COMPUTER SCIENCE:
A MATHEMATICAL INTRODUCTION

Prentice-Hall International
Series in Computer Science

C. A. R. Hoare, Series Editor

Published

BACKHOUSE, R. C., *Syntax of Programming Languages, Theory and Practice*
de BAKKER, J. W., *Mathematical Theory of Program Correctness*
BJORNER, D. and JONES, C., *Formal Specification and Software Development*
CLARK, K. L. and McCABE, F. G., *micro-PROLOG: Programming in Logic*
DROMEY, R. G., *How to Solve it by Computer*
DUNCAN, F., *Microprocessor Programming and Software Development*
ELDER, J., *Construction of Data Processing Software*
GOLDSCHLAGER, L. and LISTER, A., *Computer Science: A Modern Introduction*
HEHNER, E. C. R., *The Logic of Programming*
HENDERSON, P., *Functional Programming: Application and Implementation*
INMOS, LTD., *The Occam Programming Manual*
JACKSON, M. A., *System Development*
JONES, C. B., *Software Development: A Rigorous Approach*
JOSEPH, M., PRASAD, V. R., and NATARAJAN, N., *A Multiprocessor Operating System*
LEW, A., *Computer Science: A Mathematical Introduction*
MacCALLUM, I., *Pascal for the Apple*
REYNOLDS, J. C., *The Craft of Programming*
TENNENT, R. D., *Principles of Programming Languages*
WELSH, J. and ELDER, J., *Introduction to Pascal, 2nd Edition*
WELSH, J., ELDER, J., and BUSTARD, D., *Sequential Program Structures*
WELSH, J. and McKEAG, M., *Structured System Programming*

COMPUTER SCIENCE:

A MATHEMATICAL INTRODUCTION

(Applied Mathematics: Modeling, Analysis, and Optimization Techniques with Applications to Software Systems)

ART LEW

Professor of Information and Computer Sciences
University of Hawaii at Manoa

Prentice/Hall International

ENGLEWOOD CLIFFS, NEW JERSEY LONDON MEXICO NEW DELHI
RIO DE JANEIRO SINGAPORE SYDNEY TOKYO TORONTO WELLINGTON

Library of Congress Cataloging in Publication Data

LEW, ART
Computer Science: A Mathematical Introduction

Bibliography: p.
Includes index
1. Electronic data processing—Mathematics.
2. Electronic digital computers—Programming.
I. Title.
QA76.9.M35L48 1984 519 84-8374
ISBN 0-13-164252-9
ISBN 0-13-164062-3 Pbk

British Library Cataloguing in Publication Data

LEW, ART
Computer science: A Mathematical Introduction
1. Electronic data processing—Mathematics
I. Title
510 QA76.9.M35
ISBN 0-13-164252-9
ISBN 0-13-164062-3 Pbk

ISBN 0-13-164252-9

ISBN 0-13-164062-3 PBK

PRENTICE-HALL INTERNATIONAL, INC., *London*
PRENTICE-HALL OF AUSTRALIA PTY., LTD., *Sydney*
PRENTICE-HALL CANADA, INC., *Toronto*
PRENTICE-HALL OF INDIA PRIVATE LIMITED, *New Delhi*
PRENTICE-HALL OF JAPAN, INC., *Tokyo*
PRENTICE-HALL OF SOUTHEAST ASIA PTE., LTD., *Singapore*
PRENTICE-HALL, INC., *Englewood Cliffs, New Jersey*
PRENTICE-HALL DO BRASIL LTDA., *Rio de Janeiro*
PRENTICE-HALL HISPANOAMERICANA, S.A., *Mexico*
WHITEHALL BOOKS LIMITED, *Wellington, New Zealand*

Printed in Great Britain by A. Wheaton & Co. Ltd., Exeter

10 9 8 7 6 5 4 3 2 1

to Richard Bellman

CONTENTS

PREFACE

This book may be characterized as one on applied mathematics. The applications are to the field of computer science, especially software design and systems programming. Our selection of mathematical topics emphasizes portions of the general fields of discrete mathematics, numerical computation, and operations research. More specifically, we shall discuss aspects of number theory, numerical approximation and analysis, logic, modern algebra, matrix algebra, graph theory, metamathematics, probability, statistics, queueing theory, and mathematical programming. The aspects of computer science to which this mathematics will be applied include programming, algorithms, languages (programming and formal), automata, computability, complexity, compilers, data structures, and operating systems.

A principal objective of this book is to provide breadth, albeit at the expense of depth. There are many books which provide more depth, covering, for example, applications of discrete mathematics to data structures, or applications of operations research to operating systems. In fact, there are numerous specialized books, each covering just one of the topics listed above. In providing a much broader treatment of both mathematics and computer science topics within one volume, we necessarily must limit the amount of detail concerning any single topic. However, enough material is included for this book to be useful as a text for a variety of different courses ranging from applied discrete mathematics to systems programming theory. The book is also suitable for private study, or as a work of reference.

We have chosen to introduce most topics at a fairly simple level, so that the book will be accessible both (i) to readers with a mathematical background but little knowledge of computer science, who may wish to learn more of a vital and dynamic application area, and also (ii) to readers with a computer programming background but little knowledge of the relevant mathematics, who may wish to learn more of the mathematical methods so very necessary for the sound *engineering* of software systems.

Our general plan is to introduce closely related mathematical topics in one chapter, followed by another chapter of applications to certain areas of computer science. Each chapter includes exercises and programming assignments, in separate sections. The former are intended to reinforce the basic concepts, while the latter are intended to illustrate concrete applications. Certain exercises and programming assignments are starred to indicate that they may require some research into the cited literature.

We will emphasize mathematical *modeling, analysis,* and *optimization* techniques. The purpose of modeling is to characterize qualitatively or compute quantitatively the *behavior* of a system. The purpose of analysis is to determine system behavior by mathematical derivations or computations, generally as a function of various system parameters. The purpose of optimization is to *control* system behavior by adjusting parameters so as to obtain desired behavior according to specified criteria. The dilemma which confronts us is that models should be complex enough to be realistic, but simple enough to be practical.

MODELING, ANALYSIS, AND OPTIMIZATION

As an illustration of the dilemma, consider the familiar example of a physical system consisting of an object having mass m, suspended by a spring with constant k, and damped by a factor b. The behavior of this system, insofar as our interests are concerned, is *defined* as the object's height $h(t)$ as a function of time t, after it has been displaced (from an equilibrium position) by a certain amount h_0, held still, and released at time 0. Our intuition tells us that the object will oscillate up and down for a while before coming to rest at the equilibrium position. How much the object oscillates and how quickly it comes to rest depend upon the parameters m, k, and b. Suppose the cost of a spring depends upon k or its stiffness. Then, an example of an optimal control problem is to select the least costly spring (e.g., by choosing k for a fixed m and b) such that the object will come to rest within a specified length of time.

In physics, the foregoing system is *modeled* mathematically by a differential equation of the form

$$c_1h''+c_2h'+c_3h=0.$$

By this, we mean that the behavior of the system $h(t)$ is defined as the solution of the differential equation, given the initial conditions $h(0)=h_0$ and

$h'(0)=0$. Of course, knowledge of the differential equation model is not sufficient; we must have *analytical* tools for solving the equation, such as provided by calculus. Unfortunately, if the model is too complex, the requisite tools may be impractical or nonexistent. On the other hand, if the model is oversimplified, the resulting solution may not precisely describe actual system behavior. For example, the differential equation model does not apply to a system which comes *precisely* to rest after a finite amount of time. Nevertheless, the model is useful because the behavior it describes is close enough for most needs, but more importantly because a more complex model which does apply would be too difficult to analyze. This illustrates our dilemma.

When confronted with such dilemmas in this book, we shall seek a balance between complexity, realism, and practicality. Readers who require more realistic models for their specific needs will also require more complex analytical tools than we provide. There are several alternatives when mathematical analysis is intractable. For example, for the spring system, we may adopt experimentation or simulation methods, using one of the following types of models.

Physical model. We may construct a model using an actual object, spring, and damping mechanism. (In 'wind-tunnel' experiments, scaled models are used instead.)

Analogue model. We may construct a model of a different kind of system whose behavior is known to be analogous. For example, an electric circuit (consisting of a resistor, capacitor, and inductor) is known to have a differential equation model of the same form as the spring.

Computational model. We may construct a model which is better suited to a numerical solution using a digital computer—e.g., by approximating the continuous differential equation by a discrete difference equation.

In these modeling methods, the behavior of the system is determined by monitoring or measuring $h(t)$. Experimentation and simulation are especially useful when dealing with computer operating systems.

In summary, the models introduced in this book are discrete in nature, which accounts for our choice of discrete mathematics as a major concern. In addition, recognizing the importance of stochastic models, we emphasize probability and statistics, especially as analytical tools. Of course, since the ultimate objective of analysis is design, preferably *optimal* design, optimization techniques are also covered. Finally, since operations research techniques are numerical, many aspects of numerical computation are of relevance.

Systems programmers should have a good understanding of the problems of numerical computation, not only because they may be in a position to assist others, but also because they may be confronted by numerical problems for their own purposes. For example, compiler writers must utilize logarithms in order to generate object code for simple arithmetic expressions which include exponentiation. They also may be asked to provide programming languages having a variety of different numerical data types (e.g., complex numbers, extended-precision, matrices). Of course, software system designers may utilize some of these data types themselves, e.g., for the simulation and analysis of operating systems.

OUTLINE

Chapter 1. In the introductory chapter, we present three subjects which are basic to any study of computer science. First, we discuss *numbers,* which are of fundamental importance because computers may be regarded as number processing machines. Character-string information can, of course, be coded as sequences of numbers. The concepts discussed here are fundamental to an understanding of numerical computation.

In the second section, we briefly discuss functions and algorithms, primarily to introduce some of the terminology used subsequently.

In the third section, we summarize the basic concepts of *logic,* upon which all mathematical proofs are based. Logical operations are generally included in every programming language, of course. However, the application of logic to the proof of correctness of computer programs is our main objective.

Chapter 2. It is typical for applied mathematics in the physical sciences to be based upon calculus, where differential-equation models of continuous systems are prevalent. In the computer sciences, however, models of discrete systems are required, due primarily to the discrete (bit) nature of digital computers. Discrete systems may be modeled as *sets* of objects, together with various relations, functions, or *operations* which apply to these objects. We discuss various properties associated with operations; depending upon the nature of these operations, different classes of *algebraic systems* are distinguished. Two well-known examples of algebraic systems are Boolean algebra (the algebra of sets and logic) and linear algebra (the algebra of matrices, discussed more fully in Chapter 3).

Chapter 3. In the first section, we continue our discussion of numerical computation. We shall emphasize *sequences* and their convergence or periodicity, concepts related to the evaluation of mathematical functions (e.g., logarithms), and to the study of computational complexity. In the next section, we discuss *matrix algebra.* Matrices are important to systems programmers because of their common occurrence as a programming language data type. They are also important to software designers who may use them, for example, to represent the graph models of Chapter 4, or to analyze the Markov chains of Chapter 6.

In the final section, we introduce the algebra of *strings,* specifically as it relates to languages. Formal language theory provides both a means for defining a language L, and a mechanism for recognizing whether x is in L for an arbitrary x.

Chapter 4. A *directed graph* (*digraph,* for short) is one of the foremost algebraic systems used in computer science. Digraphs are important both as a theoretical model and as a concrete data structure. *Trees,* which form a special class of digraphs, are also of great importance in both respects. We discuss the use of digraphs and trees as models emphasizing various characterizations and representations.

This chapter is a key one, for which much of the preceding material was preparatory, and upon which much of the subsequent material relies.

Chapter 5. We begin this chapter by describing algorithms for analyzing digraph and tree models. Of special importance are algorithms for determining their properties, for restructuring them, and for traversing them.

Subsequently, we discuss applications of digraphs and trees to programming, languages, and automata. With respect to programming, the concepts of structured programming and complexity are emphasized. With respect to languages, the major application is to compiling; for example, we discuss the construction of a parse tree. Automata are important as models of computers. However, we shall see in later chapters that the concept of a *state transition system* is much more broadly used.

Chapter 6. With this chapter, we begin our discussion of operations research topics by summarizing the elements of probability and statistics which we shall need. We include a discussion of combinatorics, which is traditional for discrete mathematics texts. However, the concepts of a random variable and a probability distribution are much more important for our purposes.

The last two sections introduce two major statistical models—Markov chains and queueing systems— both of which can be applied to the analysis of software. Markov chains are especially useful for the modeling and analysis of state transition systems. Queueing theory is developed to handle problems involving waiting lines, which arise whenever there are more users than servers.

Chapter 7. Application of the mathematics of earlier chapters to some of the problems of information coding and structuring is surveyed. First, we introduce the problem of data compression, and then discuss methods of detecting and correcting errors in data. The group theory introduced in Chapter 2 is applied here.

We then analyze the performances of various information storage and retrieval procedures. Of special note is the comparison of linear search, binary search, and hashing.

Finally, we discuss some graph related topics, including data flow analysis, graph modeling, database systems, and buffering.

Chapter 8. We discuss optimization techniques in this chapter. We emphasize the *dynamic programming* methodology because of its generality. We show how dynamic programming can be used to solve two specific problems—*optimal routing,* that of finding the shortest path between two points in a digraph, and *optimal touring,* that of finding the shortest cyclic path. Solutions to these two problems are of special importance because many software design problems can be formulated in such terms.

Chapter 9. In the first section, we discuss optimization problems on digraphs and trees. Many applications relate to the determination of optimal binary trees. Scheduling using PERT and determination of maximal flows in networks are also discussed.

The two concluding sections survey one of the major areas to which optimization techniques have been applied, that of *operating systems design.* Since operating systems are responsible for the allocation of computer resources (such as processing time and storage space), it is very natural to wish allocation decisions to be made optimally.

AUDIENCE

The prerequisites for the material in this book include 'mathematical maturity', as required to follow mathematical expositions, and computer programming experience. The former may be gained from a single college-level course in differential calculus, linear algebra, set theory, probability, or symbolic logic. Specific knowledge of any of these mathematical topics is not required, since we introduce them all herein as needed; theoretical details which are not necessary for their application are omitted. The minimum level of programming expertise we expect is about that required to sort an array or invert a matrix in a 'conventional' programming language. By conventional, we mean one with assignment, conditional (IF), and iteration (FOR or DO) statements, such as BASIC; however, we do not assume prior knowledge of this or any specific language. While we include examples in a variety of other programming languages, they are for the most part self-explanatory; these examples are intended primarily for motivation, not explanation, hence they may even be skipped without much loss. Some familiarity with the programming concepts of functions, data structures, and operating systems is also expected.

This book should be of value to a wide audience with diverse backgrounds. Readers with the minimum prerequisites may find it necessary to consult supplementary material for elaboration of some topics and for additional illustrative examples; these readers, of course, have the most to gain from this book. Readers with more extensive backgrounds may find certain portions repetitious or even oversimplified, but those portions of the book which are not familiar should more than compensate; we expect most readers will be in this category. Even readers who are familiar separately with each of the mathematical and computer science concepts covered may benefit from the many illustrations of their interrelationships; these readers may wish to read the Postscript now for a summary.

We have included enough material for use as a textbook for a full-year course. For use for a single semester course, instructors will have to be selective. Topics can be emphasized or omitted depending upon the contents of other available courses, upon the students, and upon the interests of the instructors. The choice will be easier if students have homogeneous backgrounds. For example, if all students have some prior knowledge of discrete mathematics, then this material can be skipped (or left for students to read on their own), and operations research and applications can be emphasized instead. On the other hand, if all students have primarily computer programming backgrounds, then the mathematics can be emphasized at the expense of the applications. An important consideration is where the course fits within one's curriculum, specifically with respect to prerequisites and sequels. In fact, a course based on this book would be useful as a single prerequisite to several offerings, or as a sequel which consolidates many others. Instructors may also choose between the exercises or programming assignments or both, depending upon the nature of the course. Except for the few cases where the features of a particular language provide an interesting illustration, the assignments are language-independent.

Numerous references to the literature are given, and as a guide they are classified as one of three types. Type **M** references are primarily mathematical in nature, with no particular computer science application as motivation. Type **C** references are primarily computer oriented, with an emphasis on programming rather than mathematical analysis. Type **A** references, which may have elements of both other types, illustrate applications of mathematics to computer science. While some of these references (such as those on numerical computation) are much more mathematical than others, or can justifiably be classified as more than one type, we have made an arbitrary choice.

ACKNOWLEDGEMENTS

I gratefully acknowledge the support of the University of Hawaii at Manoa, without which this book would not have been possible. Some of the examples in this book are based on my research, much of which was supported by the U.S. Army Research Office; I also gratefully acknowledge this research support. Among the publications deriving from this support is the article "Optimal resource allocation and scheduling among parallel processes," in *Lecture Notes in Computer Science, Vol. 24,* Springer-Verlag, Berlin, 1975 (Reference no. [9.37]). Springer-Verlag kindly granted permission to include excerpts from this article.

Previous drafts of this book have been used for several years at the University of Hawaii, and for one summer at the Northwest Telecommunication Engineering Institute in Xi'an. I wish to thank the numerous students who helped transform these drafts into a publishable product, and to apologize to them for having to suffer through the process.

I am particularly grateful to M. Clint and A. Whittle for their many helpful suggestions. This book is much improved due to their efforts.

For their assistance in the publication of this book, my thanks also go to H. Hirschberg, R. Decent, and J. Murray of Prentice-Hall.

I am especially indebted to R. Bellman, W. W. Peterson, and C. A. R. Hoare for their advice, support, and encouragement over the years.

Finally, the role of my family (Lucille, Anthony, Scott, and Carey) should not be minimized. They contributed, at the least, by providing welcome relief from what seemed at times a daunting task.

I welcome any correspondence from readers. Your critical comments, descriptions of what you found most useful (course outlines?), lists of corrections, and suggestions for improvements would be greatly appreciated. Please write to me, c/o Department of Information and Computer Sciences, University of Hawaii, Honolulu, Hawaii 96822 (USA).

GLOSSARY OF SYMBOLS

$+$	ordinary addition, plus sign
$\oplus$	Boolean (mod 2) sum, symbolic sum, exclusive-or, disjoint union, linear combination
$-$	ordinary subtraction, minus sign (negation), set complement
$\ominus$	unary minus
$\times$	ordinary multiplication, Boolean product, Cartesian product
$\otimes$	symbolic product
$*$	star closure, Kleene closure
$^+$	positive transitive closure
$\cdot$	ordinary multiplication, concatenation, complex product
$\circ$	general operator, scalar product, composition (of functions)
$/$	ordinary division, fraction, insertion (reduction)
$\uparrow$	exponentiation
$\sqcap$	meet
$\sqcup$	join
$.$	decimal point, radix point
$'$	prime, successor function, inverse, complement, derivative
$''$	double prime, string quotation marks, second derivative
$=$	equal
$\neq$	not equal
$<$	less than, strict partial order
$\leq$	less than or equal, partial order, lexicographical order
$>$	greater than
$\geq$	greater than or equal
$\lessdot$	less than in precedence
$\doteq$	equal in precedence
$\gtrdot$	greater than in precedence
$\equiv$	equivalent, congruent

$\not\equiv$	not equivalent
$\approx$	same order of magnitude
$\simeq$	asymptotically equal
$\propto$	proportional
$:$	comparison
$:=$	assignment
$\not R$	not *R*-related
$\cup$	union
$\cap$	intersection
$\emptyset$	empty set
ε	null string
$\in$	member of
$\notin$	not member of
$\subseteq$	subset of, subgraph of
$\supseteq$	superset of
$\not\subseteq$	not subset of
$\#$	size (of set), cardinality
2^S	power set of *S*
$\rightarrow$	implies, produces, converges, maps into
$\xrightarrow[\alpha]{}$	makes transition upon reading α
$\leftrightarrow$	equivalent
$\Rightarrow$	derives immediately, functionally implies
$\Leftrightarrow$	adjacent
$\Rightarrow^*$	derives
$\Leftrightarrow^*$	equivalent
$\vdash$	moves immediately to
$\vdash^*$	moves to
$\vee$	or, logical sum
$\wedge$	and
&	and
$\neg$	not
$\forall$	for every
$\exists$	there exists
$\nexists$	there does not exist
$\mid$	mark, alternation, APL mod fct
$\mid\ \mid$	absolute value, length
$\pm$	plus-or-minus
∞	infinity
∂	partial derivative
!	factorial
∇	gradient
T	"TRUE"
F	"FALSE"

\$	immediate successor
$\oplus$	vector/matrix sum
$\ominus$	vector/matrix difference
$\otimes$	vector product
$\circledast$	matrix product
$\odot$	inner product
*	multiplication in programming languages, ring multiplication
* *	exponentiation in programming languages
o	little-oh function
O	big-oh function
Ⓥ	logical matrix sum
Ⓐ	logical matrix product
( )	binomial coefficient
Σ	generalized n-ary sum
Π	generalized n-ary product, partition, permutation, transition probability matrix
$\{\ldots\}$	set
$(\ldots)$	ordered *n*-tuple, vector
$[\ldots]$	equivalence class, interval
$\langle\ldots\rangle$	floating-point number, sequence
$\mathscr{A}$	algebraic system, action
$\mathscr{C}$	category
$\mathscr{D}$	domain of function
$\mathscr{E}$	set of events
$\mathscr{F}$	family
$\mathscr{I}$	set of natural numbers
$\mathscr{P}$	precondition, property, process
$\mathscr{Q}$	postcondition
$\mathscr{R}$	relation, range of function
e	base of Napierian logarithm
i	$\sqrt{-1}$ (imaginary no.)
$\int$	integration
$\sqrt{}$	square root
$\frac{1}{2}$	fraction
δ	state transition function
ϵ	small quantity, error
θ	vector identity, zero matrix
λ	labeling function, arrival rate
μ	mean, service rate
ν	nullity
π	pi (=3.14159)
ρ	rank, utilization factor
σ	string, standard deviation

τ	translation
ϕ	illegal or dummy label
Θ	operation
K	cut basis
Λ	loop basis, language
Φ	initial natural number
iff	if and only if
w.r.t.	with respect to

1 BASIC CONCEPTS

1.1 NUMBERS

To begin this study of mathematics relevant to computer science, we introduce the concept of 'numbers'. This is only natural since computers are most commonly thought of as numerical machines, rather than as the abstract symbol processors which they fundamentally are. Furthermore, a number is a mathematical concept which is universally known, and should require no definition. However, the theory of numbers is actually a complex field of study, and numerous concepts from this theory should be well understood by computer users.

There are many classes of numbers (such as integer, rational, real, complex) and many ways in which to represent each number (binary, decimal, fixed-point, floating-point, etc.). For the moment, we limit ourselves to a discussion of the most elementary kind of number. Firstly, we shall need the intuitive concept that a *set* of numbers *contains* (is a collection of) distinct unequal numbers, each of these numbers being called a *member* of the set. (This is a dictionary-like, rather than formalized, definition; more likely than not, a set is what you think it is.) Also loosely speaking, we say that a *sequence* or *string* of numbers is a set whose specified ordering is significant.

1.1.1 Natural Numbers

A *natural* (or *whole*) *number* is the type of number to which we are first introduced as children: e.g., 'zero', 'one', 'two', 'three', 'ten', 'one-million', etc. These are the numbers by which we count distinguishable objects; the word 'natural' is frequently omitted when the context is clear. In the theory of numbers, all natural numbers are defined by *Peano's axioms* (see Mendelson [1.17]):

1. There exists a natural number, denoted Φ.
2. Corresponding to each natural number (x) there exists another unique natural number (x'), called the *successor* of x.
3. Φ is not the successor of any natural number.
4. $x'=y'$ implies $x=y$.
5. If $\mathcal{J}$ is a set of natural numbers such that
   (a) $\mathcal{J}$ contains Φ, and
   (b) $\mathcal{J}$ contains x' whenever it contains x,
   then $\mathcal{J}$ contains all the natural numbers.

The fifth axiom is used to define $\mathcal{J}$ by induction. We remark that the distinguished natural number Φ in the above is usually denoted 0; sometimes Φ is denoted by 1 when we wish to consider only 'positive' natural numbers. While Φ may be regarded as the 'first' (or 'zeroth') number in $\mathcal{J}$, there is no 'last' number in $\mathcal{J}$ (since every number has a successor). In other words, $\mathcal{J}$ is an *infinite* set. However, by definition, each number is itself finite.

Discussion Our definition of natural numbers is, perhaps, much more formal than is customary for some readers. Such formal definitions are of fundamental importance, however, because they are precise and unambiguous, and because many conclusions can be proven directly from them. Admittedly, the definition may not be clear to everyone because English, which is a highly ambiguous language, was used. In any event, in this book, we will define many software terms with like formality.

As an example of a conclusion which follows from Peano's axioms, we first define the *predecessor* of a natural number x, denoted **pred**(x),† to be that natural number p whose successor is x; symbolically, $p'=x$. From the axioms, we may conclude that Φ has no predecessor, but that every other natural number has a unique predecessor. Furthermore, we may define the sum of two natural numbers a and b, denoted $a+b$, as follows:

$$\text{(i)}\quad a+b=a'+\textbf{pred}(b),\ \text{for } b\neq\Phi, \qquad (1.1.1)$$
$$\text{(ii)}\quad a+\Phi=a.$$

Applying (i) iteratively until (ii) is applicable, we have, for example,

$$2+3=\Phi''+\Phi'''=\Phi'''+\Phi''=\Phi''''+\Phi'=\Phi'''''+\Phi=\Phi'''''=5.$$

(*Note*: For *positive* natural numbers, when Φ is taken to be 1 instead of 0, we would replace (ii) by $a+\Phi=a'$.) It can be proven from the axioms that the

†Readers are assumed to be familiar with functional notation, at least to the extent that it is used in programming languages. (For a review, see Section 1.2.1.)

sum as defined above satisfies the usual laws of arithmetic—e.g., associativity: $(a+b)+c=a+(b+c)$, and commutativity: $a+b=b+a$.

A significant implication of the foregoing is that, in order to perform addition, it is not necessary to know how to add directly. It is sufficient to be able to determine the successor and predecessor of any natural number, and whether or not two natural numbers are equal. These are far simpler tasks. The latter facility is necessary to decide whether to use (i) or (ii) above (i.e., we must decide whether $b=\Phi$); it also enables us to determine predecessors in terms of successors. Given addition, other arithmetic operations, such as subtraction and multiplication, can then be defined.

For later reference, we state the following:

Theorem of computability Every computable function can be constructed in terms of the *successor function* and the *equality test*.

A practical illustration of this is exemplified in the LISP language, whose full computational power can be built up from these two primitives; see Section 2.3.5.

Notations The *value* of a number is unique, but the *notation* or *symbol* used to represent that number is not. We illustrate some notational systems for the natural numbers, Φ, Φ', Φ'', Φ''', etc., in Table 1.1.1.

A	*B*	*C*	*D*	*E*	*F*
	0		0		zero
|	1	0	1	I	one
||	10	1	2	II	two
|||	11	00	3	III	three
||||	100	01	4	IV	four
|||||	101	10	5	V	five
||||||	110	11	6	VI	six
|||||||	111	000	7	VII	seven

Table 1.1.1
Notations for Numbers

A. A *monadic* number system uses a single character, here a 'mark' (|), to denote the natural numbers. The value of a symbol depends upon its length. Φ is denoted by the absence of marks (sometimes denoted ε instead†), Φ' by one mark, Φ'' by two marks, etc. (Of course, any character may be used for the mark.)

†ε is not to be regarded as an actual character; this is analogous to the difference in programming languages between the literal string " " and "", where the former contains a space character whereas the latter contains no characters at all.

B. A *binary* (or 'base 2') number system uses two characters, 0 and 1 (called the binary digits, or *bits* for short). It is a 'positional' number system in that the position of a character in a symbol determines the value of the symbol; in particular, adding leading zeroes does not affect values, (e.g., 00011=11).

C. A *dyadic* number system also uses two characters, again 0 and 1. It is partly a positional system, and partly similar to the monadic system in that the value of a symbol depends upon its length. Φ is denoted by the absence of characters (or by ε); Φ' and Φ'' are denoted by one character; Φ''', Φ'''', Φ''''', Φ'''''' are denoted by two characters; etc. Symbols of the same length have the same ordering as their binary number interpretation.

D. A *decimal* (or 'base 10') number system uses ten characters (the decimal digits), and is also a positional number system. This is the conventional number system used in the modern world, and should require no elaboration. It is the number system used in this book except when stated otherwise.

E. The *Roman* number system is mentioned as another example of one that is only partly positional.

F. Finally, each number also may be denoted by a word or words in some natural language (e.g., English).

We say that numbers, in Notations A through E respectively, are *strings* of characters chosen from the sets {1}, {0,1}, {0,1}, {0,1,2,3,4,5,6,7,8,9}, {I,V,X,L,...}; observe that other sets of *digits* or *numerals* may serve to define an indefinite number of other notations, only a few of which are useful. Note that 00011 and 11 in Notation B and 00 in Notation C all represent the same number ('three'); nevertheless, we will say that 00011, 11, and 00 are distinct *numbers* even though they have the same value. (Others would say that 00011, 11, and 00 are different 'numerals', but are the same number.)

Binary-coded Decimals Generally, people use decimal, but computers can only 'understand' binary. In order for the two to communicate, there must be some mechanism to convert numbers from one base to the other, and preferably the mechanism should be automated (i.e., part of the computer system). For hardware design reasons, the basic conversion mechanism is always built into peripheral (I/O) units rather than the central processing unit (CPU) of a computer; a CPU has no way of representing even a single decimal digit in other than binary form, and I/O units are designed to convert *single* decimal digits (from, say, punched cards or on-line terminals) to their binary equivalents and vice versa, essentially one at a time. A multi-digit (unsigned) decimal number is therefore converted to a *BCD* number (i.e., a string of *binary-coded decimals*): e.g.,

$$319_{10}=3_{10},\ 1_{10},\ 9_{10}=11_2,\ 1_2,\ 1001_2=001100011001_2.$$

Here, the subscripts are used to indicate the base in which the numbers are expressed; by convention, the subscripts themselves are expressed as decimal numbers. To mark the beginning or end of each binary number, use of commas, or use of a fixed-length representation (here, 4 bits), are commonly adopted. The 4-bit convention is called the *packed* BCD notation of a number. If the hardware for a particular I/O unit converts decimal numbers according to some other convention, e.g., to an *extended* BCD† number, it is generally easy to convert the extended number to packed form; in fact, some computers have a special machine language instruction which performs this conversion.

1.1.2 Numerical Data Types

From Peano's axioms and the rudiments of set theory, one can define other types of numbers (integers, rationals, reals, etc.) and derive their usual arithmetic and analytical properties (see Rudin [1.22]). Although certain aspects are of quite practical relevance, such a mathematical development is well beyond the scope of this book. Informally, we say that an *integer* is a natural number with a *sign*, either *positive* (+) or *negative* (−), except for the special integer *zero*, which is usually considered unsigned. The sign of integer x, denoted **sgn**(x), is often represented numerically:

$$\mathbf{sgn}(x)=\begin{cases} +1 & \text{if } x>0 \\ 0 & \text{if } x=0 \\ -1 & \text{if } x<0. \end{cases}$$

The *magnitude* or *absolute value* of x, denoted $|x|$ or **abs**(x), is such that $x=\mathbf{sgn}(x)\times\mathbf{abs}(x)$.

A *rational* is the ratio p/q of two integers, p and q, where $q\neq 0$; it has a *whole* part and a *fractional* part, either of which may be zero. For example, the rational 5/4 has whole part=1 and fractional part=0.25. An *irrational* is a number which is not rational (e.g., $\sqrt{2}$, or the transcendental numbers π and e). A *real* number is either a rational or an irrational. A *complex* number is an ordered pair of real numbers, the first called the *real* part and the other the *imaginary* part.

Reals Real numbers can be represented in either fixed-point or floating-point form. A *fixed-point* (decimal) number has an explicit decimal point; i.e., it has the form

$$\mathrm{d}_n\mathrm{d}_{n-1}\ldots\mathrm{d}_1\mathrm{d}_0.\mathrm{d}_{-1}\mathrm{d}_{-2}\ldots\mathrm{d}_{-m}$$

where each d_i is a decimal digit. Note that fixed-point numbers are positional in that the value of a number depends not only on the digits in the number, but also upon the positions of the digits relative to the decimal point. If

$$N=\mathrm{d}_n\mathrm{d}_{n-1}\ldots\mathrm{d}_1\mathrm{d}_0.\mathrm{d}_{-1}\mathrm{d}_{-2}\ldots\mathrm{d}_{-m},$$

†Two such extended BCD conventions are known as the 8-bit EBCDIC and 7-bit ASCII codes. See Mackenzie [1.16]. EBCDIC numbers are also called *unpacked* or *zoned*.

we call the integer $d_n d_{n-1} \ldots d_1 d_0$ the *whole part* of N, denoted **whole**(N), and we call the real number $.d_{-1}d_{-2}\ldots d_{-m}$ the *fractional* part of N, denoted **fract**(N). The sign of **whole**(N) and **fract**(N) are both taken to be the sign of N. For example, **whole**(+1.5)=+1, **whole**(−1.5)=−1, and **fract**(−1.5)=−0.5.

The numbers which arise in numerous applications may vary greatly in magnitude. For example, in a single problem we might have numbers as large as 6.0225×10^{23} (Avogadro's number) and others as small as 1.602×10^{-19} (charge of an electron, in coulombs), which we have expressed in their usual 'scientific notation'. The first number can also be written in the form 0.60225×10^{24} and the second as 0.1602×10^{-18}, where the decimal point is at the left of the leading or 'high-order' nonzero digit. In general, we will use this second form, which we call *floating-point* notation. In programming languages, these numbers are often denoted by .60225E24 and .1602E−18, respectively.

In general, a *floating-point* number N has the form

$$N = .f \times 10^e$$

where f (not $.f$) is called the *fraction* and e the *exponent*; f and e are both integers, and the sign of f is taken to be the sign of N. If the integer f has L digits, we say that N is an L-place floating-point number. The number is *normalized* if the leftmost or leading digit of f is nonzero; there are different conventions regarding e for normalized zeroes ($f=0$), ranging from permitting e to be any number to requiring e to be one specific number (say, 0). If the decimal point is always at the left of the fraction, and the base 10 is understood, then all numbers may be characterized by integers f and e alone; we then write $N = \langle f, e \rangle$. For example, $12.34 = \langle 1234, 2 \rangle = \langle 12340, 2 \rangle = \langle 01234, 3 \rangle$; the fraction depends upon the number of places assumed, and on whether the number is normalized. We also say that $\langle 12340, 2 \rangle$ has only four *significant* digits; the 'low-order' zero is not significant since, by convention, 12.34 may represent any number from 12.335 to 12.345. Any fixed-point number

$$N = \pm d_n \ldots d_0 . d_{-1} d_{-m}$$

can also be written as the floating-point number $\langle \pm d_n \ldots d_{-m}, n+1 \rangle$. Note that the number of places L for the fraction must be at least $n+1+m$ if the number is to be represented without 'rounding' or 'truncation' (which we discuss later). The fractional part of N ($=.d_{-1} \ldots d_{-m}$) should not be confused with the fraction f ($=d_n \ldots d_{-m}$) of the floating-point equivalent.

Base r Numbers The value of an unsigned decimal fixed-point number

$$d_n d_{n-1} \ldots d_2 d_1 d_0 . d_{-1} d_{-2} \ldots d_{-m}$$

is the sum of products

$$d_n \times 10^n + d_{n-1} \times 10^{n-1} + \ldots + d_1 \times 10^1 + d_0 \times 10^0 + d_{-1} \times 10^{-1} + \cdots + d_{-m} \times 10^{-m}$$

where the digits, d_i, are chosen from the set $\{0,1,2,3,4,5,6,7,8,9\}$. It is common for high-order zeroes to be omitted. We can extend this idea to obtain a representation for numbers in any other ***base*** (or ***radix***) ***r***. Binary, octal, decimal, and hexadecimal numbers have r=2,8,10, and 16, respectively.

In general, a ***base r number*** can also be written as

$$d_n d_{n-1} \ldots d_1 d_0 . d_{-1} \ldots d_{-m}$$

where now this number is understood to be the sum

$$d_n \times r^n + d_{n-1} \times r^{n-1} + \cdots + d_1 \times r^1 + d_0 \times r^0 + d_{-1} \times r^{-1} + \cdots + d_{-m} \times r^{-m}$$

and the digits d_i are restricted to the set $\{0,1,\ldots,r-1\}$; the separating dot is now called the ***radix point***. Thus, 13.75_{10} equals

$$1 \times 10^1 + 3 \times 10^0 + 7 \times 10^{-1} + 5 \times 10^{-2} \text{ (where } n=1 \text{ and } m=2\text{)},$$

which also equals 1101.11_2 and $\mathrm{D.C}_{16}$.† The analogous floating-point notation is

$$.d_n \ldots d_0 \ldots d_{-m} \times r^{n+1} \text{ or } \langle d_n \ldots d_0 \ldots d_{-m}, n+1 \rangle_r;$$

for example, $13.75_{10} = \langle 1375,2 \rangle_{10} = \langle 110111,100 \rangle_2$. We write $\langle f,e \rangle_r$, or just $\langle f,e \rangle$ when r is understood, to mean $.f \times r^e$. Sometimes it will be convenient to express the integers f and e in a base other than r; in print, f is usually expressed in base r, while e is expressed in base 10, so that $\langle 110111_2, 100_2 \rangle_2$ is written $\langle 110111,4 \rangle_2$.

1.1.3 Application: Programming Language Facilities

The hardware of almost all computers is designed to support binary integers. In addition, computers intended for scientific applications support floating-point binary (or hexadecimal) reals, and computers intended for business applications support (BCD) decimal integers. 'Support' is in the form of machine-language arithmetic operations on these numerical data types.

Data Types General-purpose programming languages may or may not incorporate facilities for the above three data types, depending upon the hardware of the computer on which the languages are implemented. However, even if the hardware does not support a data type, compilers can in principle be written to support it by software means; in practice, this may be too inefficient.

†For $r>10$, additional symbols must be used to denote 'digits' beyond those used for decimal numbers; upper-case letters are commonly used. For example, $\mathrm{A}_{16}=10_{10}$, $\mathrm{B}_{16}=11_{10}, \ldots, \mathrm{F}_{16}=15_{10}$. Recall the convention of using decimal subscripts to indicate base.

Example PL/I language implementations generally support all three data types (as well as floating-point decimal reals and fixed-point decimal reals). FORTRAN language implementations generally support only binary integers (INTEGER) and floating-point binary reals (REAL). COBOL language implementations generally support binary integers (COMPUTATIONAL); fixed-point decimal numbers (represented as character-strings) may be restricted for use to display purposes only. Many versions of COBOL also permit decimal integers to be used for computational purposes (i.e., in arithmetic expressions); some versions of COBOL also support floating-point binary reals. FORTRAN and PL/I both support complex numbers. A common situation is summarized in Table 1.1.3.

data type	*COBOL*	*FORTRAN*	*PL/I*
binary integer	COMPUTATIONAL	INTEGER	FIXED BINARY
binary floating-point (single-precision)	COMP–1	REAL	REAL FLOAT
complex (default is real)	n/a	COMPLEX	COMPLEX
decimal fixed point (packed)	COMP–3 (integer only)	n/a (integer or real)	FIXED DECIMAL
decimal fixed point (zoned)	DISPLAY	n/a (except as Hollerith data)	CHARACTER

Table 1.1.3
Programming Language Data Types

It should be noted that non-numerical (list or string processing) languages with no explicit numerical data types or arithmetic capabilities can provide these facilities implicitly, at least for positive integers. A positive integer n can be represented by a list or string of n items, i.e., using a monadic number system. The successor function can then be implemented by inserting an item in a list. Addition of two integers can be accomplished by combining two lists or strings. An example of numerical processing utilizing these implicit facilities is given in Section 2.3.5.

Conversion Frequently, we may wish to convert numbers of one type to another. Of course, this is not always valid, e.g., a complex number with a nonzero imaginary part has no equivalent real representation. One useful type of conversion is from real to integer. In most programming languages, this conversion is done automatically when using assignment statements or built-in library functions. For example, the functions IFIX (in FORTRAN) and TRUNC (in PL/I) perform the conversion by 'truncating' the fractional part of a real number, i.e., they implement the **whole**(N) function for any real N. Some programming languages provide instead a **floor** (and **ceiling**) function, which gives the largest integer less (greater) than or equal to any real number. The **floor** function is called **INT** (in BASIC), **ENTIER** (in ALGOL60), and **FLOOR** (in PL/I). For example, **floor**(+1.5)=+1, **floor**(−1.5)=−2, **ceiling**(+1.5)=+2 and **ceiling**(−1.5)=−1. Note that **whole** (N) and **floor**(N) are equal only for nonnegative or integer N; we will use **int**(N) to denote the function for nonnegative N. (We will not define **int**(N) for negative N, in which case we will use **whole** or **floor** instead).

To better understand the computational facilities offered in programming languages, it is necessary to know how numerical data of different types can be represented or stored in computer hardware, and how arithmetic operations upon the numbers can be performed.

1.1.4 Computer Hardware Concepts

Although computers actually operate in a binary mode, let us assume for the moment that we have a decimal computer. We will first describe internal computer representations of integer numbers, i.e., how they are *stored* in our decimal computer.

We will assume that our decimal computer has a *storage unit* (or *main memory*) which provides storage spaces for decimal digits, physically grouped into *cells* or *words*, each a given number of digits long. The space per cell is called the computer's *word-length* (measured in digits), and the total number of cells is called the computer's *size* (measured in words). Associated with each word is a number, called the *storage address*, by which the word can be referenced. For example, a storage unit with a word-length of 11 and a size of 400 has space for 4400 digits, with addresses numbered from 000 to 399.

Observe that we have used the term 'digit' here as a unit for measuring space, rather than to refer to a notational character. In one decimal digit-unit, we can store exactly one of the ten decimal digit-characters. The plus and minus signs, rather than being distinct eleventh and twelfth characters, can be represented by one of the ten decimal characters, provided its position in each word is known (e.g., by the convention that it appears first).

Representations If we want to allow for both positive and negative integers, we must use one digit, of the eleven we are allowed for our number, to store the sign. We will use a 1 in the high-order digit of a storage word to denote a minus sign and 0 to denote a plus sign. The remaining 10 digits will be used to store the magnitude of the integer. For example, the integer −53 would be stored as

1 0 0 0 0 0 0 0 0 5 3 ,

whereas +53 would be stored as

0 0 0 0 0 0 0 0 0 0 5 3 .

Since 10 digits are allowed for the magnitude of our integers, we can store any integer in the 'range'

$$-9{,}999{,}999{,}999 \text{ to } +9{,}999{,}999{,}999$$

or

$$-(10^{10}-1) \text{ to } 10^{10}-1.$$

If we tried to store an integer less than $-10^{10}+1$ or greater than $10^{10}-1$, an *underflow* or *overflow*, respectively, would result.

This *sign-magnitude* representation of integers is a natural one to use, but has several disadvantages which make it rather unattractive to computer designers. One objection is that, while negation of an integer involves simply changing or 'inverting' the sign digit, addition of two integers requires operations which depend first on a test of the signs of the numbers. For example, the addition of +12 and −34 requires extra steps compared to the addition of +12 and +34; it must first be noticed that −34 is negative, so that +12 is *subtracted* from +34 (yielding +22) and the difference is then negated (yielding the answer −22). Another objection is that +0 and −0, while equal in value, have two different internal representations.

For binary computers, integers are commonly represented in *2's-complement* form. With the left bit reserved for the sign as before, positive integers in the range 1 to $(2^{L-1}-1)$ can be represented in an L-bit word. Negative integers are represented in 'complemented' form, i.e., $-x$ is represented by 2^L-x (computed as an unsigned number). For example, for $L=4$, +7 is represented by 0111_2, −7 is represented by $2^4-7=9=1001_2$, and −1 is represented by $2^4-1=15=1111_2$. Note that negative integers in the range $-(2^{L-1})$ to −1 can be so represented. There is now a unique representation for zero, 0000_2; 1000_2 represents −8 (rather than −0). The other objection mentioned above has also been answered: integers can be added without regard to their signs. For example,

$$(12_{10}) + (-34_{10})=(00001100_2)+ (11011110_2)=(11101010_2)=(-22_{10}).$$

We emphasize that, using 2's-complement, integers in the range

$$-(2^{L-1}) \text{ to } +(2^{L-1}-1)$$

can be represented in an L-bit word; there are 2^L distinct integers in this range.

Floating-point reals are represented by a pair of integers, one for the exponent and the other for the fraction. The base and the implied location of the radix point are generally fixed. The two integers are often represented in sign-magnitude form.

Arithmetic Computer hardware is designed to perform single arithmetic calculations in much the same way people do. For example, numbers may be added one digit at a time, from right to left, with carries if necessary; multiplication involves successive additions. Knowledge of how computers perform arithmetic is important to software specialists for two major reasons. First, a clear understanding of computer arithmetic is necessary for programmers who deal with numerical calculations in order to avoid certain kinds of errors. Second, programmers may be called upon to augment the arithmetic capabilities of a computer or language, e.g., for extended precision, different radices, or complex numbers.

The basic arithmetic operations (+, −, ×, /) implemented in typical computer hardware differ from those learned by children in two important respects: in the representation of numbers, and in the fixed *precision* of all operations. By the first, we refer to internal forms of numbers (e.g., in binary rather than decimal), and by the second, we refer to the use of fixed-length numbers regardless of their magnitude. Thus, while we would write 987+65=1052, the sum of a 3-digit and a 2-digit number yielding a 4-digit decimal number, computers having an 8-bit word length would have to add full 8-bit binary numbers including any leading zeroes: e.g.,

$$00000101_2+00000011_2=00001000_2.$$

We remark that computer systems comprise several interrelated functional units, one of which is called the *arithmetic unit* (AU). (Actually, there may be several AUs, e.g., one for integers and one for reals.) It is the AU which performs the four basic arithmetic operations. These operations are binary† in the sense that each operation uses two numbers (called the *operands*) to compute a third number (called the *result*). Conceptually, it is perhaps easiest to view the functioning of the AU as follows:

(Step 1) The computer system transfers operands (or at least their locations) to the AU.
(Step 2) The AU is instructed which operation to perform.
(Step 3) The AU performs the operation, saving the result usually in a special place called an *accumulator*.
(Step 4) The AU sets certain flags, e.g., to signal the system that it has successfully completed the operation, or to indicate its status.

†We emphasize that it is the operation that is binary, not (necessarily) the radix of the operands. Operations having two operands are sometimes called *dyadic* instead of binary to avoid this confusion.

So-called 'one-address' computers have instructions of the form 'ADD X', which instructs the AU to add the value of X to the accumulator, whereas 'two-address' computers have instructions of the form 'ADD X, Y', which instructs the AU to add the values of X and Y and to place the result in the accumulator (or possibly in X). 'Three-address' computers have a fifth step, where the computer system transfers the result from the accumulator to a specified storage address. See Krutz [1.14] for further information on computer arithmetic and other computer design concepts.

1.1.5 Computational Errors

One of the first surprises which awaits a computer programmer is that computer arithmetic is not exact: for example, he/she may find that

$$10\times 0.10 \neq 1.00$$

(i.e., ten dimes do not equal a dollar). The main reason for this is that some numbers have no *exact* (conventional) computer representation.

The basic difficulty is that numbers must necessarily be stored in a finite amount of storage, using (say) at most L digits. Limitation to L digits means first of all that at most r^L distinct numbers are representable (where r is the radix of the computer), and secondly that each representable number can have at most L significant digits.

Precision The *precision* of the computer representation of a number is the number of its significant digits. For standard (full-word) integers, the precision can be up to $L-1$ digits (where L is the word-length of the computer, and one digit is reserved for the sign). For floating-point numbers, the precision can be up to the length of the fraction. Computer hardware and software are designed to operate on numbers up to some fixed precision,† which is sufficient for most, but not all problems. In order to calculate with numbers having *extended* precision, i.e., precision greater than that provided in a given computer system, a number may be represented as an array of digits, and arithmetic operations on these arrays may be provided by software. The size of the computer and computation time are the only limits on the precision achievable in this manner. (π, for example, has been computed to well over 500 000 significant decimal digits.) Interested readers should consult Knuth [1.13].

Range The L-digit limitation on integers not only limits precision but also limits the *range* of values which can be represented. Recall, for example, that

†The precision of the IBM 370 is about 10 decimal digits for integers, 7 for single-precision reals, and 16 for double-precision reals.

2's-complement integers can range in value from -2^{L-1} to $+(2^{L-1}-1)$, a total of 2^L possible values. To represent larger numbers, such as 10^L, either extended precision is necessary, or floating-point notation must be used. If the latter is used, there is still a limitation on the magnitude of the exponent, but this limit is generally considerably greater; exceeding this limit results in floating-point overflow or underflow.† Furthermore, large integers may not be exactly representable in floating-point form because the precision of the fraction is limited. Finally, we note that an extended-precision floating-point number may be represented as a pair of full-word (or extended precision) integers, and floating-point arithmetic operations on these pairs may be provided by software.

Round-off Errors A computer can (conventionally) only recognize the number 2/3 as the result of dividing 2.0 by 3.0, or

$$.6666666666\ldots \times 10^0.$$

To store this number exactly in a decimal-based computer, in either fixed-point or floating-point forms, requires infinite storage for the infinite succession of 6's. In practice, the computer must 'truncate' or 'round' the representation of the number at some point since it may have a precision of (say) only four digits. Thus 2/3 would be represented by $\langle 6666,0\rangle_{10}$ or $\langle 6667,0\rangle_{10}$ in floating-point form. Similarly, π is approximately equal to $\langle 3142,1\rangle_{10}$ rounded, and e is approximately equal to $\langle 2718,1\rangle_{10}$. The error incurred by the computer's ability to handle only finite length fractions is called ***round-off error***. We note that normalization of floating-point numbers is necessary to reduce such errors, since leading zeroes are not significant. Although the round-off error for a single number may be very small, these errors can accumulate or 'propagate' during an extensive computation and render the outcome meaningless. The study of computational errors is a major undertaking in itself, which is commonly considered a part of the field of ***numerical analysis***. We refer interested readers to the texts of Isaacson and Keller [1.11] and Blum[1.3].

Recall that a rational number was defined to be the ratio of two integers. Numbers such as $\sqrt{2}$, $\sqrt{3}$, π, and e are irrational. Mathematically, the distinction between rationals and irrationals is very important, but insofar as computers are concerned rationals and irrationals are indistinguishable. For example, 22/7 and π may both be represented by $\langle 3142,1\rangle_{10}$. We will say that a number is ***rational in base r*** if it can be represented exactly as a base r number in the fixed-point form

$$d_n d_{n-1} \ldots d_0 . d_{-1} \ldots d_{-m}$$

†In the IBM 370, integers may range up to 2^{31} ($<10^{10}$); the limit on floating-point reals is about $16^{63}=2^{252}$ ($<10^{76}$).

where each d_i is chosen from the digit-set $\{0,1,\ldots,r-1\}$ and m is finite. Thus while 2/3 is rational in base 3 ($=0.2_3=\langle 2,0\rangle_3$), it is irrational in bases 10 and 2. Of great importance is the existence of numbers that are rational in base 10 but not in base 2; e.g., 1/10.† Since most humans use decimal numbers and most computers use binary numbers, round-off errors inevitably arise simply by their translation from base 10 to base 2. We also observe that if we have a computer which operates in base r and a given number is rational in base r, it is still necessary for the fraction and exponent to be small enough in order that the number can be stored exactly. For instance, the binary number 10011.1101_2 will be stored exactly in a binary computer allowing fractions of length 9, but will be rounded-off in a binary computer allowing fractions of only length 8.

In conclusion, we formally distinguish between round-off errors incurred by ***rounding*** and by ***truncation***. Let $x=\langle f,e\rangle_r$ be a positive L-place floating-point number, where $f=d_1d_2\ldots d_L$ has an implied radix point at its left. Then, for $1\le p\le L$, TRUNC$(x,p)=\langle d_1d_2\ldots d_p,e\rangle_r$ is *x truncated* to p places. However, ROUND(x,p), *x rounded* to p places, is equal to TRUNC(x,p) if $d_{p+1}<r/2$; if $d_{p+1}\ge r/2$, then 1 is added to the fraction of TRUNC(x,p), yielding $\langle d_1d_2\ldots d_p+1,e\rangle_r$. This last term is equal to $\langle 1d_2'\ldots d_p',e+1\rangle_r$, where each $d_i'=0$, if the added 1 results in a 'carry'. For example, if $x=\langle 9999996,3\rangle$, then TRUNC$(x,6)=\langle 999999,3\rangle$ and ROUND$(x,6)=\langle 100000,4\rangle$. We note that x differs from TRUNC(x,p) by $\langle d_{p+1}\ldots d_L,\ e-p\rangle$, but differs from ROUND$(x,p)$ by the smaller of $\langle d_{p+1}\ldots d_L,\ e-p\rangle$ and $\langle r^{L-p}-d_{p+1}\ldots d_L,\ e-p\rangle$. Hence while truncation is simpler to implement, rounding is more accurate (and facilitates formal analyses). For a negative number x, we define TRUNC$(x,p)=-$TRUNC$(-x,p)$ and ROUND$(x,p)=-$ROUND$(-x,p)$. We also define TRUNC$(x,p)=0$ for $p<1$ and TRUNC$(x,p)=x$ for $p>L$. Then TRUNC$(x,e)=$**whole**(x) for any $x=\langle f,e\rangle$.

Nonassociativity Computer arithmetic (addition and multiplication) does not satisfy the 'associativity law' since it is possible that

$$(a+b)+c\neq a+(b+c)$$
$$(a\times b)\times d\neq a\times(b\times d)$$

for some values of a, b, c, and d. The limitation on the range of numbers is one reason. For example, if N is the largest representable number (integer or floating-point), then the foregoing inequalities hold for $a=N$, $b=N$, $c=-N$, $d=0$ since the left-hand calculations result in overflows, whereas the right-hand calculations do not. Furthermore, if a and b are floating-point numbers with very small exponents and opposite signs, and if $c=-b$ and $d=1/b$, then evaluation of the left-hand expressions results in underflows whereas evaluation of the right-hand ones does not; note that if the result of an underflow-

†Hence, $10\times.10$ may equal .9999 instead of 1.000 when computing in binary.

ing calculation is set to zero, as is commonly done, then all precision is lost (relative to the exact answer a).

Nonassociativity of computer arithmetic may also be due to round-off errors in floating-point calculations. For example,

$$(a+b)+c \neq a+(b+c)$$

for $a=\langle 100000,0\rangle$, $b=\langle -999999,-1\rangle$, $c=\langle 900009,-1\rangle$. If the accumulator is limited to six digits, then the sum of a and b equals $\langle 000000,0\rangle$ or $\langle 100000,-5\rangle$ depending upon whether b is rounded or truncated to $\langle -100000,0\rangle$ or $\langle -099999,0\rangle$, respectively. Hence the left-hand expression equals $\langle 900009,-1\rangle$ using rounding or $\langle 900019,-1\rangle$ using truncation, whereas the right-hand expression equals $\langle 900010,-1\rangle$ exactly. In this example, loss of significance resulted from the so-called 'catastrophic subtraction' of nearly equal numbers, a common source of computational errors.

In general, the sum of a set of floating-point numbers depends upon the order in which they are added. To reduce round-off errors, it has been suggested that the sums of the positive and negative numbers be computed separately and then subtracted, and even that the numbers be added in order of increasing magnitude. Unfortunately, no such tactic is best for arbitrary sets of numbers, and if any additional overhead can be tolerated, we may as well accumulate the partial sums as a double-precision result. (This is especially recommended when summing products, e.g., of the form $x_1 \times y_1 + x_2 \times y_2 + \cdots + x_n \times y_n$.)

Error Accumulation. While a single round-off may be insignificant, errors can readily accumulate in the course of a long sequence of calculations. Loosely speaking, if the error resulting from a single arithmetic operation is ϵ, then the error resulting from N such operations may be $N \times \epsilon$, assuming that errors are not canceled out. Fortunately, some errors do tend to be canceled, but this is neither very likely nor predictable. The significance of low-order digits should always be viewed with suspicion. As an example, consider the repeated addition to itself of $x=0.10_{10}=\langle 19999\text{A},0\rangle_{16}$ rounded or $\langle 199999,0\rangle_{16}$ truncated to six hexadecimal digits: if x is added five times, we obtain $\langle 800002,0\rangle_{16}$ rounded or $\langle 7\text{FFFFD},0\rangle_{16}$ truncated, but if added ten times, we obtain $\langle 100000,1\rangle_{16}$ or $\langle \text{FFFFFA},0\rangle_{16}$, respectively. The rounded answer is exact because the accumulated error (at that point, equal to $\langle 000004,0\rangle$) happened to be rounded off. (We remark that in a binary rather than hexadecimal computer, the rounded result would not be exact.) It should also be noted that

$$\langle 100000,6\rangle_{16}+\langle 19999\text{A},0\rangle_{16}=\langle 100000,6\rangle_{16},$$

i.e., the sum no longer increases past a certain point; hence there is a limit to the number of times x can be added to itself after which an error in the sum equal to x is incurred upon each addition.

1.1.6 Elements of Number Theory

Let us for the moment restrict ourselves to natural numbers. A body of knowledge, known as *number theory,* concerns itself with such numbers; e.g., see Niven and Zuckerman [1.18]. We say that the number a is a *factor* (or *divisor*) of another number b (or that b is 'evenly divisible' by a) if $a \neq 0$ and there exists another number q such that $b = a \cdot q$.† Each number has 1 and itself as factors. We say that a number p is *prime* if $p > 1$ and it has no factors other than 1 and itself. (By this definition, 0 and 1 are not prime; 2, 3, 5, 7, 11,... are prime; $15 = 5 \cdot 3$ is not prime.) There are infinitely many primes. Two numbers are *relatively prime* if they have no common factors other than 1.

Unique Factorization Theorem An important result in number theory (also called the **Fundamental Theorem of Arithmetic**) is that every natural number greater than one is a product of prime factors, and hence can be written in the factored form $p_1^{n1} p_2^{n2} \cdots p_k^{nk}$, where $p_1, p_2, \ldots, p_k$ are distinct primes, and where this representation is unique except for the order of the factors (or, equivalently, it is unique provided that $p_1 < p_2 < \cdots < p_k$).

One consequence of this theorem is that every number $a > 1$ has at least one prime factor (only itself if a is prime). Of greater significance is the fact that, if p_i denotes the ith smallest prime (i.e., $p_1 = 2$, $p_2 = 3$, etc.), then every positive natural number can be represented uniquely by the sequence of exponents $\langle n_1, n_2, \ldots \rangle$, where $n_i = 0$ if p_i is not a factor; we call them *Gödel sequences*. For example, $1 = \langle 0,0,0,\ldots \rangle$, $2 = \langle 1,0,0,\ldots \rangle$, $3 = \langle 0,1,0,\ldots \rangle$, $4 = \langle 2,0,0,\ldots \rangle$, $5 = \langle 0,0,1,\ldots \rangle$, $6 = \langle 1,1,0,\ldots \rangle$, etc. Furthermore, every positive natural number less than the Nth smallest prime can be represented uniquely by a finite sequence of $N-1$ numbers $\langle n_1, \ldots, n_{N-1} \rangle$. The uniqueness is of special importance since it in turn permits us to uniquely represent arbitrary sequences of numbers by single equivalent numbers.

Gödel Numbers Let $\langle n_1, n_2, \ldots, n_k \rangle$ be a sequence of k natural numbers. The *Gödel number* associated with this sequence is $p_1^{n1} p_2^{n2} \ldots p_k^{nk}$, where p_i is the ith smallest prime. For example, we may represent $\langle 1, 0, 2, 0, 3 \rangle$ by $2^1 3^0 5^2 7^0 11^3 = 66550$, and this representation is unique. Because Gödel numbers can easily become quite large, they are not of much practical value, but are instrumental in the proofs of major results in computability theory; see Section 2.3.2 for a simple application.

†As is common in mathematical literature, we use $\cdot$ here instead of $\times$ to denote the multiplication operator. The two symbols will be used interchangeably. Furthermore, whenever an operator is omitted between two numbers, multiplication is usually implied (as in the product of factors introduced here, or in the quadratic formula in Section 1.2.2). In programming languages, multiplication is commonly denoted by the symbol $*$.

Recall that a rational number is the ratio of two integers, denoted b/a. Each rational can be reduced to the ratio of two relatively prime numbers, by cancelling (dividing each by) common prime factors. Hence, each rational (except 0) has a unique factorization $p_1^{n1}p_2^{n2}...p_k^{nk}$, where p_i is a factor of b if $n_i>0$, and of a if $n_i<0$. For example, $4/10 = 2^2 3^0 5^0 .../2^1 3^0 5^1 ... = 2^1 3^0 5^{-1} ... = \langle 1,0,-1,...\rangle$. Noninteger Gödel numbers can therefore be defined for sequences of signed integers. For example, we may represent $\langle -1,0,1\rangle$ by $2^{-1}3^0 5^1=2.5$.

Remainders Let a and b be positive natural numbers. If a is a factor of b, we may write $b=a\cdot q$ for some number q. However, if a (>0) is not a factor of b, then it may be proved that there exists one and only one pair of numbers q and r such that $b=a\cdot q+r$ and $0<r<a$; q is called the *quotient* of b 'divided by' a, and r is called the *remainder*. We denote r by b **mod** a. For example, 11 **mod** 7=4, and 7 **mod** 11=7. Note that $r=0$ if a is a factor of b. For the case $a=0$, we define b **mod** $0=b$.

If $b_1=a\cdot q_1+r$ and $b_2=a\cdot q_2+r$, $0\leq r<a$, i.e., if b_1 **mod** $a=b_2$ **mod** a, then we say that b_1 is *congruent to* b_2 *modulo* a, denoted $b_1\equiv b_2$ (**modulo** a). For example, $25\equiv 35$ (**modulo** 10). Computer arithmetic is commonly performed **modulo** m, where m is hardware dependent, so that adding two numbers x and y results not in the sum $s=x+y$, but rather $s'=(x+y)$ **mod** m; note $s\equiv s'$ (**modulo** m). A result we will use later is that if d is a factor of both a and b, then d is also a factor of b **mod** a.

Modulo Function The definition of x **mod** y as the remainder of x divided by y applies only for positive integers x and y. We generalize this 'modulo' function, now denoted **mod**(x,y), as follows:

$$\mathbf{mod}(x,y)=x-\mathbf{whole}(x/y)\cdot y \qquad \text{if } y\neq 0,$$
$$\mathbf{mod}(x,0)=x.$$

This definition applies to both positive and negative integers. For example,

$$\mathbf{mod}(7,5)=2$$
$$\mathbf{mod}(-7,5)=-2$$
$$\mathbf{mod}(7,-5)=2$$
$$\mathbf{mod}(-7,-5)=-2.$$

The same definition can be extended to arbitrary reals. For example, **mod**$(x,1)$ is the fractional part of x, previously denoted **fract**(x), for any positive or negative real; thus, $x=$**whole**$(x)+$**mod**$(x,1)$.

Example: Residue number system Suppose m_1, m_2, ..., m_k are positive integers that are relatively prime in pairs (i.e., taken two at a time). Then for any k integers a_1, a_2, ..., a_k, there exists a number x such that

$$x\equiv a_i \ (\textbf{modulo } m_i) \qquad \text{for } i=1,2,...,k.$$

(This is commonly known as the **Chinese Remainder Theorem,** another important result in number theory.) Furthermore, there is exactly one such number x satisfying $1 \le x \le m_1 m_2 \ldots m_k$, so that all numbers between 1 and the product $m_1 m_2 \ldots m_k$ have a unique representation as a sequence of k integers, denoted $\langle\langle a_1, a_2, \ldots, a_k \rangle\rangle$, where $a_i = x$ **mod** m_i for $i=1,2,\ldots,k$. The remainder a_i is also called the *residue* of x with respect to the *modulus* m_i.

For example, let $m_1=2$, $m_2=3$, $m_3=5$. Then for $a_1=1$, $a_2=2$, $a_3=3$, write

$$x \equiv a_1 \text{ (modulo } m_1), \quad x \equiv a_2 \text{ (modulo } m_2), \quad x \equiv a_3 \text{ (modulo } m_3).$$

or

$$x \equiv 1 \text{ (modulo 2)}, \quad x \equiv 2 \text{ (modulo 3)}, \quad x \equiv 3 \text{ (modulo 5)}.$$

The value of x that satisfies the above is 23 [since 23 **mod** 2=1, 23 **mod** 3=2, 23 **mod** 5=3]. By the theorem, 23 is the only solution between 1 and :30, hence 23 is uniquely represented by the sequence of three numbers $\langle\langle 1,2,3 \rangle\rangle$. Each other number between 1 and 30 has such a unique representation. For example, 13 is uniquely represented in this 'residue number system' by $\langle\langle 1,1,3 \rangle\rangle$.

The main advantage of representing numbers in a residue number system, i.e., by their remainders with respect to a 'set of moduli' $\{m_1, m_2, \ldots, m_k\}$, is that certain arithmetic operations can be performed more efficiently, e.g., by employing parallel computation. (This advantage has been exploited mostly by hardware designers.) The basic idea behind 'modular arithmetic' is that

$$\langle\langle a_1, a_2, \ldots, a_k \rangle\rangle \textbf{ op } \langle\langle b_1, b_2, \ldots, b_k \rangle\rangle =$$
$$\langle\langle (a_1 \textbf{ op } b_1) \textbf{ mod } m_1,\ (a_2 \textbf{ op } b_2) \textbf{ mod } m_2, \ldots,\ (a_k \textbf{ op } b_k) \textbf{ mod } m_k \rangle\rangle$$

where **op** is $+$, $-$, or $\times$, but not $/$. See Knuth [1.13] for details.

1.1.7 Application: Euclidean Algorithm

We prove here the main results upon which the Euclidean algorithm for computing greatest common divisors is based. Let m and n be positive integers, and define **gcd**(m,n) as their greatest common divisor; an integer d is a common divisor of m and n if $m=\alpha \cdot d$ and $n=\beta \cdot d$ for some integers α and β, and d is greatest if every other common divisor of m and n also divides d. Obviously, **gcd**(m,n)=**gcd**(n,m). We leave as exercises proofs that (i) if $m=n$, then **gcd**$(m,n)=m=n$, and (ii) if $m>0$ and $n=0$, then **gcd**$(m,0)=m$.

We wish first to prove that

$$\textbf{gcd}(m,n) = \textbf{gcd}(m-n,n) \tag{1.1.7(a)}$$

for $m>n$. Let d=**gcd**(m,n) and d'=**gcd**$(m-n,n)$. Then $m=\alpha \cdot d$, $n=\beta \cdot d$, $m-n=\alpha' \cdot d'$, $n=\beta' \cdot d'$ for integers α, β, α', β'. Hence

$$m-n=(\alpha-\beta)\cdot d=\alpha' \cdot d', \quad n=\beta \cdot d=\beta' \cdot d',$$

and

$$m=\alpha \cdot d=\alpha' \cdot d'+\beta \cdot d=\alpha' \cdot d'+\beta' \cdot d'=(\alpha'+\beta')\cdot d'.$$

Since d' divides both m and n, and d divides both $m-n$ and n, we conclude that $d=d'$, as desired.

Suppose now that $m-n>n$. Then $\mathbf{gcd}(m-n,n)=\mathbf{gcd}(m-2\cdot n,n)$, or generalizing (for $m>n$)

$$\mathbf{gcd}(m,n)=\mathbf{gcd}(m-q\cdot n,n)=\mathbf{gcd}(r,n)=\mathbf{gcd}(n,r)$$

where q is the quotient of m divided by n and $r=m\,\mathbf{mod}\ n$. Thus,

$$\mathbf{gcd}(m,n)=\mathbf{gcd}(\text{'min-value', 'remainder'}) \qquad (1.1.7(b))$$

where 'min-value' is the smaller of m and n and 'remainder' is the remainder of the larger divided by the smaller. The Euclidean algorithm utilizes Eq.(1.1.7(b)) iteratively, stopping when either (i) or (ii) above applies. If a remaindering function (e.g., **mod**) is available, then Eq.(1.1.7(b)) can be used directly; otherwise, the remainder can be found by successive subtraction using Eq.(1.1.7(a)). The fact that m **mod** $n=(m-n)$ **mod** n for $m>n$ is useful when division is not available.

1.1.8 Application: Random Numbers

Random-number generating functions are commonly based upon the *linear congruential* method, where integers are generated iteratively (deterministically) using:

$$x_{i+1}=(a\times x_i+c)\ \mathbf{mod}\ m \qquad \text{for } i=0,1,2,\ldots\ .$$

The parameters a, c, and m are integer constants which are chosen so that the sequence x_1, x_2, x_3, ... satisfies desirable statistical properties, and which depend upon the radix and word-length of the computer being used.[†] The only programmer-supplied parameter is x_0, which may be chosen arbitrarily (e.g., based upon the time-of-day) to generate new sequences, or which may be a given constant to generate the same sequence again and again.

We remark that, to save time, m is usually chosen to be r^L, where r and L are the radix and word-length of the computer being used. Then if the parameters are all L-digit unsigned integers,

$$x_{i+1}=((a\times x_i)\ \mathbf{mod}\ r^L)+c$$

is simply the low-order half of the $2L$-digit product plus c; in practice, the addition may result in an overflow, which is disregarded. The low-order digits of numbers generated in this fashion tend not to be very random, so it is common to regard the numbers as fractional parts, i.e., as having an implied radix point at the left. In computers having floating-point numbers

[†]For a comprehensive treatment of random-number generating algorithms and related statistical tests, see Knuth [1.13].

with fractions shorter than full-word integers, low-order digits may be automatically dropped by right shifts if the random integers (x) are to be converted to floating-point reals (x'); e.g., if $x = \mathrm{d_1d_2d_3d_4d_5d_6d_7d_8}$, we may let $x' = \langle \mathrm{d_1d_2d_3d_4d_5d_6},0 \rangle$. Note $0 \le x' < 1$ for any random real x'. To obtain random integers between, say, 1 and 100, let

$$x'' = \textbf{int}\ (x' \times 100) + 1$$

(rather than using x **mod** $100+1$).

In conclusion, we emphasize that the numbers so generated are 'uniformly' distributed (at least insofar as simple statistical tests can determine), so that one number is as likely to be generated as any other; i.e., it is equally likely that a number will be generated between a_1 and b_1 as between a_2 and b_2, where $b_1 - a_1 = b_2 - a_2$. (Be wary, however, of the possibility that for finite sequences anything can happen; e.g., that ten successive coin flips of an honest coin might come up 'heads'.)

In practice, c is often chosen to be zero, in order to save the time required for the addition, but at the expense of shortening the 'period' before the reoccurrence of x_0 as one of the generated numbers. Recall from Section 1.1.6 that if d is a divisor of both m and x_i, then d is also a divisor of x_{i+1} and all subsequent generated numbers. As a consequence, only about $(1/d)$-th of all possible numbers would be generated. Hence x_0 should be chosen relatively prime to m. (It has also been shown that if $m = 2^L$ (with $L \ge 4$), then the multiplier should be chosen so that $a \equiv 3$ or 5 (**modulo** 8).)

Example*: *RANDU The uniform random number generating subroutine provided in the FORTRAN 'Scientific Subroutine Package' (IBM [1.10]) for the IBM 370 is as follows:

```
C.. IBM 370 SSP UNIFORM RANDOM NUMBER GENERATOR [line 1]
    SUBROUTINE RANDU(IX, IY, YFL)               [line 2]
    IY=IX*65539                                 [line 3]
    IF(IY) 5,6,6                                [line 4]
5   IY=IY+2147483647+1                          [line 5]
6   YFL=IY                                      [line 6]
    YFL=YFL*. 4656613E-9                        [line 7]
    RETURN                                      [line 8]
    END                                         [line 9]
```

(Readers are advised not to use this subroutine in practice; see Forsythe [1.5].)

Observe that

$$65539=2^{16}+3\approx\sqrt{2^{32}},\ 2147483647=2^{31}-1,$$

and

$$0.4656613\text{E}-9=2^{-31}.$$

From this, it is clear that this FORTRAN subroutine is intended for a 32-bit word computer, and will not work on computers with other word-lengths; it also assumes 2's-complement representation of integers. The subroutine has an integer input parameter (IX), and a floating-point output parameter (YFL); the remaining parameter (IY) is also an output parameter which should be used as the input for the next invocation of the subroutine. In other words, to generate an array of 100 random numbers, we would write

```
      IX=ISEED
      DO 99 I=1, 100
      CALL RANDU(IX, IY, YFL)
      IX=IY
      ARRAY (I)=YFL
   99 CONTINUE
```

where ISEED is an arbitrarily chosen starting value, which is assumed to be a positive odd integer. YFL should be a floating-point fractional number, uniformly distributed between 0 and 1.

The subroutine assumes, in line 3, that the low-order half of the double-word product IX∗ 65539 is assigned to IY, and that the high-order half of the product is disregarded whether zero or not. (The latter event is not to be regarded as an overflow error.) While IX∗ 65539 must be positive (since IX was assumed positive), the leading bit of the low-order half of the product may be either 0 or 1, so IY need not be positive. Since only positive random numbers are to be generated, a test for negative IY is made at line 4; the effect of line 5 is just to change the sign bit of IY (which is not the same as negating or taking the absolute value of IY, for 2's-complement numbers). Thus, at line 6, IY is positive. Since 65539 is odd, if IX is also odd, then IY must likewise be odd (and hence nonzero).

At line 6, the positive integer IY is converted to a normalized floating-point number; some of the low-order bits may be dropped in the process. At line 7, the floating-point number is then 'fractionalized' by, in effect, dividing it by 2^{31}, an upper bound on possible values of IY. This converts an integer-valued YFL to a fractional number, i.e., a number between 0 and 1 (exclusive), as desired. (The reader should verify that the value of YFL cannot be either of the limiting values, 0 and 1.)

1.2 FUNCTIONS AND ALGORITHMS

We have presupposed that readers have a working knowledge of the concepts of functions and operations, at least to the extent that they are found in computer programming languages. **MOD**, **ABS**, and **SGN**, for example, are functions commonly provided in programming languages; the arithmetic operations +, −, ×, and / and the relational operations <, =, and > are likewise. Many programming languages also provide the facility to define new functions and operations using *formulas* (or 'arithmetic statement functions') or, more generally, using *algorithms* (or 'subroutines').

1.2.1 Functional Notation

One apparent difference between functions and operations is notational: a reference to a function is commonly written in the form $\mathbf{F}(x,y,...)$, where $x,y,...$, are called the ***arguments*** and **F** is the name given to the function, whereas an operation (having arguments) is commonly written in the form $x \circ y$, where x and y are called the *operands* and $\circ$ is a symbol (or *operator*) used to denote the operation. This notational difference is of limited significance, however, especially since what may be a 'function' in one language may be an 'operation' in another, and vice versa. For example, the addition operation is written PLUS(X,Y) in LISP, while the remainder (modulo) function is written Y|X in APL. We shall see later that, mathematically speaking, the difference between functions and operations is in the restrictions imposed upon the arguments or operands and upon the resulting values.

Informally, we say that one property of an object or system is a 'function' of other properties if the former depends upon the latter in some fashion. For example, intelligence (I) is commonly said to be a function of heredity (H) and environment (E); symbolically, we write $I=\mathbf{F}(H,E)$, where **F** is an arbitrary name given to the function. Suppose now that I, H and E are 'measurable' (not necessarily quantitatively), and that h_1 and e_1 denote particular measured values of H and E, respectively, for some person (No. 1), h_2 and e_2 denote measured values for person No. 2, etc. Then we denote by i_1 the *value* of I or **F** associated with person No. 1, and write $i_1=\mathbf{F}(h_1,e_1)$, and similarly, $i_2=\mathbf{F}(h_2,e_2)$. If it turns out that, for two different persons, $h_1=h_2$ and $e_1=e_2$ but $i_1 \neq i_2$ (as may be the case for twins), then **F** is *not* a function. Formally, we require that functions be *single-valued*; for identical arguments, their values must be identical. Hence, the mathematical square-root operation is not a function since it has both positive and negative values; e.g., $\sqrt{4}=+2$ and -2.†

†To distinguish these cases, we define $\mathbf{psqr}(x)$ and $\mathbf{nsqr}(x)$ to be the positive and negative square roots, respectively. When the context indicates that $\mathbf{sqr}(x)$ must be a function, e.g., when referring to the programming language function SQR, then **psqr** is implied.

Functions may often be defined 'by cases'—i.e., by listing all possible arguments and specifying the value of the function associated with each argument. Functions of two arguments are commonly defined by a tabular listing, such as

F	e_1	e_2	e_3
h_1	i_1	i_3	i_2
h_2	i_4	i_2	i_1

where, for example, i_2=**F**(h_1, e_3). Such a listing is called a *Cayley table*, which may also be thought of as a *relation* which 'maps' (matches) possible arguments to possible values. In practice, such a tabular listing is only feasible if the number of cases is small. Alternatively, some functions can be defined in terms of other functions 'by formula', e.g.,

sgnxfr$(x,y)=$**abs**$(x)\times$**sgn**(y).

More complex functions, e.g., **log**(x), require definition 'by computational algorithm'. We elaborate in Section 2.3.4.

1.2.2 Algebraic Expressions

An *algebraic expression* is a generalization of the programming language element known as 'arithmetic expressions' with which we assume reader familiarity. Loosely, an *arithmetic* expression is a formula consisting of 'variables' and 'constants' connected by arithmetic operators, possibly parenthesized, and subject to certain grammatical (syntactic) rules. Other algebraic expressions are characterized by operators other than the usual arithmetic ones. For example, a logical expression contains logical operators ($\wedge$, $\vee$, $\neg$: i.e., *and, or, not*) which produce 'truth values' (**true** and **false**). The logical operators have the same meaning and precedences as in programming languages, with which we also assume reader familiarity. In a general algebraic expression, the operators can be arbitrary.

Any algebraic expression defines a function whose arguments are the variables appearing in the formula. For example, the 'quadratic formula' $(-b+$**sqr**$(b^2-4ac))/(2a)$ is an arithmetic expression, which defines a numerical-valued function **f**(a,b,c); $b^2<4ac$ is a relational expression which defines a truth-valued function **g**(a,b,c); $p \wedge \neg q \vee q \wedge \neg p$ is a logical expression which defines a truth-valued function $\alpha_1(p,q)$; A^{-1} (the inverse of matrix A) is an algebraic expression which defines a matrix-valued function **f**(A) (or **f**$(a_{11},\ldots,a_{NN})$); and (**if** $\neg p$ **then** q **else** $\neg q$) is an algebraic expression (called a 'conditional expression') which defines a truth-valued function $\alpha_2(p,q)$ having the same truth value as q if p is false, while having the opposite truth value to q otherwise.

Functional Equivalence Two algebraic expressions are said to be *functionally equivalent* if they define the same function in the sense that their values are equal for each identical set of value assignments. For example, α_1 and α_2 above are functionally equivalent, denoted $\alpha_1 \equiv \alpha_2$, since $\alpha_1(p,q) = \alpha_2(p,q)$ for each of the four possible cases, where p and q are both **true** or both **false**, or where one is **true** and the other **false**. It is common to denote the truth-value **true** by the number 1 and **false** by the number 0, or by the symbols ***T*** and ***F***, respectively. Then *evaluating* the foregoing algebraic expressions, we have

$$
\begin{aligned}
&\alpha_1(1,1)=\alpha_2(1,1)=0\\
&\alpha_1(1,0)=\alpha_2(1,0)=1\\
&\alpha_1(0,1)=\alpha_2(0,1)=1\\
&\alpha_1(0,0)=\alpha_2(0,0)=0.
\end{aligned}
$$

By 'evaluate', we mean to determine whether the expression has value 0 or 1 for particular value assignments.

Evaluation of Expressions The conventional definitions of arithmetic operations and their precedences or 'rankings' dictate how arithmetic expressions (say, $a\times-b+b\times-a$, for given numerical values of a and b) are to be evaluated. If $a=b=1$, then we write tabularly

	a	$\times$	$-$	b	$+$	b	$\times$	$-$	a
[Step 0]	(1)			(1)		(1)			(1)
[Step 1]			(−1	)				(−1	)
[Step 2]	(	−1		)		(	−1		)
[Step 3]	(				−2				)

or more compactly

a	$\times$	$-$	b	$+$	b	$\times$	$-$	a	[expression]
1	−1	−1	1	−2	1	−1	−1	1	[values]
0	2	1	0	3	0	2	1	0	[step no.]

Step 0 is the substitution of values for variable-symbols, Step 1 is the computation of highest ranking operations ($-b=-1$, $-a=-1$), Step 2 is the computation of the next highest ranking operations ($1\times-1=-1$, twice), and Step 3 is the computation of the final result ($-1+-1=-2$). This same procedure can be used to evaluate logical or more general algebraic expressions. We illustrate this further in Section 1.3.

In general, evaluation of expressions or more complex functions is governed by certain sequencing rules. For example, in the above, the operations of Step 1 must be performed before those of Step 2. However, two operations in each step may be performed in either order. Such 'ambiguity' has the advantage of providing greater flexibility, but we will consider only 'deterministic' algorithms here.

1.2.3 Algorithms

Computers are often said to be 'information processors' in that they process or transform specified input information into output information satisfying specified requirements. The transformation may be regarded as a function $OUT=\mathbf{F}(IN)$, where the function **F** is determined by the 'algorithm' (set of processing instructions) which the computer executes. The task of the computer programmer is to devise the algorithm given a specification of the function (usually in terms of input-output relationships); the algorithm must be expressed in a language which the computer (as well as other programmers) can understand and effectively process. Hence, an algorithm must be unambiguous: each of its instructions must be rigorously defined, and the sequence in which instructions are to be executed must also be well-defined. Algorithms must satisfy several finiteness constraints. Instructions must require only a finite amount of space (both to represent and to execute) and time. Furthermore, an algorithm must consist of a finite number of well-defined instructions, and each instruction must be executed only a finite number of times.

We conclude that a computer program containing an infinite loop (e.g., of the form ℓ:GOTOℓ) does not qualify as an algorithm, at least as a 'correct' one. In other words, an *algorithm* is an abstraction of what we wish computer programs to be: namely, a set of processing instructions which is well-defined and which requires a finite amount of space and time. From another point of view, algorithms can be *implemented* by computer programs, but computer programs can also implement *non*algorithms; sometimes we will say (rather loosely) that a nonalgorithm is an incorrect algorithm.

We say that a program or algorithm correctly implements a function **F** if its input-output relationships are identical, i.e., if the algorithm, given an input i, produces $\mathbf{F}(i)$ as its output for each i for which $\mathbf{F}(i)$ is defined. In practice, the algorithm should produce an error message as its output if $\mathbf{F}(i)$ is not defined. Two algorithms are said to be *functionally equivalent* if they implement the same function.

Algorithmic Languages Any language in which an algorithm can be expressed may be regarded as an 'algorithmic' language. Examples include 'conventional' programming languages, pictorial languages (e.g., flowcharts), tabular languages (e.g., decision tables), or narrative languages (such as English). Since any algorithmic language for which a suitable processor (compiler) can be designed is a *programming* language, we use the term 'conventional' when we wish to limit our discussion to languages of the kind considered by Sammet [1.23]. Decision tables provide an example of an unconventional programming language; see Lew [1.15].

We express, without comment, the Euclidean algorithm for finding the greatest common divisor of two positive integers in a conventional (Pascal-like) programming language, in a flowchart (Figure 1.2.3), and in a repetitive[†] decision table (Table 1.2.3):

```
if x>0 and y>0
  then
    begin
      while x≠y do
        begin
          if x>y then x:=x−y;
          if y>x then y:=y−x
        end;
      put x
    end;
```

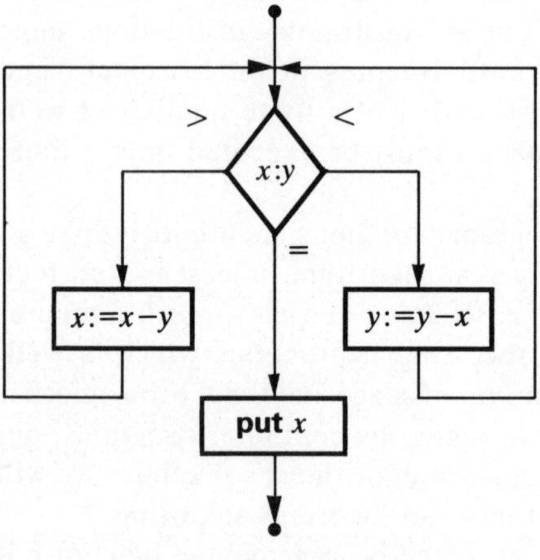

Figure 1.2.3 Flowchart

$x>0$	T T T
$y>0$	T T T
$x : y$	> < =
$x:=x-y$	× − −
$y:=y-x$	− × −
put x	− − ×
exit	− − ×

Table 1.2.3 Decision Table

1.2.4 Data Structures

A computer *data structure* is a collection of data items (usually numbers or character strings) that are related in some way, at the least by a common name. Data structures may be classified in many ways—e.g., according to how the data items are physically stored within a computer's storage units, or

[†]We adopt the convention that execution of a decision table is repeated unless an explicit **exit** action is encountered.

according to how individual data items can be located by data processing programs. We adopt the latter classification scheme here; some data structures may be classified in more than one way.

In an *ordinal* data structure or *list*, data items are logically (not necessarily physically) ordered so that there is a first, second, third, etc., item, and a position number or 'ordinate' is used to specify the location of each item. Usually, an item in the structure can only be accessed by accessing the preceding item first. (Note we say that the nth data item is the 'immediate predecessor' of the $(n+1)$th data item, and the 'immediate successor' of the $(n-1)$th.) An ordinal data structure whose data items are physically stored by increasing storage address according to their logical ordering is called *sequential*.

In a *linked* data structure, data items are connected by *pointers*, i.e., by portions (fields) within each 'data-node' which specify the locations of other items in the structure. A pointer specifying a location that is regarded as illegal is called a *null* pointer. If the data-node containing data-item x also contains a pointer specifying the location of the data-node containing data-item y (more briefly, if item x points to item y), then we say that y is an 'immediate successor' of x, and x is an 'immediate predecessor' of y. We remark that the 'location' specified by a pointer may be a direct memory address, an ordinal position number, a key (see below), or an 'indirect' memory address (i.e., the location of another pointer, which in turn specifies the 'location' of the desired data item).

In a *mapped* data structure, the location of each data item is found by evaluation of a function mapping some property or 'key' associated with an item to its location; the function can be specified either as a table or as an algorithm. A table that specifies the locations of data items is commonly called a *directory* (or *index*); an algorithm for calculating locations is termed a *direct-addressing function*, if each key is mapped to some unique location, or a *hash function*, otherwise. The remark about the 'location' specified by a pointer (in the preceding paragraph) applies also to locations specified by maps.

It is important to recognize that information is conveyed, not just by data items as individuals, but also by whether and how they are related to each other. For example, a name and its associated address, when considered independently, do not convey the information that they are attributes of the same object. We remark that a data structure generally has one distinguished data item which serves as its representative; we call this data item, or a pointer thereto, the *base* of the data structure. The 'name' of the data structure is usually associated with the address of its base. Formally, items within a data structure are accessed by using a function **f**(*item-id*)=*loc*, where *item-id* identifies the item—i.e., is its ordinate or its key—and *loc* is its storage location. Mechanisms for locating data items in a variety of data structures are provided by all programming languages. For further information, see Knuth [1.12] or Standish [1.24].

1.3 LOGIC AND PROOFS

Recall the tabular approach to evaluating algebraic expressions introduced in Section 1.2.2. To evaluate (say) the logical expression or 'proposition'

$$\alpha[p,q]=p \wedge \neg q \vee q \wedge \neg p,$$

we write

p	$\wedge$	$\neg$	q	$\vee$	q	$\wedge$	$\neg$	p	[expression]
1	0	0	1	0	1	0	0	1	[values: $p=q=1$]
0	2	1	0	3	0	2	1	0	[step no.]

i.e., the value of the logical expression $\alpha[p,q]$ is 0 for the value assignments $p=q=1$. (We have adopted here the convention found in most programming languages: namely, that $\neg$ has higher precedence than $\wedge$, and $\wedge$ than $\vee$.†) For all possible combinations of values of p and q, we have

p	$\wedge$	$\neg$	q	$\vee$	q	$\wedge$	$\neg$	p	[expression]
0	0	1	0	0	0	0	1	0	[$p=0$, $q=0$]
0	0	0	1	1	1	1	1	0	[$p=0$, $q=1$]
1	1	1	0	1	0	0	0	1	[$p=1$, $q=0$]
1	0	0	1	0	1	0	0	1	[$p=1$, $q=1$]
0	2	1	0	3	0	2	1	0	[step no.]

This is called the ***truth table*** for the given logical expression; the step numbers are commonly omitted, as they are implicitly determined by the conventional precedences of the logical operations.

Logical Operations The logical operations ***or*** ($\vee$), ***and*** ($\wedge$ or &), and ***not*** ($\neg$) are commonly defined by Cayley tables as given below.

$\vee$	0	1
0	0	1
1	1	1

$\wedge$	0	1
0	0	0
1	0	1

$\neg$	0	1
	1	0

†Subsequently, $\lessdot$, $\doteq$, and $\gtrdot$ will be used to denote precedences, so that $\neg \gtrdot \wedge$ and $\vee \lessdot \wedge$, and each operation $\doteq$ itself.

Their equivalent truth-table definitions are shown below:

p	$\vee$	q
0	0	0
0	1	1
1	1	0
1	1	1
0	1	0

p	$\wedge$	q
0	0	0
0	0	1
1	0	0
1	1	1
0	1	0

$\neg$	p
1	0
0	1
1	0

The *exclusive-or* of p and q, denoted $p \oplus q$, is defined by

$\oplus$	0	1
0	0	1
1	1	0

or

p	$\oplus$	q
0	0	0
0	1	1
1	1	0
1	0	1
0	1	0

Note that it is functionally equivalent to the expression $p \wedge \neg q \vee q \wedge \neg p$.

The binary operation *implies*, denoted $\rightarrow$, is defined by

$\rightarrow$	0	1
0	1	1
1	0	1

or

p	$\rightarrow$	q
0	1	0
0	1	1
1	0	0
1	1	1
0	1	0

From the truth table

$\neg$	p	$\vee$	q
1	0	1	0
1	0	1	1
0	1	0	0
0	1	1	1
1	0	2	0

we see that $\neg p \vee q$ is functionally equivalent to $p \to q$; hence $\to$ may be defined in terms of $\neg$ and $\vee$. We call p the *antecedent* and q the *consequent* of the *(material) implication* $p \to q$. Note that the statement 'p implies q' is true 'by default' whenever p is false, regardless of the truth or falsity of q.

We next define the *(material) equivalence* of p and q, denoted $p \leftrightarrow q$, by

$\leftrightarrow$	0	1
0	1	0
1	0	1

or

p	$\leftrightarrow$	q
0	1	0
0	0	1
1	0	0
1	1	1
0	1	0

From the truth table

$(p$	$\to$	$q)$	$\wedge$	$(q$	$\to$	$p)$
0	1	0	1	0	1	0
0	1	1	0	1	0	0
1	0	0	0	0	1	1
1	1	1	1	1	1	1
0	1	0	2	0	1	0

we see that $(p \to q) \wedge (q \to p)$ is functionally equivalent to, and hence may also define, $p \leftrightarrow q$.

We emphasize that material equivalence ($\leftrightarrow$) is an operation defined between truth values whereas functional equivalence ($\equiv$) is a relation between logical expressions. Two variables are said to be functionally and materially equivalent if they must always have the same value. In analogy, we say expression α functionally implies expression β, denoted $\alpha \Rightarrow \beta$, if $(\alpha \to \beta) = 1$ for all value assignments. We then say that β is a *necessary* condition for α, and that α is a *sufficient* condition for β.

In conclusion, we extend our definition of precedences to include the operations $\leftrightarrow$, $\to$, and $\oplus$ as follows:

$$\leftrightarrow \lessdot \to, \qquad \to \lessdot \oplus, \qquad \oplus \doteq \vee, \qquad \vee \lessdot \wedge, \qquad \wedge \lessdot \neg.$$

1.3.1 Tautologies

The *propositional calculus* is concerned in essence with the evaluation of logical expressions, usually with the intent of determining whether an expression $\alpha[x_1, ..., x_K]$ is equal to 1 for all possible value assignments. If so, the expression is called a *tautology* (or *valid*), and we write $\alpha \equiv$**true**. For example,

$$p \oplus q \leftrightarrow p \wedge \neg q \vee q \wedge \neg p,$$
$$\neg p \vee q \leftrightarrow p \rightarrow q,$$

and

$$(p \rightarrow q) \wedge (q \rightarrow p) \leftrightarrow (p \leftrightarrow q)$$

are tautologies (from above). As another example, the logical expression below, commonly known as *modus ponens*, is a tautology:

$(p$	$\rightarrow$	$q)$	$\wedge$	p	$\rightarrow$	q
0	1	0	0	0	1	0
0	1	1	0	0	1	1
1	0	0	0	1	1	0
1	1	1	1	1	1	1
0	1	0	2	0	3	0

(This is called an *implicational tautology* because it is an implication $\rightarrow$ that is always equal to 1; the preceding examples are *equivalence tautologies* because they are equivalences $\leftrightarrow$ that are always equal to 1.) We also say that an expression α is a *contradiction* (or *inconsistent*) if it is always equal to 0, and write $\alpha \equiv$**false**. We note that α is a contradiction if and only if $\neg \alpha$ is a tautology. Expressions which are neither tautologies nor contradictions are called *contingencies*. For example, $p \wedge \neg p$ is a contradiction, $p \rightarrow p$ [the *law of identity*] and $p \vee \neg p$ [the *law of the excluded middle*] are tautologies, and $p \rightarrow q$ is a contingency. Tables 1.3.1(a) and (b) summarize a number of other important tautologies, each of which can be proven by truth-table evaluations.

[R1] (modus ponens)	$(p \rightarrow q) \wedge p \rightarrow q$
[R2] (modus tollens)	$(p \rightarrow q) \wedge \neg q \rightarrow \neg p$
[R3] (disjunctive syllogism)	$(p \vee q) \wedge \neg p \rightarrow q$
[R4] (hypothetical syllogism)	$(p \rightarrow q) \wedge (q \rightarrow r) \rightarrow (p \rightarrow r)$
[R5] (dilemma)	$((p \rightarrow q) \wedge (r \rightarrow s)) \wedge (p \vee r) \rightarrow (q \vee s)$
[R6] (absorption)	$(p \rightarrow q) \rightarrow (p \rightarrow p \wedge q)$
[R7] (simplification)	$p \wedge q \rightarrow p$
[R8] (addition)	$p \rightarrow p \vee q$
[R9] (contradiction–I)	$(\neg p \rightarrow p) \rightarrow p$
[R10] (contradiction–II)	$(\neg p \rightarrow q) \rightarrow ((\neg p \rightarrow \neg q) \rightarrow p)$
[R11] (contradiction–III)	$q \wedge (q \wedge \neg p \rightarrow r) \wedge (q \wedge \neg p \rightarrow \neg r) \rightarrow p$

Table 1.3.1(a)

[A1] (associativity)	$p \wedge (q \wedge r) \leftrightarrow (p \wedge q) \wedge r$
	$p \vee (q \vee r) \leftrightarrow (p \vee q) \vee r$
[A2] (commutativity)	$p \wedge q \leftrightarrow q \wedge p$
	$p \vee q \leftrightarrow q \vee p$
[A3] (distributivity)	$p \wedge (q \vee r) \leftrightarrow (p \wedge q) \vee (p \wedge r)$
	$p \vee (q \wedge r) \leftrightarrow (p \vee q) \wedge (p \vee r)$
[A4] (De Morgan's law)	$\neg(p \wedge q) \leftrightarrow \neg p \vee \neg q$
	$\neg(p \vee q) \leftrightarrow \neg p \wedge \neg q$
[A5] (involution)	$\neg(\neg p) \leftrightarrow p$
[A6] (idempotence)	$p \wedge p \leftrightarrow p,\ p \vee p \leftrightarrow p$
[A7] (implication)	$p \rightarrow q \leftrightarrow \neg p \vee q$
[A8] (equivalence)	$(p \leftrightarrow q) \leftrightarrow (p \rightarrow q) \wedge (q \rightarrow p)$
[A9] (contraposition)	$p \rightarrow q \leftrightarrow \neg q \rightarrow \neg p$
[A10] (exportation)	$(p \wedge q) \rightarrow r \leftrightarrow p \rightarrow (q \rightarrow r)$

Table 1.3.1(b)

1.3.2 Deduction

In general, it is desirable to use known properties of the various logical operations to prove or disprove that an arbitrary logical expression is a tautology without resorting to a truth-table evaluation. (The latter is not only regarded as inelegant, but also is impractical for complex multi-variable expressions. If there are r variables, then the truth table has 2^r rows—the number of 'r-arrangements' of the two truth values—see Section 6.2.3.) We do so by treating the known tautologies as axioms, and using certain 'rules of inference' to deduce the validity of a given expression, if possible. The main rule of inference is the *substitution rule*: if logical expressions α and β are functionally equivalent, then we may replace any occurrence of α by β in any other logical expression without affecting the latter's truth values. (In the special case where α is a variable, this rule permits us to make a 'change of variable' by replacing *each* occurrence of α by β, which may be a materially equivalent variable or an arbitrary expression.) We remark that if $\alpha_1 \equiv \alpha_2$ and $\alpha_2 \equiv \alpha_3$, then $\alpha_1 \equiv \alpha_3$ (so that we may replace α_1 by α_3).

Examples To prove contraposition ([A9] of Table 1.3.1(b)) assuming other tautologies as axioms, we use the definition of $\rightarrow$ ([A7] of Table 1.3.1.(b))

$$p \rightarrow q \equiv \neg p \vee q.$$

Substituting $\neg q$ for p and $\neg p$ for q, we obtain

$$\neg q \rightarrow \neg p \equiv \neg(\neg q) \vee (\neg p)$$

and, by [A5] and [A2],

$$\neg q \to \neg p \equiv q \vee \neg p \equiv \neg p \vee q.$$

Substituting $\neg q \to \neg p$ for $\neg p \vee q$ in [A7], we have [A9]

$$p \to q \equiv \neg q \to \neg p.$$

To prove [A10] assuming other tautologies as axioms, we proceed as follows:

$$\begin{aligned} (p \wedge q) \to r &\equiv \neg(p \wedge q) \vee r && \text{by [A7]} \\ &\equiv (\neg p \vee \neg q) \vee r && \text{by [A4]} \\ p \to (q \to r) &\equiv p \to (\neg q \vee r) && \text{by [A7]} \\ &\equiv \neg p \vee (\neg q \vee r) && \text{by [A7]} \end{aligned}$$

where [A10] follows by [A1].

Modus Ponens In many cases, we are not interested in whether a logical expression is a tautology, but instead are interested only in whether a logical expression is true for some specific value assignments, the latter being specified by a set of 'premisses'. A *premiss* is a logical expression α_i that is assumed to be true. Premisses may impose certain restrictions on the variables which they contain. For example, if $p \to q$ is always true, then $p=1, q=0$ is not a permissible value assignment; if $p \to q$ and p are both assumed to be true, then q must equal 1.

A logical expression β that can be deduced from a set of premisses $\{\alpha_1, \ldots, \alpha_n\}$ is called a *conclusion* (of the set). Let

$$A = (\alpha_1 \wedge \alpha_2 \wedge \cdots \wedge \alpha_n).$$

If the logical expression $A \to \beta$ can be shown to be a tautology, then a true antecedent A (as assumed) implies that the consequent β is true (as desired), and we may write $A \Rightarrow \beta$. This statement is actually a 'rule of inference', a special case of *modus ponens*,

[R1] $((\alpha \to \beta) \wedge \alpha) \to \beta,$

which we have previously shown to be a tautology. Any implication tautology can be used as a rule of inference in a similar fashion. For example, *modus tollens* [R2] permits us to deduce that β is true if the premisses $\alpha_1 = (\neg \beta \to q)$ and $\alpha_2 = (\neg q)$ are true. Note that we have implicitly used the *conjunction* rule, that $\alpha_1, \ldots, \alpha_n$ are all true iff $A = \alpha_1 \wedge \cdots \wedge \alpha_n$ is also true, and the previously stated substitution rule. It should finally be emphasized that *a false* (inconsistent) *antecedent conveys no information about a consequent.*

We remark that the tautologies in Tables 1.3.1(a) and (b) are not independent since some can be deduced from others. This redundancy is helpful in that proofs of other logical expressions may become much simpler if there is a wider choice of permissible substitutions (equivalence tautologies) and rules of inference (implicational tautologies). Note that each equivalence also specifies two implications.

Proof by Contradiction A common approach to proving a logical expression p to be true is by *contradiction*. $p \wedge \neg p$ is a contradiction by the law of the excluded middle. Hence, the simplest form of a proof by contradiction is to show that $\neg p \rightarrow p$. If it is assumed that p is not true, and if it can be shown that this premiss implies that p is true, then the assumption must be false. In other words,

[R9] $(\neg p \rightarrow p) \rightarrow p$

is a tautology.

If q is a known consequent of $\neg p$, then p may be proved by showing that $\neg p \rightarrow \neg q$ is also true. No statement can imply both q and $\neg q$, hence $\neg p$ must be false, or p must be true. In other words,

[R10] $(\neg p \rightarrow q) \rightarrow ((\neg p \rightarrow \neg q) \rightarrow p)$

is a tautology.

An even more complex form of a proof by contradiction is based on the tautology

[R11] $q \wedge (q \wedge \neg p \rightarrow r) \wedge (q \wedge \neg p \rightarrow \neg r) \rightarrow p.$

In other words, if q is true (the conjunction of all premisses and axioms), and if q and $\neg p$ together imply both r and $\neg r$ (the contradiction), then p must be true.

In summary, a proof of p by contradiction involves assuming $\neg p$ is true, and inferring a contradiction (i.e., p, $q \wedge \neg q$, or $r \wedge \neg r$, respectively). If $\neg p \rightarrow$ **false**, then $p \equiv$ **true**.

In the event that p is an implication, i.e., a logical expression of the form $q \rightarrow r$, we have

$$\neg(q \rightarrow r) \equiv \neg(\neg q \vee r) \equiv (\neg(\neg q) \wedge \neg r) \equiv (q \wedge \neg r).$$

Hence, if we assume $\neg r$, and show that $\neg r \wedge q$ leads to a contradiction, then $q \rightarrow r$ is true. In other words,

$$\neg(q \wedge \neg r) \rightarrow (q \rightarrow r)$$

is a tautology. If r is false, then q must be false also in order for the implication to be true. Of course, if r can be shown to be a tautology, then the implication must be true since $r \rightarrow (q \rightarrow r)$ is a tautology.

Finally, in dealing with implications, it is often convenient to define the *contrapositive* of $q \rightarrow r$ as $\neg r \rightarrow \neg q$, the *converse* of $q \rightarrow r$ as $r \rightarrow q$, and the *inverse* of $q \rightarrow r$ as $\neg q \rightarrow \neg r$. Note that by the contraposition law ([A9] of Table 1.3.1(b)), an implication is functionally equivalent to its contrapositive, and its converse is functionally equivalent to its inverse.

Syllogisms A 'general' *syllogism* may be defined as an argument having two premisses (P_1 and P_2) and a conclusion C, the objective being to prove that the logical expresson

$$P_1 \wedge P_2 \rightarrow C$$

is true. (Many of the tautologies given in Tables 1.3.1(a) and 1.3.1(b) are syllogisms by this definition.) By the definition of material implication, the expression is true if either P_1 or P_2 is false. The main problem then is to show that C is true if P_1 and P_2 are both true.

An important special case is when P_1, P_2 and C are each implications:

[R4] $(p \rightarrow q) \wedge (q \rightarrow r) \rightarrow (p \rightarrow r)$;

this is called a *hypothetical* syllogism, and as noted above is a tautology (the **syllogism law**). A useful alternative is

$$(p \rightarrow (q \rightarrow r)) \rightarrow ((p \rightarrow q) \rightarrow (p \rightarrow r)),$$

which states that if $p \rightarrow q$ is true, then $q \rightarrow r$ need not be true in order for $p \rightarrow r$ to be true; instead, $q \rightarrow r$ need only be implied by p.

Another important class of syllogisms, a slight variation of *modus ponens*, is

[R3] $(p \vee q) \wedge \neg p \rightarrow q$,

also known as a *disjunctive* syllogism.

Finally, a *categorical* syllogism is an argument in which the premisses and conclusions are statements about sets[†] of the following forms:

all members of S_1 are members of S_2;
no member of S_1 is a member of S_2;
some member of S_1 is a member of S_2;
some member of S_1 is *not* a member of S_2.

Traditionally, *Venn-diagram* representations of sets are used to test the validity of categorical syllogisms; see Copi [1.4]. However, a more general approach utilizes the *predicate calculus* (as opposed to the *propositional calculus*), where logical expressions are 'quantified': the four statements above would be written

[†]In set notation (see Chapter 2), these statements may be written as

$S_1 \cap (-S_2) = \emptyset$
$S_1 \cap S_2 = \emptyset$
$S_1 \cap S_2 \neq \emptyset$
$S_1 \cap (-S_2) \neq \emptyset$

respectively.

$(\forall x)$ (x is in $S_1 \rightarrow x$ is in S_2)
$(\forall x)$ (x is in $S_1 \rightarrow x$ is not in S_2)
$(\exists x)$ (x is in $S_1 \wedge x$ is in S_2)
$(\exists x)$ (x is in $S_1 \wedge x$ is not in S_2)

where $\forall x$ means 'for every x' and $\exists x$ means 'there exists an x' (in a universal set); $\nexists x$ means 'there does *not* exist an x.' [$\forall$ is called the *universal quantifier*, and $\exists$ is called the *existential quantifier*; the right-hand side of a quantified statement is called a *predicate*, sometimes denoted **P**(x).] Deduction involving quantified statements is more complex than before, and will not be discussed here; see Quine [1.19]. We remark that the complexity increases even further when the sets, predicates, or functions are also quantified.

1.3.3 Mathematical Induction

We have seen above several ways in which logical expressions can be proven to be true (or false). One way is by truth-table evaluation, and another is by making use of equivalence and implicational tautologies. Suppose now that we wish to prove a proposition **P** about all the natural numbers, where for each $n=0, 1, 2, \ldots$, **P**(n) is either true or false. If **P** is defined for only a finite subset of $\mathcal{J}$, then **P**(n) can be proven by enumeration (or 'exhaustion'), i.e., for each value of n for which **P** is defined. However, if **P** is defined for all the natural numbers, such enumeration is infeasible. An alternative is to use *mathematical induction*:

Principle of mathematical induction Let **P**(n) be a proposition that is defined for each natural number. If **P**(0) (or **P**(1)) is true, and if **P**($i+1$) is true whenever **P**(i) is true, then **P**(n) is true for all natural numbers $n \geq 0$ (or $n \geq 1$).

(This principle follows from Peano's 'induction' axiom, where $\mathcal{J}$ is the set of natural numbers for which **P**(n) is true.) We remark that 'mathematical' induction is not the same as 'empirical' induction, which infers on the basis of a limited number of samples (say, **P**(1), **P**(2),..., **P**(k) are true) that a proposition (say, **P**($k+1$)) is true. Inference of this sort is discussed in Chapter 6.

Example: Let **P**(n) be the statement

$$\Sigma_{i=1}^{n} i = n(n+1)/2.$$

This is true for $n=1$ since

$$1 = 1(2)/2.$$

Consider now $\mathbf{P}(n+1)$:

$$\begin{aligned}1+2+\cdots+(n+1)&=(n+1)(n+2)/2\\&=(n^2+3n+2)/2.\end{aligned}$$

If $\mathbf{P}(n)$ is true, then

$$\begin{aligned}(1+2+\cdots+n)+(n+1)&=(n(n+1)/2)+n+1\\&=n^2/2+n/2+n+1\\&=(n^2+3n+2)/2.\end{aligned}$$

Hence, by the principle of mathematical induction, $\mathbf{P}(n)$ is true for all $n\geq 1$. As exercises (and for later reference), the reader should verify that

$$\Sigma_{i=1}^{n} i^2=n(n+1)(2n+1)/6,$$

and that

$$\Sigma_{i=1}^{k} i\, 2^i=2+2^k(2k-2).$$

1.3.4 Application: Program Correctness Proofs

For computer scientists, a major application of logic is its use in proving propositions about programs—specifically, that a program is correct. For example, consider the following program:

```
input n;
i:=1;
x:=1;
while i≠n do begin
  i:=i+1;
  x:=x+i; end;
exit
```

We claim that this program adds the first n positive numbers; i.e., that, for $n\geq 1$, $x=1+2+\cdots+n$ upon exit. To prove this, we proceed as follows.

First, we attach as labels to each program statement (or block of statements) an *assertion* specifying the 'state' (values of variables) whenever that statement is to be executed. Each assertion is called the *precondition* of the statement which it labels, and is a *postcondition* of each statement whose execution can precede it. For each statement, its precondition, $\mathscr{P}$, together with the actions specified in the statement, $\mathscr{A}$, must imply its postcondition(s), $\mathscr{Q}$, we call this a *specification*, and denote it

$$\mathscr{P}\{\mathscr{A}\}\mathscr{Q}.$$

Labeling the sample program

P_0: **input** n;
P_1: $i:=1$; $x:=1$;
P_2: **while** $i \neq n$ **do begin**
  P_3: $i:=i+1$;
  P_4: $x:=x+i$;
P_5: **end**;
P_6: **exit**

we have the following specifications

P_0 {**input** n} P_1
P_1 $\{i:=1;\ x:=1\}$ P_2
P_2 {**while**...**do**...} P_6
P_3 $\{i:=i+1\}$ P_4
P_4 $\{x:=x+i\}$ P_5

Our objective now is to determine a set of assertions for which these specifications all hold, such that P_6 is our desired conclusion

$$P_6=[x=\Sigma_{j=1}^{n} j \ \&\ n \geq 1].$$

We emphasize that we need both the set of assertions, and also a means of verifying each specification. For the former, we must generally use our knowledge of how the program works; we remark that it does so by computing partial sums

$$\Sigma_{j=1}^{i} j.$$

For the latter, we require rules or axioms which will permit us to infer that $\mathcal{P}\{\mathcal{A}\}\mathcal{Q}$, for each class of statements $\mathcal{A}$. For example, for simple assignment statements of the form $x:=y$, we have as the Axiom of Assignment

$$\mathbf{P}(y)\{x:=y\}\mathbf{P}(x);$$

that is, if $\mathbf{P}(y)$ holds and x is set to y, then $\mathbf{P}(x)$ holds. In fact, we say that this axiom *defines* the assignment statement. We shall need the following corollary:

$$\mathbf{P}(x)\{x:=x+y\}\mathbf{P}(x-y).$$

To handle iteration statements, we observe that, for a statement S within a loop, its postcondition must imply its precondition upon each iteration; we call such a common assertion the loop ***invariant***, denoted I. (In the example, P_3 and P_5 must both equal I.) In other words, the invariant I of a '**while** B **do** S' loop must be such that $I\ \&\ B\{S\}I$. It can be shown (by induction on the number of iterations) that, provided the loop terminates,

$I \,\&\, B\,\{S\}I$ implies
$I\,\{$**while** B **do** $S\}I \,\&\, \neg B$.

This is the inference rule or axiom which defines the **while** statement.

We return now to the problem of determining the required assertions for the sample program. Recalling our earlier remarks, the invariant should be based upon the computation of the partial sum: thus,

$$I=[x=\Sigma_{j=1}^{i}\, j].$$

The other assertions are more obvious:

$$F_0=[n\geq 1]$$
$$P_1=[n\geq 1]$$
$$P_2=[n\geq 1 \;\&\; i=1 \;\&\; x=1]$$
$$P_4=[x=\Sigma_{j=1}^{i-1} j].$$

Note that P_2 implies $I \,\&\, n\geq 1$. For these assertions, we conclude from the axioms that the specifications

$P_0\,\{$**input** $n\}\,P_1$
$P_1\,\{i:=1;\; x:=1\}\,P_2$
$I \,\&\, n\geq 1\,\{$**while**...**do**...$\}\,I \,\&\, i=n \,\&\, n\geq 1$
$I\,\{i:=i+1\}\,P_4$
$P_4\,\{x:=x+i\}\,I$

all hold; hence the program is correct provided it terminates. Termination of the loop is guaranteed since i is initialized to 1 and can only be incremented by 1 within the loop until $i=n$, which is ensured by the condition $n\geq 1$.

This section is based on the work of Hoare [1.9]. For further discussion and examples, see Gries [1.8] and Reynolds [1.21].

EXERCISES

1. Suppose that Φ is 1 instead of 0. Show that $2+3=5$.
2. Show, by example, that the sum of natural numbers (as defined in Section 1.1.1) satisfies the associative law of arithmetic. For example, show that $(2+3)+4=2+(3+4)$.
3. Make a study of programming languages available to you. Describe the data types implemented in these languages.
4. Make a study of computers available to you. Describe how different data types are represented in these computers. What are the ranges of their possible values?
5. Make a study of computers available to you. Which arithmetic operations are performed by hardware instructions for various data types?
6. Determine the largest and smallest single-precision floating-point numbers representable in an actual computer (e.g., in the IBM 370).

7. For the IBM 370, (a) what is the maximum number of significant decimal digits that can be possessed by a full-word integer, (b) what is the maximum number of significant decimal digits that can be possessed by a normalized single-precision floating-point number? (The answers were given in the text as 10 and 7, respectively; show why!)
8. Give an example of a number rational in base 10 and base 2, but not in base 3.
9. Let $x=\langle f,e\rangle$ be a positive L-place floating-point number. Show that **TRUNC** $(x,e)=\mathbf{int}(x)$, in the cases where $e<1$ and $e>L$.
10. Show that $S=((1/N)\Sigma_{i=1}^{N}x_i^2)-((1/N)\Sigma_{i=1}^{N}x_i)^2$ can be negative for a set of numbers $X=\{x_1,x_2,\ldots,x_N\}$ because of round-off errors. You may do so by example, i.e., by exhibiting a set X such that evaluation of S results in a negative number. (*Note*: the square root of this quantity, if exact, would be the standard deviation of the set of numbers.)
11. Find the Gödel sequence for each number from 1 to 20.
12. For $m_1=2$, $m_2=3$, $m_3=5$, find the residue number system representation for each integer x between 1 and 30.
13. Show that (i) if $m=n$, then $\mathbf{gcd}(m,n)=m=n$, and (ii) if $m>0$ and $n=0$, then $\mathbf{gcd}(m,0)=m$.
14. Suppose random integers are generated by the formula

$$x_{i+1}=a\cdot x_i \ \mathbf{mod}\, m$$

where $\mathbf{gcd}(x_0,m)=d$. For x_0 in the range $0\le x_0\le m$, what fraction of this range of integers can actually be generated:
(*Hint*: as stated in the text, the answer is 'about' $1/d$; you are to obtain the precise fraction.)
15. In the random number generating subroutine **RANDU**, the statement IY=IY+2147483647+1 appears.
    (a) Why can't this statement be rewritten IY=IY+2147483648?
    (b) Why can't this statement be rewritten IY=IY−(−2147483648)?
    (c) Can IY ever become zero (when the seed is nonzero)?
16. Show that, if d is a factor of a and b, then d is also a factor of b **mod** a.
17. Evaluate tabularly, for $A=B=C=D=1$, the arithmetic expressions:
    (a) $A+(B*C)-D$
    (b) $(A+B)*(C-D)$.
18. Make a study of programming languages available to you. Compare their expression evaluation rules, especially with respect to precedences and left-to-right or right-to-left orderings. (For example, is $--A**B**C$ evaluated identically in all languages?)
19. Express the algorithm for finding greatest common divisors in (a) BASIC, (b) FORTRAN, (c) COBOL, (d) PASCAL, and (e) an assembly language.
20. Trace Table 1.2.3 for $x=100$ and $y=64$.
21. Make a study of programming languages available to you. Compare their data structure facilities. (Show how data structures not supported directly by a language can be supported by the user instead.)
22. Show, using the truth table procedure of Section 1.3, that
    (a) $(p\vee q)\wedge\neg q\rightarrow\neg p$ (b) $\neg(p\wedge q)\leftrightarrow\neg p\vee\neg q$.
23. Show, using other tautologies as axioms, that
    (a) $(p\vee q)\wedge\neg p\rightarrow q$ (b) $\neg(p\vee q)\leftrightarrow\neg p\wedge\neg q$.

24. Show that implication is not associative, i.e., that

$$(p \rightarrow q) \rightarrow r \not\equiv p \rightarrow (q \rightarrow r).$$

25. Which of the implicational tautologies of Table 1.3.1(a) are also equivalences?

26. Prove that
(a) $\Sigma_{i=0}^{H} 2^i = 2^{H+1} - 1$.
(b) $\Sigma_{i=1}^{K} i\ 2^i = 2 + 2^K(2K-2)$.
(c) $\Sigma_{i=1}^{n} i^2 = n(n+1)(2n+1)/6$.

27. Prove that

$$\Sigma_{j=i+1}^{W} 1 + \Sigma_{j=1}^{W-i} 1 = 2(W-i).$$

28. Prove that the **gcd** program of Section 1.2.3 (in any one of the given forms) is correct.

PROGRAMMING ASSIGNMENTS

1. Implement the **gcd** function in terms of the equality test and successor function using LISP, SNOBOL, PL/I, or any other language with recursion facilities.
2. Write a program (a) which inputs a binary string with an embedded radix point (e.g., 101.101) and outputs its decimal equivalent (e.g., 5.625); (b) which inputs a decimal fixed-point number (e.g., 5.625) and outputs its binary equivalent (e.g., 101.101).
3. Determine experimentally, to the nearest power of 10, the largest integer and the largest and smallest floating-point reals allowed [in the computer of your choice]. (Beware of infinite looping: overflows and underflows are not fatal errors in some systems.)
4. Write programs which convert from 'vectors' $\langle n_1, n_2, \ldots, n_k \rangle$ to Gödel numbers, and vice versa. (Limit yourself to numbers less than or equal to 100; 97 is the 25th prime.)
5. Write programs whose execution result in the following errors:
   (a) integer overflow
   (b) floating-point overflow
   (c) floating-point underflow
   (d) division-by-zero
   (e) significance
   If an error is not fatal, what results?
6. Are **sgn**, **abs**, **sgnxfr**, TRUNC, FLOOR, ROUND, **mod**, **sqr**, complex arithmetic, and a random-number generating function all available in your computer system? If so, exercise them. If not, program your own version of them.
7. Write a program which computes $n!$, for $n=1$ to 20, *exactly*! (20! is a 19-digit integer.)
8. Write a program which plots the curve $y=x^2$, for $x=0$ to 1 in steps of 0.02, in a 51×51 grid (or Cartesian space), where the point $(x=0, y=0)$ is in the lower left corner, and the point $(x=1, y=1)$ is in the upper right corner.

9. Write a program which illustrates the nonassociativity of computer arithmetic. (What other anomalies can you illustrate?)

10. Write a program which computes π to 25 significant digits.

***11.** The RANDU subroutine has been discredited because of high correlations between successive numbers. One problem is the unfortunate choice of multiplier, and another is the use of single-precision arithmetic. Write a better subroutine.

2 ALGEBRAIC SYSTEMS

2.1 SETS

A typical dictionary definition of *mathematics* will state that it is the study of 'quantities' (numbers) and 'relations' (structure). A mathematical study then requires that there be distinguishable things to count or relate. We shall call these things *objects*, and say that a collection of objects is a *set*. Examples of objects include numbers, words, and sets; an object which is not a set is called a *primitive* (or *atom*), a collection (set) of sets is called a *family*. The objects belonging to a set are called its *members* (or *elements*). Membership (belonging, containment) of an object x *in* a set S is denoted $x \in S$; nonmembership is denoted $x \notin S$. By axiom (the **law of the excluded middle**), any object x either belongs or does not belong to any set S.

A set is denoted, in 'list' notation, by placing its distinct members within braces in arbitrary order, each member appearing only once. Thus, the set D of decimal digits is $\{2,3,5,7,0,1,9,8,6,4\}$. The 'listing ordering' is irrelevant: the above set D is identical to $\{0,1,2,...,9\}$. It is common to use ellipses (...) to indicate omission of some (hopefully obvious) members in the listing. More formally, we say that two sets (A and B) are *equal* (denoted $A=B$) iff they contain precisely the same members; this definition of set equality is also known as the **axiom of extension**, and is necessary because equality and identity are not always synonymous. For example, if we interpret $x \in S$ to mean that 'x is the child of S', then the axiom would say that M=F iff M and F have the same children; thus, nonidentical parents (a Mother and Father) may be equal. For sets (and objects), however, we shall by axiom consider equality to be the same as identity of members. (Inequality of A and B is denoted $A \neq B$.)

The number of distinct members in a set S is called its *size*, denoted $\#(S)$. For the above example, $\#(D)=10$. A set whose size is zero contains no members, and is called the *empty set*; it is denoted by $\emptyset$, or sometimes by $\{\}$. By the axiom of extension, the empty set is unique in that all sets with no members are identical (and equal).

We observe that $\emptyset$ and $\{\emptyset\}$ are not the same: $\emptyset$ is the empty set and contains no members, whereas $\{\emptyset\}$ is nonempty and contains one member which happens to be a set. A third set which contains the preceding two sets as its members is denoted $\{\emptyset, \{\emptyset\}\}$, and by generalizing an infinite number of such sets may be defined. Each such set may denote (and, in fact, define) a natural number: e.g., $0=\emptyset$, $1=\{\emptyset\}$, $2=\{\emptyset,\{\emptyset\}\}$, $3=\{\emptyset,\{\emptyset\},\{\emptyset,\{\emptyset\}\}\}$, etc. There are an infinite number of such sets, but each such set contains only a finite number of members.

When a set S has an infinite number of members, we shall say that $\#(S)$, no longer a natural number, is the *cardinality* of S. (While the cardinality of a finite set may be defined as its size, which is a natural number, ordinarily the term is reserved to apply to infinite sets.) Infinite sets may have different cardinalities; for example, the cardinality of the set of all integer numbers is the same as the cardinality of the set of all rational numbers, but is less than the cardinality of the set of all real numbers. (This will be discussed further in Section 2.3.2.)

Since sets may contain sets as objects, the questions of whether a set may contain itself, or whether a set may contain all sets (including itself), arise. To allow such sets would lead to paradoxes of the sort, 'I am lying!': if a person making this statement is lying, then he is not, and vice versa. To preclude such problems, we shall by axiom say that $S \notin S$ for any set S under consideration in our *universe of discourse* $U_{\mathbf{D}}$; by definition, $U_{\mathbf{D}}$ contains all the objects or sets other than $U_{\mathbf{D}}$ about which statements may be made. The postulate that $U_{\mathbf{D}}$ cannot contain itself ($U_{\mathbf{D}} \notin U_{\mathbf{D}}$) is of fundamental importance, as we shall see; it may also be considered a restriction on the definition of a set.

The **axiom of specification** states that, to every set A and to every truth-valued statement $\mathbf{P}(x)$ about an object x, there exists a set B whose members are exactly those members of A for which $\mathbf{P}(x)$ is true; this set B, in 'formula' notation, is written

$$B=\{x \in A \mid \mathbf{P}(x)\}$$

or $\{x \mid \mathbf{P}(x)\}$ when A is understood. The axiom permits us to construct sets (B) out of other ones (A) by asserting something about the members. For example, suppose $\mathbf{P}_1(x)$ and $\mathbf{P}_2(x)$ are the statements '$x=x$' and '$x \neq x$', respectively; then

$$B_1=\{x \in A \mid x=x\}=A$$

and

$$B_2=\{x \in A \mid x \neq x\}=\emptyset.$$

Consider now the case $A=U_{\mathbf{D}}$, whence $B=\{x \in U_{\mathbf{D}} \mid \mathbf{P}(x)\}$. The restriction $U_{\mathbf{D}} \notin U_{\mathbf{D}}$ implies that x in the formula cannot equal $U_{\mathbf{D}}$, so that $P(U_{\mathbf{D}})$ need not be defined. Let $\mathbf{P}_0(x)$ be the statement '$x \notin x$', and $B_0=\{x \in U_{\mathbf{D}} \mid x \notin x\}$. We observe that B_0 cannot be in $U_{\mathbf{D}}$ (in fact, $B_0=U_{\mathbf{D}}$); the assumption otherwise (i.e., that $B_0 \in U_{\mathbf{D}}$) leads to a contradiction ($B_0 \notin U_{\mathbf{D}}$), known as the *Russell paradox*. (Another way of avoiding the paradox is to permit the universe of discourse to contain itself while disallowing the existence of B_0; we shall later, in effect, restrict $U_{\mathbf{D}}$ to be a 'power set', q.v.) For details, see Halmos [2.11].

2.1.1 Ordered Sets

We have stated earlier that the members of a set have no inherent ordering. However, for many applications, it is convenient to impose some ordering by designating one member as first, a different member as second, yet another different member as third, etc. We then say that the set is *ordered*. By convention, in the list notation for ordered sets, the first member would be written leftmost in the braces, the second member to its immediate right, etc.; in other words, for ordered sets, this 'listing ordering' is significant. Thus

$$A=\{2,3,5,7,0,1,9,8,6,4\} \qquad \text{and} \qquad B=\{0,1,2,\ldots,9\},$$

while equal as sets, are not equal as ordered sets. If

$$A=\{a_1,a_2,\ldots,a_n\}$$

is an ordered set,

$$A^{\mathbf{R}}=\{a_n,a_{n-1},\ldots,a_1\}$$

is called the *reverse* of A. Furthermore, if $\{a_1,a_2,a_3,\ldots\}$ is a given ordered set, we sometimes write

$$a_1<a_2,\ a_2<a_3,\ a_3<a_4,\ \ldots$$

(or more briefly $a_1<a_2<a_3<\ldots$), where '$<$' is understood to be with respect to the given ordering; in other words, with respect to the ordered set A, we have $9<8$! Finally, when the ordering is to be emphasized, it is common to use parentheses rather than braces to denote ordered sets, e.g.,

$$A=(2,3,5,7,0,1,9,8,6,4) \qquad \text{and} \qquad A^{\mathbf{R}}=(4,6,8,9,1,0,7,5,3,2).$$

2.1.2 Multisets

A *multiset* is a collection of objects whose members need not be distinct. Thus $A=\{3,2,2,1\}$ is not the same multiset as $B=\{3,2,2,2,1\}$, although both are equal (as sets!) to $C=\{3,2,1\}$. The number of repetitions of each distinct member of a multiset is called the *multiplicity* of that member; the member 2 has multiplicity 2, 3, and 1, respectively, in A, B, and C. If the members of a multiset all have a multiplicity of one, it is said to be *without repetition*; a multiset *with repetition* may have any number of repeated members. The *size* of a multiset is the sum of the multiplicities of each distinct member; the *index* of a multiset is the number of distinct members. (Hence the sizes of A, B, and C are 4, 5, 3, respectively; their indices are all 3.) A multiset without repetition has size equal to its index, and of course may also be considered a set. An *ordered* multiset is one with a designated ordering; otherwise, like sets, multisets are considered *unordered*.

Classification Let S be a set. (Unless explicitly stated otherwise, sets are generally considered neither ordered nor multisets.) Multisets of elements of S are classified according to whether they are considered ordered or unordered, and whether they are with or without repetition. There are four classes:

(i) a *combination* is an unordered multiset without repetition (i.e., a set whose members are selected from S).
(ii) a *permutation* is an ordered multiset without repetition (i.e., a set whose members are selected from S and arranged in a designated fashion).
(iii) a *selection* is an unordered multiset with repetition.
(iv) an *arrangement* is an ordered multiset with repetition.

We may also think of a combination as just a subset of S, or as a selection without repetition; a permutation as an ordered subset of S, or as an ordered combination, or as an arrangement without repetition; a selection as a combination with repetition; and an arrangement as a permutation with repetition, or as an ordered selection. For more information, see Liu [2.15].

Sometimes the terms r-combination, r-permutation, r-selection, and r-arrangement are used when the respective multiset of S has size r. The term 'permutation of S' is then reserved to mean an N-permutation of S, where $N=\#(S)$, and we say an r-permutation is a permutation of a subset of S. Note that $r \leq N$ for r-permutations and r-combinations since there can be no repetition, but r is unlimited when repetition is permitted. Finally, parentheses rather than braces are used to denote permutations and arrangements, to emphasize the significance of their listing ordering.

Example Let $B=\{0,1\}$. Then the 2-selections of B are $\{0,0\}$, $\{0,1\}$, and $\{1,1\}$; the 2-selection $\{1,0\}$ is the same as $\{0,1\}$. There is only one 2-combination of B—namely, $\{0,1\}$ (or $\{1,0\}$); $\{0,0\}$ and $\{1,1\}$ have repetitions. The 2-arrangements of B are (0,0), (0,1), (1,0), and (1,1), and the 2-permutations of B are (0,1) and (1,0); here the different orderings matter. (It should be noted that the r-arrangements of B correspond to the r-bit binary integers.)

Now let $S=\{0,1,2\}$. Then the 2-combinations of S are $\{0,1\}$, $\{0,2\}$, $\{1,2\}$, the 2-permutations of S are (0,1), (0,2), (1,2), (1,0), (2,0), (2,1), the 2-selections of S are $\{0,1\}$, $\{0,2\}$, $\{1,2\}$, $\{0,0\}$, $\{1,1\}$, $\{2,2\}$, and the 2-arrangements of S are (0,1), (0,2), (1,2), (1,0), (2,0), (2,1), (0,0), (1,1), (2,2).

In Chapter 1, we implicitly used the formula for the number of different r-arrangements of a set S of size N. Specifically, the number of possible L-digit binary integers was given as 2^L. In Section 2.2.1, we show that the number of r-arrangements of a set of size N equals N^r. In Section 6.2.3, we give formulas for the number of possible r-permutations, r-combinations, and r-selections. We often make use of the fact that there are $N!$ permutations of a set of size N, e.g., of the set $\{1,2,\ldots,N\}$.

2.1.3 Subsets

An equivalent definition of the equality of two sets is that each member of one is also a member of the other. However, we say B is a *subset* of (is *included* in) A, denoted $B\subseteq A$, if each member of B is also a member of A, but not necessarily vice versa. B is called a *proper* subset of A if $B\subseteq A$ and there is a member of A not in B. If $B\subseteq A$, we also say that A is a *superset* of (*includes*) B, denoted $A\supseteq B$. If B is not a subset of A, we write $B\nsubseteq A$. Two sets are *incomparable* if $A\nsubseteq B$ and $B\nsubseteq A$.

A *minimal* set in U, with respect to a property of sets $\mathcal{P}$, is a set A in U having property $\mathcal{P}$ such that every nonempty proper subset of A does not have property $\mathcal{P}$. A *maximal* set (in U w.r.t. $\mathcal{P}$) is a set A in U having property $\mathcal{P}$ such that every proper superset of A in U does not have property $\mathcal{P}$.

Properties of Set Inclusion Some important properties of 'set inclusion' ($\subseteq$) which follow directly from its definition include: for any subsets A and B of U,

[P1] $\emptyset\subseteq A$ and $A\subseteq A$.
[P2] $A=B$ if $A\subseteq B$ and $B\subseteq A$.
[P3] If $A\subseteq B$ and $B\subseteq C$, then $A\subseteq C$.
[P4] A is a proper subset of B iff $A\subseteq B$ and $A\neq B$.
[P5] If $x\in A$, then $\{x\}\subseteq A$.

Note the difference between set inclusion ($A\subseteq A$ is always true) and set membership ($A \in A$ is never true). The property $\emptyset \subseteq A$ can be verified by noting that the defining condition holds 'vacuously': each member of $\emptyset$ is a member of A since there is no member of $\emptyset$ which is not a member of A.[†] This argument may be less comforting than a proof by contradiction: suppose $\emptyset \nsubseteq A$, i.e., there is a member of $\emptyset$ not in A; then $\emptyset$ would not be empty, which contradicts its definition, therefore $\emptyset \subseteq A$. Note also that $\emptyset \subseteq \emptyset$.

We remark that the set U in the above is arbitrary; since all sets under consideration are subsets (not members) of U, we call U the *universal set* (of our discussion). We note that $\emptyset \subseteq U$ and $U \subseteq U$. It is the latter property which makes it convenient to regard all sets as subsets of U rather than as members of $U_{\mathbf{D}}$. Generally, we will restrict the members of U to be primitives.

Example*: *Sets of numbers Let R be the set of all real numbers, R_1 be the set of all rationals, I be the set of all integers, and $\mathcal{J}$ be the set of all natural (nonnegative integer) numbers. Then

$$\mathcal{J} \subseteq I \subseteq R_1 \subseteq R.$$

Let $[a,b]$ denote the set of all integer or real numbers x such that $a \leq x$ and $x \leq b$; we call this set the *interval* (of integers or reals) from a to b. Then $[a_1,b_1] \subseteq [a_2,b_2]$ iff $a_2 \leq a_1$ and $b_1 \leq b_2$. (We remark that a binary computer with word length L can represent 2's-complement integers only in the interval $[-2^{L-1}, 2^{L-1}-1]$.) $[a,b]$ is sometimes called a 'closed' interval to distinguish it from the 'open' interval or *segment* (a,b), defined as the set of numbers x such that $a < x$ and $x < b$; note that the end-points a and b belong to the interval $[a,b]$, but not to the segment (a,b).

2.1.4 Set Operations

Set inclusion is a 'relation' which any two sets may or may not satisfy. Given two sets, we may perform any of a number of different 'operations' so as to produce a new set. (We shall formalize the concepts of relation and operation later.) Let A and B be sets (subsets of U). Then the *union* (or *disjunction*) of the sets, denoted $A \cup B$, is the set $\{x \in U \mid x \in A \text{ or } x \in B\}$; the term *or* is used in the 'inclusive' sense, meaning x is a member of either A or B or both. The *intersection* (or *conjunction*) of the sets, denoted $A \cap B$, is the set $\{x \in U \mid x \in A \text{ and } x \in B\}$; *and* means that x must belong to both sets A and B.

[†]We shall often define a property $\mathcal{P}$ as holding if 'α *implies* β'; in the event that α is always false, the implication is true by default, hence property $\mathcal{P}$ holds. In other words, whenever α is true, β must also be true in order for property $\mathcal{P}$ to hold; but if α is never true, then $\mathcal{P}$ holds whether β is true or false. We also say $\mathcal{P}$ holds vacuously when there can be no counter-example showing a case where α does not imply β. (See also the discussion of 'material implication' [$\rightarrow$] in Section 1.3.) In the foregoing example, $\mathcal{P}$ is the property $B \subseteq A$, α is the statement $x \in B$, and β is the statement $x \in A$; for $B = \emptyset$, $x \in \emptyset$ is never true, hence the property $\emptyset \subseteq A$ holds.

The *relative complement* of B with respect to A, denoted $A-B$, is the set $\{x\in A \mid x\notin B\}$; if A is understood (say, when $A=U$) and $B\subseteq A$, then we write $-B$, called the *absolute complement* of B. The *disjoint union* (or *symmetric difference*) of A and B, denoted $A\oplus B$, is the set $(A-B)\cup(B-A)$. We remark that newer languages, such as Pascal, have a set data type and associated operations.

Power set We define the *power set* of a set A, denoted 2^A, as the set $\{x \mid x\subseteq A\}$; 2^A is the set of all subsets of A. (Sets A and B in the above are members of 2^U; in effect, we have restricted U_D here to be 2^U.)

An interesting property of the power set operation, which explains its notation,is that

$$\#(2^A)=2^{\#(A)}.$$

(Proof of this is left as an exercise.) For example, $2^{\emptyset}=\{\emptyset\}$ has $2^0=1$ member, $2^{\{a\}}=\{\emptyset,\{a\}\}$ has $2^1=2$ members, $2^{\{a,b\}}=\{\emptyset,\{a\},\{b\},\{a,b\}\}$ has $2^2=4$ members, etc. Note that since $\emptyset$ and A are subsets of A for any set A, i.e., $\emptyset\subseteq A$ and $A\subseteq A$, then by definition $\emptyset$ and A are members of 2^A, i.e., $\emptyset\in 2^A$ and $A\in 2^A$ (whereas $\emptyset$ need not be a member of A, and A is never a member of A, $A\notin A$). It should also be emphasized that members of A (except if $\emptyset\in A$) are never members of 2^A. We note especially that $U\in 2^U$ but $U\nsubseteq 2^U$; however, $\{U\}\subseteq 2^U$.

Properties Two important properties of the $\cup$, $\cap$, and $\oplus$ operations are that they are each associative and commutative: that is, for all sets A, B, and C,

$$\begin{aligned}(A\cup B)\cup C&=A\cup(B\cup C)\\ (A\cap B)\cap C&=A\cap(B\cap C)\qquad\text{[associativity]}\\ (A\oplus B)\oplus C&=A\oplus(B\oplus C)\end{aligned}$$

and

$$\begin{aligned}A\cup B&=B\cup A\\ A\cap B&=B\cap A\qquad\text{[commutativity]}\\ A\oplus B&=B\oplus A.\end{aligned}$$

These properties essentially follow from the associativity and commutativity of *or* and *and* in the definitions, which we postulate. Some identities (statements which are true for all sets $A\in 2^U$) involving $\emptyset$ and U are:

$$\begin{aligned}&A\cup\emptyset=A, &&A\cup U=U\\ &A\cap\emptyset=\emptyset, &&A\cap U=A\\ &A-\emptyset=A, &&A-U=\emptyset\\ &\emptyset-A=\emptyset, &&U-A=-A\end{aligned}$$

$$-\emptyset=U \quad , \qquad -U=\emptyset$$
$$A\oplus\emptyset=A, \qquad A\oplus U=U-A, \qquad A\oplus A=\emptyset.$$

A in the above may also be $\emptyset$ or U.

2.1.5 Families of Sets

Let $\mathscr{F}=\{S_1,S_2,S_3,\ldots\}$ be a family of sets. Written in the form $\mathscr{F}=\{S_i|i\in I\}$, where $I=\{1,2,\ldots\}$, we say $\mathscr{F}$ is an *indexed* set and I is a set of (numerical) *indices*. Since the union and intersection operations on two sets are associative and commutative, it is natural to generalize them to a family of sets. We define the *union of* $\mathscr{F}$, denoted $\cup_{i\in I}S_i$, as the set

$$\{x\,|\,x\in S_i \text{ for } \textit{any } i\in I\},$$

and define the *intersection of* $\mathscr{F}$, denoted $\cap_{i\in I}S_i$, as the set

$$\{x\,|\,x\in S_i \text{ for } \textit{every } i\in I\}.$$

Note that these sets are independent of any ordering of the members of $\mathscr{F}$. We may equivalently define

$$\cup_{i\in I}S_i=((S_1\cup S_2)\cup S_3)\cup\cdots,$$

and

$$\cap_{i\in I}S_i=((S_1\cap S_2)\cap S_3)\cap\cdots.$$

We also use the notations $\cup_i S_i$ and $\cap_i S_i$, when I is understood.

Two sets A and B are *disjoint* if their intersection is empty (i.e., if $A\cap B=\emptyset$); otherwise, we say that they 'overlap'. We say $\mathscr{F}$ is a family of ***mutually exclusive*** sets if each pair of sets in $\mathscr{F}$ is disjoint. It is not sufficient for the intersection of $\mathscr{F}$ to be empty for mutual exclusion to hold.

Let the 'disjoint union of $\mathscr{F}$' be defined by

$$\bigoplus_i S_i=((S_1\oplus S_2)\oplus S_3)\oplus\cdots,$$

where the ordering of the sets S_i in $\mathscr{F}$ is irrelevant. We then define a *linear combination* of members of $\mathscr{F}$ as the disjoint union of a subset of $\mathscr{F}$, and say that $\mathscr{F}$ is an *[linearly] independent set* if no member of $\mathscr{F}$ is a linear combination of the other members of $\mathscr{F}$.

We say $\mathscr{F}$ *covers* (or is a *cover* of, or *spans*) A if $A\subseteq\cup_i S_i$. A family of sets $\mathscr{F}=\{S_1,S_2,\ldots\}$ is a minimal cover of B if $B\subseteq\cup_i S_i$ but $B\nsubseteq\cup_i S_i'$ for any proper subfamily $\{S_1',S_2',\ldots\}\subseteq\mathscr{F}$; we note it need not be a unique cover.

A *partition* of a set A, denoted $\Pi(A)$, is a family of pairwise disjoint nonempty subsets of A whose union equals and hence also covers A; i.e., $\Pi(A)=\{S_1,S_2,\ldots\}$, where $S_i\cap S_j=\emptyset$, $S_i\neq\emptyset$, and $\cup_i S_i=A$. The members S_i are called the *blocks* of $\Pi(A)$, and the size of $\Pi(A)$ is called its *rank* (or *index*). (A 'bipartition' has a rank equal to two.)

Examples As a concrete illustration of the various definitions, let $A=\{a,b\}$, $B=\{b,c\}$, $C=\{c,d\}$. Then

$$\begin{array}{ll}
A\cup B=\{a,b,c\} & A-B=\{a\} \\
A\cap B=\{b\} & B-A=\{c\} \\
A\cap B\cap C=\emptyset & -A=\{c,d\} \\
A\oplus B=\{a,c\} & -B=\{a,d\} \\
A\oplus B\oplus C=\{a,d\} & -\emptyset=\{a,b,c,d\}. \\
2^A=\{\emptyset,\{a\},\{b\},A\} &
\end{array}$$

In the absolute complements, the 'understood' set is not $U_{\mathbf{D}}$, but is $U=\{a,b,c,d\}$, the set of primitives. $U_{\mathbf{D}}$, *of the above discussion*, is the set

$$\{a,b,c,d,\{a,b\},\{b,c\},\{c,d\},\{a,b,c\},\{b\},\{a\},\{c\},$$
$$\{a,d\},\{a,b,c,d\},\{a,c\},\{\emptyset,\{a\},\{b\},\{a,b\}\},\emptyset\}.$$

We also note that $\{A,C\}$ is a bipartition of U, and is also a minimal cover. A and C are disjoint sets, but A and B are not; the family of sets $\mathcal{F}=\{A,B,C\}$ is not mutually exclusive although its intersection is empty. The family $\mathcal{F}$ is an independent set; moreover, $\{A,B,C\}$ is a cover of U, but is not minimal. $\{\{a,b,c\},\{c,d\}\}$ is a minimal cover, but is not a partition of U.

2.2 RELATIONS AND ORDERS

Almost all the entities arising in the study of computer software systems, including computer programs, languages, data structures, and operating systems, can be abstractly (mathematically) defined as 'algebraic systems': i.e., as sets of objects together with certain 'relations' and 'operations' associated with the sets. We formalize these concepts in this and subsequent sections.

2.2.1 Ordered Pairs

In Section 2.1.1, we defined an 'ordered set' as a set whose members are ordered in some fashion. Here, we define the concept of ordering in terms of (unordered) sets. The *ordered pair* of two objects a and b, denoted (a,b), is defined as the set (of sets) $\{\{a\},\{a,b\}\}$. The ordered pair differs in one fundamental way from the sets $\{a,b\}$ or $\{\{a\},\{b\}\}$: the listing order counts! In particular, $(b,a)=\{\{b\},\{a,b\}\}\neq(a,b)$ if $a\neq b$. Orderings of more than two objects may then be defined in terms of ordered pairs. The *ordered triple* of three objects a, b, and c, denoted (a,b,c), is defined as the 'nested' ordered pair $((a,b),c)$. The *ordered n-tuple* $\alpha=(a_1,a_2,\ldots,a_n)$ is defined analogously as $(((a_1,a_2),a_3),\ldots)$. We call n the length of α and call a_i the ith *coordinate* of α. Two ordered n-tuples are *like* if their lengths are the same. Two ordered n-tuples are *equal* if and only if they are like and their coordinates are respectively equal. We remark that nothing has been said about the individual members of a coordinate which is also a set.

Cartesian Product We consider now a generalization of the concept of ordered pairs to pairings of members of sets. The *Cartesian product* of two nonempty sets A and B, denoted $A\times B$, is defined as the set of ordered pairs $\{(a,b)\,|\,a\in A,\ b\in B\}$. We also say that $\emptyset\times A=A\times\emptyset=\emptyset$. The Cartesian product of n sets, $A_1,A_2,\ldots,A_n$, denoted $A_1\times A_2\times\cdots\times A_n$, is defined as $(((A_1\times A_2)\times A_3)\times\cdots)\times A_n$; its members are evidently ordered n-tuples. Hence we may write

$$A_1\times A_2\times\cdots\times A_n=\{(a_1,a_2,\ldots,a_n)\,|\,a_i\in A_i\}.$$

Note that $A_1\times A_2\times A_3=(A_1\times A_2)\times A_3\neq A_1\times(A_2\times A_3)$; i.e., the Cartesian product is not associative.

The *n-fold product* of a single set A, denoted A^n, is the Cartesian product of n sets identical to A, i.e., $A^n=\{(a_1, a_2, \ldots, a_n)\,|\,a_i\in A\}$, the set of ordered n-tuples whose coordinates are members of a common set A. A^n may also be defined iteratively by $A^{i+1}=A^i\times A$, $i=1,2,\ldots,n-1$, where $A^1=A$ by definition; A^0 is undefined. (Note that $A^{i+1}\neq A\times A^i$.) A member of A^n is called a *string* of length n of elements of A (or more briefly, a string *over* A); we discuss strings further in Section 2.4.5. When A is a set of numbers, a string is often called an *n-vector* (or, just *vector* when n is understood).

A member of A^n is also an n-arrangement of members of A. From the obvious fact (sometimes called the **law of products**) that

$$\#(A\times B)=\#(A)\cdot\#(B)$$

we obtain the more general formula

$$\#(A_1\times A_2\times\cdots\times A_n)=\Pi_{i=1}^{n}\#(A_i),$$

where Π denotes the generalized product. Therefore, in the special case where each $A_i=A$, the number of strings of length n over A is given by

$$\#(A^n)=[\#(A)]^n.$$

A change of notation then shows that the number of r-arrangements of a set of size N equals N^r, as previously stated.

Example Let $A=\{a,b\}$ and $B=\{b,c,d\}$. Then

$$A\times B=\{(a,b),\ (a,c),\ (a,d),\ (b,b),\ (b,c),\ (b,d)\}$$

and

$$A^2=\{(a,a),\ (a,b),\ (b,a),\ (b,b)\}.$$

Note that $\#(A)=2$, $\#(B)=3$, $\#(A\times B)=6$, and $\#(A^2)=4$.

We emphasize that, in the following, when we refer to a 'pair' of sets, we mean the ordered pair, not the Cartesian product of the sets. The same applies to n-tuples of sets.

Application*: *Boolean n-vectors Let $B=\{0,1\}$. Then the n-fold product B^n is the set of ordered n-tuples whose coordinates are either 0 or 1. Each member $\beta=(b_1,b_2,\ldots,b_n)\in B^n$ is called a *Boolean n-vector*. Often, the parentheses and commas are omitted in the notation: e.g., $(1,0,0,1,0,1)\equiv 100101$. We shall call the latter the 'binary number' representation for Boolean n-vectors. We may immediately conclude that there are 2^n possible Boolean n-vectors.

Consider now an ordered set $A=\{a_1,a_2,\ldots,a_n\}$. For any subset S of A, let $\beta_S\in B^n$ be defined such that $b_i=1$ if $a_i\in S$, and $b_i=0$ if $a_i\notin S$. This is called the 'characteristic vector' of S. From this representation, it is clear that $\#(2^A)=2^{\#(A)}$, as stated in Section 2.1.4.

Application*: *Cells Let I denote the set of all integers, and recall (from Section 2.1.3) the definition of an interval $[a,b]$ of integers from a to b,

$$[a,b]=\{x\in I\,|\,a\leq x \text{ and } x\leq b\}.$$

Noting that intervals are subsets of I, it is natural to generalize the concept to subsets of the k-fold product I^k. I^k itself is called a discrete Cartesian 'coordinate system' or 'space'. Let

$$E=\{[a_1,b_1],[a_2,b_2],\ldots,[a_k,b_k]\}$$

be an ordered set of k intervals; the associated *k-cell C* is the set of k-tuples,

$$C=[a_1,b_1]\times[a_2,b_2]\times\cdots\times[a_k,b_k],$$

which is equal to the set

$$\{(x_1,\ldots,x_k)\,|\,a_i\leq x_i\leq b_i,\ i=1,2,\ldots,k\}.$$

(An interval is a 1-cell.) We call E the *extent* of C, and k its *dimension*. Note that the size of $[a,b]$ is $(b-a+1)$, and that

$$\#(C)=(b_1-a_1+1)\cdot(b_2-a_2+1)\cdot\cdots\cdot(b_k-a_k+1).$$

Examples Some examples of ordered pairs of integers are (0,0), (1,2), (−1,1), and (4,3); these are all members of $I^2=I\times I$. The intervals $[a,b]$ associated with each (a,b) are: $[0,0]=\{0\}$, $[1,2]=\{1,2\}$, $[-1,1]=\{-1,0,1\}$, and $[4,3]=\emptyset$. The 2-cell with extent $\{[1,2],[-1,1]\}$ is the Cartesian product:

$$[1,2]\times[-1,1]=\{(1,-1),(1,0),(1,1),(2,-1),(2,0),(2,1)\}.$$

2.2.2 Relations

A *relation* (or 'mapping'), from a set A *to* a set B (or *in* a set S), is defined as a subset of $A\times B$, or equivalently as a member of $2^{A\times B}$ (or $2^{S\times S}$); that is, any set R of ordered pairs defines a relation. This concept can be generalized to sets of ordered n-tuples, which define 'n-ary' relations; we generally restrict ourselves to binary ($n=2$) relations unless explicitly stated otherwise. The empty set is called the *null* relation, and $A\times B$ (or $S\times S$) the *universal* relation. The *identity* relation in a set S, denoted I, is the set $\{(a,a)\,|\,a\in S\}$. If R is a relation, we usually write aRb if $(a,b)\in R$, and $a\not R b$ if $(a,b)\notin R$. The *domain* of R is the set $\{a\in A\,|\,(a,b)\in R$ for some b in $B\}$, and the *range* of R is the set $\{b\in B\,|\,(a,b)\in R$ for some a in $A\}$; these sets are also called the *projections* of R onto the first and second coordinates, respectively.

A relation $R\subseteq A\times B$ has a convenient pictorial representation. Let the members of A and B be depicted by points (the labeled circles in Figure 2.2.2(a)). Then each member $r=(a,b)$ in R can be depicted by an arrow from point a in A to point b in B. Figure 2.2.2(a) depicts the relation $R_0=\{(1,2), (2,1), (2,2), (2,3)\}$, where $A=\{1,2\}$ and $B=\{1,2,3\}$; note that all the arrows go from the left part to the right part of the figure, so this diagram is called the 'bipartite' representation of R_0 $[\subseteq A\times B]$. When $A=B$ $(=S)$, it is customary to depict each member of S only once, so that a member $r=(s,s)$ of R in S is depicted by an arrow from point s to itself. Figure 2.2.2(b) so depicts the relation $R_0=\{(1,2), (2,1), (2,2), (2,3)\}$, where $S=\{1,2,3\}$; this pictorial representation of a relation R in S is called the *graph* of R (cf. Chapter 4).

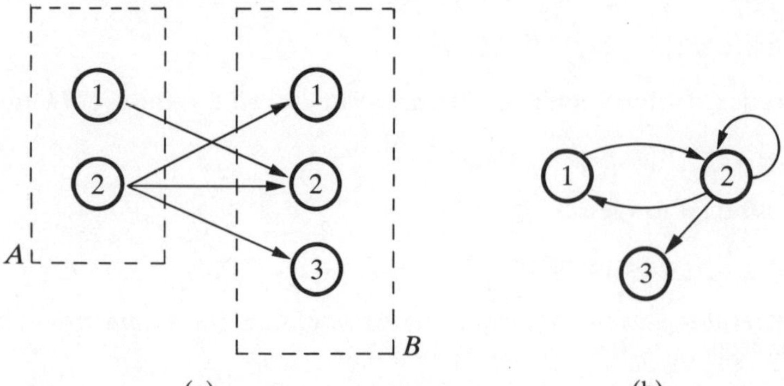

Figure 2.2.2

Characterizations A relation R in a nonempty set S ($R \subseteq S \times S$) is said to be:

reflexive	if	xRx
symmetric	if	xRy implies yRx
transitive	if	xRy and yRz implies xRz
asymmetric	if	xRy implies $y \not{R} x$
antisymmetric	if	xRy and yRx implies $x=y$ (or $x \equiv y$)
irreflexive	if	$x \not{R} x$.

($\equiv$ is used instead of $=$ when we wish to emphasize that it does not mean identical.) The above statements must be true for *all* x, y, z in S. Hence relations may be neither reflexive nor irreflexive, neither symmetric nor asymmetric, both symmetric and antisymmetric, or both asymmetric and antisymmetric. Note that the null relation ($R=\emptyset$) is symmetric, transitive, asymmetric and antisymmetric because the defining conditions hold 'vacuously' (see the footnote in Section 2.1.3); it is also irreflexive. The universal relation ($R=S \times S$) is reflexive, symmetric, and transitive.

In the graph of R in S, an arrow from a point to itself is called a 'reflexive loop'. If there is an arrow from s_1 to s_2, and another arrow from s_2 to s_3, we say that there is an 'indirect path' from s_1 to s_3 (via s_2). To determine whether or not a relation R in S satisfies the properties defined above, it is often easiest to simply inspect its graph. A relation is reflexive if *every* point in its graph has a reflexive loop, is irreflexive if *no* point has a reflexive loop, and is neither reflexive nor irreflexive if at least one but not all points have a reflexive loop. A relation is symmetric if *every* arrow connecting different points in its graph is part of a 'symmetric loop' (true vacuously if there are no arrows connecting different points), antisymmetric if *no* arrow in its graph is part of a symmetric loop, asymmetric if both antisymmetric and irreflexive. A relation is transitive if, in its graph, whenever there is an indirect path from point s to point s', there is also an arrow (a direct path) from s to s'; a relation is *in*transitive otherwise.

Example The relation R depicted in Figure 2.2.2(b) is neither reflexive, symmetric, transitive, asymmetric, antisymmetric, nor irreflexive. Adding (1,1) and (3,3) to R would make it reflexive. Adding (3,2) to R, or deleting (2,3) from R, would make it symmetric. Adding (1,3) to and deleting (2,1) from R would make it transitive. Deleting (1,2) [or (2,1)] and (2,2) from R would make it asymmetric. Deleting (1,2) or (2,1) from R would make it antisymmetric. Deleting (2,2) from R would make it irreflexive.

Composite Relations Let $R, R_1, R_2, \ldots,$ be relations in a set S. We say:

R_1 *includes* R_2 if xR_2y implies xR_1y (i.e., $R_1 \supseteq R_2$).
R_1 is the *inverse* (or *converse*) of R_2 [denoted R_2^{-1}]

if xR_2y iff yR_1x.
R is the *composition* of R_1 and R_2 [denoted R_1R_2]
if xRy iff xR_1z and zR_2y for some z.

The nth *power* of R, R^n, for $n>0$, is defined iteratively: $R^1=R$, $R^2=RR$, $R^3=R^2R=RR^2$, or in general, $R^n=R^{n-1}R=RR^{n-1}$ for $n>1$ (or for $n\geq 1$, if we define R^0 as the *identity* relation I). I is defined such that xIy iff $x=y$. Hence $R^0R=RR^0=R$ for any R. We emphasize that the nth power of a relation R is not the same as the n-fold product of the set R (see Exercise 30). Finally, define

R^+, the (*positive*) *transitive closure* of R,
such that xR^+y if xR^ny for some $n>0$
R^*, the *star* (or *reflexive transitive*) *closure* of R,
as $R^0 \cup R^+$.

It can be proven that R^+ is transitive, and in fact is the smallest transitive relation which includes R; furthermore, R^* is both transitive and reflexive, and is the smallest such relation which includes R.

Informally, in the graph of R in S, R includes R' if R' contains a subset of the arrows of R. The graph of the inverse of R has arrows all reversed in direction from those of R. The graph of the composite relation R_1R_2 has an arrow from x to y iff there is an arrow from x to some point z in the graph of R_1 as well as an arrow from that z to y in the graph of R_2. The graph of R^2 has an arrow (direct path) from x to y iff there is an indirect path from x to y via some point z in the graph of R. (In the above, x, y, and z need not be different points.) The graph of R^n, $n>0$, has an arrow from x to y iff there is a 'path of length n' from x to y in the graph of R. (A 'path of length n' will be formally defined in Chapter 4. For the moment, the pictorial notion of following n arrows from point to point in the graph of R in S will be relied upon; a path of zero length is not defined here.) The graph of R^0 in S consists of a reflexive loop for each point in S; hence there is a path of length one from each point to itself in the graph of R^0. The graph of R^+ has an arrow from x to y if there is a path (of any nonzero length) from x to y in the graph of R. R^* includes R^+, but has in addition a reflexive loop for each point whether or not R does. R^* is also called the *reachability* relation associated with R in S.

Example Suppose $S=\{1,2,3\}$, $R=\{(1,2), (2,1), (2,2), (2,3)\}$, as depicted in Figure 2.2.2(b). Then $S\times S\supseteq R\supseteq\emptyset$. Furthermore,

$R^{-1}=\{(2,1), (1,2), (2,2), (3,2)\}$
$R^2=\{(2,1), (1,2), (2,2), (2,3), (1,3), (1,1)\}$
$R^0=I=\{(1,1), (2,2), (3,3)\}$
$R^+=\{(2,1), (1,2), (2,2), (2,3), (1,3), (1,1)\}$
$R^*=\{(2,1), (1,2), (2,2), (2,3), (1,3), (1,1), (3,3)\}$
$R^3=R^4=R^5=\ldots=R^2$.

2.2.3 Equivalence Relations

A relation in a set S which is reflexive, symmetric, and transitive is called an *equivalence relation*, denoted $\equiv$. Given S and $\equiv$, for each $x \in S$, we call the set $\{y \in S \mid x \equiv y\}$ the *equivalence class* containing x, denoted $[x]$; $[x]$ contains all and only members of S that are equivalent to x. Note $x \equiv y$ if and only if $[x]=[y]$. Also, if $x \not\equiv y$, then $[x] \cap [y] = \emptyset$, i.e., equivalence classes associated with non-equivalent members of S are disjoint sets. In fact, if $S=\{x_1,x_2,\ldots,x_n\}$, then $\{[x_1],[x_2],\ldots,[x_n]\}$ is a partition of S; the *index* of an equivalence relation is the size (or rank) of the partition.

Example The identity (or 'equality') relation, $I=\{(x_i,x_i) \mid x_i \in S\}$, is an equivalence relation, where $[x_i]=\{x_i\}$, with index n. The universal relation

$$S \times S = \{(x_i,x_j) \mid x_i \in S,\ x_j \in S\}$$

is an equivalence relation, where $[x_i]=S$ for each i, and its index is 1. Finally, let

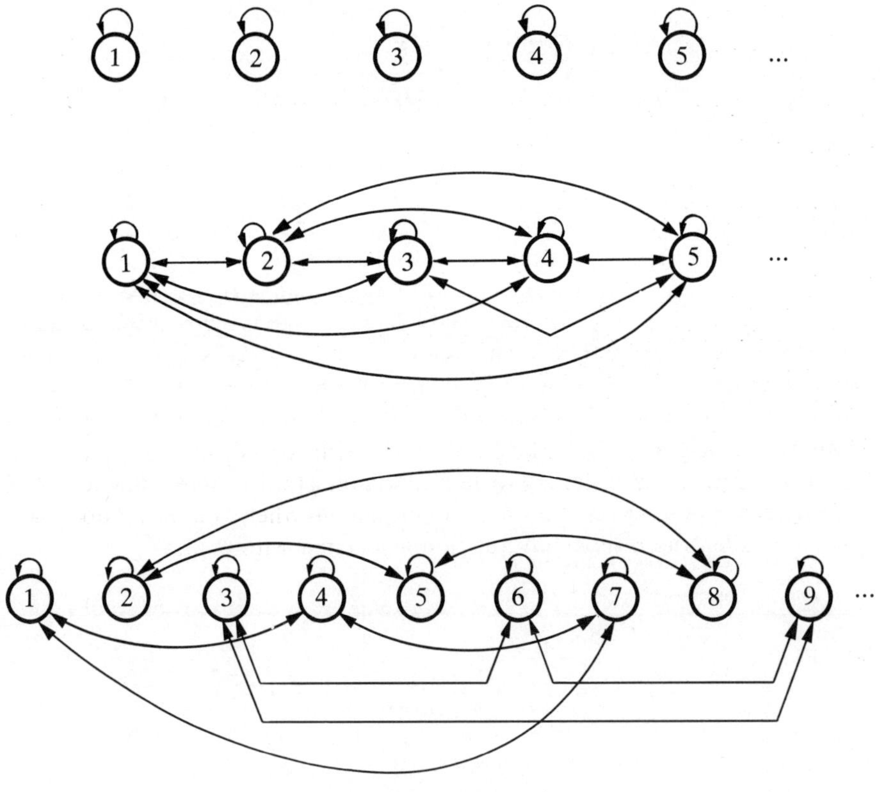

Figure 2.2.3

$S=\{1,2,\ldots\}$ be the set of natural numbers, and define C_m by letting $(x,y)\in C_m$ if x **mod** $m=y$ **mod** m; C_m is then an equivalence relation, and we write $x\equiv y$ (**modulo** m). This is the 'congruence' relation of Section 1.1.6, and its index is m. For $m=3$, there are three equivalence classes: $\{1,4,7,10,\ldots\}$, $\{2,5,8,11,\ldots\}$, and $\{3,6,9,12,\ldots\}$.

In the graph of an equivalence relation R in S, the equivalence class containing a point $x\in S$ is the set of all points y (necessarily including itself, because of reflexivity) such that a path from x to y (or y to x, because of symmetry) is in the graph of R. Points x and y are in different equivalence classes iff there is no path from one to the other in the graph. Hence it is always possible to draw the graph of an equivalence relation such that arrows connecting points in one equivalence class do not cross arrows connecting points in another, so that in a visual inspection of the graph the equivalence classes can be easily distinguished and counted. The equivalence classes are the 'disjoint pieces' in the graph. The index of R is the number of such pieces. For example, the graphs of I, $S\times S$, and C_3 are sketched in Figure 2.2.3.

2.2.4 Partially Ordered Sets

A relation (in S) which is transitive, reflexive, and antisymmetric is called a *weak partial order* (in S), and is commonly denoted $\leq$; the pair $(S,\leq)$ is then called a (*weak*) *partially ordered set*. A relation which is transitive and irreflexive (hence necessarily asymmetric) is called a *strict partial order*, and is commonly denoted $<$. A *total order* R in S is a partial order (weak or strict) such that either xRy or yRx for every distinct pair x and y in S; (S,R) then is often called a 'linearly' ordered set. If (S,R) is a linearly ordered set, it is customary for the listing ordering, $S=\{s_1,s_2,s_3,\ldots\}$, to be chosen to coincide with the ordering relation R, so that s_1Rs_2, s_2Rs_3, ... ; this assumes that S has a 'least' number, s_1, in a sense to be defined below. (Linearly ordered sets always have least members if they are finite, but not necessarily if they are infinite.)

We remark that for every weak partial order ($\leq$) we can define a strict partial order ($<$), by

$$x\leq y \quad \text{and} \quad x\neq y \quad \text{implies} \quad x<y;$$

and for every strict partial order ($<$) we can define a weak partial order ($\leq$), by

$$x<y \quad \text{or} \quad x=y \quad \text{implies} \quad x\leq y.$$

In other words, given $<$, $\leq$ is the set $<\cup I$; and, given $\leq$, $<$ is the set $\leq-I$ (where I is the identity relation). Also, we may denote $\nleq$ by $>$ and $<$ by $\ngeq$. For brevity, we usually use the term 'partial order' to refer to the weak relation, with the understanding that statements made about it have analogues for the strict case.

Let $(S,\leq)$ be a partially ordered set, with $<$ defined as the associated strict partial order. We say that a member b in S is a *successor* of a in S if $a<b$, and is an *immediate* successor if in addition there does not exist a member c in S such that $a<c<b$. We note that this defines another relation in S,

$$\mathbf{\$}=\{(a,b)\,|\,b \text{ is an immediate successor of } a\}.$$

The transitive closure of this immediate successor relation is called the *successor* relation, i.e. $\mathbf{\$}^+=<$; the respective inverse relations are the *immediate predecessor* and *predecessor* relations.

A member s_0 in S is 'the' *least* (*greatest*) member of S if it is a predecessor (successor) of every other member of S: i.e., $s_0\leq x$ ($x\leq s_0$) for every x in S. A member s_0 in S is 'a' *minimal* (*maximal*) member of S if it has no predecessor (successor): i.e., $x\leq s_0$ ($s_0\leq x$) implies $x=s_0$. A partially ordered set always has one or more minimal (maximal) members, but has a least (greatest) member iff it is also the unique minimal (maximal) member.

A member s_0 in S is a *lower* (*upper*) *bound* of a subset E of S iff $s_0\leq x$ ($x\leq s_0$) for every x in E. If the set of lower (upper) bounds of E has a greatest (least) member, then that member is called the *infimum* (*supremum*) of E. The infimum of E (denoted **inf** E) is also called the 'greatest lower bound' of E (**glb** E); the supremum of E (denoted **sup** E) is also called the 'least upper bound' of E (**lub** E). We note that the bounds of $E\subseteq S$ need not be members of E.

Topological Order In the above, we said that $S=\{s_1,s_2,\ldots,s_N\}$ is a 'linearly' ordered set if s_iRs_{i+1} holds for each i for some total order relation R. We wish to extend this ordering concept for S to the case where R is not a total order. In essence, we wish the ordered set $S=\{s_1,s_2,\ldots,s_N\}$ to be such that the order in which members are listed in the set S is 'consistent' with the given relation R in the sense that no member of S can be both a predecessor and successor of any other member of S. Formally, we say $S=\{s_1,s_2,\ldots,s_N\}$ is a *topologically ordered* (or *sorted*) set, with respect to a relation R, if s_iRs_j implies $i<j$ for each distinct pair s_i and s_j. Of course, the reverse of S, $\{s_N,s_{N-1},\ldots,s_1\}$, is also a topologically ordered set (with respect to R^{-1}). Clearly, if s_iRs_j and s_jRs_i both hold for $i\neq j$, then S cannot be topologically ordered (nor can any permutation of S be topologically sorted). We wish to determine what conditions R must satisfy to ensure that S can be topologically ordered. From the above, we conclude that antisymmetry of R is a necessary condition, but for sufficiency the transitive closure of R must also be antisymmetric. Note that both weak and strong partial orders satisfy this condition. We discuss an algorithm for producing a topologically ordered set given a relation R in Section 5.2.5. Briefly, given S, where S is not topologically ordered, and a

partial order R in S, we can produce a topologically ordered set S' by successively finding and deleting a minimal member of S while placing it in S' in order.

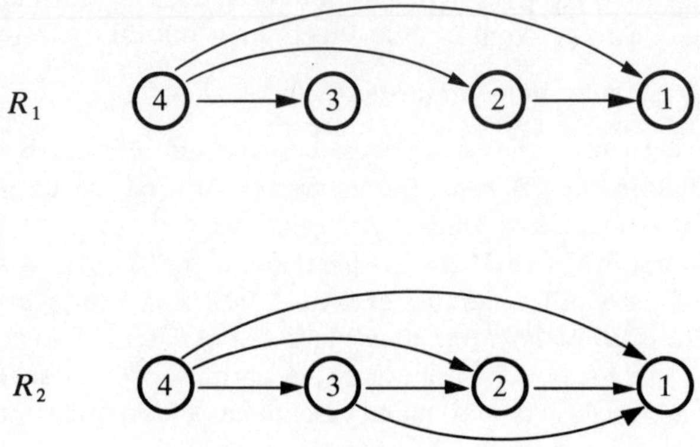

Figure 2.2.4

Examples Consider the relations depicted in Figure 2.2.4. The first one,

$$R_1=\{(4,3),\ (4,2),\ (4,1),\ (2,1)\},$$

is transitive and irreflexive, hence is a strict partial order. It is not a total order since, for example, 2 and 3 are not related. The corresponding weak partial order is

$$\{(4,3),\ (4,2),\ (4,1),\ (2,1),\ (1,1),\ (2,2),\ (3,3),\ (4,4)\}.$$

The immediate successor relation is

$$\$=\{(4,3),\ (4,2),\ (2,1)\}.$$

Note that **\$** is not transitive, hence is itself not a partial order; moreover, $\$^+=R_1$. Finally, there is a least (unique minimal) member of S (namely, 4), but two maximal members (namely, 1 and 3) (hence no greatest member); 1, 2, and 3 are upper bounds of $E=\{4\}$, but E has no supremum.

Consider next

$$R_2=\{(4,3),\ (4,2),\ (4,1),\ (3,2),\ (3,1),\ (2,1)\}.$$

R_2 is a total order. Let $E=\{1,2\}$. Then $\{4,3\}$ is the set of lower bounds of E, and 3 is the greatest of these lower bounds, hence **inf** E $(=3)\notin E$. Suppose $R_2'=R_2\cup I$, and $E=\{1,2\}$. Then 2 is also a lower bound of E, hence **inf** $E(=2)\in E$.

Finally, R_1^+ and R_2^+ are both asymmetric, hence a topological sort of the set $\{1,2,3,4\}$ can be found. $\{4,3,2,1\}$, $\{4,2,3,1\}$, and $\{4,2,1,3\}$ are topologically sorted with respect to R_1, but only $\{4,3,2,1\}$ is topologically sorted with respect to R_2.

Application: Set Inclusion Let us consider again a universal set U (of primitives). We observe that '$\subseteq$' is a relation, in our now formal sense, where $A\subseteq B$ means that $(A,B)\in$'$\subseteq$', for any pair of sets A and B in 2^U (the set of subsets of U). [Rather circularly, we can write '$\subseteq$'$\subseteq 2^U\times 2^U$!] From its definition in Section 2.1.3 (cf. Properties P1–P3), the subset relation is easily seen to be reflexive ($S\subseteq S$), transitive ($S_1\subseteq S_2$ and $S_2\subseteq S_3$ implies $S_1\subseteq S_3$) and antisymmetric ($S_1\subseteq S_2$ and $S_2\subseteq S_1$ implies $S_1=S_2$). Hence it is also a weak partial order; 'proper subset' is the associated strict partial order, and 'superset' ($\supseteq$) is its inverse relation. (We also observe that set membership, '$\in$', may be regarded as a bipartite relation.)

2.2.5 Application: Lexicographical Order

Consider N linearly ordered sets (S_1,R_1), (S_2,R_2), ..., (S_N,R_N). Let $A_i=S_1\times S_2\times\cdots\times S_i$, for $i=1,2,\ldots,N$, and let $A=\cup_i A_i$. We now proceed to define a relation R in A. For $a=(a_1,a_2,\ldots,a_r)$, $r\leq N$, and $b=(b_1,b_2,\ldots,b_s)$, $s\leq N$, let aRb if one of the following conditions holds:

(A) a is equal to b (i.e., $r=s$ and $a_i=b_i$ for each i)
(B) a is a 'prefix' of b (i.e., $r<s$ and $a_i=b_i$ for $1\leq i\leq r$)
(C) $a_i=b_i$ for $1\leq i\leq k-1$, but $a_k\neq b_k$ for some k $(1\leq k\leq r)$ and $a_k R_k b_k$.

R defined in this fashion is a total order in A, called the *lexicographical order*, and so we write $a\leq b$ if aRb ($a>b$ otherwise). It is common for the sets S_i to be identical (say, all equal to S), in which case S is called the 'character set' and A is a set of 'character strings' or 'words' (of length less than or equal to N).

Example: Alphabetical Order Let each $S_i=\{a,b,\ldots,z\}$, the set of letters. Define R_i, for each i, so that $aR_ib,\ldots,yR_iz$. Then $(a)\leq(a,d)\leq(a,d)\leq(a,d,d)\leq(a,t)\leq(b,e)$. We observe that $(a,d)\leq(a,d)$ since they are equal; $(a)\leq(a,d)\leq(a,d,d)$ by the 'prefix' condition; $(a,d,d)\leq(a,t)$ by the third condition (where $k=2$, dR_2t); and $(a,t)\leq(b,e)$ by the third condition (where $k=1$, aR_1b). This example illustrates that lexicographical order corresponds to alphabetical (dictionary) order, in the set of words. In the set of strings of digits, however, our intuition is not as reliable. For example, if each $S_i=\{0,1,\ldots,9\}$, and $0R_i1,\ldots,8R_i9$, then $(2,3,4)\leq(5,6)$, whereas as numbers $56\leq 234$.

Software Implications The lexicographical order relation permits us to extend the notion of 'alphabetical' order to apply to arbitrary character strings. However, not all programming languages permit direct comparison of character strings for other than equality, and the ordering of the character set (letters, digits, punctuation, etc.) is not standardized, being dependent upon computer hardware, for example. Furthermore, the ordering between letters and digits in the common ASCII and EBCDIC codes is reversed. Another problem is that, in some systems, short strings may be 'padded' [with blanks on the right] when compared with long strings.

A major application relates to a linear ordering of the elements of an n-dimensional array (see Section 2.5.1). Lexicographical ordering of data is required in order to use binary search. This is discussed in Section 7.2.

2.2.6 Sequences

We have earlier defined a 'string' or 'vector' of fixed length n, of elements of a set S, as a member of the n-fold product S^n, i.e., as an n-tuple $(a_1,a_2,\ldots,a_n)$, each a_i in S. A *sequence* [*in* S] is a generalization to the case where the length is not necessarily fixed (finite). (It is common to use angle brackets instead of parentheses to denote sequences—e.g., $\langle a_1,a_2,\ldots,a_n\rangle$ for finite sequences, or $\langle a_1,a_2,\ldots\rangle$ for infinite ones; sometimes, we will start with a_0 instead of a_1.) If R is a total order in S, then the sequence $\langle a_1,a_2,\ldots,a_n\rangle$ is said to be *increasing* if $a_i R a_{i+1}$ for $i=1,\ldots,n-1$. If $x=\langle a_1,a_2,\ldots,a_n\rangle$ is a sequence in S, and $\langle i_1,i_2,\ldots,i_m\rangle$ is an increasing sequence of integers such that $1\leq i_1<i_2<\ldots<i_m\leq n$, then $x'=\langle a_{i_1},a_{i_2},\ldots,a_{i_m}\rangle$ is called a *subsequence* of x; moreover, a subsequence x' is a *substring* of x if $i_{j+1}=i_j+1$, i.e., if $x'=\langle a_i,a_{i+1},\ldots,a_{i+m-1}\rangle$ is a 'consecutive' subsequence of x. A *run* is a maximal increasing substring.

Recurrence Relations Given a sequence of numbers $\langle a_0,a_1,a_2,\ldots\rangle$, an equation relating a number a_n, for any $n>n_0$, to r of its predecessors in the sequence is called a *recurrence relation* (of order r); the values $a_0,\ldots,a_{n_0}$ are called *boundary conditions*. There must be at least r boundary conditions, i.e., $n_0+1\geq r$ must hold. In fact, equality commonly holds, with a_n related to its r *immediate* predecessors, e.g.,

$$a_n=\mathbf{F}(a_{n-1},a_{n-2},\ldots,a_{n-r},n),$$

where the function $\mathbf{F}$ may also depend on n. An example of a special case where $\mathbf{F}$ does not depend on n, and $r=2$, is the recurrence relation

$$a_n=a_{n-1}+a_{n-2}, \qquad n>1,$$

with boundary conditions $a_0=0$, $a_1=1$; this 'defines' the **Fibonacci sequence** of numbers $\langle 0,1,1,2,3,5,8,13,\ldots\rangle$. An example of a special case where $\mathbf{F}$ does depend on n, and $r=1$, is the recurrence relation

$$a_n=a_{n-1}\cdot n, \qquad n>0$$

with boundary condition $a_0=1$; this defines the sequence of factorials $\langle 1,1,2,6,24,120,\ldots\rangle$, which we discuss further in Section 2.3.4. (See also

Section 3.1.4 for a discussion of the case where each a_n is a function.)

Recurrence relations involving two (or more) subscripts define arrays of the form

$$\begin{bmatrix} a_{00} & a_{01} & a_{02} & \cdots \\ a_{10} & a_{11} & a_{12} & \cdots \\ a_{20} & a_{21} & a_{22} & \cdots \\ \cdots & \cdots & \cdots & \cdots \end{bmatrix}$$

where a_{nm} is defined in terms of a_{ij}, $i \leq n$, $j \leq m$. For example, the recurrence relation

$$a_{n,m} = a_{n-1,m-1} + a_{n-1,m}, \qquad n > 0, \quad m > 0$$

with boundary conditions $a_{i,0} = 1$ for $i \geq 0$, and $a_{0,j} = 0$ for $j > 0$, defines the **Pascal triangle**

$$\begin{bmatrix} 1 & 0 & 0 & 0 & 0 & \ldots \\ 1 & 1 & 0 & 0 & 0 & \ldots \\ 1 & 2 & 1 & 0 & 0 & \ldots \\ 1 & 3 & 3 & 1 & 0 & \ldots \\ 1 & 4 & 6 & 4 & 1 & \ldots \\ \ldots & \ldots & \ldots & \ldots & \ldots & \ldots \end{bmatrix}$$

(The nonzero numbers in this array are also known as 'binomial coefficients'; see Section 6.2.3.)

Other important examples of recurrently defined sequences are 'arithmetic progressions'

$$\langle a_0, a_0 + d, a_0 + d + d, a_0 + d + d + d, \ldots \rangle$$

and 'geometric progressions'

$$\langle a_0, a_0 k, a_0 k^2, a_0 k^3, \ldots \rangle.$$

The former is defined by the recurrence relation

$$a_n = a_{n-1} + d, \qquad n > 0$$

and the latter by

$$a_n = a_{n-1} \cdot k, \qquad n > 0$$

both with a single boundary condition a_0. Often we may wish to determine one member of the sequence without computing all preceding members. In this event, a 'closed' formula rather than a recurrence relation is needed—for example,

$$a_n = a_0 + d \cdot (n-1) = (a_0 - d) + d \cdot n = c + d \cdot n$$

and

$$a_n = a_0 \cdot k^{(n-1)} = (a_0 \cdot k^{-1}) \cdot k^n = b \cdot k^n$$

respectively, for the above progressions. (In both cases, $n>0$; for $n=0$, $a_n=a_0$.) Given such a closed formula, it is common to denote the sequence $\langle a_1,a_2,\ldots\rangle$ using set notation, e.g., $\{c+d\cdot n\,|\,n>0\}$, where the set is taken to be ordered by increasing value of the 'index' n. See also the discussion of 'generating functions' in Knuth [2.14].

A *periodic* sequence $\langle a_0,a_1,a_2,\ldots\rangle$, of 'period' $r>1$, is one defined by a recurrence relation where

$$a_n=a_{n-r} \qquad \text{for} \quad n>n_0\geq r-1.$$

In other words, the function $\mathbf{F}(a_{n-1},a_{n-2},\ldots,a_{n-r},n)$ depends only on a_{n-r}. If $n_0>r-1$ (usually much greater), we say the sequence is 'ultimately' or 'asymptotically' periodic. For example,

$$\langle 1,2,3,4,5,6,7,8,9,10,8,9,10,8,9,10,\ldots\rangle$$

is an ultimately periodic sequence whose period $r=3$ and where $n_0=9$. In conclusion, we note that if $r=1$, then the sequence is constant; such sequences are 'trivially' periodic in a sense, but are usually disregarded by requiring that $r>1$ (and that $a_0\neq a_1$).

2.3 FUNCTIONS

Functions were discussed informally in Section 1.2. Formally, a *function* **F** *on* (or *of*) a set A *into* a set B is a relation (i.e., a set of ordered pairs, from A to B) which is single-valued, i.e., where for each a in A there may be at most one b in B such that $(a,b)\in\mathbf{F}$; this unique b is called the *image* of the *argument* a, or the *value* of **F** 'for' (or 'at') a, and we write $b=\mathbf{F}(a)$ or $\mathbf{F}[a]$. We also write $\mathbf{F}$: $A\rightarrow B$ (instead of $\mathbf{F}\subseteq A\times B$ or $\mathbf{F}\in 2^{A\times B}$, as for multi-valued relations) to denote sets which are functions; we also say '**F** *maps* A (in)to B.' **F** may be regarded as a function of n arguments by letting

$$A=A_1\times A_2\times\cdots\times A_n;$$

a single argument is then an ordered n-tuple of n subarguments. Of special interest is the case where the sets A_i and B are identical; we then call **F** an 'operation' (discussed further in Section 2.4).

We define the *domain* of **F** as the set $\mathscr{D}_{\mathbf{F}}=\{a\in A\,|\,\mathbf{F}(a) \text{ is defined}\}$, and the *range* of **F** as $\mathscr{R}_{\mathbf{F}}=\{\mathbf{F}(a)\,|\,a\in\mathscr{D}_{\mathbf{F}}\}$. Thus, $\mathbf{F}=\{(a,\mathbf{F}(a))\,|\,a\in\mathscr{D}_{\mathbf{F}}\}$, where **F** by itself denotes the set, and $\mathbf{F}(a)$ denotes function values. We say a is a *fixed point* of **F** if $a=\mathbf{F}(a)$. If $\mathscr{R}_{\mathbf{F}}\subseteq\mathscr{D}_{\mathbf{F}}$, we say the domain is *closed* under **F**. If $\mathbf{F}$: $A\rightarrow B$, and $X\subseteq A\subseteq Y$, then $\mathbf{F}'$: $X\rightarrow B$ defined by $\mathbf{F}'(a)=\mathbf{F}(a)$ for any $a\in X$ is called the *restriction* of **F** to X, and $\mathbf{F}''$: $Y\rightarrow B$ defined by $\mathbf{F}''(a)=\mathbf{F}(a)$ for any $a \in A$, with $\mathbf{F}''(a)$ arbitrary for $a \notin A$, is called an *extension* of $\mathbf{F}$ to Y.

A function $\mathbf{F}$: $A \to B$ is said to be

total	if $\mathscr{D}_{\mathbf{F}}=A$, otherwise *partial*
onto	(or a 'surjection') if $\mathscr{R}_{\mathbf{F}}=B$
one-to-one	(1–1, or an 'injection') if $a \neq b$ implies $\mathbf{F}(a) \neq \mathbf{F}(b)$
invertible	(or a 'bijection') if 1–1, onto, and total.

Unless otherwise stated, functions are ordinarily considered total but not necessarily onto. If $\mathbf{F}$ is invertible, then the *inverse* function $\mathbf{F}^{-1}$: $B \to A$, defined by $\mathbf{F}^{-1}(b)=a$ if $\mathbf{F}(a)=b$, is also invertible (and, in fact, $(\mathbf{F}^{-1})^{-1}=\mathbf{F}$); we then also say that there is a 1–1 *correspondence* between A and B, and that A and B have the same size or cardinality.

Example Let R_1 denote the set of nonnegative reals, and R denote the set of all reals. The (positive) square-root function defined on R_1, **sqr**: $R_1 \to R_1$, is total, onto, 1–1, hence invertible; if $x=\mathbf{sqr}(y)$, then $y=\mathbf{sqr}^{-1}(x)=x^2$. Extending the map to R, **sqr**: $R \to R$, the same function is partial (undefined for negative arguments). If negative roots are allowed, then **sqr** would no longer be single-valued; we would then call it an 'operation' instead of a function.

Labeling Functions It is often convenient to associate with each member a_i of a set $A=\{a_1,a_2,\ldots\}$ a member ℓ_j of another set $L=\{\ell_1,\ell_2,\ldots\}$. To do so, we formally define a *labeling* function λ: $A \to L$, where L is called the set of labels, A is called a *labeled set*, and $\lambda(a_i)=\ell_j$ is the label of a_i. If the function is 1–1, we call the labels (or labeling) *distinctive*: each member of a_i then has a unique label different from the label of a_j, $i \neq j$, so that the members of A may be referenced by just their labels without ambiguity. (However, if λ is not onto, there may be 'undefined' labels.) If a labeling is not distinctive, we regard those members of A with a common label as 'equivalent' (with respect to the labels). A (numerical) *sequencing* of a set A is a distinctive labeling of A with $L=\{1,2,\ldots,\#(A)\}$; if $\lambda(a_i)=j$, it is then common to denote a_i by j or by $A(j)$. We recall that an ordered set $\{a_1,a_2,\ldots\}$ may be regarded as a set with an implied sequencing, where the label of a_i is i, called its *ordinate* (or *relative*) position in the listing ordering of members. We remark, in conclusion, that an infinite sequence $\mathbf{A}=\{a_1,a_2,\ldots\}$ in S can be regarded (and is sometimes defined) as a 1–1 function $\mathbf{A}$: $\mathscr{I} \to S$, where $\mathscr{I}=\{1,2,\ldots\}$. Infinite sets are said to be *countable* if they can be sequenced (numerically labeled) in this sense; we discuss this further in Section 2.3.2. A set, in addition to this implied sequencing, may have many other labelings or sequencings.

Special Functions Let $\mathbf{f}$: $\mathscr{I} \to \mathscr{I}$, where $\mathscr{I}$ is the set of natural numbers. A *constant* function is defined by

$$\mathbf{const}_k(i)=k, \qquad \text{for every} \quad i \in \mathscr{I},$$

where k is a fixed number (or 'constant'); the special case $\mathbf{const}_0$ is denoted **zero**. The *identity* function is defined by

$$\mathbf{ident}(i)=i, \qquad \text{for every} \quad i\in\mathcal{J}.$$

The *successor* function is defined by

$$\mathbf{succ}(i)=i+1, \qquad \text{for every} \quad i\in\mathcal{J}.$$

The *predecessor* function is defined by

$$\mathbf{pred}(i)=i-1, \qquad \text{for } i>0, \quad \text{and} \quad \mathbf{pred}(0)=0.$$

If $A=A_1\times\cdots\times A_n$ is an n-ary relation, we say $\mathbf{ident}_i$:$A\to A_i$, defined by

$$\mathbf{ident}_i(a_1,\ldots,a_n)=a_i$$

is a *selection* or *generalized-identity* function; it 'selects' the ith of its n arguments as its value.

A total function on a set A into the set of truth values {**true, false**} or $\{1,0\}$ is called a *predicate* (on A). If S is a subset of U, the *characteristic function* of S (with respect to U), denoted $\mathbf{C}_S$, is a predicate which maps members of U to $\{0,1\}$: i.e., for any $x\in U$,

$$\mathbf{C}_S(x)=\begin{cases} 1 \text{ if } x\in S \\ 0 \text{ if } x\notin S. \end{cases}$$

Note that if $\mathbf{f}$: $A\to B$, then the characteristic function of $\mathcal{D}_{\mathbf{f}}$ (w.r.t. A) maps each member of A to 1 iff $\mathbf{f}$ is total.

2.3.1 Composition of Functions

Recall that $R=R_1R_2$ is the composition of relations R_1 and R_2 if xRy iff xR_1z and zR_2y for some z. Similarly, the composition of functions $\mathbf{f}_1$: $A_1\to B_1$ and $\mathbf{f}_2$: $A_2\to B_2$ is a function from A_1 to B_2 if the range of $\mathbf{f}_1$ is a subset of the domain of $\mathbf{f}_2$ (i.e., if $B_1\subseteq A_2$). We write $\mathbf{f}=\mathbf{f}_2\circ\mathbf{f}_1$ and $\mathbf{f}(x)=\mathbf{f}_2(\mathbf{f}_1(x))$ for $x\in A_1$, $\mathbf{f}_1(x)\in A_2$, and $\mathbf{f}(x)\in B_2$; regarded as the composition of relations, the *composite function* $\mathbf{f}_2\circ\mathbf{f}_1$ equals $\mathbf{f}_1\mathbf{f}_2$. Note that, when defined (i.e., when $B_2\subseteq A_1$), the 'reversed' composite function $\mathbf{f}_1\circ\mathbf{f}_2$ need not be equal to $\mathbf{f}$. Examples of composite functions appear below, and an important application is discussed in Section 3.2.8.

Application: Indexing Functions Let $S=\{a_1, a_2,\ldots,a_N\}$ be an ordered set, let the ordered N-tuple $\Pi(S)=(a_{k_1},a_{k_2},\ldots,a_{k_N})$ be a permutation of S, and let $\mathbf{P}$: $\mathcal{J}_N\to S$ where $\mathcal{J}_N=\{1,2,\ldots,N\}$ and $\mathbf{P}(j)=a_{k_j}$, the jth coordinate of $\Pi(S)$. Define $\mathbf{I}$: $\mathcal{J}_N\to\mathcal{J}_N$ by letting $\mathbf{I}(k)$ be the location (coordinate position) in $\Pi(S)$ of a_k; $\mathbf{I}(k)=j$ specifies that a_k is the jth coordinate of $\Pi(S)$. $\mathbf{I}$ is called an *indexing function*, and the composite function $(\mathbf{P} \circ \mathbf{I})$: $\mathscr{I}_N \to S$ is such that $\mathbf{P}(\mathbf{I}(k)) = a_k$. Let i_k denote $\mathbf{I}(k)$. Then, in terms of ordered pairs, the functions

$$\mathbf{I} = \{(k, i_k) \mid k \in \mathscr{I}_N\},$$
$$\mathbf{P} = \{(i_k, a_k) \mid k \in \mathscr{I}_N\},$$

and

$$\mathbf{P} \circ \mathbf{I} = \{(k, a_k) \mid k \in \mathscr{I}_N\}.$$

Example If $S = (\alpha, \beta, \gamma)$ and $\Pi(S) = (\beta, \gamma, \alpha) = (\mathbf{P}(1), \mathbf{P}(2), \mathbf{P}(3))$, then $\mathbf{I} = \{(1,3),(2,1),(3,2)\}$, $\mathbf{P} = \{(3,\alpha),(1,\beta),(2,\gamma)\}$, and $\mathbf{P} \circ \mathbf{I} = \{(1,\alpha),(2,\beta),(3,\gamma)\}$. Note also that if we define $\mathbf{A}$: $\mathscr{I}_N \to S$ by $\mathbf{A}(k) = a_k$, then $(\mathbf{A} \circ \mathbf{I}^{-1})$: $\mathscr{I}_N \to S$ is such that $\mathbf{A}(\mathbf{I}^{-1}(j)) = \mathbf{P}(j)$. Denoting $\mathbf{I}^{-1}$ by k, we may write $\mathbf{P}(j) = a_{k_j}$, as above. (An application of the foregoing to the use of 'dope vectors' appears in Section 4.3.5.)

Related to the concept of composition of functions is that of combination of functions. If $\mathbf{f}_1 : A_1 \to B_1$ and $\mathbf{f}_2 : A_2 \to B_2$, then we may define $\mathbf{f} : A_1 \times A_2 \to B_1 \times B_2$ as follows:

$$\mathbf{f}(x_1, x_2) = (\mathbf{f}_1(x_1), \mathbf{f}_2(x_2)).$$

This should be contrasted with the function $\mathbf{f}' : A_1 \times A_2 \to B_1 \times B_2$, where $\mathbf{f}_1 : A_1 \times A_2 \to B_1$ and $\mathbf{f}_2 : A_1 \times A_2 \to B_2$, defined as follows:

$$\mathbf{f}'(x_1, x_2) = (\mathbf{f}_1(x_1, x_2), \mathbf{f}_2(x_1, x_2)).$$

In general, we say $\mathbf{f}$ is a *combination* (or *product*) of $\mathbf{f}_1$, $\mathbf{f}_2$, ..., $\mathbf{f}_n$ if $\mathbf{f}(x) = (\mathbf{f}_1(x), \mathbf{f}_2(x), \ldots, \mathbf{f}_n(x))$; we write $\mathbf{f} = \mathbf{f}_1 \times \mathbf{f}_2 \times \cdots \times \mathbf{f}_n$.

The functions **zero**, **succ**, and $\mathbf{ident}_i$ for any i, are called the 'primitive' functions. We show, in Section 2.3.6, how a major class of functions can be defined in terms of these primitive functions by permitting recursion. First, we show how functions can be composed and combined to obtain other functions:

$$\mathbf{const}_3(i) = 3 = \mathbf{succ}(\mathbf{succ}(\mathbf{succ}(\mathbf{zero}(i))))$$
$$\mathbf{incr}(i) = i + 3 = \mathbf{succ}(\mathbf{succ}(\mathbf{succ}(i)))$$
$$\mathbf{ident}(i) = \mathbf{ident}_1(i)$$
$$\mathbf{pred}(i+1) = \mathbf{ident}(i)$$
$$\mathbf{exch}(i,j) = (j,i) = (\mathbf{ident}_2(i,j), \mathbf{ident}_1(i,j))$$

Two other 'functionals' (used here to mean, loosely, functions of functions) are worth special mention. First, if $\mathbf{f}$: $A \times A \to A$, we define the *insertion* of $\mathbf{f}$ (denoted $/\mathbf{f}$), a function from $\cup_{i \geq 1} A^i$ to A, as follows:

$$/\mathbf{f}(a_1, a_2, \ldots, a_n) = \mathbf{f}(a_1, /\mathbf{f}(a_2, \ldots, a_n)) \qquad \text{if} \quad n \geq 2,$$
$$/\mathbf{f}(a_1) = a_1.$$

Second, if $\mathbf{f}$: $A \to B$, we define the *apply-to-all* of $\mathbf{f}$ (denoted $\alpha\mathbf{f}$), a function from A^n to B^n, as follows:

$$\alpha\mathbf{f}(a_1,a_2,\ldots,a_n)=(\mathbf{f}(a_1),\mathbf{f}(a_2),\ldots,\mathbf{f}(a_n)) \qquad \text{if} \quad n \geq 1.$$

These functionals are used in LISP and APL, languages which have some of the 'functional programming' features advocated by Backus [2.1]. Of related interest is the discussion of 'operators' given in Iverson [2.13]. We remark that insertion is called *reduction* in APL, and apply-to-all is called *map* (or some variant) in LISP.

2.3.2 Application: Countability

A set S was said to be *countable* (or 'countably infinite') if there is a 1–1 correspondence between S and the set $\mathcal{I}=\{1,2,3,\ldots\}$; in other words, there must exist an invertible function from S to $\mathcal{I}$ (or $\mathcal{I}$ to S). This holds when the members of S can be listed in a sequential order such that each member is uniquely determined (distinctively labeled) by its integer position number in the ordering; in other words, each member of S can be 'counted' in a systematic fashion. In a sense then, each countable set has the same number of members as $\mathcal{I}$; we call the cardinality of $\mathcal{I}$, or any countable set, ***aleph-null***. In contrast, a set S is finite (of size N) iff there is a 1–1 correspondence between S and $\mathcal{I}_N=[1,N]$, the set of integers from 1 to N. Observe that $\mathcal{I}$ in the above may be taken as the set $\{0,1,2,\ldots\}$ (or $\mathcal{I}_N$ as $[0,N-1]$), for if $\mathbf{f}:S \to \{0,1,2,\ldots\}$ is 1–1, then $\mathbf{g}:S \to \{1,2,3,\ldots\}$, defined by $\mathbf{g}(s)=\mathbf{f}(s)+1$, is also 1–1; note that $\mathbf{g}^{-1}(i)=\mathbf{f}^{-1}(i-1)$.

$\mathcal{I}$ itself is evidently countable. Subsets of $\mathcal{I}$ are either countable or finite; infinite subsets $\mathcal{I}'$ of $\mathcal{I}$ are countable. The latter can be demonstrated by listing the members of $\mathcal{I}' \subseteq \mathcal{I}$ in the same order as they appear in a listing of $\mathcal{I}$. Hence, for any infinite set S to be countable, it suffices that there exist a 1–1 correspondence between S and *any* infinite subset of integers. We shall show below that $2^{\mathcal{I}}$, the set of all subsets of $\mathcal{I}$, is not countable. Thus there is a hierarchy: some sets contain a finite number of elements, other sets contain a countably infinite number of elements, and still other sets contain uncountably many elements.

Example*: *Countability of n-tuples An important example of a countable set is the n-fold product $\mathcal{I}^n$ (for any finite n); in other words, we claim that there are 'countably many' n-tuples. If $(i_1, i_2,\ldots,i_n)$ is regarded as the coordinate of a point in an n-dimensional Cartesian space, that there are countably many such points can be shown by listing these points sequentially in (say) an outward 'spiral' from the point (0,0,...,0). For example, for $n=2$, we may list the points in the sequence (0,0), (1,0), (0,1), (2,0), (1,1), (0,2), (3,0), ... Alternatively, we may define $\mathbf{G}:\mathcal{I}^n \rightarrow \mathcal{I}$ by $\mathbf{G}(i_1,i_2,\ldots,i_n)=p_1^{i_1}p_2^{i_2}\ldots p_n^{i_n}$, where p_k is the kth smallest prime. Recalling our discussion of Gödel numbers, the function $\mathbf{G}$ is 1–1 and onto an infinite subset $\mathcal{I}'$ of $\mathcal{I}$, hence $\mathcal{I}^n$ is countable. Since a rational number is a ratio of integers, corresponding to a member of $\mathcal{I}\times\mathcal{I}$, this also shows that there are countably many rationals. (However, the set of real numbers is not countable.)

Example*: *Uncountability of $2^{\mathcal{I}}$ In Section 2.1.4, we stated that the power set $2^{\mathcal{I}}$, defined as $\{S \mid S\subseteq\mathcal{I}\}$, has size $2^{\#(\mathcal{I})}$. When $\mathcal{I}$ is the set of natural numbers, $2^{\mathcal{I}}$ is infinite. In fact, we show here that $2^{\mathcal{I}}$ is uncountable, i.e., that there does not exist a 1–1 correspondence between $\mathcal{I}$ and $2^{\mathcal{I}}$. Let us assume otherwise, i.e., that

$\mathbf{f}:\mathcal{I}\rightarrow 2^{\mathcal{I}}$ is a 1–1 *onto* function.

We shall show that this leads to a contradiction (cf. Section 1.3.2). Note that $\mathbf{f}(x)$ is a set; i.e., for any $x\in\mathcal{I}$, $\mathbf{f}(x)\in 2^{\mathcal{I}}$ or $\mathbf{f}(x)\subseteq\mathcal{I}$. Now define

$$A=\{x\in\mathcal{I} \mid x\notin\mathbf{f}(x)\};$$

note $A\subseteq\mathcal{I}$ and $A\in 2^{\mathcal{I}}$. Furthermore, let $a\in\mathcal{I}$ be such that $\mathbf{f}(a)=A$, which must exist by the onto assumption. Now on one hand, if $a\in A$, then $\mathbf{f}(a)=A$ contradicts $a\notin\mathbf{f}(a)$, which is implied by the definition of A; hence $a\notin A$. On the other hand, if $a\notin A$, then $\mathbf{f}(a)=A$ contradicts $a\in\mathbf{f}(a)$, which is also implied by the definition of A; hence $a\in A$. We conclude that the onto assumption cannot hold, hence $2^{\mathcal{I}}$ is uncountable. (It is uncountability of a set which we will use in Section 5.4.4 to show the existence of functions which are not 'computable'.)

2.3.3 Predicates and Equations

A predicate $\mathbf{P}$: $A\rightarrow\{\boldsymbol{T},\boldsymbol{F}\}$ partitions A into two portions: that subset of A containing members x of A for which $\mathbf{P}(x)$ is true (denoted just $\mathbf{P}(x)$, for short), and that subset for which $\mathbf{P}(x)$ is false (denoted $\neg\mathbf{P}(x)$). (These subsets are disjoint and span A since $\mathbf{P}$, as a function, is single-valued and assumed total.) A predicate thus defines the set $\{x\in A \mid \mathbf{P}(x)\}$ and its complement, which we note correspond to the 'formula' notation for sets introduced in Section 2.1.

Common examples of predicates are 'relational' expressions such as

$$b^2-4ac<0$$
$$\mathbf{d}(x,y)\leq\mathbf{d}(x,z)+\mathbf{d}(z,y)$$
$$x^2+y^2=z^2$$

which are true if the given relation, from the set $\{<,\leq,=,\neq,\geq,>\}$, is true for specified values of the variables. Relational expressions involving the equality relation (=) are called *equations,* and the other examples above are called *inequalities*. Predicate expressions involving other relations, such as $\in$, $\subseteq$, $\equiv$, are also common. Predicate expressions can also be combined logically (using the logical opeıations) to form new ones.

For a fixed n-variable predicate expression, the domain of the predicate can be regarded as a Cartesian product $A_1 \times A_2 \times \cdots \times A_n$, where the ith variable assumes values from A_i. For example, the 3-variable predicate expression or equation $x^2+y^2=z^2$ can be regarded as a function $\mathbf{P}: \mathcal{J}^3 \to \{\boldsymbol{T}, \boldsymbol{F}\}$, where $\mathcal{J}$ is the set of positive integers. Members $(x,y,z) \in \mathcal{J}^3$ for which $\mathbf{P}(x,y,z)=\boldsymbol{T}$ include (3,4,5) and multiples thereof. When the domain is finite, its members can (in principle if not in practice) be listed one at a time, and the function evaluated 'by cases', as in a *truth table* (cf. Section 1.3).

Given an equation $\mathbf{P}(x)$, we generally wish to determine the members of $\{x \in A \mid \mathbf{P}(x)\}$, the set of *solutions* to the equation for a specified domain A. For example, the set of integer solutions to the equation $x^2+y^2=z^2$,

$$\{(x,y,z) \in \mathcal{J}^3 \mid x^2+y^2=z^2\},$$

includes (3,4,5), (5,12,13), ... However, if R denotes the set of reals, the set of real solutions to the same equation,

$$\{(x,y,z) \in R^3 \mid x^2+y^2=z^2\},$$

also includes $(1,1,\sqrt{2})$, ... If z is a positive constant, usually denoted r instead, then

$$\{(x,y) \in R^2 \mid x^2+y^2=r^2\} = \{(r \textbf{ sin } \theta, r \textbf{ cos } \theta) \mid 0 \leq \theta \leq 2\pi\},$$

where θ is an angle expressed in radians. Some equations may have no solution, at least in the specified domain. For example, $\{x \in R \mid x^2=-1\}=\emptyset$; however, if C denotes the set of complex numbers and $\mathrm{i}=\sqrt{-1}$, then $\{x \in C \mid x^2=-1\}=\{\pm \mathrm{i}\}$. Furthermore, we remark that, according to **Fermat's last theorem** (as yet unproven),

$$\{(i,j,k) \in \mathcal{J}^3 \mid i^n+j^n=k^n\}=\emptyset$$

for any integer $n>2$.

Given an inequality $\mathbf{P}(x)$, we generally are not interested in particular members ('solutions') of $\{x \in A \mid \mathbf{P}(x)\}$ for a specified domain A, but rather are interested in characterizing the domain for which the inequality is true or false. For example, given the inequality $x^2 \geq 0$ (defined only for $x^2 \in R$), the domain for which the inequality is true is the set of reals R; hence $\{x \in R \mid x^2<0\}=\emptyset$. On the other hand, the domain for which $x^2 \geq 0$ is false is the set of imaginary numbers.

In conclusion, we emphasize that the variables in predicates (equations and inequalities) need not have numerical values; they may in fact have functions as values. For example, consider the equation

$$\mathbf{f}(a+b)=\mathbf{f}(a)\times\mathbf{f}(b)$$

where **f**, the unknown, is a function **f**: $R\rightarrow R$ (where R is the set of real numbers) and + and × denote ordinary addition and multiplication (of reals). (Recall that the function **f** is a set, in fact a subset of $R\times R$; the set of all possible such subsets is $2^{R\times R}$.) The set of solutions to the equation can be expressed as follows:

$$\{\mathbf{f}:R\rightarrow R\,|\,\mathbf{f}(a+b)=\mathbf{f}(a)\times\mathbf{f}(b)\}.$$

One member of this set is the 'exponential' function $\mathbf{f}(a)=10^a$ (since $10^{a+b}=10^a\times10^b$). Equations whose solutions are functions are also known as *functional equations*.

2.3.4 Arguments and Recursion

A function **f**: $A\rightarrow B$ has a unique value $\mathbf{f}(a)\in B$ for each single *argument* $a\in A$. If $A=\{a_1,a_2,\ldots\}$, then as a set of ordered pairs we write

$$\mathbf{F}=\{(a_1,\mathbf{f}(a_1)),(a_2,\mathbf{f}(a_2)),\ldots\}.$$

Recall from above that a single argument may be an ordered n-tuple of n subarguments. Later, we discuss the case where arguments themselves may be functions.

In computer programming languages, a distinction is made between function 'definitions', function 'references', and function 'evaluations'. A function *definition* is an association of a name with a specification (by cases, by formula, or by computational algorithm) of the set of ordered pairs constituting the function, and with an implicit or explicit specification of its domain (i.e., its set of 'legal' arguments). A *reference* to a defined function is the use of its name, say, in a formula or a computational algorithm. An *evaluation* of a referenced function **F** is the determination of the value of **F** for a particular argument a; this argument a (or its subarguments) is sometimes called the *actual* parameter(s) for a particular evaluation of **F**, as opposed to the *formal* parameter(s) specified in formula or algorithmic definitions of **F**. The actual parameter is a member of the domain A; the formal parameter is a symbolic variable.

Example Consider the 'factorial' function **FACT**: $\mathcal{J}\rightarrow\mathcal{J}$ where $\mathcal{J}$ is the set of natural numbers. We write $n\geq0$ to mean $n\in\mathcal{J}$. The definition of **FACT** by cases, i.e., as a specified set of ordered pairs, is

FACT$=\{(0,1),(1,1),(2,2),(3,6),(4,24),(5,120),\ldots\}$.

Since this is an infinite ('open-ended') set, a 'closed' *formula* giving the values of the function is generally preferred; one such formula is:

$$\mathbf{FACT}(n)=\Pi_{i=1}^{n} i, \quad n \geq 0.$$

Thus, a formula definition of the function is **FACT**$=\{(n,\Pi_{i=1}^{n} i) \mid n \geq 0\}$. If the generalized product is not defined in our language (as is the case with most programming languages), the **FACT** function must then be defined by a computational *algorithm* (i.e., by a 'program') utilizing only ordinary multiplication: for example, in a PL/I-like language:

```
FACT: PROC(N);
        F=1;
        DO I=1 TO N;
        F=F * I; END;
        RETURN(F);
END;.
```

In both the formula and algorithmic definitions of **FACT**, n is the formal parameter. Evaluation of **FACT** for an actual parameter, (say) 5, yields the result **FACT**$(5)=120$, or $(5,120)\in$**FACT**.

Recursive Functions A *recursive* function[†] is one whose formula or algorithmic definition references itself by name. For example, the 'factorial' function **FACT**: $\mathcal{J}\rightarrow\mathcal{J}$, defined previously by the generalized product formula

$$\mathbf{FACT}(n)=\Pi_{i=1}^{n} i$$

may instead be defined recursively by

$$\mathbf{FACT}(n)=n\times\mathbf{FACT}(n-1), \quad n\geq 1$$
$$\mathbf{FACT}(0)=1.$$

Note that the evaluation of **FACT**, say, for $n=5$, requires its evaluation for $n=4$. Furthermore, every recursive equation requires some 'boundary' or 'termination' condition (e.g., **FACT**$(0)=1$) to terminate the recursion. Some programming languages (PL/I, LISP, SNOBOL) incorporate the facility to define such recursive functions; other languages (FORTRAN, COBOL) do not. The facility requires a mechanism (usually 'stacking') which saves the states of previous computations when the function is evaluated again for another argument. (We remark that we have already used the notion of

[†]This should not be confused with the definitions of 'primitive recursive' or 'partial recursive' functions given in Section 2.3.6.

recursive functions in defining

SUM(A,B)=SUM(SUCC(A),PRED(B))

in Section 1.1.1.) For further discussion, see Barron [2.2] and Wand [2.19].

The set **FACT**={(0,1),(1,1),(2,2),(3,6),...} can also be defined as follows:

$$A_0=\{(0,1)\}$$
$$A_{i+1}=A_i \cup \{(i+1, j\times(i+1)) \mid (i,j)\in A_i\},\ i\geq 0$$
$$\textbf{FACT}=\cup_i A_i.$$

In general, if S is a set, $A_0\subseteq S$, and **f**: $S\to S$, then

$$A_{i+1}=A_i \cup \{\mathbf{f}(x) \mid x\in A_i\},\ i\geq 0 \qquad (2.3.4)$$

is said to *inductively* define the set $S=\cup_i A_i$; this set S is 'closed' with respect to **f** in the sense that if $x\in S$ then $\mathbf{f}(x)\in S$, and furthermore each member of S either is in A_0 or is $\mathbf{f}(x)$ for some x in S. Note that Eq. (2.3.4) may also be regarded as a recurrence relation which defines the sequence $\langle A_0, A_1, \ldots\rangle$. As another example, let $A_0=\{\Phi\}$ be a set consisting of a distinguished object Φ, and let $\mathbf{f}(x)$ denote the 'successor' of x. Then

$$A_0=\{\Phi\}$$
$$A_1=A_0\cup\{\mathbf{f}(x) \mid x\in A_0\}=\{\Phi,\mathbf{f}(\Phi)\}$$
$$A_2=A_1\cup\{\mathbf{f}(x) \mid x\in A_1\}=\{\Phi,\mathbf{f}(\Phi),\mathbf{f}(\mathbf{f}(\Phi))\}$$
$$\ldots$$

and $S=\cup_i A_i$ is the set of natural numbers defined in Section 1.1.1.

In conclusion, we wish to distinguish the specific terms recursive, recurrent, and inductive, and the general term iterative.† A function is said to be recursively defined if its formula or algorithmic definition references itself by name. A sequence is said to be recurrently defined if the 'next' member of the sequence is related to previous members. A set is said to be inductively defined if it is the union of the members of a recurrently defined sequence of sets. We also say that a set or function is iteratively defined if it is a member of a recurrently defined sequence. In Section 2.2.1, we considered a sequence of sets $\langle A^1, A^2, A^3, \ldots\rangle$, where A^i was defined in terms of A^{i-1}, hence each is an iteratively defined set; on the other hand, we say $\cup_i A^i$ is an inductively defined set. In Section 3.1.4, we show how common mathematical functions can be defined in terms of sequences. In the next section, we describe one well-known application of recursive functions.

†Some writers use these terms somewhat interchangeably. We have no real objection to this, our distinctions being mainly a matter of taste.

2.3.5 Application: Conditional Expressions

A *conditional expression* is a function of the form

$$(p_1 \rightarrow e_1 \,,\, p_2 \rightarrow e_2 \,,\, \dots \,,\, p_n \rightarrow e_n)$$

whose value is e_1 if p_1 is true, else is e_2 if p_2 is true,..., else is e_n if p_n is true (else is undefined if none of the 'predicates' $p_1, p_2, \dots, p_n$ are true). The ordering of the predicate tests is significant; i.e., if p_i and p_j are both true, with $i<j$, then the value of the expression is e_i, not e_j. The predicates may contain the logical operations, defined in terms of conditional expressions as follows:

$$\begin{aligned} p \wedge q &= (p \rightarrow q \,,\, \boldsymbol{T} \rightarrow \boldsymbol{F}) \\ p \vee q &= (p \rightarrow \boldsymbol{T} \,,\, \boldsymbol{T} \rightarrow q) \\ \neg p &= (p \rightarrow \boldsymbol{F} \,,\, \boldsymbol{T} \rightarrow \boldsymbol{T}) \\ p \rightarrow q &= (p \rightarrow q \,,\, \boldsymbol{T} \rightarrow \boldsymbol{T}). \end{aligned}$$

(Note that choosing p_n to be $\boldsymbol{T}$ avoids the undefined situation mentioned above.)

Assuming the equality test (=) and the successor function (′) on natural numbers as primitives, as well as the capability to evaluate recursive conditional expressions, the predecessor function **pred**(n) may then be defined as **pred2**$(n,0)$ for $n \neq 0$, where **pred2** is defined recursively by

$$\textbf{pred2}(n,m) = (m' = n \rightarrow m \,,\, \boldsymbol{T} \rightarrow \textbf{pred2}(n,m')).$$

The sum of two numbers (cf. Eq. (1.1.1)) may then be defined recursively by

$$\textbf{sum}(m,n) = (n=0 \rightarrow m \,,\, \boldsymbol{T} \rightarrow \textbf{sum}(m', \textbf{pred}(n))).$$

The inequality relation $\leq$ on numbers may be defined by

$$m \leq n = (m=0) \vee (n \neq 0 \wedge \textbf{pred}(m) \leq \textbf{pred}(n))$$

and, of course,

$$m < n = (m \leq n) \wedge (m \neq n)$$

where $m \neq n = \neg(m=n)$. These arithmetic and relational operations may in turn be used to define more complex operations.

Examples The remainder of m divided by n, denoted **mod**(m,n), may be defined recursively as follows:

$$\textbf{mod}(m,n) = (m<n \rightarrow m \,,\, \boldsymbol{T} \rightarrow \textbf{mod}((m-n),n)).$$

In other words, **mod**(m,n) can be found by successively subtracting n from m until no longer possible. The greatest common divisor of m and n, denoted **gcd**(m,n), may then be found as follows:

$$\mathbf{gcd}(m,n)=(m<n\rightarrow\mathbf{gcd}(n,m),\ \mathbf{mod}(m,n)=0\rightarrow n\ ,\ T\rightarrow\mathbf{gcd}(n,\mathbf{mod}(m,n))).$$

For example,

$$\begin{aligned}
&\mathbf{gcd}(6,22)=\mathbf{gcd}(22,6) && [\text{since } 6<22]\\
&\mathbf{gcd}(22,6)=\mathbf{gcd}(6,4) && [\text{since } \mathbf{mod}(22,6)=4]\\
&\mathbf{gcd}(6,4)\ =\mathbf{gcd}(4,2) && [\text{since } \mathbf{mod}(6,4)\ =2]\\
&\mathbf{gcd}(4,2)\ =2 && [\text{since } \mathbf{mod}(4,2)\ =0].
\end{aligned}$$

Note that this algorithm produces successive pairs of minimum values and remainders; see Section 1.1.7. (In the foregoing, subtraction $(m-n)$ was not defined; it is left as an exercise.) Further examples may be found in McCarthy [2.17] and in various textbooks on the LISP language. The use of LISP for functional programming is of special interest; see Henderson [2.12].

2.3.6 Computability

Given the composite function $\mathbf{f}_2(\mathbf{f}_1(x))$, it is often convenient to view $\mathbf{f}_1$ as the argument of the function $\mathbf{f}_2$. Furthermore, any function of n arguments $\mathbf{f}(x_1,\ldots,x_n)$ may reference another function $\mathbf{g}_i$ in its ith argument; in general,

$$\mathbf{f}(x_1,\ldots,x_n)=\mathbf{f}(\mathbf{g}_1(x_1,\ldots,x_n),\ldots,\mathbf{g}_n(x_1,\ldots,x_n)).$$

A 'general' recursive function is one which references itself in one or more of its arguments. We then write

$$\mathbf{f}(x_1,\ldots,x_n)=\mathbf{g}(\ldots,\mathbf{f}(\ldots),\ldots,\ \mathbf{f}(\ldots),\ldots) \tag{2.3.6}$$

where the composite function **g** may be **f**, and where **f** as an argument may reference other functions. For example,

$$\mathbf{FACT}(x)=\mathbf{PROD}(x,\mathbf{FACT}(\mathbf{pred}(x))),$$

for $x>0$, with termination condition $\mathbf{FACT}(0)=1$; here, $\mathbf{f}=\mathbf{FACT}$ and $\mathbf{g}=\mathbf{PROD}$ (the function which computes the product of its arguments), where the predecessor function is referenced as the argument of an argument. Composition and recursion provide means of constructing new functions from a given 'basis' set of functions.

Primitive Recursive Functions 'General' recursion, of the form shown in Eq.(2.3.6), is too general for easy analysis. *Primitive* recursion restricts the recursion to only one argument, as follows:

$$\mathbf{f}(x_1,\ldots,x_{n-1},x_n+1)=\mathbf{g}(x_1,\ldots,x_n,\mathbf{f}(x_1,\ldots,x_n))$$

with termination condition

$$\mathbf{f}(x_1,\ldots,x_{n-1},0)=\mathbf{g}_0(x_1,\ldots,x_{n-1}),$$

where $\mathbf{f}:\mathcal{J}^n\to\mathcal{J}$, $\mathbf{g}:\mathcal{J}^{n+1}\to\mathcal{J}$, and $\mathbf{g}_0:\mathcal{J}^{n-1}\to\mathcal{J}$, for $\mathcal{J}=\{0,1,2,\ldots\}$. We define the class of *primitive recursive* functions as those which can be constructed using composition and primitive recursion only, given a basis set consisting of the 'primitive' functions, i.e., **succ**, **zero**, and $\mathbf{ident}_i$ for any i. Each primitive function is itself primitive recursive. Since the primitive functions are total, so are all the primitive recursive functions.

Examples Some examples of primitive recursive functions, defined by composition, recursion, and termination conditions, are given below:

1. $\mathbf{sum}(x,y)=x+y$ (addition):
   $\mathbf{sum}(x,y+1)=\mathbf{succ}(\mathbf{sum}(x,y))$
   $\mathbf{sum}(x,0)=\mathbf{ident}(x)$
2. $\mathbf{prod}(x,y)=x\cdot y$ (multiplication):
   $\mathbf{prod}(x,y+1)=\mathbf{sum}(\mathbf{ident}_1(x,y),\mathbf{prod}(x,y))$
   $\mathbf{prod}(x,0)=\mathbf{zero}(x)$
3. $\mathbf{fact}(y)=y!$ (factorial):
   $\mathbf{fact}(y+1)=\mathbf{prod}(\mathbf{succ}(y),\mathbf{fact}(y))$
   $\mathbf{fact}(0)=\mathbf{succ}(0)$
4. $\mathbf{diff}(x,y)=x-y$ if $x\geq y$, $=0$ if $x<y$:
   $\mathbf{diff}(x,y+1)=\mathbf{pred}(\mathbf{diff}(x,y))$
   $\mathbf{diff}(x,0)=\mathbf{ident}(x)$.

Recall that $\mathbf{pred}(x+1)=\mathbf{ident}(x)=\mathbf{ident}_1(x)$. We next define a 'distance' function in terms of the 'difference' (or 'proper subtraction') function as follows:

5. $\mathbf{dist}(x,y)=\mathbf{sum}(\mathbf{diff}(x,y),\mathbf{diff}(y,x))$.

Finally, let

6. $\mathbf{null?}(x)=\mathbf{diff}(1,x)$,

which equals 1 for $x=0$, else equals 0 for $x>0$; this function is a predicate which tests whether or not its argument is zero. The 'equality test' predicate which equals 1 if its arguments are equal, else equals 0, may then be defined as follows:

7. $\mathbf{equal?}(x,y)=\mathbf{null?}(\mathbf{dist}(x,y))$.

Computable Functions There are functions which are not primitive recursive; for example, any partial function from $\mathcal{J}^n$ to $\mathcal{J}$ is not. However, the function defined by

$$\mathbf{a}(x+1,y+1)=\mathbf{a}(x,\mathbf{a}(x+1,y))$$
$$\mathbf{a}(0,y)=y+1$$
$$\mathbf{a}(x,0)=\mathbf{a}(x-1,1)$$

(known as **Ackermann's function**) is also not primitive recursive even though it is total; note that it uses a more general form of recursion, being recursive in both of its arguments. We define the class of *partial recursive* functions as those which can be obtained by adding the additional operation of *minimalization* [to the two previously permitted construction operations for primitive recursive functions (i.e., composition and primitive recursion)]. Given y), we construct the function $\mathbf{M}(x_1,\ldots,x_n)$= the minimum value in the set 0,y)=0}. Since there may be arguments for which $\mathbf{f}\neq 0$, **M** may be a partial function. A *recursive* function is defined as a partial recursive function which is also total.

Informally, a function $\mathbf{f}{:}A\rightarrow B$ is said to be *computable* if there exists an algorithm or computer program **P** which 'implements' the function in the sense that if $I\in A$ is the input of **P**, then $\mathbf{f}(I)\in B$ is the output. We say that a set is *recursive* if its characteristic function is computable, and a set is *enumerable* if it is the range of a computable function. Note that every recursive function is computable since the construction operations (composition, primitive recursion, and minimalization) can be implemented by a computer program, as can the primitive functions. That every total function which can be conceived as computable is also recursive is known as **Church's thesis**. (See Beckman [2.1], Brady [2.7], and Section 5.4 for further discussion.) This implies, of course, that every algorithm which can be expressed as a conventional computer program can also be expressed purely 'functionally'; in particular, assignment statements and goto statements are not required.

2.4 OPERATIONS

An (n-ary) *operation* Θ in a set S is a [($n+1$)-ary] relation from ordered n-tuples of members of S to S; i.e., $\Theta\subseteq S^n\times S$. In the event the relation is single-valued, i.e., if Θ is a function, we say the operation is 'deterministic'; otherwise, we say it is 'nondeterministic'. For $n=0$, 1, 2, and 3, we say that the operation is *null, unary, binary,* and *ternary*, respectively. A null operation maps domain S^0 (which need not be defined) into S; no matter how S^0 is defined, the members of the range C are called *constants,* or *distinguished* members, of S (C is a subset of S). A unary operation in S is a relation from S to S; a deterministic unary operation is a single-valued relation in S (i.e., a function). Unless otherwise specified, e.g., for sets of constants (regarded as null operations) and for multi-valued relations (regarded as unary operations), operations are generally assumed to be deterministic. (An example of a nondeterministic operation is the unary square-root operation in the set of reals, which has both positive and negative roots.) We say that an operation is *closed* in S, or that S is a 'closed set under the operation', if the operation is a total function. (The square-root operation is not closed in the set of reals, since it is undefined for negative reals, but it is closed in the set of complex numbers.)

A (deterministic) binary operation Θ is a function on $S \times S$ into S. In this special case, instead of writing $\Theta(a,b)$, we introduce a special symbol, say $\circ$, to denote the operation or operator, and write $a \circ b = c$ if $((a,b),c) \in \Theta$; a and b are called the *operands* and c the *result* of the operation $\circ$. This notation, where the symbol denoting an operation appears in between its (two) operands, is called *infix*; *prefix* and *postfix* notations have the operation preceding and following the operands, respectively. The latter two notations, unlike infix, are generalizable to other than binary operations. Note that $\circ$ is closed in S, or S is closed under $\circ$, iff $a \circ b$ is defined and is in S for each pair a, b in S.

In the case where $\Theta \subseteq S^n \times S$ is a possibly multi-valued relation, i.e., when Θ is a nondeterministic operation (and hence is not a function), it may be converted into a function, $\Theta': S^n \rightarrow 2^S$, by letting

$$\Theta'(s_1,\ldots,s_n) = \{s \in S \mid ((s_1,\ldots,s_n),s) \in \Theta\}.$$

However, Θ' would not then be an operation on S; of course, we may define Θ'': $(2^S)^n \rightarrow 2^S$, a single-valued operation on 2^S, but only a partial function with domain $\hat{S}^n$, where $\hat{S} = \{\{s\} \mid s \in S\}$.

2.4.1 Properties of Operations

Suppose we have a binary operation in S, denoted $\circ$. (We will usually assume that binary operations are deterministic and closed, unless otherwise stated.) We further say that the operation $\circ$

is *associative* iff $a \circ (b \circ c) = (a \circ b) \circ c$
is *commutative* iff $a \circ b = b \circ a$
has a *nullifier* z iff $z \circ a = z = a \circ z$
has an *identity* e iff $e \circ a = a = a \circ e$
has *inverse* elements iff $(\exists a')a' \circ a = e = a \circ a'$,
  where e is the identity [a' is defined as the inverse of a]
has the *involution* property iff $(a')' = a$,
  where a' is the inverse of a
is *idempotent* iff $a \circ a = a$
has the *(left) cancellation* property iff
  $c \circ a = c \circ b$ implies $a = b$ (whenever $c \neq e$)

for all a, b, c in S. The nullifier z and identity element e, and the inverse elements a' of each a, must also be in S and be uniquely defined (i.e., with respect to $\circ$, S may contain only one nullifier and identity, and a may have only one inverse a').

We remark that an operation may have only a *left* identity (e_1) or a *right* identity (e_2), defined by

$$(\exists e_1)e_1 \circ a = a, \qquad (\exists e_2)a \circ e_2 = a,$$

respectively; the operation has an (ordinary) identity iff it has both a left and right identity and these are identical. Similarly, an operation may have only left inverse or right inverse elements, defined by

$$(\exists a')a' \circ a = e, \qquad (\exists a')a \circ a' = e,$$

respectively. The existence of inverse elements also permits the definition of a unary operation, called *inversion*, $\mathbf{I}$: $S \rightarrow S$, such that $(a, a') \in \mathbf{I}$ iff a' is the inverse of a. In this event, we also say (somewhat informally) that the operation $\circ$ has an inverse, when in fact it is the elements which have inverses. We note also that inversion, as a function, equals its inverse—i.e. $\mathbf{I} = \mathbf{I}^{-1}$.

Examples The addition operation ($+$: $S \rightarrow S$) in the set of positive integers is binary, deterministic, closed, associative, and commutative, but has no identity. In the set of nonnegative integers, addition has the additional number 0 (zero) as an identity, but has no inverse. In the set of all integers, the negative numbers are inverses of the positive ones and vice versa; zero is its own inverse. The involution and cancellation properties hold, but idempotence does not. See Section 3.1.1 for further discussion of the arithmetic operations.

2.4.2 Operation Pairs

Suppose we have two binary operations in S, denoted $\sqcap$ and $\sqcup$. We say they satisfy the stated property,

consistency, iff $a \sqcap b = a$ is equivalent to $a \sqcup b = b$

absorption, iff $a \sqcap (a \sqcup b) = a \sqcup (a \sqcap b) = a$

(mutual) *distributivity*, iff $a \sqcap (b \sqcup c) = (a \sqcap b) \sqcup (a \sqcap c)$
and $a \sqcup (b \sqcap c) = (a \sqcup b) \sqcap (a \sqcup c)$

(*universal*) *boundedness*, iff $(\exists B_1)$ $a \sqcap B_1 = B_1$, $a \sqcup B_1 = a$
and $(\exists B_2)$ $a \sqcup B_2 = B_2$, $a \sqcap B_2 = a$

complementarity, iff $(\exists a')$ $a \sqcap a' = B_1$, $a \sqcup a' = B_2$,
where B_1 and B_2 are universal bounds
(a' is defined as the complement of a)

DeMorgan's law, iff $(a \sqcap b)' = a' \sqcup b'$
and $(a \sqcup b)' = a' \sqcap b'$,
where a' is the complement of a

for all a, b, c in S. The bounds B_1 and B_2, and the complements a' for each a in S, must also be in S. We emphasize that the complement of a, although also denoted a', is *not* the inverse of a with respect to either $\sqcap$ or $\sqcup$. Note also that if $\sqcap$ and $\sqcup$ are not commutative, then we must distinguish between left and right distributivity.

Example Each of the foregoing properties holds for union and intersection of sets, and for the logical *and* and *or* operations. See Section 2.5.3 for additional details. Only one of the two distributive laws holds for the arithmetic operations (+ and ×); we leave as an exercise consideration of the other properties.

2.4.3 Morphisms

A 'morphism' is a function $\mathbf{F}: A \to B$ such that some property (or set of properties) $\mathscr{P}$ associated with A is also associated with B; we say then that $\mathbf{F}$ 'preserves' $\mathscr{P}$. For example, suppose $\mathscr{P}$ is the 'semigroup' property that a closed, associative, binary operation, denoted $\circ$, is defined on the sets (where context distinguishes $\circ$ on A from $\circ$ on B). Then we say that $\mathbf{F}$ is a *homomorphism* if

$$\mathbf{F}(x \circ y) = \mathbf{F}(x) \circ \mathbf{F}(y)$$

for all x, y in A. If a homomorphism $\mathbf{F}$ is both 1–1 and onto, so that $\mathbf{F}$ is invertible, then we say $\mathbf{F}$ is an *isomorphism*.

Considering the preservation of a relation, we say $\mathbf{F}: A \to B$ is an *order morphism* (or 'order-preserving') if

$$(x,y) \in R(\text{in } A) \text{ implies } (\mathbf{F}(x),\mathbf{F}(y)) \in R(\text{in } B).$$

If, in addition, R in both A and B are strict partial orders, we then say that $\mathbf{F}$ is a *monotone* function. Monotone functions may be ('monotonically') increasing, decreasing, non-increasing, or non-decreasing, depending upon the relations.

A *category* $\mathscr{C}$ is a set of objects (which may in turn be sets) together with, for each ordered pair of these objects, a set of morphisms which can be composed (i.e., if $\mathscr{C} = \{A, B, C, \ldots\}$, and $\mathbf{f}: A \to B$ and $\mathbf{g}: B \to C$ are morphisms, then $\mathbf{f} \circ \mathbf{g}$ is also a morphism), where the composition of morphisms is associative and has an identity. The development and application of categorical algebra to computer science is relatively recent; we refer interested readers to Eilenberg [2.8].

2.4.4 Algebraic Systems

We now formally define an *algebraic system* as a set S together with a collection of operations in S, $\{\Theta_1,\Theta_2,...,\Theta_k\}$; we usually denote this algebraic system by the $(k+1)$-tuple $(S, \Theta_1,...,\Theta_k)$. By convention, n-ary operations should be single-valued for $n>1$, but we permit null and unary operations to be nondeterministic. Alternatively, we may define an algebraic system as a set together with distinguished members, relations, and (then necessarily single-valued) operations. Often, it is convenient to regard S as the union of distinguished (not necessarily disjoint) subsets, $S=S_1\cup S_2\cup\cdots\cup S_i$, and to write $(S_1,...,S_i, R_1,...,R_j, \Theta_1,...,\Theta_k)$ to denote the algebraic system. The simplest non-trivial algebraic system consists of a set S together with a relation R in S, (S,R), and is called a *directed graph*. This class of algebraic systems is of fundamental importance, and deserves special treatment (see Chapter 4). Other algebraic systems which we discuss herein include:

numbers and the arithmetic operations
sets and the set operations
truth-values and the logical operations
matrices and the matrix operations
vectors and the vector operations
strings and the string operations.

In the next section, we introduce, as an illustration, an algebraic system $(S, \circ,e)$ having a single associative operation $\circ$ with an identity e; such algebraic systems are called *monoids*.

2.4.5 Application: Strings

Given a finite nonempty set of symbols A, called the *alphabet* (or 'character set'), a *string* x (*over* A) of integer length $n\geq 1$, of members of A, was defined in Section 2.2.1 as a member of A^n (the n-fold product of A), i.e., $x=(x_1,x_2,...,x_n)$, each $x_i\in A$. When no ambiguity results, we usually write x without parentheses and commas, $x=x_1x_2...x_n$; then

$$A^1=\{x_1|x_1\in A\}=A,\ A^2=\{x_1x_2|x_1\in A,\ x_2\in A\},$$

and

$$A^n=\{x_1x_2...x_n|x_i\in A\}.$$

A string of length zero, called the *null* (or *empty*) string, is denoted ε and defines $A^0=\{\varepsilon\}$ ($\neq\emptyset$!).

Recall (from Section 1.1) that there is an infinite number of numbers, but that each number is itself finite. Similarly, there is an infinite number of strings (at least one of every length n), but each string is itself of finite length. (That is, there is no infinitely long string.[†]) The set of all strings over A, denoted A^*, is equal to the infinite union $A^0 \cup A^1 \cup A^2 \cup \cdots$; let $A^+ = A^* - \{\varepsilon\}$ denote the set of all non-null strings. For example, if $A = \{0,1\}$, then $A^+ = \{0,1\}^+ = \{0, 1, 00, 01, 10, 11, 000, 001, 010, 011, ...\}$, the set of all binary numbers (including leading zeroes); $\{0,1\}^* = \{0,1\}^+ \cup \{\varepsilon\}$. We emphasize that any set of strings over A is a subset of A^*, or a member of 2^{A^*}.

Two strings $x = x_1 ... x_r$ and $y = y_1 ... y_s$ in A^* are *equal* (or 'match') if $r = s$ and $x_i = y_i$ for each i. For unequal strings x and y, we may write $x < y$ or $y < x$ with respect to a lexicographical order (see Section 2.2.5) if A is linearly ordered: for $x \neq y$, $x < y$ if $r < s$ and $x_i = y_i$ for $1 \leq i \leq r$, or if $x_k \neq y_k$ for some k $(1 \leq k \leq r)$ with $x_i = y_i$ for $1 \leq i \leq k-1$ and x_k less than y_k in the ordering of A. $(A^*, \leq)$ then is a lexicographically ordered set.

We now define an operation on A^* called *concatenation*: the concatenation of strings $x = x_1 ... x_r$ and $y = y_1 ... y_s$, denoted $x \cdot y$ (or xy), is the string $x_1 ... x_r y_1 ... y_s$. We emphasize that an r-tuple concatenated with an s-tuple yields an $(r+s)$-tuple. For x and y in A^*, we say that x is a *substring* of y if $y = uxv$ for some u and v in A^*. If $x = \varepsilon$ (i.e., $r = 0$), then $xy = y$; similarly, $x\varepsilon = x$. Concatenation is a closed, associative, binary operation, and has the null string ε as its identity. Hence the algebraic system $(A^*, \cdot, \varepsilon)$ is a monoid.

We next extend the definition of $\cdot$, which was defined between two individual strings, to a 'generalized concatenation' operation, which is defined between two *sets* of strings. If X and Y are subsets of A^*, i.e., members of 2^{A^*}, then we define the *complex product* (or *composition*) of X and Y, denoted $X \cdot Y$ (or XY), as the set of strings $\{xy \mid x \in X, y \in Y\}$. (Note that $X \cdot Y$ is not the same as the Cartesian product $X \times Y$; e.g., the complex product is associative, but the Cartesian product is not.) If either X or Y is empty, then so is XY; however, if X (or Y) $= \{\varepsilon\}$, the (nonempty) set containing only the null string, then $XY = Y$ (or X, respectively). In other words, the complex product has the set $\{\varepsilon\}$, not $\emptyset$, as its identity; hence $(2^{A^*}, \cdot, \{\varepsilon\})$ is a monoid. If X contains only a single member, $X = \{a\}$, we write aY or Ya (instead of XY or YX, respectively). If $X \subseteq A^*$ and $n > 0$, then the nth (*complex*) *power* of X is given by $X^n = X \cdot X^{n-1} = X^{n-1} \cdot X$, where $X^0 = \{\varepsilon\}$. (Note that the nth power of X is not the same as the n-fold product of X.) X^*, also called the *Kleene closure* of X, is the set $\bigcup_{n \geq 0} X^n$. We may regard Kleene closure as a unary operation $* : 2^{A^*} \to 2^{A^*}$. If the set X of strings over A has only one member, $X = \{x\}$ where $x = a_1 ... a_k$, then we denote X^n by $(a_1 ... a_k)^n$ and $\bigcup_{n \geq 0} (a_1 ... a_k)^n$ by $(a_1 ... a_k)^*$; if $k = 1$, we write a_1^n and a_1^*, omitting the parentheses.

In the context of strings, we call the union of two sets of strings X and Y

[†]There are, however, infinitely long *sequences*.

over A (i.e., $X \cup Y$) the *alternation* of X and Y, denoted $X|Y$. It is clear that $(2^{A^*}, |, \emptyset)$ is also a monoid. In addition, alternation is commutative, but the complex product is not. Furthermore, the complex product distributes over alternation (on both the left and right):

$$X \cdot (Y|Z) = (X \cdot Y)|(X \cdot Z)$$
$$(X|Y) \cdot Z = (X \cdot Z)|(Y \cdot Z).$$

In summary, $(2^{A^*}, |, \cdot, *, \emptyset, \{\varepsilon\})$ is an algebraic system defined on sets of strings over a given A, where $\emptyset$ is the identity associated with $|$ and $\{\varepsilon\}$ is the identity associated with $\cdot$. We call this an 'associative algebra'; the non-commutativity of $\cdot$ is one property which distinguishes this system from the major classes of algebraic system discussed in Section 2.5.

Any set $L \subseteq A^*$ of strings over A is called a *language*. When L is an infinite set, some means of characterizing the language using a finite number of symbols (i.e., in a 'closed' form rather than by listing each number of L) is desirable. In Section 3.3, we discuss 'grammars' which may be used to define given languages, and later describe algorithms which may be used to test whether an arbitrary string is in a given language.

2.4.6 Application: Algebraic Expressions

Let $\mathscr{A} = (S, \Theta_1, \ldots, \Theta_k)$ be an algebraic system. If Θ_i is unary, we write $\Theta_i a$ [or $a\Theta_i$] to denote the result of the operation on the operand a; if Θ_i is binary, we write $a\,\Theta_i b$; if Θ_i is n-ary, we write $\Theta_i[a_1, \ldots, a_n]$. An *algebraic expression* (with respect to $\mathscr{A}$), abbreviated a.e., is defined inductively as follows (using prefix notation, except for infix binary operations):

(A) Each member of S is an a.e.
(B) If α is an a.e., then so is $(\Theta_i\alpha)$, for each unary operation Θ_i.
(C) If α and β are a.e.s, then so is $(\alpha\,\Theta_i\beta)$, for each binary operation Θ_i.
(D) If $\alpha_1, \alpha_2, \ldots, \alpha_n$ are a.e.s, then so is $(\Theta_i[\alpha_1, \alpha_2, \ldots, \alpha_n])$, for each n-ary operation Θ_i.

Defined in this fashion, algebraic expressions are completely parenthesized, and the parentheses explicitly dictate the order in which the operations are to be performed. We state this as the first of three a.e. evaluation rules.

Rule 1 In an algebraic expression, operations are performed in order from the innermost pair of parentheses to the outermost pair.

(For expressions involving an n-ary operation Θ_i, all operations within the square brackets must be performed before Θ_i; hence replacement of the brackets by parentheses would not lead to ambiguity.) The result of performing all the operations appearing in an algebraic expression (in the indicated order) must be a member of S, which is called the *value* of the a.e. A convenient generalization of the concept of algebraic expressions is to the case where symbols ('variables') may denote members of S (called 'constants'). Let V be a set of symbols distinguishable from those of $\mathscr{A}$. Then we add to the foregoing definition the following:

(A′) Each member of V is an a.e.

A (generalized) a.e. is sometimes called a 'formula' or 'form'. Suppose an a.e. α contains k distinct variables, $v_1,\ldots,v_k$: we denote it $\alpha[v_1,\ldots,v_k]$. Then the a.e. may be regarded as a function α: $S^k \to S$; if to each variable v_i, we assign as its value a member a_i of S, then $\alpha[a_1,\ldots,a_k]$ is the value of α with v_i replaced by a_i. Two a.e.s, α_1 and α_2, containing the same variables are said to be *functionally equivalent*, denoted $\alpha_1 \equiv \alpha_2$, if

$$\alpha_1[a_1,\ldots,a_k] = \alpha_2[a_1,\ldots,a_k]$$

for all identical value assignments; by 'identical', we mean that each distinct variable v_i in both α_1 and α_2 must be assigned the same value a_i.

Let $(S,\Theta_1,\ldots,\Theta_k)$ be an algebraic system having only binary and unary operations. It is often convenient to define a partial order (in this context, called a *precedence* relation) on the set of operations so that algebraic expressions can be written as free of parentheses as possible. For example, consider the set of arithmetic operations $\{+,-,\times,/,\uparrow,\ominus\}$ on numbers, where $\ominus$ denotes the 'unary minus' operation (i.e., negation or complementation) and $\uparrow$ denotes exponentiation; algebraic expressions with respect to these operations are called 'arithmetic expressions'. A universally adopted convention, by mathematicians and by programming-language designers (and of which we have implicitly assumed reader knowledge), is that $\times$ and / have higher precedence (or 'rank') than + and −: we denote this

$$+ \lessdot \times, \quad + \lessdot /, \quad - \lessdot \times, \quad - \lessdot /,$$

where $\lessdot$ denotes the 'less than' (in precedence) relation.

Rule 2 Within any matching pair of parentheses, the operations are performed in decreasing order of precedence.

Hence

$$(a + b \times c - d) \equiv (a + (b \times c) - d)$$

and

$$(a \times b + c/d) \equiv ((a \times b) + (c/d)).$$

Some operations may have the same precedence (denoted $\doteq$): e.g.,

$+ \doteq -$, and $\times \doteq /$, and any operation $\doteq$ itself. In this event, we adopt a *left-to-right* convention.

Rule 3 Within any pair of parentheses, (binary) operations having the same precedence are performed from left to right.

Hence

$$(a-(b\times c)+d) \equiv ((a-(b\times c))+d)$$

rather than $(a-((b\times c)+d))$. Precedence relations involving $\uparrow$ and $\ominus$ have no universally accepted convention, nor is the left-to-right convention always adopted for $\uparrow$. We adopt the latter, and in addition use

$$\times \lessdot \uparrow, \quad / \lessdot \uparrow, \quad \uparrow \lessdot \ominus.$$

Hence

$$\begin{aligned}&(a+b-c\times d/e\uparrow \ominus f\uparrow g/h\times i-j+k)\equiv\\ &\quad ((((a+b)-((((c\times d)/((e\uparrow(\ominus f))\uparrow g))/h)\times i))-j)+k).\end{aligned}$$

In conclusion, we remark that $-$ and $\ominus$ are usually denoted by the same symbol ($-$) and are distinguished by context.

Example For the string operations defined in the preceding section for the algebraic system $(2^{A^*}, |, \cdot, *, \emptyset, \{\varepsilon\})$, we adopt the precedence conventions:

$$\cdot \gtrdot | \quad \text{and} \quad * \gtrdot \cdot$$

so that $X\cdot Z|Y\cdot Z=(X\cdot Z)|(Y\cdot Z)$ and $X\cdot Y^* = X\cdot (Y^*)$.

Algebraic expressions consisting of symbols and the operations $|$, $\cdot$, and $*$ are sometimes called ***regular*** expressions, and denote sets of strings; see Section 3.3.4. For example, the expression $(1\cdot(0|1)^*)$ denotes the set $\{1X\}$, the set of all positive binary numbers without leading zeros, where $X=\{0,1\}^*$.

2.5 CLASSES OF ALGEBRAIC SYSTEMS

The simplest meaningful algebraic system is a pair (S,R), where S is a set and R is a relation in S (or a unary operation in S); we call this a *directed graph* (see Chapter 4). More complex algebraic systems are classified according to the properties which their operations possess (as defined in Sections 2.4.1 and 2.4.2). A pair $(S, \circ)$, where S is a set and $\circ$ is a single-valued binary operation which is closed and associative, is called a *semigroup*. A semigroup having a distinguished member of S as an identity e is called a *monoid*, denoted $(S, \circ, e)$. A monoid whose operation has an inverse is called a *group*. A group $(S, \circ, e)$, where $\circ$ is also commutative, is called a *commutative* (or *Abelian*) group.

A set S with two binary operations, denoted $+$ and $*$, such that $(S,+)$ is a commutative group, such that $(S, *)$ is a monoid, and such that $*$ distributes over $+$ on the left and right—i.e.,

$$a * (b+c)=a * b+a * c$$
$$(b+c) * a=b * a+c * a$$

—is called a *ring*. A ring is *commutative* if $*$ is. The 'additive' identity of $(S,+)$ is denoted by 0; the 'multiplicative' identity of $(S, *)$ is denoted by 1. If $(S-\{0\}, *, 1)$ is a commutative group, so that $c * a=c * b$ implies $a=b$ for any $c \neq 0$ (the cancellation law), then the ring $(S,+, *, 0, 1)$ is called a *field*, with the members of S called *scalars*. In other words, a field is a commutative ring whose multiplication operation $*$ would be invertible if it were not the case that the element 0 had no inverse; to emphasize this exception, we denote $*$ by $\times$ for fields.

An algebraic system $(S, +, *, 0, 1)$ satisfying all the definitions of a ring, except that $+$ need not have an inverse relation (i.e., $(S,+,0)$ is not a group), and in addition satisfying

$$a+1=1,$$
$$a * 0=0 * a=0,$$

is called a *semiring*. Note that the additive identity 0 is a nullifier for $*$, and that the multiplicative identity 1 is a nullifier for $+$, properties which we have called 'universal boundedness'; this does not hold for rings. In summary, $(S, +, *, 0, 1)$ is a semiring if $(S, +, 0)$ and $(S, *, 1)$ are monoids, $+$ is commutative, $*$ distributes over $+$, and universal boundedness holds.

We note that it is the associativity of $+$ and $*$ for rings, fields, and semirings which permits us to write

$$\sum_{i=1}^{n} x_i=x_1+\cdots+x_n$$

$$\prod_{i=1}^{n} x_i=x_1 * \cdots * x_n$$

without ambiguity. In other words, the *generalized n*-ary sum and product operations can be computed using the binary operations in any order.

2.5.1 Vector Spaces

Let $F=(S,+,\times,0,1)$ be a field (of scalars), and let V be any set (whose members are called *vectors*) with a binary operation called *vector sum*, denoted $\oplus$, such that $(V, \oplus)$ is a commutative group whose identity is denoted θ. Let $\circ: S \times V \to V$ be a total function, called the *scalar product*. If the scalar product is associative in the sense that

$$(a \times b) \circ x = a \circ (b \circ x), \qquad a \in S, \quad b \in S, \quad x \in V,$$

if in addition

$$1 \circ x = x, \qquad x \in V,$$

and furthermore if the distributive laws

$$a \circ (x \oplus y) = a \circ x \oplus a \circ y$$
$$(a + b) \circ x = a \circ x \oplus b \circ x$$

hold, then $(V, \oplus, \circ, \theta)$ is called a *vector space* 'over' F.

A 'vector product' from $V \times V$ to V is ordinarily not defined. However, a function $\mathbf{P}$: $V \times V \to S$ which satisfies

(a) $\mathbf{P}(x,x) > 0$ if $x \neq \theta$, $\mathbf{P}(\theta,\theta) = 0$ (positive definiteness)
(b) $\mathbf{P}(x,y) = \mathbf{P}(y,x)$ (symmetry)
(c) $\mathbf{P}(a \circ x \oplus b \circ y, z) = a \times \mathbf{P}(x,z) + b \times \mathbf{P}(y,z)$ (linearity)

is called an *inner product*, a very common and useful operation. A unary operation $\mathbf{T}$: $V \to V$, defined on a vector space, is said to be a *linear transformation* if

$$\mathbf{T}(a \circ x \oplus b \circ y) = a \circ \mathbf{T}(x) \oplus b \circ \mathbf{T}(y)$$

for all a, $b \in S$ and x, $y \in V$.

Let $(V, \oplus, \circ, \theta)$ be a vector space over F. We say $(V', \oplus, \circ, \theta)$ is a *subspace* of $(V, \oplus, \circ, \theta)$ if $V' \subseteq V$ and $(V', \oplus, \circ, \theta)$ is a vector space. A *linear combination* of k vectors $\{v_1, v_2, \ldots, v_k\}$ is defined as the sum

$$a_1 \circ v_1 \oplus a_2 \circ v_2 \oplus \cdots \oplus a_k \circ v_k$$

where $a_1, a_2, \ldots, a_k$ are scalars. A set $\{v_1, \ldots, v_k\}$ is *linearly dependent* iff there are scalars, $a_1, \ldots, a_k$, not all zero, such that

$$a_1 \circ v_1 \oplus a_2 \circ v_2 \oplus \cdots \oplus a_k \circ v_k = \theta;$$

the set is *linearly independent* if it is not linearly dependent. A set V' of vectors in V *spans* a vector space $(V, \oplus, \circ, \theta)$ if every vector in V can be expressed as a linear combination of vectors in the set V'. If two sets of linearly independent vectors span the same (vector) space, then the numbers of vectors in the two sets are equal, and this number is called the *dimension* of the space. Any set of k linearly independent vectors which spans a space of dimension k is said to be a *basis* of the space.

The *range* of a linear transformation $\mathbf{T}: V \to V$ is the set $\{\mathbf{T}(x) \mid x \in V\}$; the *null space* of $\mathbf{T}$ is the set $\{x \in V \mid \mathbf{T}(x) = \theta\}$. Both the range and the null space of $\mathbf{T}$ are subspaces: the *rank* of $\mathbf{T}$ is the dimension of the range, and the *nullity* of $\mathbf{T}$ is the dimension of the null space. It can be shown that the sum of the rank and the nullity of $\mathbf{T}$ equals the dimension of the vector space V. See Halmos [2.10] for details.

Application: Linear Algebra Let S be a set. We define an *n-vector* to be a member of the n-fold product $V = S^n = S \times S \times \cdots \times S$, for some finite n; n is called its *extent*. (We have previously called n-vectors 'n-tuples'.) Consider now the m-fold product $M = V^m = S^n \times S^n \times \cdots \times S^n$. A member of M will be called an $(m \times n)$ *matrix*, and (m,n) is called its *extent*. (A member of M is an m-tuple, each coordinate of which is an n-tuple.) We show below that vectors and matrices may be regarded as ordered multisets, called *arrays*, of n and $m \cdot n$ members of S, respectively. In particular, we say that a vector is a *one-dimensional* ('singly subscripted') array, and that a matrix is a *two-dimensional* ('doubly subscripted') array; *k-dimensional* ('k-tuply subscripted') arrays are defined analogously.† A *nonsubscripted* ('0-tuply subscripted') array is defined as a single member of S.

Let A be a multiset of members of S (considered unordered), I be a set of integers, and I^k be the k-fold product of I. We say A is a ***k-dimensional array*** if there is a distinctive (1–1) labeling function $\lambda: A \to I^k$; the label of a member of A is called its *subscript*. It is customary to restrict the range of λ to a k-cell of integers $C \subseteq I^k$, so that $\lambda: A \to C$ is onto and invertible. (Note $\#(A)$ must equal $\#(C)$.) There is then a 1–1 correspondence between members of the array and k-tuples of integers. An array can therefore be considered ordered by the lexicographical ordering of the subscripts; that is, if C is the lexicographically ordered set $\{c_1, c_2, \ldots\}$, then we may define the ordered multiset $A = \{a_1, a_2, \ldots\}$ so that $\lambda(a_i) = c_i$ or $\lambda^{-1}(c_i) = a_i$. Usually, we denote λ^{-1} [: $C \to A$] by A, so that $A(c_i) = a_i$; thus an array name followed by a parenthesized subscript denotes (or 'selects') a member or *element* of the array.‡ A, used in this way, will be called an *array map* from subscripts (in C) to values (in S).

Suppose $A = \{a_1, a_2, \ldots, a_6\}$ is an ordered multiset of members of S. If C is the 1-cell $[0,5]$, we may regard A as a 6-vector $V = (a_1, a_2, \ldots, a_6) \in S^6$, with $a_1 = A(0)$, $a_2 = A(1)$,..., $a_6 = A(5)$. If C is the 2-cell $[1,2] \times [-1,1]$, we may regard A as a (2×3) matrix $M = ((a_1, a_2, a_3), (a_4, a_5, a_6)) \in S^3 \times S^3$, with $a_1 = A(1,-1)$, $a_2 = A(1,0)$, $a_3 = A(1,1)$, $a_4 = A(2,-1)$, $a_5 = A(2,0)$, $a_6 = A(2,1)$.

†The dimension (number of subscripts) of an array should not be confused with the dimension of a vector space.

‡Sometimes, we write $a_{ij\ldots}$ to mean $A(i,j,\ldots)$.

(We recall that $\{(1,-1),(1,0),(1,1),(2,-1),(2,0),(2,1)\}$ is $[1,2]\times[-1,1]$ ordered lexicographically.) The matrix M can be displayed pictorially as shown:

$$M=\begin{bmatrix} a_1 & a_2 & a_3 \\ a_4 & a_5 & a_6 \end{bmatrix}=\begin{bmatrix} A(1,-1) & A(1,0) & A(1,1) \\ A(2,-1) & A(2,0) & A(2,1) \end{bmatrix}.$$

In this form, we say that the two components of M (each of which in turn has three components) correspond to rows, and so lexicographical ordering is also called *row order*. We also say that A is the row-order representation of M; the *column-order* representation of M is the ordered multiset $\{a_1,a_4,a_2,a_5,a_3,a_6\}$.

Linear algebra is concerned with arrays as algebraic systems, i.e., sets of arrays, and various operations on arrays. In particular, if S is a set of 'scalars' having certain well-defined operations, + and $\times$, where $F=(S, +, \times, 0, 1)$ is a field, then generalizations of these operations to arrays may be defined and consequences of these definitions derived. If $A=\{a_1,a_2,\ldots,a_N\}$ and $B=\{b_1,b_2,\ldots,b_N\}$ are both k-dimensional arrays (of members of S) having the same extent, we say they are *like* (or 'comformable') arrays; like arrays are said to be *equal* iff $a_i=b_i$, for each i. We will also use θ to denote the array whose elements are all the scalar 0.

We define the *sum* of like arrays A and B, denoted $A\oplus B$, to be the like array $C=\{c_1,c_2,\ldots,c_N\}$ where $c_i=a_i+b_i$; the *difference*, $C=A\ominus B$, is defined by $c_i=a_i-b_i$. Addition and subtraction, as operations on sets of like arrays, have the conventional properties associated with + and − in the set of scalars. However, the product of arrays is not defined by $c_i=a_i\times b_i$ for various reasons, not the least of which is that we would like to multiply unlike arrays. The *scalar product* of a scalar s and an array A, denoted $s\circ A$, is an array B, $b_i=s\times a_i$. With these definitions, the set of arrays forms a vector space $(S^N, \oplus, \circ, \theta)$ over F. We also define the *inner product* of arrays A and B, denoted $A\odot B$, as the scalar $\Sigma_{i=1}^N a_i\times b_i$.

Linear algebra, in the special case ($k=2$) where the arrays are (two-dimensional) matrices, is of great importance and will be discussed further in Section 3.2.

Application: Boolean Arithmetic If B is the set $\{0,1\}$, we define an algebraic system $(B,\oplus,\times,0,1)$ where $\oplus$ and $\times$ are operations on B defined as follows:

$$\begin{array}{ll} 0\oplus 0=0, & 0\times 0=0 \\ 0\oplus 1=1, & 0\times 1=0 \\ 1\oplus 0=1, & 1\times 0=0 \\ 1\oplus 1=0, & 1\times 1=1 \end{array}$$

for $x,y\in\{0,1\}$; we call $x\oplus y$ the ***Boolean sum*** of x and y, and $x\times y$ the ***Boolean product*** of x and y.

The distinguishing feature of the Boolean arithmetic operations is that $1 \oplus 1 = 0$ (rather than 2, which is not in B); $\oplus$ is sometimes called the 'modulo 2' sum. Note that 0 and 1 are still the additive and multiplicative identities, respectively, and that $\oplus$ has an inverse while $\times$ does not. Restricted to $B - \{0\} = \{1\}$, $\times$ has an inverse; 1 is then both the additive and multiplicative inverse of itself. Note also that $\oplus$ and $\times$ are both commutative, and that $\times$ distributes over $\oplus$. Hence $(B, \oplus, \times, 0, 1)$ is a field. (We emphasize again that the Boolean arithmetic operations should not be confused with the 'logical' operations, discussed in Section 1.3; in particular, the *logical sum*, denoted $\vee$ instead of $\oplus$, is defined such that $1 \vee 1 = 1$.)

For $B = \{0, 1\}$, B^n is the set of Boolean n-tuples or n-vectors introduced in Section 2.2.1. Define the Boolean vector sum $\oplus: B^n \times B^n \to B^n$ such that

$$(x_1, x_2, \ldots, x_n) \oplus (y_1, y_2, \ldots, y_n) = (x_1 \oplus y_1, \ldots, x_n \oplus y_n)$$

where $x_i \oplus y_i$ is the Boolean sum. Denote the n-tuple of zeroes $(0,0,\ldots,0)$ by θ. Then $(B^n, \oplus, \theta)$ is a group. Each member of B^n is its own inverse (i.e., $x \oplus x = \theta$ for each $x \in B^n$). Therefore, $z = x \oplus y$ implies $x \oplus z = y$ and $z \oplus y = x$. If we define the scalar product $\circ$ by $0 \circ x = \theta$ and $1 \circ x = x$ for each $x \in B^n$, then $(B^n, \oplus, \circ, \theta)$ is a vector space over B.

2.5.2 Groups

Let $(G, \circ, e)$ be a group. We say $(G', \circ, e)$ is a *subgroup* of $(G, \circ, e)$ if $G' \subseteq G, e \in G'$, and $(G', \circ, e)$ is a group. Then for $a \in G$, the set $\{a \circ b \mid b \in G'\}$, denoted aG', is called a *(left) coset* of G' in G. Note that $eG' = G'$ (since $e \circ b = b$), $aG' = G'$ for all $a \in G'$ (since $aG' \cap bG' \neq \emptyset$ implies $aG' = bG'$), and $a \in aG'$ for all $a \in G$ (since $e \in G'$). A fundamental theorem in group theory (we call it the **group decomposition theorem**) is the following:

> For $(G', \circ, e)$ a subgroup of $(G, \circ, e)$, the family of left cosets $\{aG' \mid a \in G\}$ partitions G.

This means that the members of G may be divided into equivalence classes, each member of G belonging to one and only one coset. Thus, the coset containing a may be denoted $[a]$. If G is finite, then $\#(aG') = \#(G')$ for any a; therefore,

> $\#(G')$ evenly divides $\#(G)$, and the quotient is the rank of the partition.

This is known as **Lagrange's theorem**.

Application: Boolean Vectors Let $(B^n, \oplus, \circ, \theta)$ be the vector space over $B = \{0, 1\}$ defined in Section 2.5.1. Let $x = (x_1, \ldots, x_n)$ and $y = (y_1, \ldots, y_n)$ be

Boolean n-vectors ($x,y \in B^n$). A vector product operation $\otimes$: $B^n \times B^n \to B^n$ is not ordinarily defined. Instead, a function $\odot$: $B^n \times B^n \to B$, called the (Boolean) *inner product*, is defined such that

$$(x_1,\ldots,x_n) \odot (y_1,\ldots,y_n) = \sum_{i=1}^{n} x_i \times y_i$$

where Σ and $\times$ are the Boolean sum and product, respectively. If Σ is interpreted as the ordinary arithmetic sum, then $\Sigma x_i \times y_i$ equals the number of l's $x=(x_1,\ldots,x_n)$ and $y=(y_1,\ldots,y_n)$ have as common coordinates. A more useful measure is the number of coordinates in which n-vectors x and y differ, a function called the **Hamming distance**, $\mathbf{d}(x,y)$. It can easily be shown that $\mathbf{d}(x,y) = \mathbf{wgt}(x \oplus y)$, where the weight of any $x \in B^n$, $\mathbf{wgt}(x)$, is defined as the number of coordinates in x which equal 1.

Example For $n=6$, let

$$\hat{C} = \{000000,\ 001110,\ 010101,\ 011011,\ 100011,\ 101101,\ 110110,\ 111000\},$$

where we have used the binary number representation. The Hamming distances between distinct members of $\hat{C}$ are either 3 or 4; for example, $\mathbf{d}(100011, 110110) = \mathbf{wgt}(010101) = 3$, and $\mathbf{d}(011011, 101101) = \mathbf{wgt}(110110) = 4$. (The fact that $\hat{C}$ has minimal distance equal to 3 has an important application, as we shall see in Section 7.1.) The reader should verify that $(\hat{C}, \oplus, \theta)$ is a subgroup of $(B^6, \oplus, \theta)$, and calculate the left cosets of $\hat{C}$ in B^6. By the group decomposition theorem and Lagrange's theorem, we know that there are 8 distinct cosets. That member of a coset having minimal weight will be called the coset *leader*; the leaders of the 8 cosets in the example are 000000, 100000, 010000, 001000, 000100, 000010, 000001, and 100100 (or 010010, or 001001, this last coset not having a unique minimal weight member).

2.5.3 Boolean Algebra

A 'Boolean algebra' is a particular kind of algebraic system. (Recall our definition of an algebraic system as a set S together with a collection of operations $\{\Theta_1, \Theta_2,\ldots,\Theta_k\}$, where the operations may be constants, relations, unary and binary operations, etc.) A Boolean algebra has two constants (or distinguished elements), a partial ordering relation, one unary operation, and two binary operations, which together satisfy a set of properties. These properties *define* a Boolean algebra, and also serve as a redundant set of axioms from which additional properties can be derived. It is common either (a) to start with a partial order and then to relate the other operations to it, or (b) to start with two binary operations and then to relate a partial order and other operations to them.

A partially ordered set $(S, \leq)$ is a *lattice* iff every subset consisting of two members of S, $\{a,b\}$, has an infimum in S and a supremum in S. For any subset-pair $\{a,b\}$ of members of S, denote **inf**($\{a,b\}$) by $a\sqcap b$, and denote **sup**($\{a,b\}$) by $a\sqcup b$. If a lattice has minimal and maximal members, they must be (unique) least and greatest members, called (*universal*) *bounds*, and denoted B_1 and B_2:

$$B_1 \le x \le B_2 \qquad \text{for each } x \text{ in } S.$$

A lattice having such bounds is said to be *bounded*. An element x' in S of a bounded lattice which satisfies

$$x\sqcap x' = B_1 \qquad \text{and} \qquad x\sqcup x' = B_2$$

is called a *complement* of x, and if such an x' exists for each x in S, the lattice is said to be *complemented*. Note that $\sqcap$ and $\sqcup$ may be regarded as single-valued binary operations on S, which are called the *meet* and *join*, respectively. A *distributive* lattice is one where $\sqcap$ and $\sqcup$ satisfy the mutual distributivity properties. A *Boolean algebra* $(S, \le, \sqcup, \sqcap, ', B_1, B_2)$ may be defined as a lattice $(S, \le)$ which is bounded, complemented, and distributive.

A lattice may also be defined as an algebraic system $(S, \sqcup, \sqcap)$, where $\sqcup$ and $\sqcap$ are single-valued binary operations on S which are closed, associative, commutative, and satisfy the absorption property. It can be proven, from these properties, that $\sqcup$ and $\sqcap$ are also idempotent, and satisfy the consistency property. The latter implies that the two relations in S, $\{(a,b) \mid a\sqcap b = a\}$ and $\{(a,b) \mid a\sqcup b = b\}$, are identical; we denote this relation by $\le$ (i.e., $a \le b$ iff $a\sqcap b = a$ iff $a\sqcup b = b$). This notation is justified since the commutative and absorption properties imply that $\le$ is a partial order. It can be shown then that, for any subset $\{a,b\}$ consisting of two members of S, $a\sqcap b$ is **inf**($\{a,b\}$) and $a\sqcup b$ is **sup**($\{a,b\}$). (Since $a\sqcap a\sqcap b = a\sqcap b$ and $b\sqcap a\sqcap b = a\sqcap b$ [by idempotence], $a\sqcap b$ is a lower bound of a and b [i.e., $a\sqcap b \le a$, $a\sqcap b \le b$]; $a\sqcap b$ is in fact the greatest lower bound since $c \le a$ and $c \le b$ imply

$$c = a\sqcap c = a\sqcap(b\sqcap c) = (a\sqcap b)\sqcap c;$$

hence $c \le a\sqcap b$.) Therefore, $(S, \le)$ is a lattice. If, in addition, $\sqcap$ and $\sqcup$ are mutually distributive, and satisfy the universal boundedness and complementarity properties, then $(S, \le, \sqcup, \sqcap, ', B_1, B_2)$ is a Boolean algebra.

In summary, a *Boolean algebra* is an algebraic system $(S, \le, \sqcup, \sqcap, ', B_1, B_2)$, where $(S, \le)$ is a partially ordered set satisfying the properties of a bounded, complemented, and distributive lattice, $(S, \sqcup, \sqcap, B_1, B_2)$ and $(S, \sqcap, \sqcup, B_2, B_1)$ are both semirings, $'$ is a unary operation satisfying complementarity, and the distinguished elements B_1 and B_2 are universal bounds in S. The two principal examples of Boolean algebras are where $S = 2^U$, the set of all subsets of any given set U, and where S is the set of truth values $\{\boldsymbol{T}, \boldsymbol{F}\}$. We remark that algebraic expressions with respect to Boolean algebras are called 'Boolean expressions'. In conclusion, we emphasize that lattices which do not satisfy all the properties of a Boolean algebra also have useful applications; see Scott [2.18], for example.

Application*: *Set Algebra Consider the set inclusion relation ($\subseteq$) in 2^U (see Section 2.2.4). Let

$$\sqcap=\{((a,b),a)\,|\,a\subseteq b\} \text{ and } \sqcup=\{((a,b),b)\,|\,a\subseteq b\}.$$

These sets coincide, respectively, with the intersection ($\cap$) and union ($\cup$) operations (from $2^U \times 2^U$ to 2^U) defined previously, and in fact provide alternative definitions for $\cap$ and $\cup$ in terms of $\subseteq$. For $\boldsymbol{a}$ and $\boldsymbol{b}$ in 2^U, $\boldsymbol{a} \cap \boldsymbol{b}$ is the largest member of 2^U that is smaller than (a subset of) both $\boldsymbol{a}$ and $\boldsymbol{b}$, and $\boldsymbol{a} \cup \boldsymbol{b}$ is the smallest member of 2^U that is greater than (includes) both $\boldsymbol{a}$ and $\boldsymbol{b}$. We have already observed that $\cap$ and $\cup$ are closed, associative and commutative, and note that $\emptyset$ and U serve as the universal bounds (B_1 and B_2, respectively). The intersection and union operations have identity elements, which happen to be the bounds U and $\emptyset$, respectively, but do not have inverses. However, the complement of a set A, defined as $(U)-A$, satisfies the involution property $-(-A)=A$, the complementarity property $A \cap -A=\emptyset$ and $A \cup -A=U$, and DeMorgan's law. The operations individually satisfy the idempotence, substitution, and cancellation properties, and together satisfy the consistency, absorption, and distributivity properties. We conclude that $(2^U, \cup, \cap, \emptyset, U)$ is a semiring, and $(2^U, \subseteq, \cup, \cap, -, \emptyset, U)$ is a Boolean algebra. In this algebra, the precedence relations $\cup \lessdot \cap$ and $\cap \lessdot -$ are adopted by convention.

Application*: *Logic Let $(S, \leq, \sqcup, \sqcap, ', B_1, B_2)=(\{0,1\}, \rightarrow, \vee, \wedge, \neg, 0, 1)$, where the relation $\rightarrow$ is the set $\{(0,0), (0,1), (1,1)\}$, where $\vee$ and $\wedge$ are binary operations defined as follows:

$$\begin{array}{ll} 0\vee 0=0, & 0\wedge 0=0 \\ 0\vee 1=1, & 0\wedge 1=0 \\ 1\vee 0=1, & 1\wedge 0=0 \\ 1\vee 1=1, & 1\wedge 1=1 \end{array}$$

and where $\neg$ is a unary operation defined as follows:

$$\neg 0=1, \qquad \neg 1=0.$$

This defines a Boolean algebra which is called ('two-valued') *logic*.† Note that if 0 corresponds to **false** and 1 corresponds to **true**, then $\wedge$, $\vee$, and $\neg$ correspond to the conventional logical operations, which have the precedences:

$$\vee \lessdot \wedge, \qquad \wedge \lessdot \neg.$$

†There are 'multiple-valued' logics, which we will not discuss here; see Zadeh [2.20].

EXERCISES

1. Let $A=\{a,b,c\}$, $B=\{b,c,d\}$, $C=\{e\}$. Determine

$A\cup B$, $A\cap B$, $A-B$, $B-A$, $-A$, $-B$, $A\oplus B$, $A\cup C$, $A\cap C$, $A-C$, $C-A$, $-C$, $A\oplus C$, 2^A, 2^C.

***2.** The Barber cuts the hair of *all* and *only* those who do not cut their own hair. Who cuts the Barber's hair?

3. Let $S=\{0,1,2\}$. List each of the possible

(a) 3-selections
(b) 3-combinations
(c) 3-arrangements
(d) 3-permutations.

4. Let $S_1=\{a,b,c\}$, $S_2=\{b,c,d\}$, $S_3=\{c,d\}$, and $I=\{1,2,3\}$.

(a) Determine $\cup_{i\in I}S_i$, $\cap_{i\in I}S_i$, and $\oplus_{i\in I}S_i$.
(b) Is the family of sets $\mathcal{F}=\{S_1,S_2,S_3\}$ an independent set?
(c) Is $\mathcal{F}$ a minimal cover?

5. Show that $\#(2^A)=2^{\#(A)}$.

6. Prove that $\oplus_{i\in I}S_i$ is independent of the ordering of I.

***7.** (Principle of inclusion and exclusion)

(a) Show that $\#(A\cup B)=\#(A)+\#(B)-\#(A\cap B)$.
(b) Show that

$$\begin{aligned}&\#(A\cup B\cup C)=\\ &\quad \#(A)+\#(B)+\#(C)+\#(A\cap B)\\ &\quad -\#(A\cap C)-\#(B\cap C)+\#(A\cap B\cap C).\end{aligned}$$

8. (a) Why isn't (a,b,c) defined as $\{\{a\},\{a,b\},\{a,b,c\}\}$?
(*Hint*: Show that two nonidentical ordered triples would be equal under this definition.)
(b) Show that $((a,b),c)\neq(a,(b,c))$.

9. Show that the nth power of a relation R is not the same as the n-fold product of (the set) R.

10. For $A=\{a,b\}$ and $B=\{a,b,c\}$, determine $B\times A$ and B^2.

11. Determine R^{-1}, R^2, R^3, R^+ and R^* for the following relations:

(a) $R=\{(1,2),(2,1),(2,2),(2,3),(3,2)\}$ on $S=\{1,2,3\}$
(b) $R=\{(1,2),(2,3),(3,4)\}$ on $S=\{1,2,3,4\}$.

12. Characterize the following relations (i.e., are they reflexive, symmetric, transitive, asymmetric, antisymmetric, irreflexive, a partial order, an equivalence relation?):

(a) $R=\{(1,2),(2,1),(2,2),(2,3),(3,2)\}$ on $S=\{1,2,3\}$
(b) $R=\{(1,2),(1,4),(3,4)\}$ on $S=\{1,2,3,4\}$
(c) $R=\{(1,2),(2,3),(3,4)\}$ on $S=\{1,2,3,4\}$.

13. Draw the graph of the subset relation $\subseteq$ on $S=\{a,b,c\}$. Determine its immediate successor relation. Let $E=\{a,b\}$; what are the lower and upper bounds of E?

14. Given the recurrence relation

$$a_{nm}=a_{n-1,m}\cdot(n-1)+a_{n-1,m-1} \text{ for } n\geq 1,\ m\geq 1$$

with boundary conditions $a_{n1}=(n-1)!$ for $n\geq 1$, $a_{01}=0$, $a_{n0}=0$ for $n\geq 1$, and $a_{00}=1$, determine a_{nm} for $0\leq n\leq 8$, $0\leq m\leq 8$. [a_{nm} are called 'Stirling numbers of the first kind'.]

15. If $\mathbf{f}:A\rightarrow B$, with $\#(A)=a$ and $\#(B)=b$, characterize functions such that

(a) $a<b$, (b) $a>b$, (c) $a=b$, (d) $\#(\mathbf{f})$ is maximum.

16. If $\mathbf{f}:A\rightarrow B$ is a function, show that $\{(a,b)\,|\,\mathbf{f}(a)=\mathbf{f}(b)\}$ is an equivalence relation on A.

17. Prove that $\mathcal{I}\times\mathcal{I}\times\mathcal{I}$ is a countable set. (You need to define $\mathbf{F}:\mathcal{I}^3\rightarrow\mathcal{I}$ appropriately. Be explicit!) For your definition of $\mathbf{F}$, what is $\mathbf{F}^{-1}(10)$?

18. Show that composition of functions is associative but not commutative.

19. Given sets A and B,

(a) how many different functions are there from A to B?
(b) how many of these functions are total?
(c) how many of these total functions are 1–1?
(d) how many of these functions are invertible?

20. The inequalities ($\leq$ and $<$) were defined in terms of equality test, the successor function, and the logical operations ($\vee$, $\wedge$, $\neg$). Define the inequalities using conditional expressions instead of the logical operations. (For example, $m\neq n$ can be defined as $(m=n\rightarrow \boldsymbol{F},\ \boldsymbol{T}\rightarrow \boldsymbol{T})$ rather than as $\neg(m=n)$.)

21. Define subtraction $(m-n)$ utilizing conditional expressions and the functions already given.

22. Evaluate **mod**(22,6) using the given recursive definition of **mod**.

23. Suppose **gcd**(m,n) is defined by

$$(m<n\rightarrow \mathbf{gcd}(m,n-m),\ m>n\rightarrow \mathbf{gcd}(m-n,n),\ m=n\rightarrow m).$$

Evaluate **gcd**(22,6) using this definition. (In other words, the greatest common divisor of a pair of natural numbers can be found by successively subtracting the smaller of the pair.)

24. Which of the properties listed in Section 2.4.1 does the addition operation on the set of rational numbers satisfy?

25. Which of the properties listed in Section 2.4.2 do the arithmetic addition and multiplication operations on the set of reals satisfy?

26. Consider the stroke operator '$\downarrow$' defined by $0\downarrow 0=1$, $0\downarrow 1=0$, $1\downarrow 0=0$, $1\downarrow 1=0$. Which of the properties listed in Section 2.4.1 does it satisfy?

27. Construct examples showing that

(a) the complex product $S_1\cdot S_2$ is not equal to the Cartesian product $S_1\times S_2$;
(b) the nth complex power of a set is not equal to its n-fold product.

28. Completely parenthesize the set algebraic expression

$$-A \cap B \cup C \cap D \cup -E.$$

29. Let $X=\{a,b\}$, $Y=\{b,c\}$, and $Z=\{d\}$. List the members of the sets

(a) $X \cdot (Y|Z)$
(b) $(X|Y) \cdot Z$
(c) $X|Y \cdot Z^*$.

30. Show that the definition of a linearly independent set in Section 2.5.1 is consistent with that of an independent set in Section 2.1.5.

31. Prove that, for any vector space V, $0 \circ x = \theta$, for any $x \in V$.

32. Define a field F and a vector space V over F, where the vectors are Boolean n-tuples. What is the dimension of V? Give a basis for the space.

33. Prove that every field is a vector space over itself.

34. Given a set $S=\{s_1,\ldots,s_n\}$, we define a permutation of S as a 1–1 function $\mathbf{P}:S \rightarrow S$. For $S=\{a,b,c\}$, examples of permutations are the sets $\mathbf{P}_1=\{(a,b),(b,c),(c,a)\}$ and $\mathbf{P}_2=\{(a,c),(b,b),(c,a)\}$. Let $\mathcal{P}$ denote the set of all permutations of S. The product of two permutations $\mathbf{P}_1 \times \mathbf{P}_2$ is defined as the composite function $\mathbf{P}_2 \circ \mathbf{P}_1$.

(a) Show that $(\mathcal{P}, \times, e)$ is a group; what is e?
(b) For $n=3$, identify (i.e., list the members of) each subgroup of $(\mathcal{P}, \times, e)$.
(c) For one of the proper subgroups identified above, other than the trivial one containing only e, find the left cosets that partition $\mathcal{P}$.

35. For $(G', \circ, e)$ a subgroup of $(G, \circ, e)$, show that $aG'=G'$ for all $a \in G'$. (*Hint*: Show that $aG' \cap bG' \neq \emptyset$ implies $aG'=bG'$.)

***36.** Let $\mathcal{F}=\{S_1, S_2, \ldots, S_n\}$ be a family of sets, where each S_i is a subset of some universal set U. Show that if $\mathcal{F}$ is closed under both complementation and union, then it is also closed under intersection. (*Note*= Such families are called *Borel fields*.)

PROGRAMMING ASSIGNMENTS

1. Program a set of subroutines to enable set algebra computations—e.g., to calculate A∪(B∩C), the main program would have the following form:

```
CALL SET_READ(A)
CALL SET_READ(B)
CALL SET_READ(C)
CALL INTERSECT (B,C, TEMP)
CALL UNION (A, TEMP,TEMP)
CALL SET_PRINT (TEMP)
```

2. Program a set of subroutines to enable Boolean arithmetic computations.

3. Program a set of subroutines to enable operations on sets of strings—specifically, to enable the evaluation of regular expressions.

4. Write a program that determines whether or not a relation is reflexive, symmetric, transitive, irreflexive, asymmetric, or antisymmetric.

5. Write a program that tests whether or not a relation is an equivalence relation, and if so then finds the equivalence classes.

6. Write a program which reads in a set of alphanumeric character strings, and prints them out 'lexicographically'

(a) assuming letters come before digits (as in the EBCDIC code)
(b) assuming letters come after digits (as in the ASCII code).

7. Write a program which computes the first 20 Fibonacci numbers.

8. Program subroutines necessary to perform recursion in a language which has no recursion facilities (by implementing a stack, say by using array facilities). Use these subroutines to compute $n!$.

9. Write subroutines which compute $n!$ by iteration and by recursion. Compare their space and time requirements, as a function of n.

3 APPLICATIONS—NUMERICAL AND NON-NUMERICAL COMPUTATION

3.1 NUMERICAL ALGEBRA

3.1.1 Arithmetic Operations

Let R be the set of reals. Addition of two reals can be defined in terms of addition of rationals, which can in turn be defined in terms of addition of two natural numbers (see Rudin [3.21]). The addition operation $+: R \times R \rightarrow R$ so defined is a closed binary operation which is commutative and associative,has an identity (0), has inverses (the 'negations', $-x$) that satisfy involution, and satisfies the (left and right) substitution and cancellation properties. We define the subtraction operation $-: R \times R \rightarrow R$ by $x - y = x + (-y)$. Subtraction is closed, is neither commutative nor associative, has a right-identity ($a - 0 = a$, but $0 - a \neq a$), has inverses (each element is its own inverse) that satisfy involution, and satisfies the substitution and cancellation properties. We remark that in the set of natural (non-negative) numbers, subtraction is not closed since $a - b$ would not be defined for $b > a$.

The multiplication operation $\cdot : R \times R \rightarrow R$ on the set of reals, which also can be defined in terms of multiplication of rationals, is closed, associative, commutative, has an identity (1), has inverses (the 'reciprocals', $1/x$, except when $x = 0$) that satisfy involution, and satisfies the substitution and cancellation properties (the latter except when $c = 0$ in the definition). Division of reals x and y, denoted x/y, is defined if $y \neq 0$ by the real z such that $x = y \cdot z$. The exception ($y \neq 0$) means that the division operation is not closed; it has a right-identity ($a/1=a$ but $1/a \neq a$), has inverses (each element is its own inverse, except for 0) that satisfy involution, and satisfies the substitution and cancellation laws. We remark that division in the set of integers is not closed since x/y may not be an integer for arbitrary x and y (in addition to being undefined for $y=0$). Finally, we observe that, taken together, addition and multiplication satisfy only one distributivity property:

$$x \cdot (y+z) = x \cdot y + x \cdot z;$$

we say that multiplication 'distributes over' addition. Also,

$$(x-y)/z = x/z - y/z.$$

In summary, the four arithmetic operations (+, −, ·, /) defined on reals have the same properties as their analogues defined on rationals. Similar operations defined on other entities, e.g., matrices, do not necessarily have the same properties. We conclude that $(S, +, \cdot, 0, 1)$ is a field when S is the set of rationals, or the set of reals, or the set of complex numbers, but not when it is the set of integers, where +, ·, 0 and 1 have the usual interpretations. (If S is the set of all integers, we have just a commutative ring. If S is the set of only nonnegative integers, we have something less.)

3.1.2 Numerical Functions

A *numerical function* **F**: $A \to B$ is one where A and B are sets of numbers. A function **F** where only the range B is a set of numbers is called 'numerical-valued'.

Let **F**: $A \to B$ be a homomorphic numerical function. In general, the operations or relations 'preserved' (a misnomer, now) by **F** need not be identical, even if the sets A and B are identical. For example, consider a numerical-valued function where $\circ$ on a set of numbers B is addition, i.e., where

$$\mathbf{F}(x \circ y) = \mathbf{F}(x) + \mathbf{F}(y);$$

such a function is said to be *additive*. On the other hand, if $\circ$ on B is multiplication,

$$\mathbf{F}(x \circ y) = \mathbf{F}(x) \cdot \mathbf{F}(y),$$

then **F** is called *multiplicative*. If $\circ$ on A is multiplication, then an additive function is said to have the 'logarithmic' property:

$$\mathbf{F}(x \cdot y) = \mathbf{F}(x) + \mathbf{F}(y).$$

If $\circ$ on A is addition, then a multiplicative function is said to have the 'exponential' property:

$$\mathbf{F}(x+y)=\mathbf{F}(x)\cdot\mathbf{F}(y),$$

and an additive function is said to be *linear*:

$$\mathbf{F}(x+y)=\mathbf{F}(x)+\mathbf{F}(y).$$

Of related interest are numerical functions satisfying the inequality

$$\mathbf{F}(x+y)\leq\mathbf{F}(x)+\mathbf{F}(y).$$

If, in addition, $\mathbf{F}(0)=0$ and $\mathbf{F}(x)>0$ for $x\neq 0$ (a property called 'positive definiteness'), then $\mathbf{F}$ is said to measure 'distances' to the 'origin' 0. $\mathbf{F}$ is a *norm* if it also is 'homogeneous', i.e., if $\mathbf{F}(a\cdot x)=a\cdot\mathbf{F}(x)$ for positive a. The absolute value function, $\mathbf{abs}(x)$ or $|x|$, is an example of a norm. A distance function between arbitrary members ('points') x and y of a set S, denoted $\mathbf{d}(x,y)$, satisfying

(a) $\mathbf{d}(x,y)\geq 0$, and $\mathbf{d}(x,y)=0$ iff $x=y$
(b) $\mathbf{d}(x,y)=\mathbf{d}(y,x)$
(c) $\mathbf{d}(x,y)\leq\mathbf{d}(x,z)+\mathbf{d}(z,y)$ for any z

is called a *metric* on S; $(S,\mathbf{d})$ is then called a 'metric space'. (Note: (a) is a 'positive definiteness' property, (b) is a symmetry property, and (c) is called the *triangle inequality*.) For example, $\mathbf{d}_1(x,y)$, defined as $\mathbf{abs}(x-y)$ on the set of numbers, is a metric measuring the 'distance' between x and y. On the other hand, the norm $\mathbf{F}(x)$ defined as $\mathbf{d}(x,0)$ measures the distance between x and the origin 0. The *Hamming distance* between two Boolean n-tuples a and b, defined as the number of coordinates in which a and b differ, is an example of a metric on B^n (where $B=\{0,1\}$); see Section 7.1.5 for an application. Finally, we note that the definition of a metric does not require that $\mathbf{d}(x,y)=\mathbf{d}(x-y,0)$; if so, the metric is said to be *invariant* (as is $\mathbf{d}_1$, defined above). Metric spaces are discussed further in Simmons [3.22].

3.1.3 Convergence of Numerical Sequences

For infinite sequences of numbers, $\langle a_0, a_1, a_2,\ldots\rangle$ or $\{a_i\}$, it is frequently of interest to determine their 'asymptotic behavior', especially to determine limits on the values of a_i as i increases indefinitely. We say that the sequence is *bounded* above or below (by M) if $a_i\leq M$ or $a_i\geq M$, respectively, for some (necessarily finite) number M. Informally, we say that a numerical sequence $\{a_i\}$ 'converges' to a (necessarily finite) number L if $\mathbf{abs}(L-a_i)$ becomes smaller as i becomes larger.

Formally, we say that $\{a_i\}$ *converges* to L if for any measure of smallness $\epsilon>0$ that we choose, there exists a point in the sequence, say at the n_0th number a_{n0}, beyond which each difference $\mathbf{abs}(L-a_i)$ for $i>n_0$ is less than ϵ. This formal definition is less restrictive than the prior informal one, because the sequence of differences, $d_i=\mathbf{abs}(L-a_i)$, need not be decreasing; if it is decreasing, we say convergence is *monotonic*. If the sequence $\{a_i\}$ converges to L, we call L the *limit* of the sequence and write $\mathbf{lim}_{i\to\infty}a_i=L$, or more briefly $\{a_i\}\to L$. A sequence is said to *diverge* if it does not converge; we write $\mathbf{lim}_{i\to\infty}a_i=\infty$, or $\{a_i\}\to\infty$ to denote divergence. (We emphasize that the symbol ∞ here denotes divergence, not unboundedness.) A necessary condition for a sequence to be convergent is that it be bounded both above and below. Boundedness is also a sufficient condition if the sequence is monotonically increasing or decreasing. However, a bounded sequence can be divergent, e.g., if the sequence is not monotone but 'oscillates' instead. For example, the sequence $\langle -1,+1,-1,+1,\ldots\rangle$ diverges, so we write $\{(-1)^i\}\to\infty$. For details, see Rudin [3.21].

An *infinite series* is an infinite sum of the form $\Sigma\mathbf{f}(j)$, which is defined as the limit of the sequence of partial sums $S_i=\mathbf{f}(0)+\mathbf{f}(1)+\cdots+\mathbf{f}(i)$; i.e.,

$$\sum_{j=0}^{\infty}\mathbf{f}(j)=\mathbf{lim}_{i\to\infty}\sum_{j=0}^{i}\mathbf{f}(j).$$

We say the series converges if the sequence $\{S_i\}$ does.

Examples of divergent sequences include

$$\begin{aligned}
\langle 1,2,3,4,5,\ldots\rangle &= \{i\}\\
\langle 1,4,9,16,25,\ldots\rangle &= \{i^2\}\\
\langle 2,4,8,16,32,\ldots\rangle &= \{2^i\}\\
\langle 1,2,6,24,120,\ldots\rangle &= \{i!\}
\end{aligned}$$

We say that a sequence increases ('grows') *linearly* with i if it is of the form $\{c_0+c_1i\}$, *quadratically* if it is of the form $\{c_0+c_1i+c_2i^2\}$, *polynomially* if it is of the form $\{c_0+\ldots+c_ni^n\}$, *exponentially* if it is of the form $\{c^i\}$, and *factorially* if it is of the form $\{c\cdot i!\}$. Hence the four examples above grow linearly, quadratically, exponentially, and factorially, respectively.

Other examples of convergence and divergence include:

(a) $\mathbf{lim}_{i\to\infty}(1/i)=0$
(b) $\mathbf{lim}_{i\to\infty}(1-1/i)=1$
(c) $\Sigma_{j=0}^{\infty}(1/j!)=e$ (base of natural logarithm)
(d) $\mathbf{lim}_{i\to\infty}i^{\alpha}=\infty$ for $\alpha>0$
(e) $\mathbf{lim}_{i\to\infty}\beta^{i}=\infty$ for $\beta>1$
(f) $\mathbf{lim}_{i\to\infty}i!=\infty$
(g) $\mathbf{lim}_{i\to\infty}\dfrac{ai^2+bi+c}{i^2}=a$
(h) $\mathbf{lim}_{i\to\infty}(ci/i^2)=0$
(i) $\mathbf{lim}_{i\to\infty}(i^{\alpha}/\beta^{i})=0$ for $\alpha>0$, $\beta>1$
(j) $\mathbf{lim}_{i\to\infty}(\beta^{i}/i!)=0$
(k) $\Sigma_{j=0}^{\infty}x^{j}=1/(1-x)$ for $0<x<1$.

Case (g) shows that 'polynomial growth' is dominated by its leading (highest order) term. Case (h) shows that linear divergence is slower than quadratic divergence. Case (i) is of special importance since it shows that 'exponential growth' (raising a constant β to increasing powers) is faster than 'polynomial growth'; see Section 3.1.6 for implications. Cases (c) and (i) follow from the 'power series' formula for e^x (as found in mathematical handbooks, and discussed further in the next section):

$$e^x=\sum_{j=0}^{\infty}\frac{x^j}{j!} = \sum_{j=0}^{r}\frac{x^j}{j!}+\frac{x^{r+1}}{(r+1)!}+\cdots .$$

Case (c) is simply a special case. To prove case (i), we note that $e^i>i^{r+1}/(r+1)!$ for any fixed integer r, hence $e^i/i^r>i/(r+1)!$ or

$$\lim_{i\to\infty}\frac{i^r}{e^i}\left(<\frac{(r+1)!}{i}\right)=0.$$

Case (j) shows that factorial growth is even faster than exponential growth.

3.1.4 Application: Power Series

Most of the mathematical functions provided in scientific programming languages, e.g., SIN, COS, ARCTAN, are evaluated using power-series formulas, of the type found in mathematical handbooks. A *power series* in x is an arithmetic expression of the form

$$\mathbf{F}(x)=a_0+a_1x+a_2x^2+a_3x^3+\cdots$$

where the 'coefficients' $\{a_0,a_1,\ldots\}$ differ for different functions. Let

$$\mathbf{S}_i(x)=a_0+a_1x+a_2x^2+\cdots+a_ix^i,$$

the polynomial consisting of the first $i+1$ terms in the power series. Then we may regard the members of the numerical sequence $\langle\mathbf{S}_0(x),\mathbf{S}_1(x),\mathbf{S}_2(x),\ldots\rangle$ as *successive approximations* to the infinite power series. The coefficients must be chosen carefully if $\{\mathbf{S}_i(x)\}$ is to converge to $\mathbf{F}(x)$—i.e., so that

$$\mathbf{lim}_{i\to\infty}\ \mathbf{abs}(\mathbf{F}(x)-\mathbf{S}_i(x))=0$$

for any x; for practicality, i.e., for fast convergence, $\mathbf{S}_i(x)$ must be approximately equal to $\mathbf{F}(x)$ for a reasonably small i.

Let us rephrase our definition of $\mathbf{S}_i(x)$ as a recurrence relation (see Section 2.2.6):

$$\mathbf{S}_i(x)=\mathbf{S}_{i-1}(x)+a_ix^i, \qquad \text{for } i\geq 1$$
$$\mathbf{S}_0(x)=a_0.$$

In this form, we say $\mathbf{S}_i(x)$ is defined iteratively, $\mathbf{S}_1(x)$ in terms of $\mathbf{S}_0(x)$, $\mathbf{S}_2(x)$

in terms of $\mathbf{S}_1(x)$, and so forth.

An especially important class of iteratively defined functions is that where $\mathbf{S}_i(x)$ depends only upon $\mathbf{S}_{i-1}(x)$, and not upon i (or constants a_i): we then write

$$\mathbf{S}_i(x)=\mathbf{T}(\mathbf{S}_{i-1}(x)).$$

If $\{\mathbf{S}_i(x)\}$ converges to some $\mathbf{F}(x)$, we say that $\mathbf{F}(x)$ is a *fixed point* of the transformation $\mathbf{T}$; convergence usually requires that $\mathbf{T}$ satisfy certain nice properties and that $\mathbf{S}_0(x)$ be sufficiently close to $\mathbf{F}(x)$. See Blum [3.4]. The special case where $\mathbf{S}_i(x)$ depends only upon i and not upon x is also very important, since a fixed point of $\mathbf{T}$ is then a single number which can be regarded as the solution of some equation. For example, the solution of the general equation

$$\mathbf{f}(S)=0$$

can be rewritten as

$$S+\mathbf{f}(S)=S.$$

If this equation cannot be solved explicitly for S, we may be tempted to define $\mathbf{T}_\mathbf{f}$ by the recurrence relation

$$s_i=s_{i-1}+\mathbf{f}(s_{i-1})=\mathbf{T}_\mathbf{f}(s_{i-1}).$$

If $\mathbf{T}_\mathbf{f}$ has a fixed point S, then S is the desired solution, i.e., $\mathbf{f}(S)=0$. While this particular choice for $\mathbf{T}_\mathbf{f}$ is not likely to have the nice properties required for convergence, other choices may—e.g., the Newton–Raphson formula

$$s_i=s_{i-1}-\frac{\mathbf{f}(s_{i-1})}{\mathbf{f}'(s_{i-1})}=\mathbf{T}_\mathbf{f}(s_{i-1})$$

where $\mathbf{f}'$ denotes the derivative of $\mathbf{f}$. This formula is based on the definition $\mathbf{f}'(x)=(\mathbf{f}(x+\Delta)-\mathbf{f}(x))/\Delta$ as $\Delta\rightarrow 0$, with the substitutions $x=s_{i-1}$, $x+\Delta=s_i$, $\Delta=s_i-s_{i-1}$, and $\mathbf{f}(s_i)=0$. See below for an example.

Example*: *Exponentiation and Logarithms By 'exponentiation', we refer to the raising of a number a to some power b, denoted a^b or $a \uparrow b$ or $a ** b$. (We adopt the upwards arrow notation.) If b is a positive integer, then $a \uparrow b$ may be computed by successively multiplying (1, initially) by a, b times: e.g., $a \uparrow 4=(((1\cdot a)\cdot a)\cdot a)\cdot a$; if b is a negative integer, then $a \uparrow b$ may be computed by successively dividing (1, initially) by a, $-b$ times, or alternatively by computing $1\ /\ (a \uparrow -b)$. These procedures are not practical if the magnitude of b is great, and not possible if b is not an integer. In these events, exponentiation by arbitrary reals may be performed by making use of logarithms. The logarithm of a in the base r, denoted $\mathbf{log}_r a$, is defined (for $r>0$, $r\neq 1$, and $a>0$) as the solution b to the equation $r^b=a$. Note then that

$$a \uparrow b = r \uparrow (b \cdot \mathbf{log}_r a).$$

Evaluation of the functions $\mathbf{log}_r(x)$ and $r \uparrow (z)$ by using equivalent power-series formulas is fairly simple when the 'base' r is equal to e (the base of the *natural* logarithms):

$$\mathbf{log}_e(x) = 2\left(\frac{x-1}{x+1}\right) + \frac{1}{3}\left(\frac{x-1}{x+1}\right)^3 + \frac{1}{5}\left(\frac{x-1}{x+1}\right)^5 + \cdots, \qquad x>0$$

$$e \uparrow (z) = 1 + z + \frac{z^2}{2!} + \frac{z^3}{3!} + \frac{z^4}{4!} + \cdots.$$

These particular series are not commonly used in practice since they converge rather slowly; for better ones, see Cody [3.6]. Note that e can be found (or defined) by evaluating e $\uparrow$ (1). Note also that $\mathbf{log}_r(x)$ is not defined for $x\leq 0$; any attempt to compute $x \uparrow y$ in this case is regarded as an error except if y is an integer n, when we may instead use

$$x \uparrow n = (-1) \uparrow n \cdot \mathbf{abs}(x) \uparrow n.$$

Also, the LOG function of most programming languages is assumed to be with respect to the base e, and $\mathbf{log}_r(x)=\mathrm{LOG}(x)/\mathrm{LOG}(r)$; e $\uparrow (z)$ is commonly denoted EXP(z).

We observe that roots of numbers can also be computed in the way outlined above by letting y be a fraction. For example, the square root of x, denoted $\mathbf{sqr}(x)$, is equal to $x \uparrow 0.5$; the nth root of x is equal to $x \uparrow (1/n)$. There are, however, more efficient ways for computing roots, especially square roots, which can, for example, be found by solving $\mathbf{f}(S)=S^2-X=0$ using the Newton–Raphson formula:

$$s_i = s_{i-1} - \frac{\mathbf{f}(s_{i-1})}{\mathbf{f}'(s_{i-1})} = \frac{1}{2}\left[s_{i-1} + \frac{X}{s_{i-1}}\right] = \mathbf{T}_f(s_{i-1}),$$

where $\mathbf{f}'$ is the derivative of $\mathbf{f}$. Then, for any s_0, the successive approximations $\langle s_1, s_2, s_3, \ldots\rangle$ converge to the fixed point $S=\mathbf{sqr}(X)$. There are ways of computing s_0 so that convergence is faster, but any positive s_0 will do (say, $s_0=X/2$).

3.1.5 Successive Approximation Errors

In the foregoing, we introduced the concept of 'convergence', especially that of power series and of successive approximations to fixed points. There are theoretical conditions which the 'coefficients' $\{a_i\}$ or the 'transformation' $\mathbf{T}$ must satisfy in order for convergence of $\{S_i\}$ to F to occur (for any fixed x). Rather than be concerned with these theoretical conditions, we assume that sufficient conditions are satisfied, so that convergence is assured in theory, and address the computational problems instead.

One problem is that convergence can only be assured 'in the limit', i.e., if infinitely many approximations, $\{S_1, S_2, S_3, \ldots\}$, are computed. If we only compute finitely many, i.e., up to S_n for a given n, then there is no guarantee (in general) that S_n will be sufficiently close to F for our purposes. For this reason, we would like to compute successive approximations $\{S_i\}$ to F until the *absolute approximation error*, **abs**$(S_i - F)$, or *relative approximation error*, **abs**$((S_i - F)/S_i)$, is small. However, since we do not know F (we are trying to compute it), the best we can do in general is to compare consecutive approximations, i.e.,

$$\epsilon_{\mathsf{A}}(i) = \mathbf{abs}(S_i - S_{i-1})$$

or preferably

$$\epsilon_{\mathsf{R}}(i) = \mathbf{abs}((S_i - S_{i-1})/S_{i-1}).$$

If $\{S_i\}$ converges to F, we would expect $\epsilon_{\mathsf{R}}(i)$ to decrease as i increases; however, this cannot be guaranteed unless the convergence of $\{S_i\}$ to F is 'nice' and arithmetic is exact. The latter condition is necessary because it is entirely possible that round-off errors (due to inexact computer representations of numbers and arithmetic) may exceed approximation errors (due to the inability to compute more than finitely many approximations). Furthermore, as i increases, so that $\epsilon_{\mathsf{R}}(i)$ decreases, the round-off errors due to the increase in the number of arithmetic operations may also increase. We conclude that if floating-point numbers are accurate to at most 6 or 7 decimal places, then we cannot expect $\epsilon_{\mathsf{R}}(i)$ to be less than 10^{-6} for any i, and even this accuracy may be too much to expect. (Hence, if greater accuracy is required, we must use extended precision.)

In summary, successive approximations, S_i, for $i = 1, 2, 3, \ldots$, should be computed at least until $\epsilon_{\mathsf{R}}(i) < 10^{-j}$ if j decimal-place accuracy is desired. The larger j is, the longer the calculation will take. However, it cannot always be assured that $\epsilon_{\mathsf{R}}(i)$ will ever be less than 10^{-j}, so some limitation on the magnitude of i may be necessary. While a maximum execution-time limit on any given computer program serves this purpose, a programmed test for $i > L$ is preferable, hopefully where L can be chosen on the basis of theoretical considerations. ('Bisection' methods serve as an example of a case where a suitable L can be easily determined, but, in general, the problem is quite difficult. For example, see Isaacson and Keller [3.12].)

3.1.6 Asymptotic Complexity of Functions

Let $\mathbf{f}(x)$ and $\mathbf{g}(x)$ be numerical functions. We say $\mathbf{f}(x)$ is *order* $\mathbf{g}(x)$, denoted $\mathbf{f}(x) = \mathbf{O}(\mathbf{g}(x))$, if, for any $x > x_0$, $\mathbf{abs}(\mathbf{f}(x)/\mathbf{g}(x)) \leq M$, where x_0 and M are constants. In other words, past a certain point x_0, the magnitude of $\mathbf{f}(x)$ is never greater than some constant multiple of the magnitude of $\mathbf{g}(x)$. Hence we may consider $\mathbf{g}(x)$ to be a measure of the *asymptotic complexity* of $\mathbf{f}(x)$. Furthermore, $\mathbf{g}(x)$ (or, more specifically, $M \cdot \mathbf{abs}(\mathbf{g}(x))$) serves as a bound or approximation to the magnitude of $\mathbf{f}(x)$. The term 'asymptotic' is used to emphasize that we are primarily concerned with the behavior of $\mathbf{f}(x)$ for large x (i.e. for $x>x_0$). Furthermore, $\mathbf{f}(x)=\mathbf{O}(\mathbf{g}(x))$ is a non-symmetric relation, not an equation, hence we cannot write $\mathbf{O}(\mathbf{g}(x))=\mathbf{f}(x)$. If $\mathbf{f}(x)=\mathbf{O}(\mathbf{g}(x))$ and $\mathbf{g}(x)=\mathbf{O}(\mathbf{f}(x))$, then we say $\mathbf{f}$ and $\mathbf{g}$ are of the 'same order of magnitude', and write $\mathbf{f}(x)\approx\mathbf{g}(x)$.

Suppose that the domain of $\mathbf{f}$ and $\mathbf{g}$ is the set of natural numbers, so that $\mathbf{f}(n)$ and $\mathbf{g}(n)$ for $n=1, 2, \ldots,$ are sequences. We say $\mathbf{f}(n)=\mathbf{O}(\mathbf{g}(n))$ if, for any $n>n_0$, $\mathbf{abs}(\mathbf{f}(n)/\mathbf{g}(n))\leq M$, where n_0 and M are constants. Since we are only interested in asymptotic behavior, i.e. behavior for $n>n_0$, $\mathbf{g}(n)$ may be zero for $n\leq n_0$. Recalling our discussion of convergent sequences, we note that:

$$\mathbf{f}(n)=\mathbf{O}(\mathbf{g}(n)) \qquad \text{if } \mathbf{lim}_{n\to\infty}\frac{\mathbf{f}(n)}{\mathbf{g}(n)}=M$$

where M is a nonzero constant. This provides a simple sufficient test for the boundedness of $\mathbf{f}$ by $\mathbf{g}$; the condition is not necessary since, for example, the sequences may oscillate. If $M=1$, we write $\mathbf{f}(x)\simeq\mathbf{g}(x)$, and say that $\mathbf{f}$ and $\mathbf{g}$ are 'asymptotically equal'. If $M=0$, we write $\mathbf{f}(n)=\mathbf{o}(\mathbf{g}(n))$, read $\mathbf{f}$ is 'little-oh' of $\mathbf{g}$ (as opposed to 'big-oh' of $\mathbf{g}$), and say that $\mathbf{f}(n)$ is (asymptotically) 'negligible' compared with $\mathbf{g}(n)$. See Knuth [3.13] and Lewis [3.15] for further discussion.

Examples

$an^2+bn+c=\mathbf{O}(n^2)$ (with $M=a$)
$n^2=\mathbf{O}(an^2+bn+c)$ (with $M=1/a$)
$an^2+bn+c\approx n^2$
$an^2+bn+c\simeq an^2$
$bn+c=\mathbf{o}(n^2)$
$n^\alpha=\mathbf{o}(\beta^n)$ (for $\alpha>0$, $\beta>1$)
$x^r=\mathbf{o}(r!)$

There are two major applications of the concept of asymptotic complexity. One relates to the approximation of functions—e.g., of e^x by the finite series

$$\mathbf{f}_r(x)=1+x+x^2/2!+x^3/3!+\cdots+x^r/r!.$$

The approximation error,

$$e^x-\mathbf{f}_r(x)=\mathbf{O}(x^{r+1}/(r+1)!),$$

decreases as we increase the number of terms r taken in the series, as we would like; the more terms we compute, the more accurate is our approximation (at least in theory). In practice, of course, the more we compute, the more round-off errors may accumulate, which may easily offset decreases in approximation errors. The other major application relates to the analysis of algorithms: we say that an algorithm with nested loops of the form

```
DO I=1 TO N
...
DO J=1 TO N
...
END J
...
END I
```

has $\mathbf{O}(N^2)$ complexity since its execution time approximately equals aN^2+bN+c [$=\mathbf{O}(N^2)$], where a is the execution time of the statements in the inner loop, which are executed N^2 times, b is the execution time of the statements in the outer but not inner loop, which are executed N times, and c is the execution time of the statements outside the loops, which are executed only once.

3.2 MATRIX ALGEBRA

The algebra of matrices was introduced in Section 2.5.1. We recall that a matrix A is a 2-dimensional array of elements from a set S of scalars, with an associated 'array map' which maps a set of subscripts C to S. We assume, primarily for notational convenience, that an $(m \times n)$ matrix is *unity-indexed* (i.e., the subscripts belong to the 2-cell $C=[1,m]\times[1,n]$), and can be displayed as follows:

$$\begin{bmatrix} A(1,1) & A(1,2) & \dots & A(1,n) \\ A(2,1) & A(2,2) & \dots & A(2,n) \\ \dots & \dots & \dots & \dots \\ A(m,1) & A(m,2) & \dots & A(m,n) \end{bmatrix}.$$

The row-order representation of this matrix A is the sequence

$$\langle A(1,1),A(1,2),\dots,A(1,n),A(2,1),A(2,2),\dots,A(m,n-1),A(m,n)\rangle.$$

Two special cases of matrices are when m or n equals 1. An 'n-vector' can be regarded as either a $(1\times n)$ or an $(n\times 1)$ matrix, in which case it is called a 'row vector' or 'column vector', respectively. We say a matrix is *square* if $m=n$. The *main diagonal* of a square $(n\times n)$ matrix A is the sequence of elements $\langle A(1,1),A(2,2),\dots,A(n,n)\rangle$. A matrix, all of whose elements are equal to a, will be denoted $[a]$; e.g., $[1]$ denotes the $(m \times n)$ matrix all of whose elements are unity. For the special case where $a=0$, we use θ instead of $[0]$.

3.2.1 Matrix Operations

Recall from Section 2.5.1 the definitions of the sum $\oplus$ and difference $\ominus$ of like k-dimensional arrays. For the special case where $k=2$, i.e., for matrices, $C=A\oplus B$ is defined by $C(i,j)=A(i,j)+B(i,j)$, $D=A\ominus B$ is defined by $D(i,j)=A(i,j)-B(i,j)$, and the scalar product $B=s\circ A$ is defined by $B(i,j)=s\cdot A(i,j)$, where each of the matrices A, B, C, and D has the same extent, say (m,n). Furthermore, two like matrices are equal, $A=B$, iff corresponding elements are equal, i.e., $A(i,j)=B(i,j)$ for all i and j. Equality, addition, and subtraction are not defined for unlike matrices.

Let A_1 and A_2 be $(m_1 \times n_1)$ and $(m_2 \times n_2)$ matrices, respectively. The *matrix product* of A_1 and A_2, denoted $A_1 \circledast A_2$, is defined, *only* if $n_1=m_2$, as the $(m_1 \times n_2)$ matrix B, where

$$B(i,j)=\sum_{k=1}^{n} A_1(i,k)\cdot A_2(k,j),$$

and $A_1(i,k)\cdot A_2(k,j)$ is a product of scalars. (Note that $B(i,j)$ is the inner product of the ith row of A_1 and the jth column of A_2.) We say that A_1 is 'postmultiplied' by A_2, and A_2 is 'premultiplied' by A_1. It can be shown that the matrix product operation $\circledast$: $(S^n)^m \times (S^n)^m \rightarrow (S^n)^m$ is not closed, is associative whenever defined, is *not* commutative, has left-identities different from right-identities, hence no identity or inverse, and distributes over addition.

Restricted to $(n \times n)$ square matrices, $\circledast$: $S^{n^2} \times S^{n^2} \rightarrow S^{n^2}$, multiplication is closed and has an *identity* I (defined by $I(i,j)=1$ for $i=j$, else $I(i,j)=0$). Identity matrices are denoted by I for any n. The matrix θ, all of whose elements are zero, is a nullifier. Some square matrices (A) have *inverses* (denoted A^{-1}), $A^{-1}\circledast A=I=A\circledast A^{-1}$, but others do not; such matrices are called *nonsingular* and *singular*, respectively. (Whether or not a matrix is singular can be tested by computing its 'determinant', which we shall not discuss here; if nonsingular, the elements of the $(n \times n)$ inverse matrix can be found by solving a linear set of n^2 equations in n^2 unknowns.) For a square matrix A, the kth *power* of A is always defined, for $k>0$, by $A^k=A^{k-1}\circledast A$, $A^0=I$. We remark that the set of all $(n \times n)$ square matrices, together with their sum and product as defined here, form a noncommutative ring $(S^{n^2}, \oplus, \circledast, \theta, I)$.

A common unary operation, called 'matrix transposition', which maps unity-indexed $(m \times n)$ matrices to $(n \times m)$ matrices is defined as follows: the *transpose* of A, denoted A^{T}, is such that

$$A^{\mathsf{T}}(i,j)=A(j,i) \qquad \text{for} \quad 1\le i\le n, \quad 1\le j\le m.$$

A is said to be *symmetric* if $A = A^{\mathsf{T}}$. Unary operations which map unity-indexed $(m \times n)$ matrices into like matrices, called 'permutations', are defined as follows. Let $R = (r_1, r_2, \ldots, r_m)$ be a permutation of $\{1, 2, \ldots, m\}$. Then B is the *row permutation* of A with respect to R if

$$B(i,j) = A(r_i, j) \qquad \text{for} \quad 1 \le i \le m, \quad 1 \le j \le n.$$

Let $C = (c_1, c_2, \ldots, c_n)$ be a permutation of $\{1, 2, \ldots, n\}$. Then B is the *column permutation* of A with respect to C if

$$B(i,j) = A(i, c_j) \qquad \text{for} \quad 1 \le i \le m, \quad 1 \le j \le n.$$

If A is an $(n \times n)$ square matrix, then B is a *permutation* (*of corresponding rows and columns*) of A if

$$B(i,j) = A(r_i, r_j) \qquad \text{for} \quad 1 \le i \le n, \quad 1 \le j \le n$$

for some permutation $R = (r_1, r_2, \ldots, r_n)$ of $\{1, 2, \ldots, n\}$; the parenthesized qualification is implied when a row or column permutation is not explicitly specified. Since there are $n!$ permutations of $\{1, 2, \ldots, n\}$ (cf. Section 2.1.2), there are $n!$ permutations of any $(n \times n)$ matrix. Any permutation of an identity matrix is called an 'elementary' matrix; any permutation of a matrix A equals A pre- and/or post-multiplied by a sequence of elementary matrices. (We leave the demonstration of the latter statement as an exercise.)

The algebra of matrices is discussed in greater detail in Paige [3.17]. Some important applications to statistical and optimization problems, which we discuss later in Chapters 6 and 8, are introduced in Hovanessian [3.11]. See also Bellman [3.1].

Examples Let A and B be the following (2×2) matrices:

$$A = \begin{bmatrix} 1 & 2 \\ 3 & 4 \end{bmatrix}, \qquad B = \begin{bmatrix} 5 & 6 \\ 7 & 8 \end{bmatrix}.$$

Then

$$A \oplus B = \begin{bmatrix} 1+5 & 2+6 \\ 3+7 & 4+8 \end{bmatrix} = \begin{bmatrix} 6 & 8 \\ 10 & 12 \end{bmatrix}$$

$$A \oplus B = \begin{bmatrix} 1-5 & 2-6 \\ 3-7 & 4-8 \end{bmatrix} = \begin{bmatrix} -4 & -4 \\ -4 & -4 \end{bmatrix}$$

$$A \circledast B = \begin{bmatrix} 1 \cdot 5 + 2 \cdot 7 & 1 \cdot 6 + 2 \cdot 8 \\ 3 \cdot 5 + 4 \cdot 7 & 3 \cdot 6 + 4 \cdot 8 \end{bmatrix} = \begin{bmatrix} 19 & 22 \\ 43 & 50 \end{bmatrix}$$

$$I=\begin{bmatrix}1 & 0\\ 0 & 1\end{bmatrix},\qquad 5\circ A=\begin{bmatrix}5 & 10\\ 15 & 20\end{bmatrix}$$

$$A^{-1}=\begin{bmatrix}-2 & 1\\ 3/2 & -1/2\end{bmatrix},\qquad B^{-1}=\begin{bmatrix}-4 & 3\\ 7/2 & -5/2\end{bmatrix}.$$

A^{-1} can be found by solving a system of four linear equations in four unknowns. Let its elements be denoted

$$A^{-1}=\begin{bmatrix}x_1 & x_2\\ x_3 & x_4\end{bmatrix};$$

then since $A \circledast A^{-1}=I$, we have

$$\begin{bmatrix}1 & 2\\ 3 & 4\end{bmatrix}\circledast\begin{bmatrix}x_1 & x_2\\ x_3 & x_4\end{bmatrix}=\begin{bmatrix}1 & 0\\ 0 & 1\end{bmatrix}$$

or

$$\begin{array}{cccccc} x_1 & & +2x_3 & & = & 1\\ & x_2 & & +2x_4 & = & 0\\ 3x_1 & & +4x_3 & & = & 0\\ & 3x_2 & & +4x_4 & = & 1\end{array}$$

which when solved (e.g., by 'Cramer's rule' or 'Gaussian elimination')† yields $x_1=-2$, $x_2=1$, $x_3=3/2$, $x_4=-1/2$.

The transposes of A and B are

$$A^{\mathsf{T}}=\begin{bmatrix}1 & 3\\ 2 & 4\end{bmatrix},\qquad B^{\mathsf{T}}=\begin{bmatrix}5 & 7\\ 6 & 8\end{bmatrix}.$$

With respect to (2,1), a permutation of {1,2},

$\begin{bmatrix}3 & 4\\ 1 & 2\end{bmatrix}$ is the row permutation of A,

$\begin{bmatrix}2 & 1\\ 4 & 3\end{bmatrix}$ is the column permutation of A,

.

†For a computer solution, use of Gaussian elimination is far more efficient than use of Cramer's rule. See Isaacson and Keller [3.12]. This is just one of numerous examples where a good 'by-hand' procedure is a poor 'by computer' procedure.

and

$$\begin{bmatrix} 4 & 3 \\ 2 & 1 \end{bmatrix}$$ is the permutation of corresponding rows and columns of A.

The main diagonals of A and B are $\{1,4\}$ and $\{5,8\}$, respectively.

3.2.2 Stochastic Matrices

A *probability vector* is a row vector whose elements are all numbers in the interval of reals $[0,1]$ and whose sum is one: i.e., $(a_1,\ldots,a_N)$ is a probability vector if $0 \le a_i \le 1$ for each i and $\Sigma_{i=1}^{N} a_i = 1$. A *stochastic matrix* is square, with each of its rows a probability vector. It can be shown that the stochastic property of matrices is preserved under matrix multiplication, i.e., that the system of stochastic matrices is closed under multiplication. Therefore, all powers A^k of a stochastic matrix are also stochastic matrices.

A stochastic matrix A is

regular if each element of A^k is nonzero for some $k \ge 1$
periodic (with 'period' r) if $A^{k+r} = A^k$ for $k \ge 1$.

If neither regular nor periodic, A is said to be *reducible*. A reducible matrix is such that, for some permutation of its rows and columns, it (and each of its powers) has the block triangular form

$$\begin{bmatrix} B & C \\ 0 & D \end{bmatrix}.$$

A regular matrix has the property that the sequence of powers A^k converges to (say) A^* ; the matrix A^* has the property that all of its rows are equal to each other.

We shall use the concepts presented here in our discussion of Markov chains (see Section 6.3). For details, see Pearl [3.18].

Example

$$\begin{bmatrix} 0 & 1 & 0 \\ 0 & 0 & 1 \\ 1 & 0 & 0 \end{bmatrix}$$ is periodic with period 3; $$\begin{bmatrix} 0 & 1/2 & 1/2 \\ 0 & 0 & 1 \\ 0 & 0 & 1 \end{bmatrix}$$ is

reducible, with

$$B = \begin{bmatrix} 0 & 1/2 \\ 0 & 0 \end{bmatrix}, \quad C = \begin{bmatrix} 1/2 \\ 1 \end{bmatrix}, \quad D = [1];$$

and

$$\begin{bmatrix} 0 & 1/2 & 1/2 \\ 0 & 0 & 1 \\ 1 & 0 & 0 \end{bmatrix}$$

is regular and converges to

$$\begin{bmatrix} 0.4 & 0.2 & 0.4 \\ 0.4 & 0.2 & 0.4 \\ 0.4 & 0.2 & 0.4 \end{bmatrix}.$$

3.2.3 Boolean Matrices

A *Boolean* array is one in which each element is either 0 or 1. The sum and inner product of Boolean vectors (i.e., n-tuples) were defined in terms of the Boolean arithmetic operators $\oplus$ and $\times$ in Section 2.5.1. Generalizing, the *Boolean matrix sum* of A and B, $C=A\oplus B$, is defined by the Boolean sum of corresponding elements.

(a) $C(i,j)=A(i,j)\oplus B(i,j)$.

The *Boolean matrix product* of A and B, $C=A\circledast B$, is defined by

(b) $C(i,j)=[A(i,1)\times B(1,j)]\oplus\ldots\oplus[A(i,n)\times B(n,j)]$;

note that $C(i,j)$ equals the Boolean inner product of the ith row of A and jth column of B.

For some applications, it is preferable to use the logical sum [$\vee$] instead of the Boolean sum [$\oplus$] in (a) and (b) above. (The logical operations were discussed earlier in Section 2.5.3.) Doing so defines the *logical* matrix sum and product, denoted $A \textcircled{\vee} B$ and $A \textcircled{\wedge} B$, respectively. The logical matrix product is assumed in the definition of (logical) *powers* of Boolean matrices, $M^n=M^{n-1}\textcircled{\wedge}M$, $M^0=I$. An important application will be discussed in Section 4.3.2.

Example Let $A=\begin{bmatrix} 1 & 1 \\ 1 & 0 \end{bmatrix}$ and $B=\begin{bmatrix} 0 & 1 \\ 1 & 1 \end{bmatrix}$.

Then

$$A\oplus B=\begin{bmatrix} 1 & 0 \\ 0 & 1 \end{bmatrix}, \quad A\circledast B=\begin{bmatrix} 1 & 0 \\ 0 & 1 \end{bmatrix}$$

$$A\textcircled{\vee}B=\begin{bmatrix} 1 & 1 \\ 1 & 1 \end{bmatrix}, \quad A\textcircled{\wedge}B=\begin{bmatrix} 1 & 1 \\ 0 & 1 \end{bmatrix}.$$

3.2.4 Symbolic Operations

When the elements of arrays are strings rather than numbers, we may define *symbolic sum* $\oplus$ and *symbolic product* $\otimes$ of strings as follows: for any two strings x and y in A^* (the set of all strings over an alphabet A), let

$$x \oplus y = \{x, y\}$$
$$x \otimes y = x \cdot y$$

i.e., the symbolic sum of two strings is the set containing both, and the symbolic product of two strings is their concatenation (see Section 2.4.5). (Note that $\oplus$ is commutative, but $\otimes$ is not.)

Let X and Y be sets of strings. Then $\oplus$ and $\otimes$, operations on A^* above, may be extended to 2^{A^*} as follows: for $X = \{x_1, \ldots, x_r\}$, $Y = \{y_1, \ldots, y_s\}$, both finite subsets of A^*, let

$$X \oplus Y = X \mid Y \qquad \text{(the alternation of } X \text{ and } Y\text{)}$$
$$X \otimes Y = X \cdot Y \qquad \text{(the complex product of } X \text{ and } Y\text{).}$$

We emphasize that $X \otimes \emptyset = \emptyset$ whereas $X \otimes \{\varepsilon\} = X$. The *symbolic matrix sum* $\boldsymbol{\oplus}$ and *symbolic matrix product* $\circledast$ of matrices of *sets* of strings are then defined in terms of $\oplus$ and $\otimes$ in the conventional fashion, as in (a) and (b) of the preceding section.

Example Let $X = \{a, b\}$ and $Y = \{b, c, \varepsilon\}$. Then

$$X \oplus Y = \{a, b, c, \varepsilon\},$$
$$X \otimes Y = \{ab, ac, a, bb, bc, b\}.$$

If

$$A = \begin{bmatrix} \{a\} & \{a,b\} \\ \{\varepsilon, b\} & \emptyset \end{bmatrix} \quad \text{and} \quad B = \begin{bmatrix} \{a\} & \emptyset \\ \{bc\} & \{\varepsilon\} \end{bmatrix},$$

then

$$A \boldsymbol{\oplus} B = \begin{bmatrix} \{a\} & \{a,b,c\} \\ \{\varepsilon, b, bc\} & \{\varepsilon\} \end{bmatrix},$$

$$A \circledast B = \begin{bmatrix} \{aa, abc, bbc\} & \{a,b\} \\ \{a, ba\} & \emptyset \end{bmatrix}.$$

3.2.5 Software Implications

Arrays correspond to what are generally called 'subscripted variables' in programming languages. Variable names are treated as array names either by explicit declaration (e.g., using DECLARE or DIMENSION statements), or less commonly by implicit declaration (e.g., by context, such as the appearance of subscripts or the use of array functions or subroutines). Usually, an array name standing alone represents the entire set of array elements, whereas an array name followed by a parenthesized subscript(s) represents one of the elements. Many languages restrict the former usage to input and output statements, or to subprogram calls, and the latter usage to arithmetic expressions and assignment statements. In such languages, it is not possible to write 'array expressions', e.g., $A \oplus B \circledast C$, where A, B, C are arrays; instead programmers must code element-by-element calculations, or use built-in or library functions, e.g.,

```
SUM(A,PROD(B,C)).
```

Some languages, however, 'overload' the symbols +, −, *, and /, interpreting them in expressions as array operations when the operands have been explicitly declared to be arrays.

A declaration of an array in a programming language is generally accompanied by the specification of the types of elements that the array may contain, and of the number of subscripts and their permissible values. Arrays conventionally contain only one type of element (say, integers, reals, character strings, or truth values) and there may also be restrictions on the values of these elements. For example, a *Boolean* array is one where elements may have only one of two integer values: 0 and 1. Subscript specifications are made by identifying (i.e., 'declaring' or 'dimensioning') the extent of a cell $[a_1,b_1]\times\cdots\times[a_k,b_k]$. In some languages, each a_i must be 0 or 1 and only the values b_i are specified; also, k must generally be less than some given limit. The most common special case of an array is where $a_1=1$, $b_1=N$, and $k=1$, and the elements $A(1), A(2),\ldots,A(N)$ are N numbers (or strings) grouped together under a common name A; we then call A a 'standard' (unity-indexed) vector of size N.

3.2.6 Sequential Storage of Arrays

A *sequence* S is a physically sequential data structure, i.e. a list of data items, where the logical ordering of the data items corresponds to their physical locations in the storage unit of a computer. The storage unit consists of a numerically ordered set of '(storage) cells', ordered by increasing 'address'. The address of the first data item is the *base address* of S; the remaining data items are stored in cells with successively increasing addresses. A sequence is usually regarded as an ordinal data structure, where access to the individual data items must always be made consecutively, starting from the first item, to the second, to the third, and so forth. We then say that the items are processed sequentially or serially. A special case of a sequence is one whose data items are from some character set. Such a sequence is called a (character) *string*. Generally, it is possible for several characters to share one cell.

Examples Examples of sequences (of character strings) are shown in Figure 3.2.6. Figure 3.2.6(a) shows the case where each character-string data item has three cells allocated to it, despite the fact that only one needs all three. While inefficient in its use of space, the use of a fixed length makes it fairly easy to find the ith data item. Figure 3.2.6(b) shows the use of variable lengths (to save space); but to find the second data item now requires examining each character until the 'end-marker' @ is encountered. Alternatively, a 'count' of the number of characters in the string can precede each string to enable its end to be easily found, as shown in Figure 3.2.6(c). Yet another alternative is to have each cell include a flag which indicates whether or not the data item continues into the next cell; this method is called *spanning*, and is illustrated in Figure 3.2.6(d).

If the data items of a sequence are stored one to a cell, access to an arbitrary data item can be made without accessing all of its predecessors by use of a direct-addressing function: the address of the ith data item, denoted **addr**$[S(i)]$, is given by

$$\textbf{addr}[S(i)]=(\text{base address of } S)+(i-1);$$

$i-1$ is called the *displacement*. Analogous formulas (displacements) can be derived in the event that data items are stored m to a cell, or if each data item requires m cells of storage, where m is an integer. If data items are accessed using a base-displacement formula, a sequence may be regarded as a mapped data structure (i.e., as a one-dimensional array).

Arrays A k-dimensional *array* A is a set of data items, where the logical ordering of the data items corresponds to some conventional ordering of an associated set of subscripts (k-cell) C. If the data items are physically stored and accessed in this order, an array may be regarded as an ordinal data structure, but more commonly an array is regarded as a mapped data structure where the location of an arbitrary data item is found by a direct-addressing base-displacement formula:

$$\textbf{addr}[A(i_1,\ldots,i_k)]=\text{base address of } A+\text{displacement } (i_1,\ldots,i_k).$$

For example, let $C=[a_1,b_1]\times\cdots\times[a_k,b_k]$, and assume the data items of A are ordered lexicographically by subscript (i.e., in 'row order'), one to a memory cell. Then

$$\text{base address of } A=\textbf{addr}[A(a_1,\ldots,a_k)]$$

and

$$\text{displacement } (i_1,\ldots,i_k)=\sum_{j=1}^{k}[i_j-a_j]\cdot\prod_{\ell=j+1}^{k}(b_\ell-a_\ell+1).$$

100	B A N A
101	N A b b
102	b b b b
103	C O C O
104	N U T b
105	b b b b
106	G U A V
107	A b b b
108	b b b b
109	M A N G
110	O b b b
111	b b b b
112	P A P A
113	Y A b b
114	b b b b
115	P I N E
116	A P P L
117	E b b b
118	@ b b b

(a)

B A N A
N A @ C
O C O N
U T @ G
U A V A
@ M A N
G O @ P
A P A Y
A @ P I
N E A P
P L E @
@ b b b

(b)

6 B A N
A N A 7
C O C O
N U T 5
G U A V
A 5 M A
N G O 6
P A P A
Y A 9 P
I N E A
P P L E
0 b b b

(c)

100	1BAN
101	0ANA
102	1COC
103	1ONU
104	0Tbb
105	1GUA
106	0VAb
107	1MAN
108	0GOb
109	1PAP
110	0AYA
111	1PIN
112	1EAP
113	0PLE
114	@bbb

(d)

	0	1	2	3	4	5	6	7	8	9
100	B	A	N	A	N	A				
110	C	O	C	O	N	U	T			
120	G	U	A	V	A					
130	M	A	N	G	O					
140	P	A	P	A	Y	A				
150	P	I	N	E	A	P	P	L	E	
160										

(e)

Figure 3.2.6

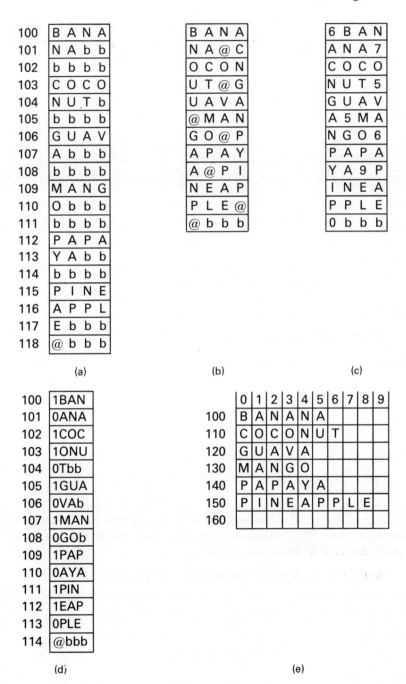

Note that the displacement does not depend upon a_1 and b_1, or the number of rows. For unity-indexed arrays (when $a_i=1$, for each i), the foregoing reduces to

base address of vector $A=\mathbf{addr}[A(1)]$
displacement $(i)=i-1$

and (for $m=b_1$, $n=b_2$)

base address of $(m\times n)$ matrix $A=\mathbf{addr}[A(1,1)]$
displacement $(i,j)=(i-1)\cdot b_2+(j-1)$

for $k=1$ and 2, respectively. If each data item occupies L cells, then the displacements should be multiplied by L.

Example Figure 3.2.6(e) illustrates a one-dimensional array A in which each character occupies one cell, and each character-string data item occupies ten cells. For this example,

$$\begin{aligned}\mathbf{addr}[A(i)]&=\mathbf{addr}[A(1)]+(i-1)\cdot 10\\&=100+i\cdot 10-10=90+i\cdot 10,\end{aligned}$$

hence $\mathbf{addr}[A(4)]=90+4\cdot 10=130$. Figure 3.2.6(e) also illustrates a two-dimensional row-order array in which the data items are the individual characters. The location of the jth character of the ith string is given by

$$\mathbf{addr}[A(i,j)]=\mathbf{addr}[A(1,1)]+(i-1)\cdot n+(j-1);$$

thus,

$$\mathbf{addr}[A(4,5)]=\mathbf{addr}[A(1,1)]+(4-1)\cdot 10+(5-1)=100+30+4=134.$$

3.2.7 Application: Directories and Hash Tables

Let $A=\{a_1,\ldots,a_M\}$ be a set of data items, K be a set of distinctive labels (here called *keys*), and λ: $A\rightarrow K$ with $\lambda(a_i)=k_j$. Suppose data item a_i is stored in computer memory location $\mathbf{L}(a_i)\in[X,Y]$, an interval of integers. Then we call the set of ordered pairs

$$\mathbf{D}=\{(k_j,\ell_i)\,|\,k_j=\lambda(a_i),\ \ell_i=\mathbf{L}(a_i),\ a_i\in A\}$$

a *directory* (or *index*). To find the location of data item a_i given its key k_j, the first component of the directory is searched for k_j; when found, the corresponding second component specifies the desired location. We note that $\mathbf{D}$ may be regarded as a direct-addressing function, $\mathbf{D}$:$K\rightarrow[X,Y]$, so that A and $\mathbf{D}$ together form a mapped data structure, called a *directoried table*. The function $\mathbf{D}$ need not be onto; in fact, the locations of the data items may be widely scattered. The directory is usually stored as two 'parallel' one-dimensional arrays, and the base of such a directoried table is the pair of bases of the arrays. The directory, as a separate entity, may be structured in a variety of other ways (e.g., linked or hashed); it is common for the directory to be sorted lexicographically in key-order in order to facilitate searching.

Example Figure 3.2.7(a) illustrates a directoried table, where the data items are character strings, and the keys are their initial two characters. The directory is shown sorted. Note that if we used only initial characters as keys, there would be duplicate keys (in other words, λ would not be 1–1). This situation is commonly disallowed, by convention, for directoried tables, but techniques described next can be used to handle it.

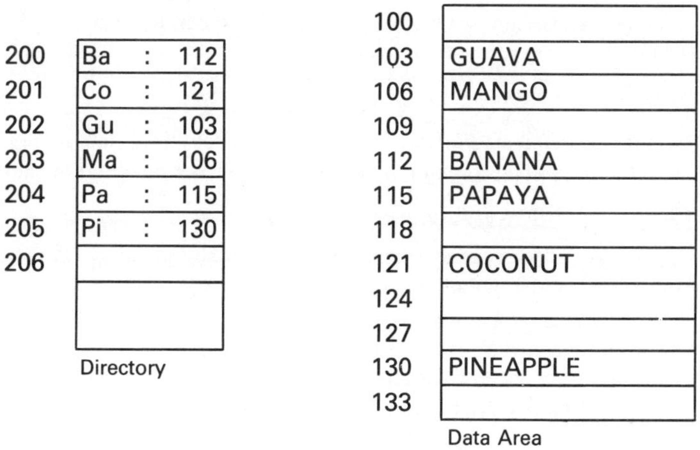

Figure 3.2.7(a)

Hash Tables Let $K=\{k_1,\ldots,k_N\}$ be a set, $\mathbf{L}=[X,Y]$ be an interval of integers, and $\mathbf{h}$: $K\rightarrow\mathbf{L}$ be a function *onto* $\mathbf{L}$. We call members of K *keys*, and members of $\mathbf{L}$ *locations* or *buckets*. We assume $N=\#(K)$ is much greater than $Y-X+1=\#(\mathbf{L})$, so that $\mathbf{h}$ cannot be a 1–1 function; many different keys may map to the same location. Suppose we now have a set of data items $A=\{a_1,a_2,\ldots,a_M\}$, $M\leq\#(\mathbf{L})$, each of which we would like to associate with a (computer memory) location at which we would store the data item. If each data item has a distinctive key (or label), i.e., λ: $A\rightarrow K$ such that $\lambda(a_i)=k_j$, then we may store a_i at location $\mathbf{h}(\lambda(a_i))$. This defines a mapped data structure, called a *hash table,* with $\mathbf{h}$ called the *hash function*. (We remark that $\mathbf{h}$ must by convention be defined by an algorithm, rather than by a directory). The base of a hash table may be regarded as the entry point of the hash-function 'subroutine' (although the data items are stored starting at location X).

One problem which may arise is that, given two different data items a_i and a_j, with $\lambda(a_i) \neq \lambda(a_j)$, it is possible that $\mathbf{h}(\lambda(a_i)) = \mathbf{h}(\lambda(a_j))$. In this event, called a *collision,* both data items would be mapped to the same location, which may not be able to accommodate them. A colliding or 'overflowing' data item, also called a *synonym,* would then have to be placed in the next available location in the table (a policy called *open addressing*), in an arbitrarily chosen location in the table, or in a separate overflow area. In the latter two cases, a pointer to the location of the overflowing data item would usually be used (a policy called *chaining*); in the former case, the table would be searched sequentially for the overflowing data item starting from its expected location.

Example Figure 3.2.7(b) illustrates a hash table, where the hash function $\mathbf{h}: K \rightarrow \mathbf{L}$ is defined as follows:

$$\mathbf{h}(k) = \mathbf{ordinate}(k) + 100$$

where $\lambda(a)$ is the initial character (letter) of the character-string data item a, and **ordinate**(k) is the ordinal position number of letter k in the alphabet. For example,

$$\mathbf{h}(\lambda(\text{PAPAYA})) = \mathbf{h}(\text{P}) = \mathbf{ordinate}(\text{P}) + 100 = 116.$$

Note that data items 'PAPAYA' and 'PINEAPPLE' collide; in the figure, they are shown sharing the same cell. If a cell can accommodate only one data item, one of the two would have to be placed elsewhere and some mechanism for finding the overflowing item would have to be made available. Using the open addressing policy, 'PINEAPPLE' would be placed at address 117, whereas using a separate overflow area, it would be placed (say) at address 200. In a retrieval process, if a data item is not found where it is expected, a search for it must be made.

101		114	
102	BANANA	115	
103	COCONUT	116	PAPAYA : PINEAPPLE
104		117	
105		118	
106		119	
107	GUAVA	120	
108		121	
109		122	
110		123	
111		124	
112		125	
113	MANGO	126	

Figure 3.2.7(b)

We discuss the analysis of various hashing strategies in Section 7.2.5. A pioneering study appears in Peterson [3.19]; see Maurer [3.16] for a survey.

3.2.8 Application: Sparse Arrays

In many applications, many elements of a large array are known to be zero, and only the nonzero elements are of interest. Such arrays are called *sparse*. Let $\mathbf{A}:C \to S$ be an array map, and C' be a nonempty proper subset of C such that $\mathbf{A}:C' \to \{0\}$. C' is a set of subscripts associated with (not necessarily all) known zero elements of the array. It is then often convenient to define the composition of two total functions, $\mathbf{A}_2 \circ \mathbf{A}_1$, where

$$\mathbf{A}_1:(C-C') \to I,\ \mathbf{A}_2:I \to S,$$

such that $\mathbf{A}_2(\mathbf{A}_1(c))=\mathbf{A}(c)$ for $c \in C-C'$; $\mathbf{A}_1(c)$ is undefined for $c \in C'$, when $\mathbf{A}(c)=0$. $\mathbf{A}_1$ and $\mathbf{A}_2$ are most commonly defined so that $I=[1,K]$, the interval of integers, where $K=\#(C-C')$. Note that $\mathbf{A}_2 \circ \mathbf{A}_1$: $(C-C') \to S$.

For example, let $C=[1,N]\times[1,N]$, and let $\mathbf{A}:C \to S$ be an $(N\times N)$ matrix such that $\mathbf{A}(i,j)=0$ whenever $i<j$. Such a matrix is called 'lower triangular' since, in its pictorial display, nonzero elements are limited to the triangular portion of the matrix at or below the main diagonal. C' is the set $\{(i,j) \in C \mid i<j\}$, and $K=\#(C-C')=N(N+1)/2$. We remark that known symmetric matrices can be represented as lower-triangular ones since the upper-triangular portion is redundant.

Methods of representing sparse arrays in computers, without wasting space for storing zeroes or other implicitly known elements, are of great importance. For a lower-triangular array A, suppose the nonzero data items are stored one to a cell row-wise—i.e., in the order

$$A(1,1),\ A(2,1),\ A(2,2),\ A(3,1),\ \ldots,\ A(4,1),\ \ldots,$$
$$A(N-1,1),\ \ldots,\ A(N-1,N-1),\ A(N,1),\ \ldots,\ A(N,N).$$

Define $\mathbf{A}_2$ such that $\mathbf{A}_2(k)$ is the kth data item in the above ordering. Then if $\mathbf{A}_1$ is defined by the displacement formula

$$\mathbf{A}_1(i,j)=\frac{(i-1)i}{2}+j$$

we have

$$\mathbf{A}_2(\mathbf{A}_1(i,j))=\mathbf{A}_2((i-1)i/2+j)=\mathbf{A}(i,j)$$

(whenever $i \geq j$) as desired. In other words, use of the map $\mathbf{A}_2 \circ \mathbf{A}_1$ defined by the special displacement formula, instead of the map $\mathbf{A}$ defined by the general displacement formula $[(i-1)N+(j-1)]$, avoids wasting space for storing about $N^2/2$ zeroes.

Use of special displacement formulas to save space in implementations of sparse arrays is applicable only when the nonzero elements occupy known positions in some systematic pattern. Furthermore, it is sometimes necessary to test whether the displacement formula is applicable at all, i.e., whether the subscript is in $C-C'$. This test may not always be as simple as above, when the defining condition for C', namely $i<j$, suffices.

A general approach to implementing sparse arrays, applicable when its zeroes are randomly distributed and even when their positions are changeable, is to use linked lists. For example, we may represent an $(M \times N)$ array by M different lists, one for each row, and in each such row list we would chain together only those nonzero elements in the row. Similarly, we may have N column lists. Each node in this data structure must include self-identifying information, i.e., its subscript values (row and column numbers). Because of the necessity to store this additional information as well as pointers, the approach saves space only for very sparse or large arrays.

For further information on hashing as well as the concepts discussed in Sections 3.2.6 to 3.2.8, see the data structure text books cited in Chapter 4.

3.2.9 Array Representations of Sets and Relations

Sets may be represented in computers by one-dimensional arrays, i.e., by ordinary vectors. If $S=\{a_1,a_2,\ldots,a_M\}$ is a set of objects, and A is a vector name, then we may associate a_i with $A(i)$, the ith element of the vector A. If the members of the set S are primitive data items (i.e., single numbers or character strings), then, in most programming languages, this association may be made by assignment statements, e.g., of the form '$A(i):=a_i$'. If the data items may themselves be sets, then more complex data structuring schemes may become necessary. Suppose each object a_i has a distinct numerical label, $\lambda(a_i)$; then a set of objects may be represented by the array of associated labels.

To perform set operations, it is convenient to deal with *sorted* vectors. If the objects in a set S have a natural (e.g., numerical or lexicographical) ordering, then the vector A is said to be sorted ***by value*** if $A(1)$ is the smallest member of S, $A(2)$ is the second smallest, ..., and $A(M)$ is the largest. If the objects in S have no natural ordering, then A is said to be sorted ***by label*** if $A(1)$ is the member of S having the smallest numerical label, etc. If the vector representation of a set is sorted, then it is not necessary, for example, to compare a given data item x with each element of the vector to determine whether or not $x \in S$; one needs only compare x [or $\lambda(x)$] with $A(i)$, in the order $i=1,2,\ldots$, until either $x=A(i)$ or $x<A(i)$. Determination of whether or

not one set is a subset of another, and construction of new sets by union, intersection, or other operations on given sets, are likewise facilitated.

If the objects in any set S are required to be members of a given finite universal set $U=\{u_1,u_2,\ldots,u_N\}$, considered ordered, then we may represent $S\subseteq U$ by an N-vector $\mathbf{C}_S$ defined by

$$\mathbf{C}_S(i)=\begin{cases}1 \text{ if } u_i\in S\\ 0 \text{ if } u_i\notin S.\end{cases}$$

We call this representation the *characteristic vector* associated with the set S (cf. Sections 2.2.1 and 2.3).

Since relations, functions, and operations are sets, they may, in principle, also be represented as vectors. However, the members of these sets are no longer primitive objects, and consequently vector representations are not advantageous. However, matrix representations may be used instead.

Let $S=\{a_1,a_2,\ldots,a_n\}$ be an ordered set (with $\lambda(a_i)=i$ or $\lambda^{-1}(i)=a_i$). A relation $R\subseteq S\times S$ can be represented by a unity-indexed $(n\times n)$ matrix $\mathbf{C}_R$ as follows:

$$\mathbf{C}_R(i,j)=\begin{cases}1 \text{ if } (a_i,a_j)\in R\\ 0 \text{ if } (a_i,a_j)\notin R.\end{cases}$$

This Boolean matrix is called the *characteristic matrix* associated with the relation R in S. A natural extension to single-valued binary operations, $\mathbf{F}$: $S\times S\rightarrow S$, is to define

$$\mathbf{C}_{\mathbf{F}}(i,j)=\begin{cases}k \text{ if } ((a_i,a_j),a_k)\in \mathbf{F}\\ 0 \text{ otherwise}.\end{cases}$$

The characteristic matrix representation of relations has several important properties. We cite, for example, the following:

(1) if R_1 is the inverse of R_2,
then $\mathbf{C}_{R_1}$ is the transpose of $\mathbf{C}_{R_2}$

(2) if $R=R_1R_2$ (the composition of the relations),
then $\mathbf{C}_R=\mathbf{C}_{R_1}Ⓐ\mathbf{C}_{R_2}$ (the *logical* matrix product).

Furthermore, whether or not a relation has certain properties (e.g., reflexivity, symmetry, transitivity) can be determined by examination of its matrix representation; a general computer program can be written to perform these examinations for arbitrary input matrices. Often, sparse matrix representations are useful, e.g., in the case of symmetric relations.

An n-ary operation $\Theta\subseteq S^n\times S$ has an especially convenient matrix representation provided S is a finite ordered set (where the members of S are represented by their ordinates). Then Θ can be represented by an n-dimensional array A where $A(x_1,x_2,\ldots,x_n)=y$ iff $((x_1,x_2,\ldots,x_n),y)\in\Theta$.

Analogous array representations can be used for functions **F**: $S_1 \times S_2 \times \cdots \times S_n \rightarrow Y$.

3.3 STRING PROCESSING

We recall (from Section 2.4.5) that a string, over a finite nonempty alphabet A, is a finite sequence of members of A, with the set of all strings (including the null string ε) denoted by A^*. Furthermore, $(A^*, \cdot, \varepsilon)$, where $\cdot$ is the concatenation operation on pairs of strings over A, is a monoid. We define a *language* as a subset of A^*, i.e., any set of strings (also called *sentences* or *words*) over A is a language. A language L is 'ε-free' if it does not include the null string, i.e. if $L \subseteq A^+$. We also recall the definitions of the complex product (generalized concatenation) and alternation (union) operations on sets of strings, X and Y in 2^{A^*}:

$$X \cdot Y = \{xy \mid x \in X, y \in Y\}; \qquad X \mid Y = X \cup Y.$$

Furthermore, $X^2 = X \cdot X$, or in general $X^n = X^{n-1} \cdot X$ for $n \geq 1$, $X^0 = \{\varepsilon\}$, and $X^* = \cup_{n \geq 0} X^n$ (which we called the *Kleene closure* of X). $(2^{A^*}, |, \cdot, *, \emptyset, \{\varepsilon\})$ is the associative algebra defined in Section 2.4.5.

For arbitrary strings x and y in A^*, we say x is a *substring* of (or 'matches') y if $y = uxv$ for some u and v in A^*: x is a *prefix* (or *initial* substring) if $u = \varepsilon$; x is a *suffix* (or *terminal* substring) if $v = \varepsilon$; x is a proper substring if $u \neq \varepsilon$ or $v \neq \varepsilon$. Let $|x|$ denote the length of the string x; then $x \neq \varepsilon$ iff $|x| > 0$. The *index* of a substring x, of $y = uxv$, equals $|u| + 1$. In general, a string x may occur more than once as a substring of y (say, $y = u_1 x v_1 = u_2 x v_2$, $u_1 \neq u_2$); in this event, the substring with the smallest index is called the *leftmost match* of x in y. (It should be noted that the null string ε occurs as a substring of every string as a prefix, as a suffix, between each adjacent pair of symbols, and in fact between other null strings; we define the leftmost match of ε in y to be a single 'null-prefix', $y = \varepsilon y$, and in general disregard other null substrings.) A prefix or suffix (of y) of length one is called the *leading* or *trailing* symbol of y, respectively.

3.3.1 Rewriting Systems

A *rewriting system* (or 'associative calculus') over a finite alphabet A is a set of strings $W \subseteq A^*$ and a (usually) symmetric relation P on A^*, $P \subseteq A^* \times A^*$; we denote this algebraic system (A^*, P, W). A member of P is called a (simple) *substitution rule* (or *production*). Given $w = \alpha\beta\gamma$ in W and (β, β') in P, we say that (β, β') is *applicable* to w, that $w' = \alpha\beta'\gamma$ is the result of a *substitution* of β' for substring β in w, and that w' results from a *rewriting* of w. If W is a proper subset of A^*, this substitution is *admissible* only if w' is in W. A 'canonical' substitution is one for the leftmost match of β in w. Finally, we say that x is *adjacent* to y ($x \Leftrightarrow y$) if $x,y \in W$ and y results from a rewriting of x by an admissible substitution, and that x is *equivalent* to y ($x \Leftrightarrow^* y$) if $x,y \in W$ and there exists a sequence of zero or more admissible substitutions such that

$$x = w_0 \Leftrightarrow w_1 \Leftrightarrow \cdots \Leftrightarrow w_r = y$$

where each $w_i \in W$; we then also say that y may be obtained by 'successively rewriting' x. Note that $\Leftrightarrow^*$ is the star closure of the relation $\Leftrightarrow$ in W. If P is symmetric, then $\Leftrightarrow^*$ is an equivalence relation which partitions W into sets of equivalent strings. For each member σ in W, the equivalence class $[\sigma]$ may be regarded as a language $L_\sigma \subseteq W$ over A which is defined by the rewriting system (A^*, P, W); for each σ in W, (A^*, P, L_σ) is also a rewriting system.

Rewriting systems are introduced in Trakhtenbrot [3.23]. An interesting application is described in Bellman [3.2].

3.3.2 Languages and Grammars

An 'unrestricted' *grammar* (also called a *semi-Thue* system) is a rewriting system (A^*, P, A^*) where P is finite, but not necessarily symmetric, and where $W = A^*$ has a distinguished *initial* string $\sigma_0 \in A^*$ and a distinguished set $T \subseteq A$ of *terminals*. $A - T$ is the set of *variable* (or *nonterminal*) symbols, and the members of P are called *productions* (or 'syntactic rules'); we denote this algebraic system (A, T, P, σ_0). If $(\beta, \beta') \in P$, we say β may *produce* β' and write $\beta \rightarrow \beta'$; β' is called the 'right-hand side' of the production. If y results from the 'application' of a production to x, we write $x \Rightarrow y$. Since $\Rightarrow$ is a relation on A^*, its positive transitive closure ($\Rightarrow^+$) and star closure ($\Rightarrow^*$) are also. If $x \Rightarrow^* y$, and $x, y \in A^*$, then we say y can be *derived* (or *generated*) from x by a sequence of productions, and that

$$x = w_0 \Rightarrow w_1 \Rightarrow \cdots \Rightarrow w_r = y$$

is a *derivation* of length r (length zero if $x = y$). Since P need no longer be symmetric, languages will be associated with strings derivable from the initial string (rather than with equivalence classes).

The *language* generated by a grammar $G = (A, T, P, \sigma_0)$ is the set, denoted $\mathbf{L}(G)$, of all strings of terminals derivable from σ_0: i.e.,

$$\mathbf{L}(G) = \{ w \in A^* \mid \sigma_0 \Rightarrow^* w,\ w \in T^* \}.$$

Strings in A^* derivable from σ_0 are called *sentential forms* of G; derivable strings in T^* are *sentences*. We define four main classes or *Chomsky* types of 'restricted' grammars, each of which defines a corresponding type of language.

Type 0 A *phrase-structure grammar* (PSG) is a grammar (A,T,P,σ_0) having no productions of the form $\varepsilon \rightarrow w$ for any $w \in A^*$, and where $\sigma_0 \in A-T$ (i.e., a single variable).

Type 1 A *context-sensitive grammar* (CSG) is a PSG restricted to ('context-sensitive') productions of the form $\alpha\sigma\gamma \rightarrow \alpha w \gamma$, where $\sigma \in A-T$, $\alpha, \gamma \in A^*$, and $w \in A^+$. In other words, the nonterminal σ can be rewritten by the nonnull string w only in the 'context' of α and γ. (Type 1′ grammars, defined by the sole restriction that all productions $\beta \rightarrow \beta'$ are 'non-abbreviating', i.e., $|\beta| \leq |\beta'|$, generate the same type of language.)

Type 2 A *context-free grammar* (CFG) is a PSG restricted to ('context-free') productions of the form $\sigma \rightarrow w$, where $\sigma \in A-T$ and $w \in A^*$. In other words, only a single nonterminal may appear on the left-hand side of a production, and it may be rewritten without regard to the context in which it is found. (The grammar is 'ε-free' if the right-hand sides of productions are restricted to be non-null, so that $w \in A^+$.)

Type 3 A *right-linear grammar* (RLG) (also called a 'regular' grammar) is a CFG restricted further to ('right-linear') productions of the form $\sigma \rightarrow u$ or $\sigma \rightarrow u\sigma'$, where $u \in T^*$ and $\sigma' \in A-T$. The right-hand side of each production is restricted to have at most one nonterminal and it must appear on the right; the left-hand side is restricted to a single nonterminal. (A CFG is 'left-linear' if $\sigma \rightarrow u\sigma'$ is replaced by $\sigma \rightarrow \sigma' u$.)

Let T be a set. A Type 0 language $L_0 \subseteq T^*$ is one which a Type 0 grammar can generate, i.e., $L_0 \subseteq T^*$ is a Type 0 language iff there exists a Type 0 grammar G_0 such that $L_0 = \mathbf{L}(G_0)$. In general, a *Type i language* $L_i \subseteq T^*$ is one which a Type i grammar can generate (for $i=0$, 1, 2, and 3). It can be shown that every ε-free Type i language (for $i=1,2$, and 3) is also a Type $i-1$ language, but not conversely; i.e., the class of Type 0 languages is the most general or inclusive, while the class of Type 3 languages is the most restrictive. The ε-free qualification is because the null string is not in any Type 1 language (since in the definition of a CSG, $w \in A^+$ rather than A^*). Usually, when a language is classified as Type i, it is implied that it is not Type $i+1$.

Finiteness of P in our definition of 'unrestricted' grammars is a significant restriction on the class of languages which can be generated. Recall that languages are, in general, *infinite* sets of strings. Nonenumerable languages are those which are nonenumerable sets; recall Section 2.3.6. If $L = \{w_1, w_2, \ldots\}$ is a nonenumerable language over A, each member w_i in L can be derived by a production $\sigma_0 \rightarrow w_i$ if an infinite number of productions are permitted. If P is restricted by definition to be finite, then only enumerable languages can be generated. It can be shown that the Type 0 restrictions do not further limit the class of languages which can be generated. In other words,

Theorem Nonenumerable languages cannot be generated by a Type 0 grammar. However, each enumerable language can be generated by a Type 0 grammar.

It can also be shown that Type 1 grammars correspond to nonempty languages which are recursive sets. See Ginsburg [3.9].

3.3.3 Examples of Grammars

Let $T=\{a,b,\ldots,z,+,\times\}$ and $L_{\text{sae}}\subseteq T^*$ be the set (language) of simple arithmetic expressions (a.e.s.) consisting of strings of letters alternating with arithmetic operators; examples of members of L_{sae} include a, $b+c$, $d\times f+g$, $h+i+j$, and $a+b\times c+d$. Let $\lambda=\{a,b,\ldots,z\}$. Since each member of λ is an a.e., we introduce the productions $\sigma_0\rightarrow\lambda$, $\lambda\rightarrow a$, $\lambda\rightarrow b,\ldots$, $\lambda\rightarrow z$. (λ and σ_0 are nonterminals with σ_0 being the 'initial' one.) Since any a.e. can be extended by appending an operator and another letter, we add the productions $\sigma_0\rightarrow\sigma_0+\lambda$ and $\sigma_0\rightarrow\sigma_0\times\lambda$. Hence the grammar $G_{\text{sae}}=(A, T, P_{\text{sae}}, \sigma_0)$, where $A=T\cup\{\sigma_0, \lambda\}$ and P_{sae} is the set of productions

$$
\begin{aligned}
\{&\sigma_0\rightarrow\sigma_0\times\lambda\\
&\sigma_0\rightarrow\sigma_0+\lambda\\
&\sigma_0\rightarrow\lambda\\
&\lambda\rightarrow a\\
&\lambda\rightarrow b\\
&\ldots\\
&\lambda\rightarrow z\},
\end{aligned}
$$

generates the language of simple a.e.s; this grammar happens to be of Type 2 (context-free).

Often it is convenient to let symbols be terminals rather than nonterminals. For example, if λ is regarded as a terminal that represents any of the characters $a, b,\ldots, z$, then the grammar $G_{\text{sae}}'=(A', T', P_{\text{sae}}', \sigma_0)$, where $A'=T'\cup\{\sigma_0\}$, $T'=\{\lambda, +, \times\}$, and P_{sae}' is the set of productions

$$
\begin{aligned}
\{&\sigma_0\rightarrow\sigma_0\times\lambda\\
&\sigma_0\rightarrow\sigma_0+\lambda\\
&\sigma_0\rightarrow\lambda\},
\end{aligned}
$$

is much simpler; it is in fact a left-linear grammar. An equivalent right-linear grammar G_{sae}'', i.e., a right-linear grammar that generates the same language L_{sae}, has the set P_{sae}'' of productions

$$
\begin{aligned}
\{&\sigma_0\rightarrow\lambda\times\sigma_0\\
&\sigma_0\rightarrow\lambda+\sigma_0\\
&\sigma_0\rightarrow\lambda\}.
\end{aligned}
$$

Now suppose that we wish to generate the language of parenthesized a.e.s, L_{pae}, where any parenthesized a.e. $[\sigma_0]$ can take the place of a single letter $[\lambda]$. To the grammar G_{sae}, we may simply add the production $\lambda \rightarrow (\sigma_0)$, as well as add the parentheses to the set of terminals T; G_{sae} so extended will be denoted G_{pae}. To the left- and right-linear grammars, where λ is a terminal, we may add the production $\sigma_0 \rightarrow (\sigma_0)$; the resulting grammars are no longer left or right linear, and in fact no such grammars exist for the more general language L_{pae}. L_{sae} is a Type 3 language while L_{pae} is Type 2. Any language requiring balanced parentheses (or other paired symbols) cannot be Type 3, as should become evident later; e.g., the language $a^* bc^* = \{a^i bc^j \mid i \geq 0, j \geq 0\}$ is Type 3 (see below), but $\{a^n bc^n \mid n \geq 0\}$ is Type 2. However, the language $\{a^n b^n c^n \mid n \geq 1\}$ is Type 1, and can be generated by the Type 1 grammar $G = (A, T, P, \sigma_0)$, where

$$A = \{\sigma_0, \beta, \gamma, x, y\} \cup T, \ T = \{a, b, c\},$$

and P is the set of productions

$$\begin{aligned}
\{&\sigma_0 \rightarrow a\sigma_0\beta\gamma \\
&\sigma_0 \rightarrow ab\gamma \\
&\gamma\beta \rightarrow \gamma y \\
&\gamma y \rightarrow xy \\
&xy \rightarrow x\gamma \\
&x\gamma \rightarrow \beta\gamma \\
&b\beta \rightarrow bb \\
&b\gamma \rightarrow bc \\
&c\gamma \rightarrow cc\}.
\end{aligned}$$

or by the Type 1′ grammar $G' = (A, T, P', \sigma_0)$, where

$$A = \{\sigma_0, \beta, \gamma\} \cup T, \ T = \{a, b, c\},$$

and P' is the set of productions

$$\begin{aligned}
\{&\sigma_0 \rightarrow a\sigma_0\beta\gamma \\
&\sigma_0 \rightarrow ab\gamma \\
&\gamma\beta \rightarrow \beta\gamma \\
&b\beta \rightarrow bb \\
&b\gamma \rightarrow bc \\
&c\gamma \rightarrow cc\}.
\end{aligned}$$

(Note that the non-abbreviating production $\gamma\beta \rightarrow \beta\gamma$ is not context-sensitive.) Since Type 1 languages cannot include the null string, the language $\{a^n b^n c^n \mid n \geq 0\}$ is evidently Type 0, which can be generated by the foregoing grammar augmented by the production $\sigma_0 \rightarrow \varepsilon$.

3.3.4 Regular Sets

A *regular set* (over A) is defined inductively as follows:

1. $\emptyset$ is a regular set.
2. $\{\varepsilon\}$ is a regular set.
3. $\{a\}$ is a regular set for each a in A.
4. If X and Y are regular sets, then so are $X|Y$, $X \cdot Y$, and X^* (and Y^*).

The last defining condition essentially states that regular sets can be obtained by taking the alternation (union), complex product (dot), or Kleene closure of other regular sets. For example, let $A=\{a, b, c\}$. Then $\emptyset$, $\{\varepsilon\}$, $\{a\}$, $\{b\}$, $\{c\}$ are regular sets. Unions of these 'primitives' yield $\{a,b\}$, $\{b,c\}$, $\{a,c\}$, and $\{a, b, c\}$, as well as each of the foregoing together with ε. Single concatenations of the primitives yield $\{ab\}$, $\{bc\}$, $\{ac\}$, $\{ba\}$, $\{cb\}$, and $\{ca\}$. Closures of the primitives include $a^* = \{\varepsilon, a, aa, aaa, \ldots\}$, b^*, and c^*. An example of a regular set constructed from a combination of alternation, composition, and closure operations is

$$[((A \cdot B)|C)^* \cdot A] | [(C \cdot A)^*]$$

where $A=\{a\}$, $B=\{b\}$ and $C=\{c\}$; this set may be expressed in the simpler form

$$\{ab, c\}^* a | (ca)^*$$

where it is assumed that $*$ has higher precedence than the implicit $\cdot$, and $\cdot$ has higher precedence than $|$. Such algebraic expressions are said to be 'regular', and every regular expression denotes a regular set.

It can be proven that every regular set is a Type 3 language. Hence, every language which can be written in the form of a regular expression—i.e., as a string of symbols connected by the operators $|$, $\cdot$, and $*$—can be generated by a right-linear grammar. For example, we have stated above that $a^* bc^*$ is a Type 3 language. It can be generated by the right-linear grammar defined by the set of productions

$$\{\sigma_0 \rightarrow a\sigma_0, \quad \sigma_0 \rightarrow b\sigma_1, \quad \sigma_1 \rightarrow c\sigma_1, \quad \sigma_1 \rightarrow \varepsilon\}.$$

3.3.5 Translations and Programmed Rewritings

A *translation* is a function τ from one language L_1 to another L_2: i.e., $L_1 \subseteq A_1^*$, $L_2 \subseteq A_2^*$, and $\tau: L_1 \rightarrow L_2$. A translation is *syntactic* if $\tau: \mathbf{L}(G_1) \rightarrow \mathbf{L}(G_2)$ for grammars G_1 and G_2. A translation is *length-preserving* if $\tau(w_1)=w_2$ implies $|w_1|=|w_2|$. A translation is *prefix-inclusive* if $\tau(x)$ is a prefix of $\tau(xw)$. A translation is *homomorphic* if $\tau(x_1x_2 \ldots x_r) = \tau(x_1)\tau(x_2)\ldots\tau(x_r)$; *isomorphic* if also a bijection. In general, we say that a translation *preserves* property $\mathcal{P}$ of L_1 if L_2 also has property $\mathcal{P}$.

A 'translator' is a system which implements a translation τ in the sense that given an 'input' string w_1, the translator produces an 'output' string w_2. Rewriting systems may serve as translators by adding determinism.

We shall say that a *programmed grammar,* with *input* string σ_0, is a grammar $PG=(A,T,P,\sigma_0)$ together with a set of deterministic generation rules which dictate the exact order in which applicable productions may be used in rewriting σ_0 successively; i.e., the derivation

$$\sigma_0 \Rightarrow \sigma_1 \Rightarrow \sigma_2 \Rightarrow \ldots \Rightarrow \sigma_k \Rightarrow \cdots$$

is uniquely determined by the generation rules. It is possible, in general, for this derivation sequence to be infinite (nonterminating) in that a production is applicable to every sentential form σ_k. However, if the derivation sequence terminates, i.e., if no production is applicable to some σ_k, we say that the programmed grammar PG translates input σ_0 into output σ_k; we then write $\tau_{PG}(\sigma_0)=\sigma_k$. We note in passing that if the output string consists only of terminals (i.e., $\sigma_k \in T^*$), then the programmed grammar PG with input $\sigma_0 \in A^*$ generates a sentence in the language $\mathbf{L}(PG)$.

Consider next the same grammar with the same deterministic generation rules, except that the initial string σ_0 is now regarded as a changeable input. $\tau_{PG}(\sigma_0)$ is considered defined only if the associated derivation sequence terminates; we may also wish to require that output strings consist only of terminals, in which case we define $\tau_{PG}'(\sigma_0)$ only if $\sigma_k \in T^*$. The programmed grammar PG thus defines partial functions τ_{PG} and τ_{PG}' from A^* to A^* , whose domains and ranges are both languages associated with PG. In other words, the total functions τ_{PG}: $\mathcal{D}_\tau \rightarrow \mathcal{R}_\tau$ and τ_{PG}': $\mathcal{D}_\tau' \rightarrow \mathcal{R}_\tau'$ are translations, where $\mathcal{D}_\tau \subseteq A^*$, $\mathcal{R}_\tau \subseteq A^*$, $\mathcal{D}_\tau' \subseteq A^*$ and $\mathcal{R}_\tau' \subseteq T^*$. (Note that these domains and ranges should not be confused with $\mathbf{L}(PG)$ for a specific σ_0.)

3.3.6 Recognizers

A grammar may be regarded as a finite set of (syntactic) rules[†] which 'defines' a language L, permits us to generate the sentences in a language, and permits us to tell (in theory at least) whether or not an arbitrary string w is in L. We need the parenthesized qualification since, in general, it may not be possible to tell using any realistic (finite) procedure. In fact, there are languages for which no defining grammar exists. We shall call a system by which we can tell, in a finite amount of time, whether an arbitrary string is in a language L, a *recognizer* for L. (We discuss subsequently two kinds of recognizers: Markov algorithms, and automata.)

[†]Languages also have 'semantic' rules, which we disregard here. For example, x/y is usually considered a syntactically correct arithmetic expression, but is semantically incorrect when $y=0$.

Translators in general, and programmed grammars in particular, may serve as recognizers. Let A^* be the set of all possible input strings. A programmed grammar PG is a recognizer of a language $L \subseteq A^*$ if $L = \mathscr{D}_\tau$, the domain of the translation function τ_{PG} (or if $L = \mathscr{D}_\tau'$, the domain of τ_{PG}'); i.e., $w \in \mathscr{D}_\tau$ iff the PG produces output. Note that a recognizer for $L = \mathscr{D}_\tau$ cannot serve as a recognizer for $A^* - L$, since $\tau_{PG}(w)$ for $w \notin L$ is undefined (i.e., the PG would never produce output in a finite amount of time).

3.3.7 Markov Algorithms

A *Markov algorithm* (MA) over an alphabet A may be regarded as a programmed rewriting system (or special case of programmed grammars): (a) whose set of productions P is finite and ordered, and which has a distinguished subset of *terminal* productions; (b) where an initial string $\sigma_0 \in A^*$ is regarded as its 'input' which is successively rewritten, generating the derivation sequence

$$\sigma_0 \Rightarrow \sigma_1 \Rightarrow \sigma_2 \Rightarrow \cdots \Rightarrow \sigma_k \cdots;$$

and (c) where (following a deterministic generation rule) the first applicable production, with respect to the ordering of P, must always be used in any rewriting, and then canonically (i.e., on the leftmost matching substring). The derivation sequence terminates and the Markov algorithm produces as *output* σ_k either (a) if a terminal production is applied in rewriting σ_{k-1} to obtain σ_k, or (b) if no production is applicable to σ_k; we also then say that the Markov algorithm (a) *halts* or (b) *blocks,* respectively. If the derivation sequence does not terminate, then the Markov algorithm is said to *loop*.

A Markov algorithm may define several partial functions from A^* to A^*. For example, we may let $\tau_1(\sigma_0) = \sigma_k$ be defined only if the Markov algorithm halts with output σ_k, and let $\tau_2(\sigma_0) = \sigma_k$ be defined only if the Markov algorithm halts or blocks with output σ_k. In either case, we may also require that σ_k consists only of terminals (i.e., $\sigma_k \in T^*$, for some $T \subseteq A$). Similarly, σ_0 may also be restricted to be a string over the same or a different subalphabet T. As with programmed grammars, Markov algorithms may serve as translators or as recognizers. It is also common to say that a Markov algorithm *accepts* σ_0 if it halts, *rejects* σ_0 if it blocks, and does not recognize σ_0 if it loops. (Alternatively, we may define a proper subset W of A^* and say that a Markov algorithm *accepts* σ_0 if it halts and produces as output a string σ_k in W, *rejects* σ_0 if $\sigma_k \notin W$, and does not recognize σ_0 if the Markov algorithm loops or blocks.)

We observe that blocking may or may not be a sign of recognition, depending upon our choice of definitions. It is possible to avoid this equivocation by restricting ourselves to *closed* Markov algorithms, which we define as those which can never block; those having a dummy terminal production, which rewrites ε and is last in the ordering of P, are examples. In theory, every Markov algorithm can be transformed to a closed one which is equivalent in that they both define the same translation function, or they both recognize or accept the same language. (This transformation is not always easy in practice.) Constructing a Markov algorithm which 'computes' a given translation is in general a nontrivial *programming* task. For further discussion, see Galler [3.8] and Brainerd [3.5].

Notation A Markov algorithm is commonly written as an ordered list of productions, each on a separate line, where the top-down listing ordering corresponds to the ordering of the set of productions P. A production (β,β') in P will be written $\beta\to\beta'$, and if a terminal production, a 'dot' will be added immediately after the arrow, $\beta\to\cdot\beta'$; optionally, we may precede each production by a label, usually its ordinate in the ordering of P. The alphabet over which the Markov algorithm is defined is implicitly defined as the set of symbols appearing in the productions.

Example*: *String Reversal As an illustration, we define a Markov algorithm which 'reverses' strings over a terminal alphabet T. If $x=x_1x_2...x_r$ is in T^*, the *reverse* of x is string $x_rx_{r-1}...x_2x_1$. Let $\{\#,\#_L,\#_R\}$ be a set of 'auxiliary' (nonterminal) symbols, and for simplicity, let $T=\{a,b\}$. The Markov algorithm is then given by the productions

(1) $\#aa\to a\#a$
(2) $\#ab\to b\#a$
(3) $\#ba\to a\#b$
(4) $\#bb\to b\#b$
(5) $\#a\#_R\to\#_Ra$
(6) $\#b\#_R\to\#_Rb$
(7) $\#a\to\#_Ra$
(8) $\#b\to\#_Rb$
(9) $\#_L\#_R\to\cdot\varepsilon$
(10) $\#_L\to\#_L\#$
(11) $\varepsilon\to\#_L$

We trace the operation of the Markov algorithm with input $\sigma_0=aab$, by showing the derivation sequence along with each applicable production number:

$\sigma_0=aab$
$\Rightarrow\#_Laab$ (11)
$\Rightarrow\#_L\#aab$ (10)
$\Rightarrow\#_La\#ab$ (1)
$\Rightarrow\#_Lab\#a$ (2)
$\Rightarrow\#_Lab\#_Ra$ (7)
$\Rightarrow\#_L\#ab\#_Ra$ (10)
$\Rightarrow\#_Lb\#a\#_Ra$ (2)

$\Rightarrow \#_L b \#_R aa$ (5)
$\Rightarrow \#_L \# b \#_R aa$ (10)
$\Rightarrow \#_L \#_R baa$ (6)
$\Rightarrow baa$ (9)
$= \sigma_{11}$

The derivation sequence terminates since production (9) is terminal.

Observe that this Markov algorithm is closed; if none of the preceding productions are applicable (as is the case initially, when no auxiliary symbol appears), then the last production must be applied to the null prefix, introducing thereby the auxiliary symbol $\#_L$. Note that $\#_L$ and $\#_R$ serve as left and right end markers, respectively, and that # is introduced by production (10) to move, one at a time, the leftmost terminal symbol to the right. Productions (1)–(4) do the moving; upon reaching the right end, productions (5)–(8) delete # and set (i.e., adjust or introduce) $\#_R$. When all the terminal symbols appear to the right of $\#_R$, then the end markers are deleted by the terminal production (9).

This example is typical of many Markov algorithm programs, in that auxiliary symbols must be introduced to keep track of states, and great care must be exercised to properly order the productions. (The reader should study the effect of permuting the productions in various ways.)

Example: Greatest Common Divisor While Markov algorithms are primarily 'string translators', they may compute numerical functions provided that numbers are suitably encoded. Let the natural number n be represented by a monadic string of length n over the alphabet $\{1\}$; e.g., $5='11111'$. We represent an ordered pair of numbers, (n,m), by the string $'n\#m'$, where '#' is an input delimiting symbol that serves to separate n from m; e.g., $(10,4)='1111111111\#1111'$. A Markov algorithm which computes or outputs the greatest common divisor of the pair of positive integers (n,m), which is input as $'n\#m'$, is given by the productions:

(1) $1a \rightarrow a1$
(2) $1\#1 \rightarrow a\#$
(3) $1\# \rightarrow \#b$
(4) $b \rightarrow 1$
(5) $a \rightarrow c$
(6) $c \rightarrow 1$
(7) $\# \rightarrow \cdot\varepsilon$

Here, $\{a,b,c\}$ is the auxiliary alphabet. We trace the operation of this Markov algorithm with input $\sigma_0='1111111111\#1111'$:

$\sigma_0 = 1111111111\#1111$
$\Rightarrow 111111111a\#111$ (2)
$\Rightarrow^* a111111111\#111$ (1) nine times
$\Rightarrow a11111111a\#11$ (2)
$\Rightarrow^* aa11111111\#11$ (1) eight times
$\Rightarrow aa1111111a\#1$ (2)

$\Rightarrow^* aaa1111111\#1$	(1) seven times
$\Rightarrow aaa111111a\#$	(2)
$\Rightarrow^* aaaa111111\#$	(1) six times
$\Rightarrow^* aaaa\#bbbbbb$	(3) six times
$\Rightarrow^* aaaa\#111111$	(4) six times
$\Rightarrow^* cccc\#111111$	(5) four times
$\Rightarrow^* 1111\#111111$	(6) four times
$\Rightarrow^* aaaa\#11$	(2,1) repeatedly
$\Rightarrow^* cccc\#11$	(5) four times
$\Rightarrow^* 1111\#11$	(4) four times
$\Rightarrow^* aa11\#$	(2,1) repeatedly
$\Rightarrow^* aa\#bb$	(3) two times
$\Rightarrow^* aa\#11$	(4) two times
$\Rightarrow^* cc\#11$	(5) two times
$\Rightarrow^* 11\#11$	(6) two times
$\Rightarrow^* aa\#$	(2,1) repeatedly
$\Rightarrow^* cc\#$	(5) two times
$\Rightarrow^* 11\#$	(6) two times
$\Rightarrow^* \#bb$	(3) two times
$\Rightarrow^* \#11$	(4) two times
$\Rightarrow 11$	(7)
$=\sigma_{99}$	

This Markov algorithm is based upon the Euclidean algorithm (cf. Section 1.1.7): production (2), when applied repeatedly, computes remainders by subtracting the smaller of m and n from both. The pairs ('min-value', 'remainder') of the Euclidean algorithm appear following applications of production (6).

Patterns Markov algorithms can be simplified by permitting a single production to represent a set of permissible substitutions; in other words, a production may be of the form $B \to B'$, where B and B' are sets of strings (rather than single strings). Such productions may be regarded as either a notational shorthand, or as productions of a generalized Markov algorithm. In either case, the string-reversal Markov algorithm can then be rewritten as follows:

(1) $\#T_1T_2 \to T_2\#T_1$
(2) $\#T\#_R \to \#_RT$
(3) $\#T \to \#_RT$
(4) $\#_L\#_R \to \cdot\varepsilon$
(5) $\#_L \to \#_L\#$
(6) $\varepsilon \to \#_L$

where $T_1=T_2=T=\{a,b\}$. We interpret production (1) as follows: if a member of the complex product $\#T_1T_2$ (say, $\#ab$) matches a substring of σ, then the production is applicable and the substring is to be replaced by the (unique) corresponding member of $T_2\#T_1$ (i.e., $b\#a$). Note that productions

(2) and (3) may be combined to yield $\#TS \rightarrow \#_R T$, where S is the ordered set $\{\#_R, \varepsilon\}$; it is convenient to write instead $\#T(\#_R|\varepsilon) \rightarrow \#_R T$, so that new sets (like S) need not be identified.

A string expression consisting of the concatenation and alternation of symbols (in an alphabet) and names (of subalphabets), and appearing on the left-hand side of a production, is called a *pattern*; the right-hand side of the production is called its *object*. A generalized production is applicable to a string σ_i if the pattern 'matches' a substring x of σ_i $(= u\,x\,v)$, in which case x is replaced by the corresponding object string x' (thus deriving $\sigma_{i+1} = u\,x'\,v$).

3.3.8 Application: SNOBOL

The SNOBOL programming language provides a convenient means of implementing Markov algorithms. While most commonly available versions of the language have more complex facilities (for arithmetic, recursion, I/O, pattern definition, etc.), we shall be concerned here only with those features which correspond to Markov algorithms. For details and many programming examples, see Griswold [3.10].

A SNOBOL *statement* has the general form

⟨label⟩ ⟨subject⟩ ⟨pattern⟩=⟨object⟩: S(ℓ_1)F(ℓ_2)

where

⟨label⟩	is the statement label (cf. production number)
⟨subject⟩	specifies the name of the string to be rewritten
⟨pattern⟩	specifies the set of strings to be matched
⟨object⟩	specifies the string to be substituted for a substring of the subject matching the pattern
⟨ℓ_1⟩	is the label of the production to be tested next if this production is applicable (i.e., if there is a Successful match)
⟨ℓ_2⟩	is the label of the production to be tested next if this production is not applicable (i.e., if a match Fails to be found)

A SNOBOL *program* consists of an ordered set of statements; the last statement of the program must have the special label 'END'. The generalized string-reversal Markov algorithm may be written in SNOBOL form, as follows:

```
P1  SIGMA (# T1 T2)=(T2 # T1)  : S(P1)F(P2)
P2  SIGMA (# T (#R|))=(#R T)   : S(P1)F(P3)
P3  SIGMA (#L #R)=             : S(END)F(P4)
P4  SIGMA #L=(#L #)            : S(P1)F(P5)
P5  SIGMA ()=#L                : S(P1)F(ERROR)
END
```

We have omitted definitions for SIGMA, T1, T2, and ERROR, and have assumed that #, #L, and #R are treated as special characters.

Observe that the ⟨subject⟩ part of SNOBOL statements permits us to rewrite several different strings within the same program, and the $\mathsf{S}(\ell_1)$ and $\mathsf{F}(\ell_2)$ clauses permit us greater flexibility in ordering productions. (Markov algorithms incorporating an $\mathsf{S}(\ell_1)$ clause are said to be 'labeled'.) These generalizations are for convenience, but do not increase the class of problems whose solutions are computable.

3.3.9 Application: Parsing

An important string processing application is 'parsing', which, informally, is the task of determining whether (and how) a sentence is grammatically correct. For example, the Markov algorithm:

$$
\begin{array}{rrcl}
(1) & \lambda\ \Delta & \rightarrow & \mathsf{F}\ \Delta \\
(2) & \mathsf{T{\times}F}\ \Delta & \rightarrow & \mathsf{T}\ \Delta \\
(3) & \mathsf{F}\ \Delta & \rightarrow & \mathsf{T}\ \Delta \\
(4) & \mathsf{T}\ \Delta\times & \rightarrow & \mathsf{T}\times\Delta \\
(5) & \mathsf{E{+}T}\ \Delta & \rightarrow & \mathsf{E}\ \Delta \\
(6) & \mathsf{T}\ \Delta & \rightarrow & \mathsf{E}\ \Delta \\
(7) & \mathsf{(E)}\ \Delta & \rightarrow & \mathsf{F}\ \Delta \\
(8) & \perp\mathsf{E}\perp\Delta & \rightarrow\cdot & G_{\text{pae}} \\
(9) & \Delta\alpha & \rightarrow & \alpha\ \Delta \\
(10) & \varepsilon & \rightarrow & \Delta
\end{array}
$$

is a parser; it reduces an input string σ_0 to the symbol G_{pae} if σ_0 is a legitimate parenthesized arithmetic expression (according to grammar G_{pae} of Section 3.3.3), otherwise it loops. Production (10) introduces a marker Δ, and production (9) moves it past any terminal symbol α when applicable. To illustrate the operation of this Markov algorithm, the following trace is exhibited:

$$
\begin{array}{ll}
\sigma_0 = \perp\lambda+(\lambda)\times\lambda\perp & \\
\Rightarrow\Delta\perp\lambda+(\lambda)\times\lambda\perp & (10) \\
\Rightarrow\perp\Delta\lambda+(\lambda)\times\lambda\perp & (9) \\
\Rightarrow\perp\lambda\Delta+(\lambda)\times\lambda\perp & (9) \\
\Rightarrow\perp\mathsf{F}\Delta+(\lambda)\times\lambda\perp & (1) \\
\Rightarrow\perp\mathsf{T}\Delta+(\lambda)\times\lambda\perp & (3) \\
\Rightarrow\perp\mathsf{E}\Delta+(\lambda)\times\lambda\perp & (6) \\
\Rightarrow\perp\mathsf{E}+\Delta(\lambda)\times\lambda\perp & (9) \\
\Rightarrow\perp\mathsf{E}+(\Delta\lambda)\times\lambda\perp & (9) \\
\Rightarrow\perp\mathsf{E}+(\lambda\Delta)\times\lambda\perp & (9)
\end{array}
$$

$$
\begin{array}{ll}
\Rightarrow \perp \mathsf{E}+(\mathsf{F}\Delta)\times\lambda\perp & (1)\\
\Rightarrow \perp \mathsf{E}+(\mathsf{T}\Delta)\times\lambda\perp & (3)\\
\Rightarrow \perp \mathsf{E}+(\mathsf{E}\Delta)\times\lambda\perp & (6)\\
\Rightarrow \perp \mathsf{E}+(\mathsf{E})\Delta\times\lambda\perp & (9)\\
\Rightarrow \perp \mathsf{E}+\mathsf{F}\Delta\times\lambda\perp & (7)\\
\Rightarrow \perp \mathsf{E}+\mathsf{T}\Delta\times\lambda\perp & (3)\\
\Rightarrow \perp \mathsf{E}+\mathsf{T}\times\Delta\lambda\perp & (4)\\
\Rightarrow \perp \mathsf{E}+\mathsf{T}\times\lambda\Delta\perp & (9)\\
\Rightarrow \perp \mathsf{E}+\mathsf{T}\times\mathsf{F}\Delta\perp & (1)\\
\Rightarrow \perp \mathsf{E}+\mathsf{T}\Delta\perp & (2)\\
\Rightarrow \perp \mathsf{E}\Delta\perp & (5)\\
\Rightarrow \perp \mathsf{E}\perp\Delta & (9)\\
\Rightarrow G_{\text{pae}} & (8)\\
=\sigma_{22} &
\end{array}
$$

This example is adapted from Floyd [3.7].

EXERCISES

1. Show that the Hamming distance is a metric on B^n.

2. Determine

(a) $\mathbf{lim}_{i\to\infty}\, i^{\alpha}$ for $\alpha>0$
(b) $\mathbf{lim}_{i\to\infty}\, \beta^{i}$ for $|\beta|\leq 1$
(c) $\mathbf{lim}_{i\to\infty}\, b^{i}$ for a negative integer b
(d) $\mathbf{lim}_{i\to\infty}\, \alpha^{1/i}$ for $\alpha>0$

***3.** Show that the sequence $\langle a_1, a_2, \ldots\rangle$ converges iff $\mathbf{lim}_{m,n\to\infty}(a_m-a_n)=0$.

4. Show that

(a) $e^n=\mathbf{O}(e^{n+1})$
(b) $e^n=\mathbf{o}(10^{n-1})$

5. Show that $\mathbf{lim}_{n\to\infty}\ |\mathbf{f}(n)/\mathbf{g}(n)|=M\neq 0$ is not a necessary condition for $\mathbf{f}(n)=\mathbf{O}(\mathbf{g}(n))$.

6. For

$$A=\begin{bmatrix}4&2&3\\0&4&2\\0&0&5\end{bmatrix},\qquad B=\begin{bmatrix}1&0&0\\2&4&0\\3&5&6\end{bmatrix},$$

determine $A\oplus B$, $A\ominus B$, and A^{-1}.

7. For the Boolean matrix

$$M=\begin{bmatrix}0&1&0\\0&0&1\\1&0&0\end{bmatrix},$$

determine the logical powers M^2 and M^3.

8. For the matrices of strings

$$A=\begin{bmatrix}a&b&c\\d&e&f\\g&h&i\end{bmatrix},\quad B=\begin{bmatrix}a&d&g\\b&e&h\\c&f&i\end{bmatrix},$$

determine the symbolic matrix product $A\oplus B$.

9. For $U=\{1,2,\ldots,10\}$, give matrix representations for the following:

(a) The set of primes in U.
(b) The relation $\{(i,j)\in U\times U \mid i \text{ divides } j\}$.

10. For

$$Q=\begin{bmatrix}0&1/3&1/3\\0&0&1/2\\0&0&0\end{bmatrix},$$

determine $(I-Q)^{-1}$.

***11.** Show that any permutation of a matrix A equals A pre- and/or post-multiplied by a sequence of elementary matrices.

12. Give the displacement formula for a 'column-order' zero-indexed matrix, whose data items occupy L cells each.

13. For

$$P=\begin{bmatrix}0&1/2&1/2\\0&0&1\\0&0&0\end{bmatrix},$$

compute the matrix powers P^k for $k=1,2,\ldots,12$.

14. For the same data as shown in Figure 3.2.7(b),

(a) define a hash function for which there will be no collisions;
(b) show the table resulting from the use of the hash function

$$\mathbf{h}(k)=\mathbf{ordinate}(k)\,\mathbf{mod}\,13+100.$$

15. Give a Type 3 grammar which generates the regular set $\{ab,c\}^*\ a\,|\,(ca)^*$.

16. Give a Type 2 grammar which generates the context-free language $\{a^nbc^n \mid n\geq 0\}$.

17. Give a Type 1 grammar which generates the language $\{ww \mid w\in T^+\}$, where $T=\{0,1\}$.

18. Give a Type 3 grammar which generates the set of legal variable names in some dialect of FORTRAN and BASIC.

19. Show that the Type 1′ grammar $G''=(A,T,P,\sigma_0)$, where $A=\{\sigma_0,\gamma\}\cup T$, $T=\{a,b,c\}$, and P is the set of productions $\{\sigma_0\rightarrow abc,\ ab\rightarrow aabb\gamma,\ \gamma c\rightarrow cc,\ \gamma b\rightarrow b\gamma\}$, generates the language $\{a^nb^nc^n\,|\,n\geq 1\}$.

20. Using the Type 1 grammar given in Section 3.3.3, show the derivation of the string $a^4b^4c^4$.

21. Extend the Type 2 arithmetic expression grammar G_{pae} to include (a) exponentiation ($\uparrow$), which should have greater precedence than $\times$, and should be 'right-associative'; (b) unary minus ($\ominus$), which should have greater precedence than $\uparrow$; (c) subscripted variables.

22. Give a Type 1 grammar which generates the language

$$\{awbwc\,|\,w\in T^+\}$$

for any $a,b,c\in T^*$, $T\subseteq A$. (*Note*: this is a generalization of Exercise 17. An important implication is that programming languages which require declaration of variables before their use cannot be context-free.)

23. Trace the operation of the 'parsing' Markov algorithm for the input string $\perp\lambda+\lambda\times\lambda+\lambda\perp$. What happens for input strings not in $\mathbf{L}(G_{\text{pae}})$?

24. Consider the following Markov algorithm:

$$0C\rightarrow\cdot 1$$
$$1C\rightarrow C0$$
$$\#0\rightarrow 0\#$$
$$\#1\rightarrow 1\#$$
$$\#\rightarrow C$$
$$\varepsilon\rightarrow\#$$

(a) Trace its operation for 000, 001, 010, 011,....
(b) What does it compute, in general?

25. Construct a Markov algorithm that:

(a) adds a 1 to a (possibly empty) string of 1's,
(b) tests whether or not two strings of 1's, separated by an '=', are equal in length; if so, output '=', otherwise, output ε.

26. Construct a Markov algorithm which recognizes strings in the language $\{(^n\lambda)^n\,|\,n\geq 1\}$.

PROGRAMMING ASSIGNMENTS

1. Write a program which computes e (=2.718...) to 8 decimal places.
2. Write a program which solves a system of N linear equations in N unknowns.
3. Write a subroutine that inverts a matrix; a flag should be set in case a matrix has no inverse. (Warning: Matrix inversion is a process that is especially susceptible to round-off errors. For the purpose of this assignment, you need not

worry about obtaining great accuracy, but then you shouldn't use this subroutine in practice. See numerical analysis textbooks for ways to increase accuracy if necessary.)

4. Program a set of subroutines to enable symbolic matrix calculations.

5. (a) Implement a generalized matrix multiplication subroutine, which is called with matrix names (identifying their base addresses) and their 'dimensions' (i.e., the extents of their associated cells of subscripts). For example, to multiply (3×4) and (4×5) matrices A and B, we would use CALL MATMUL(A,B,3,4,5,C), where C is the (3×5) result. The subroutine should represent all matrices as vectors.

(b) Implement generalized matrix addition, subtraction, and inversion subroutines in a like fashion.

6. Write a program which computes the matrix power P^k, for any matrix P and nonnegative integer k.

7. Write a program which generates sentential forms of a right-linear grammar. You should, of course, set some length limitation. (*Hint*: represent the grammar as a linked list structure.)

8. Implement the 'parsing' program of Section 3.3.9 in SNOBOL.

9. Write an interpreter in a language other than SNOBOL for a Markov algorithm language (i.e., for a SNOBOL subset) sufficient to execute the Markov algorithm programs of Section 3.3.7.

4 DIGRAPHS AND TREES

4.1 DIGRAPHS—BASIC DEFINITIONS

An algebraic system $G=(S,R)$ consisting of a nonempty set S and a relation R in S ($R \subseteq S \times S$) is given a special name: *directed graph*, or *digraph*, for short. The members of S are called *nodes* (or *vertices*), and the members of R are called *branches* (or *arcs*). Digraph G is said to be *finite* if S is; *trivial* if $\#(S)=1$. G is *labeled* if S and/or R is are.

Nontrivial finite digraphs have a convenient pictorial representation: for example,

$$G_1=(S,R)=(\{a,b,c,d\},\{(a,a),(a,b),(b,a),(b,c),(d,d)\})$$

is depicted in Figure 4.1. Observe that each node $s \in S$ is represented by a point (or small circle, labeled s), and a branch $r=(a,b) \in R$ is represented by a directed line or arrow from a to b, with the direction indicated by an arrowhead at node b). A digraph $G=(S,R)$ is in fact the 'graph of the relation R in S' introduced in Section 2.2.2; we call R the 'incidence' relation of G.

Branch $r=(a,b)$ is said to be *incident from* node *a to* node b, or incident *with a* and b; also, a and b are said to be *adjacent* nodes, with a called the *initial* node (denoted **init**(r)) and b the *terminal* node (denoted **term**(r)) of r, respectively. If a and b coincide, we say the branch is *reflexive* (or is a 'reflexive loop'). Two branches incident with a common node are said to be *adjoining*; also *collinear* if one branch is incident from and the other branch is incident to the common node, or *opposing* otherwise (i.e., if the common node is the initial or terminal node of both branches).

The fact that lines, rather than planes, connect node points is sometimes emphasized by calling the digraph *linear*; these lines need not be straight. If the directions of the branches are ignored (i.e., if the arrowheads are omitted, as in graph G_2 of Figure 4.1), we have an *undirected graph*. (Duplication of lines between any pair of nodes is also eliminated.†) We call the undirected graph so obtained the *underlying graph* associated with the digraph. For example, graph G_2 is the underlying graph associated with digraph G_1.

A digraph $G=(S,R)$ is reflexive, symmetric, transitive, asymmetric, antisymmetric, and irreflexive, respectively, if the relation R is. (Digraph G_1 is neither reflexive, symmetric, transitive, asymmetric, antisymmetric, nor irreflexive.) We also say that a node a is reflexive if there is a reflexive branch (a,a), and we say a pair of branches (a,b) and (b,a) is symmetric (or is a 'symmetric loop').‡ A digraph is reflexive if each node is reflexive, irreflexive if no node is reflexive, symmetric if each branch is one of a symmetric pair, antisymmetric if there are no symmetric pairs of distinct branches (but reflexive branches are allowed), and asymmetric if there are no symmetric pairs or reflexive branches.

It is often convenient to associate with an undirected graph a 'symmetric-equivalent' digraph obtained by replacing each line in the graph by a symmetric pair of branches in the digraph. Thus, the (undirected) underlying graph associated with an unsymmetric digraph $G=(S,R)$ has as its symmetric-equivalent, the digraph $G'=(S,R \cup R')$, where R' is the inverse of the relation R; G' is also called the *symmetrized* digraph of G. (G_3 of Figure 4.1 is the symmetrized digraph of G_1.)

A *subgraph* of $G=(S,R)$ is defined as a digraph $G'=(S',R')$, where $S'\subseteq S$, $R'\subseteq R$, and $R'\subseteq S'\times S'$; we then write $G'\subseteq G$. If G is a subgraph of G'', we say G'' is a *supergraph* of G, and write $G''\supseteq G$. G' is a 'proper' subgraph if $G'\neq G$, and is a 'trivial' subgraph if $\#(S')=1$. A *node-subgraph* of (S,R) is a digraph (S',R'), where $S'\subseteq S$ and $R'=\{r\in R \mid r\in S'\times S'\}$; we also say that (S',R') is the subgraph of (S,R) 'induced' by the subset of nodes S'. A *branch-subgraph* of (S,R) is a digraph (S',R') where $R'\subseteq R$ and $S'=\{a\in S \mid (a,b)$ or (b,a) is in R' for some b in $S\}$; we also say that (S',R') is the subgraph of (S,R) 'induced' by the subset of branches R'. A *spanning* subgraph of $G=(S,R)$ is a subgraph $G'=(S,R')$, where $R'\subseteq R$.

Example In Figure 4.1, G_4 is a node-subgraph and G_5 is a branch-subgraph of G_1, both proper. G_6 is a proper subgraph of G_1, which is neither a node-subgraph nor a branch-subgraph, but is a spanning subgraph of G_1.

†In a generalization of a digraph, called a *multigraph* (or a 'k-graph'), more than one (but at most k) branch from a node to another is allowed. R then is a multiset, hence it cannot strictly be a relation. Some writers use the term 'linear' to mean $k=1$.

‡Sometimes a symmetric pair will be depicted by a single double-headed branch (one arrowhead at each end) to avoid clutter.

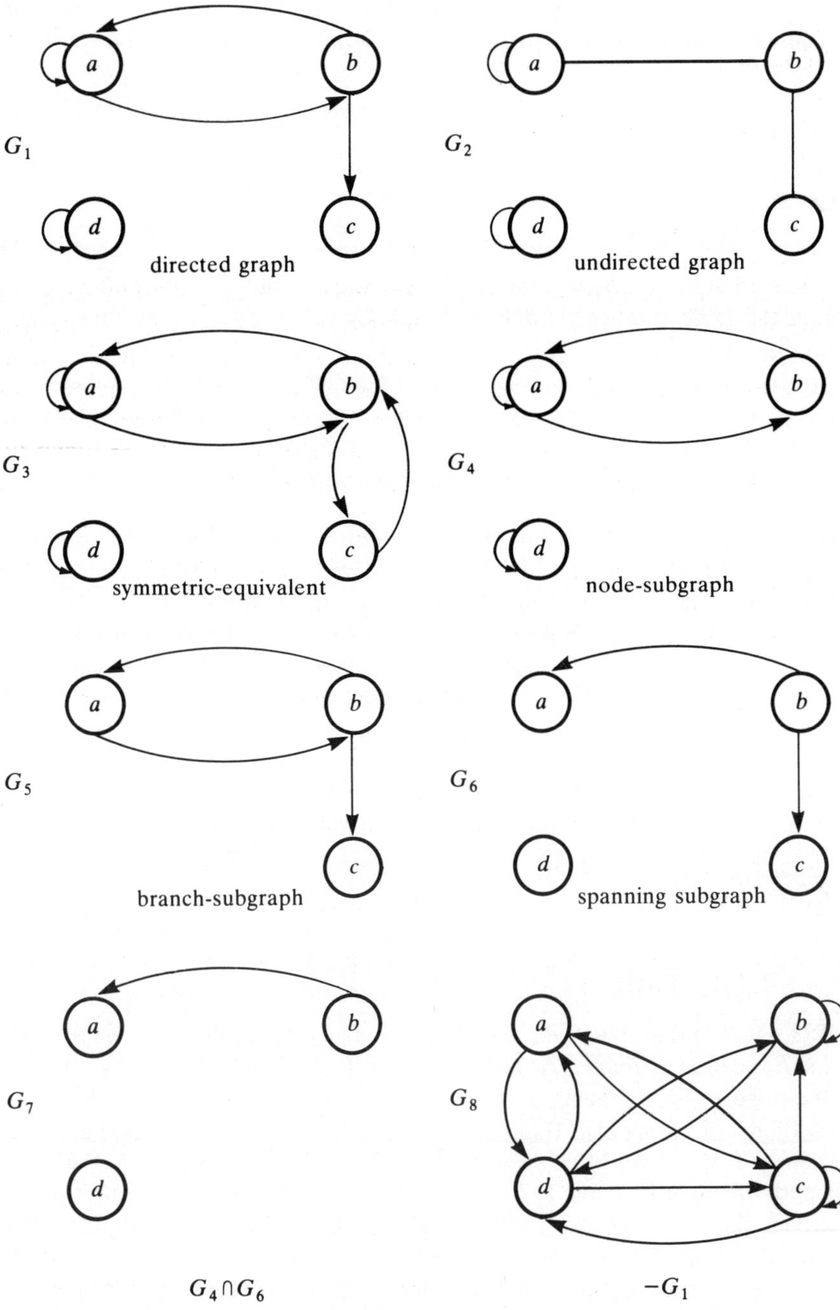

a
b
c
d
G_1
directed graph
G_2
undirected graph
G_3
symmetric-equivalent
G_4
node-subgraph
G_5
branch-subgraph
G_6
spanning subgraph
G_7
G_8
$G_4 \cap G_6$
$-G_1$

Figure 4.1

The *union* of two digraphs (S_1,R_1) and (S_2,R_2) is $(S_1 \cup S_2, R_1 \cup R_2)$. The *relative complement* of subgraph $G'=(S,R')$ with respect to digraph $G=(S,R)$ is the subgraph $G-G'=(S,R-R')$. The *absolute complement* of $G=(S,R)$ is $-G=(S,S\times S-R)$. The *intersection* of (S_1,R_1) and (S_2,R_2) is $(S_1 \cap S_2, R_1 \cap R_2)$.

Example In Figure 4.1, $G_1=G_4 \cup G_6$, $G_7=G_4 \cap G_6$, and G_8 is the absolute complement of G_1.

Two digraphs (S_1,R_1) and (S_2,R_2) are (structurally) *isomorphic* iff they have the same number of nodes and there exists an order morphism (order-preserving bijection) **F** from S_1 to S_2 such that $(a,b)\in R_1$ iff $(\mathbf{F}(a),\mathbf{F}(b))\in R_2$. In other words, there are 1–1 correspondences between their nodes and between their branches such that incidences are preserved. In pictorial terms, two digraphs are isomorphic if they can be drawn identically except for possibly different labelings of their nodes and branches.

The *in-degree* of a node a is the size of the set $\{r\in R \mid \mathbf{term}(r)=a\}$ (i.e., the number of branches incident to the node a). The *out-degree* of a node a is the size of the set $\{r\in R \mid \mathbf{init}(r)=a\}$ (i.e., the number of branches incident from the node a). The (*total*) *degree* of a node is the sum of its in- and out-degrees. A node whose in-degree is zero but out-degree is nonzero is called a *source* node; a node whose out-degree is zero but in-degree is nonzero is called a *sink* node. A node whose (total) degree is zero, excluding any branch to itself, is called an *isolated* node.

Example Referring again to Figure 4.1, node b of G_8 has in-degree 3, out-degree 2, and total degree 5. G_8 has no source, sink, or isolated nodes. In G_6, node b is a source node, and nodes a and c are sink nodes; node d is an isolated node, but is neither a source nor sink.

4.1.1 Paths

If $(a,b)\in R$ in digraph (S,R), then a is called an *immediate predecessor* of b, and b is called an *immediate successor* of a. If $(a,b)\in R^+$, the positive transitive closure of R (cf. Section 2.2.2), then a is called a *predecessor* of b, and b is called a *successor* of a. If $(a,b)\in R^*$ (recall $R^* = R^0 \cup R^+$), we say that there is a *path* from a to b. More specifically, if $(a,b)\in R^k$, $k\geq 0$, we say that there is a path of *length k* from a to b. (Note that, between any node a and itself, a 'trivial' path of length zero always exists by definition, but a path of length one exists only if the reflexive branch (a,a) is in R.)

We formally define a *path* (of *length k*) as a sequence of k branches

$$\langle (a_{i_1},b_{i_1}),\ (a_{i_2},b_{i_2}),\ldots,(a_{i_k},b_{i_k})\rangle,$$

each branch in R, such that $b_{i_j}=a_{i_{j+1}}$, for $1 \leq j \leq k-1$; we then say that the $k+1$ nodes $\{a_{i_1}, b_{i_1}, b_{i_2}, \ldots, b_{i_k}\}$ are ***traversed*** by the path from initial end-node a_{i_1} to terminal end-node b_{i_k}. If the traversed nodes are all distinct (i.e., different from each other), then the path is said to be ***simple***.† If there exists a sequence of nodes $\langle b_{i_0}, b_{i_1}, \ldots, b_{i_k} \rangle$ such that the nodes b_{i_0} and b_{i_1}, b_{i_1} and $b_{i_2}, \ldots$, and $b_{i_{k-1}}$ and b_{i_k} are (pairwise) adjacent—i.e., either $(b_{i_j}, b_{i_{j+1}})$ or $(b_{i_{j+1}}, b_{i_j})$ are branches—then we call the sequence of branches a ***semi-path***. Note that a path follows the directions of arrows along a sequence of collinear branches, whereas a semipath ignores the directions of arrows along a sequence of adjoining (collinear or opposing) branches.

A path from a node to itself (i.e., where the 'end' nodes, a_{i_1} and b_{i_k}, are identical) is called a *cycle* (or *circuit*); or a *semicycle* if a semipath. A cycle traversing distinct nodes, except for the end nodes, is said to be *simple*. (The cycle is simple, not the path.) A digraph is *acyclic* if it contains no cycle; it is *cyclic* if it contains at least one cycle, and 'trivially' cyclic if each such cycle consists of a single reflexive branch.

Two paths are ***branch-disjoint*** if they contain no common branch, and are ***node-disjoint*** if they traverse no common node. However, we also say, in the context of simple paths from given nodes a to b, that two paths are node-disjoint if a and b are the only nodes they both traverse. An *Eulerian* path (or cycle) in a digraph G is a path (or cycle, respectively) containing each branch of G exactly once. A *Hamiltonian* path (or cycle) in G is a path (or cycle, respectively) traversing each node of G exactly once (except for the end node of a cycle, which is traversed exactly twice). Not all digraphs contain Eulerian or Hamiltonian paths.

The *distance* between two nodes a and b, denoted $\mathbf{d}(a,b)$, is the number of branches in a path of shortest length from a to b; such a shortest path is called a *geodesic* from a to b. If no path exists between a and b, then $\mathbf{d}(a,b)$ is infinite; $\mathbf{d}(a,a)=0$ for all a. A longest simple path in digraph G is a *diametric* path of G, and its length is the *diameter* of G.

Examples Refer to Figure 4.1. In G_8, $\langle (b,b), (b,d), (d,a), (a,d), (d,c) \rangle$ is a path of length 5 from node b to node c; it traverses the nodes in the sequence $\langle b,b,d,a,d,c \rangle$, hence the path is not simple. A simple path from b to c is $\langle (b,d), (d,c) \rangle$. There is no path of length one from b to c, but there is such a semipath, namely, $\langle (c,b) \rangle$; $\langle (b,d), (d,c), (c,b) \rangle$ is a simple cycle; and $\langle (c,d), (d,b), (c,b) \rangle$ is a simple semicycle. Of the other digraphs shown, only G_6 and G_7 are acyclic. In G_5, $\langle (b,a), (a,b), (b,c) \rangle$ is an Eulerian path and $\langle (a,b), (b,c) \rangle$ is a Hamiltonian path. In G_8, $\langle (b,d), (d,a), (a,c), (c,b) \rangle$ is a Hamiltonian cycle, but there is no Eulerian cycle. In G_8, $\mathbf{d}(b,c)=2$ while $\mathbf{d}(c,b)=1$. (Therefore, $\mathbf{d}$ is not a metric.) One of many diametric paths in G_8 is $\langle (a,d), (d,c), (c,b) \rangle$, hence the diameter of $G_8=3$.

†Some writers call such paths 'elementary', and use the term 'simple' to refer to paths which do not include the same branch twice.

4.1.2 Connectivity

If there is a path from node a to node b, $(a,b) \in R^*$, we say that b is *reachable* from a. By definition, a node is always reachable from itself by a path of length zero, whether or not by any other. (If we wish to exclude this case, we say b is 'nontrivially' reachable from a if $(a,b) \in R^+$.) If a is reachable from b and b is reachable from a, we say that the pair of nodes is *strongly connected*. If a is reachable from b, but b is not reachable from a, we say that the pair of nodes is *unilaterally* connected. If there is at least a semipath from a to b, hence also from b to a, we say that the pair of nodes is *weakly connected*, and 'strictly' so if neither strongly nor unilaterally connected. If nodes a and b are not even weakly connected, the pair is said to be *disconnected*.

A digraph is strongly connected, or weakly connected, if each pair of its nodes is strongly or weakly connected, respectively. If some pair of its nodes is disconnected, then so is the digraph. A digraph is 'weakly' *complete* if each pair of distinct nodes is adjacent, and 'strongly' complete if each pair of distinct nodes is connected by symmetric branches. Note that if a strongly complete digraph (S,R) is also reflexive, then $R=S\times S$.

Let $G=(S,R)$ be a digraph. The maximal connected node-subgraphs of a digraph G are called its *pieces* (or *components*); a connected subgraph G' of G is 'maximal' if G' is not a proper subgraph of another connected node-subgraph [i.e., if every proper node-subgraph of G' is disconnected]. A connected digraph consists of only one piece. (An *isolated* piece consists of a single isolated node, which is strongly connected whether reflexive or not.) A *cut* of a connected digraph $G=(S,R)$ is a minimal disconnecting set of branches C, 'disconnecting' in that its complement $G'=(S,R-C)$ is a disconnected spanning subgraph of G, and 'minimal' in that no proper subset of C is disconnecting; the removal of C divides G into two pieces (whose union of nodes is S). A minimal disconnecting set of branches is a *directed cut* if the branches are 'unidirectional' with respect to the two pieces, i.e., each branch is incident from a node in a given piece to the other piece. If a cut consists of only one branch, that branch is called an *isthmus*.[†] An *articulation set* is a 'disconnecting' set of nodes A, i.e., a set of nodes whose complement $S-A$ induces a disconnected node-subgraph G' of G; an *articulation point* is an articulation set of size one. A digraph which contains no articulation point is said to be *biconnected*.

Examples In Figure 4.1, digraph G_8 is strongly connected. Digraph G_5 is only weakly connected since nodes a and b are not reachable from node c. Digraphs G_1, G_3, G_4, G_6, and G_7 are disconnected because of the isolated node d. The two pieces

[†] Some writers define an isthmus or 'bridge' as a branch whose removal divides G into two *non-isolated* pieces.

of G_4 are each strongly connected, taken separately; only the isolated piece containing node d is strongly connected in G_6 and G_7, the other piece being weakly connected. None of the digraphs of Figure 4.1 are complete, even weakly, because of the isolated node in most cases; G_5 would be weakly complete if (a,c) were added, and G_8 would be weakly complete if (a,b) were added. Strong completeness would also require symmetry; for example, if node d were omitted, G_4 would be strongly complete, while G_7 would only be weakly complete. G_5 has two cuts, $C_1=\{(a,b), (b,a)\}$ and $C_2=\{(b,c)\}$; their union is not a cut because it is not minimal. C_2 is a directed cut, but C_1 is not. (b,c) is also an isthmus, whose removal divides G_5 into two pieces, one of which is isolated. Node b is an articulation point of G_5.

4.1.3 Orientation

Paths, cycles, and directed cuts in a digraph have an *orientation,* the common direction of the branches they contain. On the other hand, semipaths, semicycles, and (non-directed) cuts have no such common direction. However, it will become necessary to define a reference orientation for these latter sets of branches, and to distinguish those branches having this reference orientation from those branches having the opposite orientation.

A semipath from one node (b_{i_0}) to another (b_{i_k}) has an orientation associated with the sequence of branches or traversed nodes $\langle b_{i_0}, b_{i_1}, \ldots, b_{i_k} \rangle$. We say that a branch $(b_{i_j}, b_{i_{j+1}})$ is a *collinear* branch in (has the same orientation as) the semipath, but that a branch $(b_{i_{j+1}}, b_{i_j})$ is an *opposing* branch in (has the opposite orientation as) the semipath. A path has no opposing branches.

The orientation of a semicycle regarded as a semipath from a node to itself may be defined in the same fashion. However, it is often inconvenient to distinguish one node of a semicycle to be its initial and terminal end node. Instead we may distinguish one branch of the semicycle, and define the orientation of the semicycle such that the distinguished branch is collinear. (Alternatively, in a pictorial representation of a digraph, provided the lines of a semicycle do not cross, the orientation of the semicycle may be defined as 'clockwise' or 'counterclockwise' with respect to a point inside; this definition, of course, is less general because of the proviso.)

Recall that a cut divides a digraph into two pieces, say, X and Y. If one of the branches in a cut is distinguished, we may define the orientation of the cut in terms of the direction of the distinguished branch, which is incident from a node in (say) piece X to a node in piece Y. A branch in the cut is said to be *collinear* if from X to Y, or *opposing* if from Y to X. A directed cut has no opposing branches.

Usually, the distinguished branch in a semicycle or a cut may be chosen arbitrarily, as the main objective is to partition its branches into two opposing sets, each having branches with the same orientation. The same partition would result for any choice of distinguished branch. (For definiteness, we may by convention distinguish the branch having the lowest ordinate with respect to a given ordering of the set of branches in the digraph.)

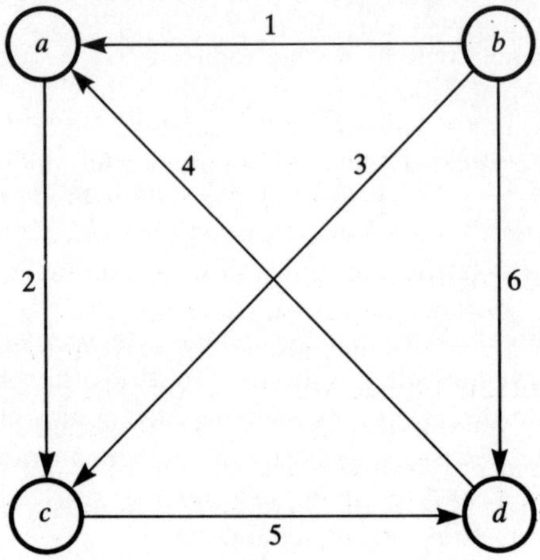

Figure 4.1.3

Example Consider the digraph of Figure 4.1.3, where we have given the branches numerical labels for identification purposes. $\langle 3,2,1\rangle$ is an example of a semipath from node b to itself; it is also a semicycle. With respect to the given sequence of branches, branch 3 is collinear and branches 1 and 2 are opposing; if branch 1 is distinguished as the reference branch of the semicycle, then branch 2 is collinear and branch 3 is opposing. $\{1,3,6\}$ is an example of a directed cut, while the cut $\{2,3,5\}$ is not directed. Distinguishing branch 2 of the latter cut, branch 3 is collinear and branch 5 is opposing.

4.1.4 Other Characterizations

Let $\Pi(S)=\{S_1,S_2,\ldots,S_K\}$ be a partition of the nodes of a digraph $G=(S,R)$, and let G_i be the node-subgraph induced by S_i. The branches of G not contained in any of the subgraphs G_i are called the *links* with respect to the partition. The *reduced* digraph resulting from 'merging' the nodes of G_i is the digraph $G'=(S',R')$ where

$$\begin{aligned} S' &= (S-S_i)\cup\{s_i\},\ s_i\notin S \\ R' &= \{(a,b)\in R \mid (a,b)\in S'\times S'\} \\ &\quad \cup\{(a,s_i)\mid (a,b)\in R \text{ for some } b \text{ in } S_i,\ a \text{ in } S-S_i\} \\ &\quad \cup\{(s_i,b)\mid (a,b)\in R \text{ for some } a \text{ in } S_i,\ b \text{ in } S-S_i\}. \end{aligned}$$

Observe that the nodes in S_i of the subgraph G_i have been replaced by a single newly introduced 'merged node' s_i, which is connected to other nodes of G by the links incident with any node in S_i (with duplications eliminated); more briefly, we say the nodes in S_i are replaced by the new node s_i. The digraph G' may be further reduced by merging other blocks of the partition. The minimal reduced digraph resulting from merging each block of the partition is called the *condensation* of G with respect to $\Pi(S)$; it consists of the merged-nodes $\{s_1,s_2,\ldots,s_K\}$ connected by the links.

A strongly connected *region* (*SCR*) of a digraph $G=(S,R)$ is a subgraph of G which is strongly connected; an SCR is maximal (then called an *MSCR*, or a *strong component*) if it is not a subgraph of another SCR. (An isolated node is a MSCR.) A finite digraph uniquely contains a finite number of MSCRs $\{G_1,G_2,\ldots,G_N\}$ which are pairwise disjoint: i.e., if $G_i=(S_i,R_i)$ and $G_j=(S_j,R_j)$, then $S_i \cap S_j=\emptyset$ (hence also $R_i \cap R_j=\emptyset$), for $i \neq j$. A strongly connected digraph contains a single MSCR—itself. The condensation of G with respect to the partition of nodes $\{S_1,S_2,\ldots,S_N\}$ associated with the MSCRs is called the *condensed* graph of G. The condensed graph consists of merged nodes replacing each MSCR of G connected by links, and is acyclic.

Examples Consider the graph G shown in Figure 4.1.4(a). It has two maximal strongly connected regions, shown in Figure 4.1.4(b). The reduced graph resulting from merging the nodes of the larger MSCR is shown in Figure 4.1.4(c); in addition, merging the other MSCR then yields the condensed graph of G, as shown in Figure 4.1.4(d). For contrast, the condensation of G with respect to the partition of nodes $\{\{1,2\},\{3,4\},\{5,6\}\}$ is shown in Figure 4.1.4(e).

A *clique* of a digraph is the set of nodes of a maximal strongly complete subgraph. A (proper) *subclique* is the set of nodes of a nonmaximal strongly complete subgraph. A *weak clique* is the set of nodes of a maximal (weakly) complete subgraph. A set of nodes $A \subseteq S$ of a digraph $G=(S,R)$ is *independent*† if no node in A is adjacent to any other node in A. It can be shown that A is a clique of G iff A is a maximal independent set of the absolute complement of G; A is a weak clique of G iff A is a maximal independent set of the absolute complement of the symmetrized digraph of G. (All cliques are also weak cliques, but not vice versa; furthermore, cliques need not be disjoint, nor all of the same size.) A partition of the nodes of G into K disjoint independent subsets of nodes is a *coloration* of G; the minimum K for which a partition is a coloration is the *chromatic number* of G. A set of nodes $A \subseteq S$ of $G=(S,R)$ is *dominating* if every node in $S-A$ is adjacent to at least one node in A. (It can be shown that every maximal independent set is a minimal dominating set.)

†Independence in this context should not be confused with the concept of linear independence of vectors (introduced in Section 2.5.1).

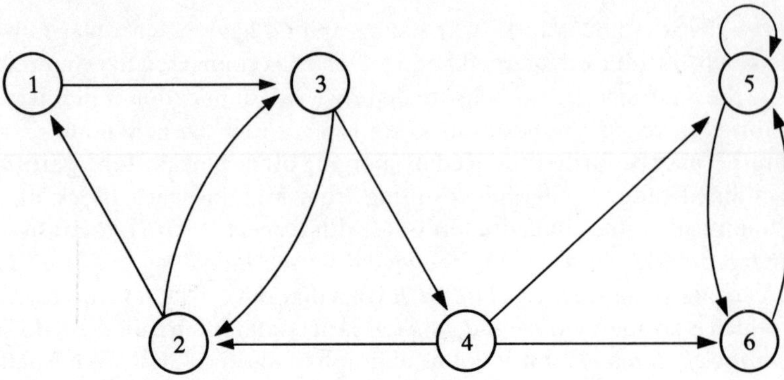

Figure 4.1.4(a)

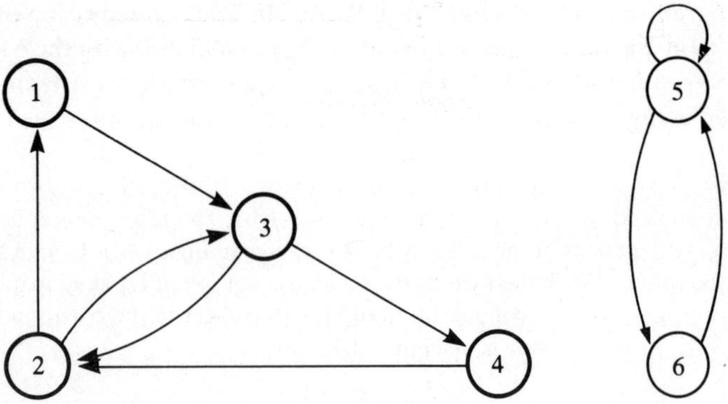

Figure 4.1.4(b)

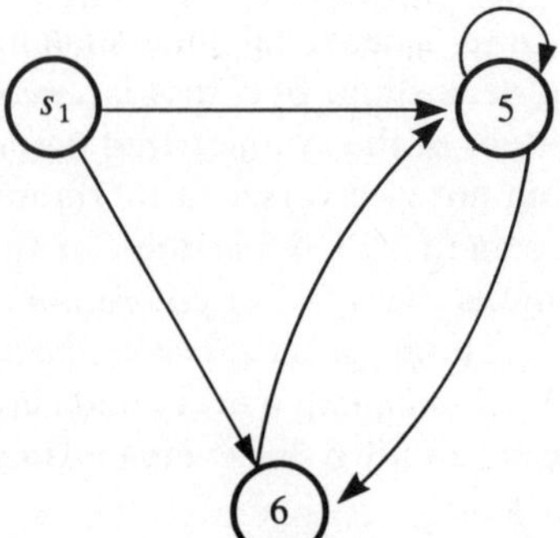

Figure 4.1.4(c)

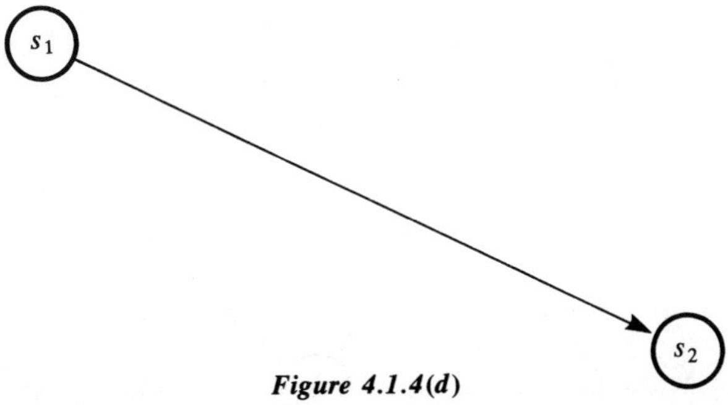

Figure 4.1.4(d)

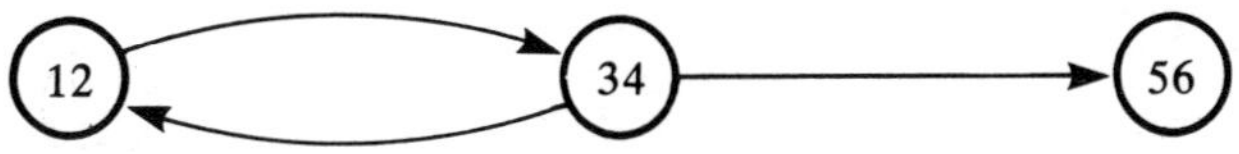

Figure 4.1.4(e)

Examples In Figure 4.1, $\{a,c,d\}$ and $\{b,d\}$ are the (strong) cliques of G_8; in addition, $\{c,d\}$ is a (strong) subclique and $\{b,c,d\}$ is a weak clique. The maximal independent sets of G_1 (the complement of G_8) are $\{a,c,d\}$ and $\{b,d\}$. Two possible colorations of G_1 are $\{\{a,c,d\},\{b\}\}$ and $\{\{a,c\},\{b,d\}\}$; the chromatic number of G_1 is 2. $\{a,c,d\}$ and $\{b,d\}$ are also dominating sets of G_1. $\{a,b\}$ is the only nontrivial (strong) clique of G_1; $\{b,c\}$ is a weak clique. $\{a,b\}$ is a maximal independent set and a dominating set of G_8, as are $\{c\}$ and $\{d\}$. $\{\{a,b\},\{c\},\{d\}\}$ is a coloration of G_8, whose chromatic number is 3.

A *node base* B of $G=(S,R)$ is a minimal subset of its nodes such that every node in S is reachable from some node in B. If a node base B has size 1, then the node in B is called an *out-root*, and G is said to be *out-rooted*. (G is *in-rooted* if there is at least one node, called the *in-root*, which is reachable from every other node; G is *rooted* if either out-rooted or in-rooted.) The subgraph of G *generated* by a node a in G is the subgraph induced by the nodes reachable from a. A digraph is the union of subgraphs generated by the nodes in a node base.

In conclusion, a digraph is said to be *planar* if it can be mapped (drawn) on a plane such that no two branches intersect. One application of this concept, of course, is to the drawing of program flowcharts, where intersections are generally to be minimized. A test for planarity is provided by **Kuratowski's theorem:** a digraph is planar iff its underlying (undirected) graph does not contain a subgraph which is isomorphic to either the 'star graph' or the 'utility graph' of Figures 4.1.4(f) and (g), respectively.

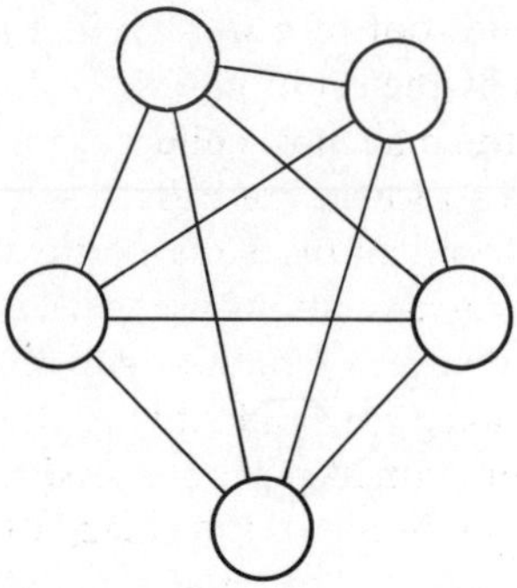

Figure 4.1.4(f)

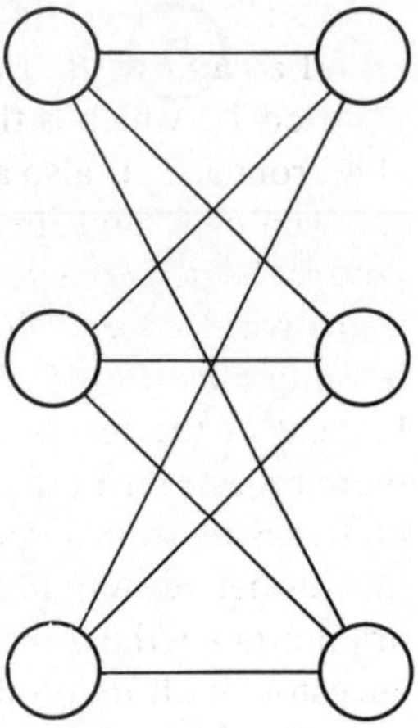

Figure 4.1.4(g)

Examples Node bases of G_1 of Figure 4.1 are $\{a,d\}$ and $\{b,d\}$. Node bases of G_8 are $\{a\}, \{b\}, \{c\}$, and $\{d\}$. Each of the nodes of G_8 may be considered an out-root; G_1 is not a rooted digraph. G_8 is planar since the branches connecting nodes *a* and *c* can be redrawn to curve around node *b* so as not to intersect the branches connecting nodes *b* and *d*.

4.2 TREES—BASIC DEFINITIONS

A *tree* is a rooted digraph having no semicycles. It must therefore be unilaterally connected, having no reflexive or symmetric branches. A *free* tree is a weakly connected digraph having no semicycles, but which is not necessarily rooted. A tree is also said to be 'out-oriented' if it is out-rooted, and 'in-oriented' if in-rooted. For convenience, we generally limit our discussion below to out-oriented trees, with the understanding that analogous statements hold for in-oriented trees. The subgraph of a tree generated by its (out-)root *R* is the tree itself. We note that *R* is the terminal node of no branch in the tree, and that every other node is the terminal node of exactly one branch. Examples of trees are shown in Figure 4.2. When the orientation is not important, the arrowheads may be omitted as long as the root is identified, usually by being drawn at the top.

Let T be a (out-oriented) tree with root R. A node adjacent to R is called a *child* of R. Each child of R, say x, is the root of a *subtree* of T, denoted T_x, which is the node-subgraph induced by the set of nodes reachable from x. T_x is also a tree with root x, hence x too may have children; the children of x, and these children's children, and so forth, are called, as a group, the *descendants* of x. In other words, 'descendant of' is the positive transitive closure of the 'child of' relation. The inverse relations are 'parent of' and 'ancestor of'. Nodes are 'siblings' if they have a common parent. Instead of the terms child, descendant, parent, ancestor, and sibling, the more prosaic terms immediate successor, successor, immediate predecessor, predecessor, and co-successor, respectively, may be used; an advantage of the former terminology is that the same terms apply to the nodes if the orientation of the tree is reversed (i.e., the root of an in-rooted tree is still the ancestor of all its predecessor nodes).

The *degree* of a node in a tree is the number of its children (i.e., its out-degree, for out-oriented trees). A *leaf* (also called a 'terminal' node) is a node with zero degree; a node with a nonzero degree is called *internal*. A tree is t*-ary* if the degree of each of its nodes is at most t. A 2-ary tree is also called a *binary* tree. A t-ary tree is *full* if the degree of each of its nodes is either t or 0. If a node in a t-ary tree has i children, then we say the node has $t-i$ *null leaves*. Every leaf in a t-ary tree has t null leaves; in a full tree, only the leaves have null leaves. Sometimes it is convenient to represent each null leaf by a special (distinguishable) node; doing so results in a full 'augmented' tree whose leaves are these special nodes.

An *ordered* tree is one where the children of each node are considered ordered, so that each internal node has a first child, second child,..., and last child. The first child of node x, say node y, is called its *left* one, the second child of x, say z, is the *right* sibling of y, the third child of x is the right sibling of z, and so forth; the subtree T_y is called the first (or left) subtree of x, the subtree T_z is the second subtree of x, etc. An ordered t-ary tree is said to be *positional* if full or if null leaves are distinguished in the ordering. For example, if a node in a positional 3-ary tree has only two non-null children, then its null leaf may take the place of either its first, second, or third child; or in a positional binary tree, a node having degree 1 has either a left or right child, and its null leaf is then either on the right or left, respectively.

The *depth* of a node x, denoted $\mathbf{depth}(x)$, in a tree is the length of the path from the root R to the node $(=\mathbf{d}(R,x))$. The 'average depth of a tree' is the average of the depths of its (non-null) leaves. The root has a depth of zero; the depth of a null leaf equals its depth in the augmented tree. The *height* of a node x, denoted $\mathbf{height}(x)$, is the length of a longest path from x to a (non-null) leaf. The 'height of a tree' (= the height of its root) is the length of a longest path in the tree $(=\mathbf{max}_x\mathbf{d}(R,x))$. The height of a leaf is zero; the height of a null leaf is undefined. The term 'level' (of a node x) is

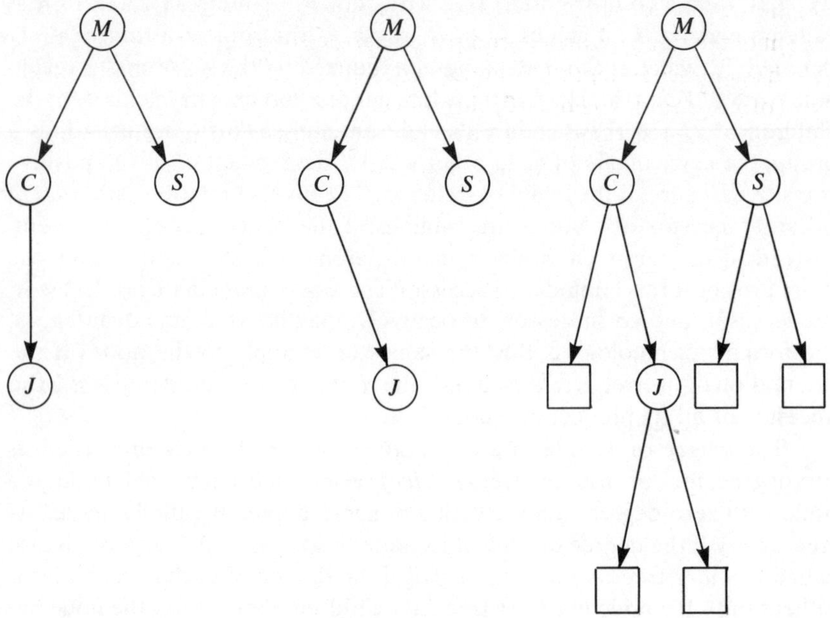

(a) (b) (c)

Figure 4.2

used as a measure of both depth and height, depending upon whether the tree is processed from the root to the leaves ('top-down'), or vice versa ('bottom-up'). [In the literature, **level**(x) has been defined variously as **depth**(x), **depth**(x)+1, **height**(x)+1, and **height**(R)−**height**(x).]

Examples Consider the directed graph shown in Figure 4.2(a). It is an out-rooted tree whose root is labeled M. M has two children, C and S. Its left subtree is the tree rooted at C; this subtree consists of the two nodes C and J, and the connecting branch. The degree of M equals 2, the degree of C equals 1, and the degrees of J and S equal 0. Hence J and S are leaves, and M and C are internal nodes. Regarded as a binary tree, the tree is not full. Node C has one null leaf, and leaves J and S have two null leaves each. The null leaf of node C is shown on the left in the positional tree of Figure 4.2(b). Figure 4.2(c) shows the augmented tree, where square nodes represent the null leaves. The depths of nodes M, C, S, and J are 0, 1, 1, and 2, respectively; the average depth of the tree equals 1.5. The heights of nodes M, C, S, and J are 2, 1, 0, and 0, respectively. The height of the tree is 2.

A tree is said to be *complete* if it is full and its leaves all have a depth equal to the height of the tree; its null leaves all have depth equal to the height of the tree plus one. A complete t-ary tree of height H has one node at depth 0, t nodes at depth 1, t^2 nodes at depth 2, ..., t^H nodes at depth H, for a total of $\Sigma_{i=0}^{H} t^i$ nodes. (For $t=2$, this sum equals $2^{H+1}-1$.) Figures 4.2(d) and (e) show complete binary and 3-ary trees of height 2 having 7 and 13 nodes, respectively. Since complete trees can have only certain fixed numbers of

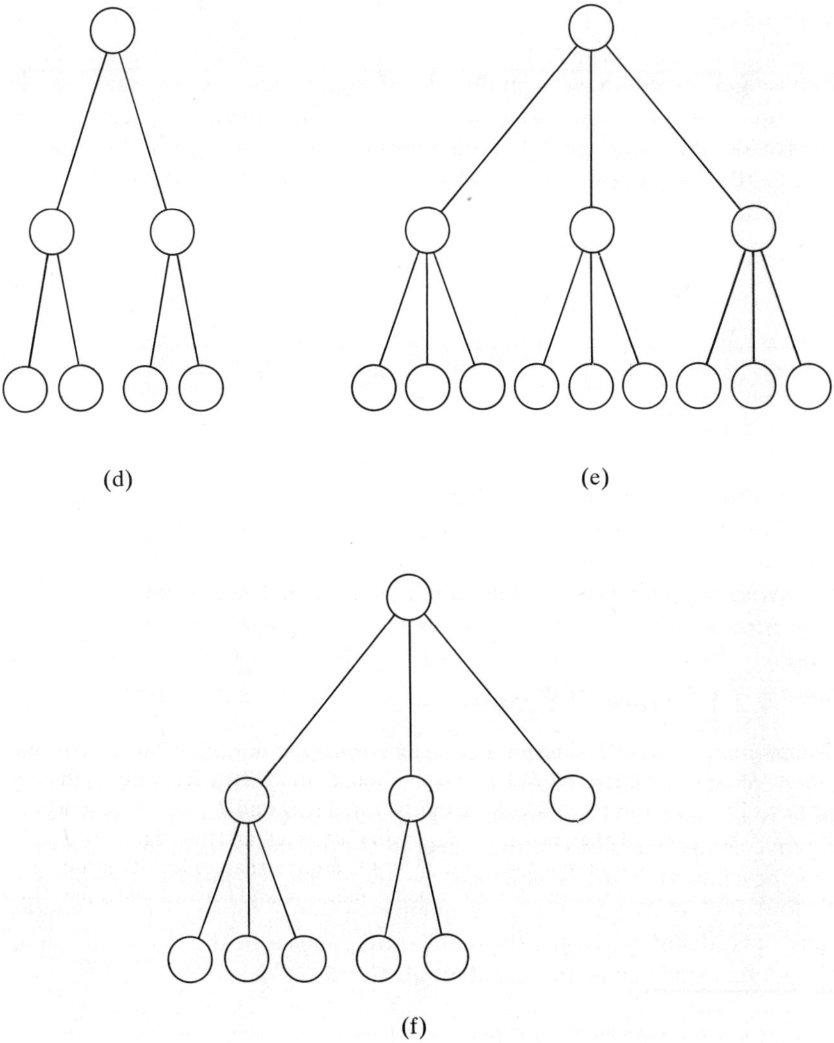

Figure 4.2

nodes, it is often convenient to define a *canonical* t-ary tree having an arbitrary number of nodes N as one which is 'almost complete' in the sense that it has t^i nodes at depth i except possibly for $i=H$; at depth H, there are $N-\Sigma_{i=0}^{H-1}t^i$ leaf nodes. Moreover, there are $\Sigma_{i=0}^{H}t^i-N$ null leaves at depth H; if the canonical tree is a positional one, these null leaves all appear on the right, by convention.† Note that a canonical tree having $N=\Sigma_{i=0}^{H}t^i$ nodes has no null leaves at depth H, i.e., the tree is complete. An example of a canonical 3-ary tree having 9 nodes is shown in Figure 4.2(f); it has $\Sigma_{i=0}^{2}3^i-9=4$ null leaves at depth 2.

Application Trees have a multitude of applications, as we shall see in subsequent sections and chapters. We mention one obvious application here to provide some motivation. In some programming languages, such as PL/I and COBOL, 'hierarchical' variables are allowed, whose declaration is of the form:

```
1  M
     2  C
          3  J
     2  S
```

Here, M is the main variable, which has two subvariables, C and S; and C in turn has one subvariable J. This hierarchical variable may be represented by a tree, as shown in Figure 4.2(a). Data values for such variables are commonly associated with only the leaves (J and S in the above).

4.2.1 Trees of Graphs

A (*spanning*) *tree of* (vs. contained *in*) a connected digraph G is a spanning subgraph of G which is also a free tree. A *chord* of G with respect to a tree T of G is a branch in G but not in T. The *cotree* of G with respect to T is the relative complement of T with respect to G (i.e., the branches of the cotree are the chords); cotrees need not be trees. The *cyclomatic number* (or *nullity*) of G, denoted $\nu(G)$, is the number of chords for any T of G; and the *rank* of G, denoted $\rho(G)$, is the number of branches in any T of G. It can be shown that, for any connected digraph $G=(S,R)$,

†Some writers extend the definition of a 'complete' tree having any number of nodes N to mean such a positional canonical tree; we prefer to restrict the definition of 'complete' to the case given where the tree is full, and to assume that canonical trees are positional unless otherwise stated.

$$\nu(G)=\#(R)-\#(S)+1,$$
$$\rho(G)=\#(S)-1.$$

If $G=(S,R)$ is not connected, consisting of p pieces (disjoint connected subgraphs), $G_1,\dots,G_p$, then

$$\nu(G)=\nu(G_1)+\dots+\nu(G_p)=\#(R)-\#(S)+p,$$
$$\rho(G)=\rho(G_1)+\dots+\rho(G_p)=\#(S)-p.$$

A digraph G is called a *forest* if each piece of G is a free tree, and a spanning forest and co-forest may be defined analogously.

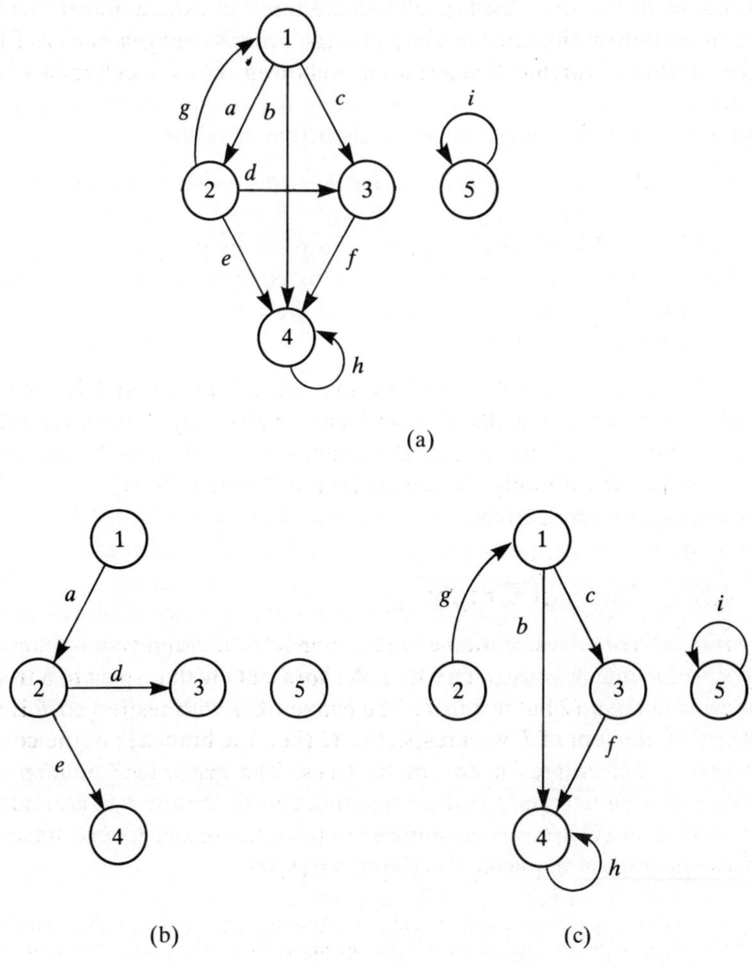

Figure 4.2.1

Example Figure 4.2.1(a) shows a digraph G consisting of two pieces. A spanning forest is shown in Figure 4.2.1(b). The chords are $\{b,c,f,g,h,i\}$; the co-forest is shown in Figure 4.2.1(c). The cyclomatic number and the rank of G are given below:

$$\nu(G)=\#(R)-\#(S)+p=9-5+2=6$$
$$\rho(G)=\#(S)-p=5-2=3.$$

4.2.2 Tree Traversals

An 'Eulerian' semipath in a tree is a semicycle from the root to itself containing each branch of the tree exactly twice (or equivalently containing each branch in the symmetrized digraph of the tree exactly once). Such a path traverses or 'visits' each node of the tree a number of times equal to its degree plus one. If the tree is ordered, we adopt the convention that the children of each node are to be traversed in left-to-right order. Hence we define a recursive *canonical traversal* algorithm as follows:

(A) On the first visit to a node x (initially the root), follow the branch leading to its first (left) child and visit this child; if node x has no first child (i.e., if x is a leaf), revisit (return to) its parent.

(B) On each subsequent visit (i.e. return) to a node x, follow the branch to its next child and visit this child; if node x has no next child, revisit (return to) its parent. (If the node x has no parent, it must be the root, hence stop.)

Note that this algorithm requires that we keep track of each visit to a node, in order to determine where to proceed; implementation of the algorithm is facilitated by using a stack, as we shall see.

An Eulerian semipath in a tree traverses each node, except for leaves, more than once; a semipath which traverses each node in an arbitrary tree exactly once does not exist. Let $S=\langle a_1,a_2,\ldots,a_L\rangle$ be the 'canonical Eulerian sequence' of nodes as traversed by the canonical algorithm, where a_1 and a_L are both the root. Define $T=\langle a_{i_1},a_{i_2},\ldots,a_{i_N}\rangle$ as a subsequence of S (i.e., where $i_j<i_{j+1}$) containing each node in the tree exactly once, hence N is the number of nodes in the tree; any such sequence will be called a *traversal sequence* of the nodes. We may define several 'canonical' traversal sequences—in particular, (a) *preorder*: where the nodes in T are chosen according to their first (leftmost) appearance in S, and (b) *postorder*: where the nodes in T are chosen according to their last (rightmost) appearance in S. These two traversal sequences can be obtained by adding to the canonical traversal algorithm a substep which appends to the sequence T a node (a) upon the first visit to the node (in Step A), or (b) prior to returning to its parent (in Step B), respectively. Note that, in the event that the tree is positional, when a null leaf of node x is encountered, the null leaf should be

visited (rather than returning to the parent of x); a return to x would be made immediately upon visiting the null leaf. In the case of (positional) binary trees, a third traversal sequence, (c) *inorder,* can be obtained by appending to T a node upon the first return to the node (from its left child). We also extend the definitions of the various sequences to ordered forests by concatenating the sequence corresponding to each tree in the forest.

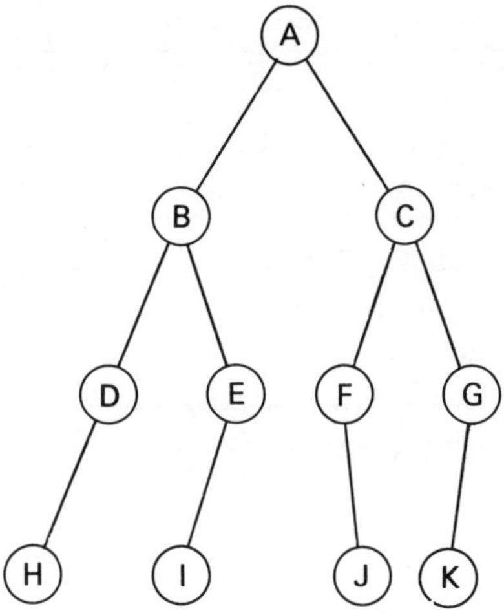

canonical Eulerian sequence:
ABDHDBEIEBACFJFCGKGCA

preorder traversal-sequence:
ABDHEICFJGK

postorder traversal-sequence:
HDIEBJFKGCA

inorder traversal-sequence:
HDBIEAFJCKG

Figure 4.2.2(a)

Example Figure 4.2.2(a) illustrates a nonpositional tree and its corresponding canonical Eulerian sequence, its preorder and postorder traversal sequences, and its inorder traversal sequence for the tree regarded as positional (i.e., taking into account null leaves); Figure 4.2.2(b) illustrates all but inorder for a forest.

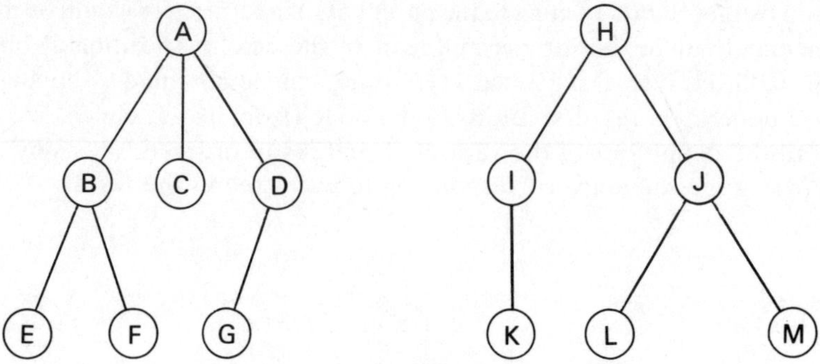

canonical Eulerian sequence:
ABEBFBACADGDAHIKIHJLJMJH
preorder traversal-sequence:
ABEFCDGHIKJLM
postorder traversal-sequence:
EFBCGDAKILMJH

Figure 4.2.2(b)

4.2.3 Binary Tree Transformation

Binary trees are of special importance because of their numerous applications. We shall first show that every tree can be represented by a binary tree. Let T be an ordered (out-rooted) tree. We construct a positional binary tree B, having the same nodes as T, as follows:

(a) If y is the first child of x in T, then let y be the left child of x in B; if x has no first child (i.e., is a leaf) in T, then the left child of x in B is a null leaf.
(b) If z is the right sibling of y in T, then let z be the right child of y in B; if y has no right sibling (i.e., is a rightmost child, or the root) in T, then the right child of y in B is a null leaf.

If T is an ordered forest, i.e., the trees comprising the forest are ordered, then the correspondingly ordered roots are considered siblings, and the first root in T has the second root in T as its right child in B, etc. Binary trees can be transformed into general trees or forests by reversing the procedure. Of course, a binary tree T can also be transformed into another (different) binary tree B by using the above procedure. If T is considered positional, then its null leaves may appear as internal nodes of B.

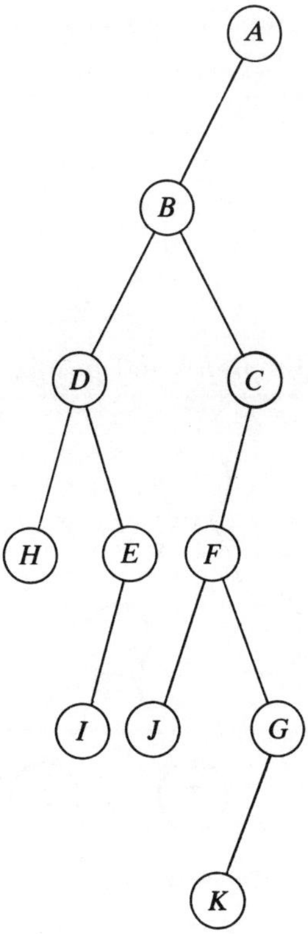

Figure 4.2.3(a)

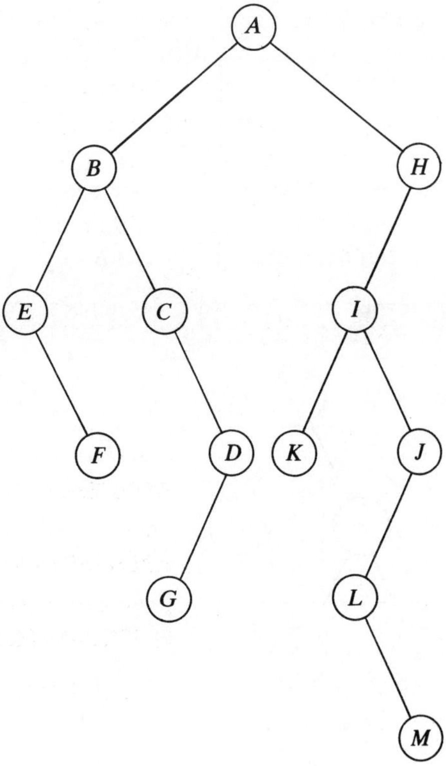

Figure 4.2.3(b)

Example Figures 4.2.3(a) and 4.2.3(b) illustrate the binary trees obtained by transforming the tree of Figure 4.2.2(a) and the forest of Figure 4.2.2(b), respectively.

4.2.4 Application: Algebraic Expressions

A common application of binary trees is their use in representing algebraic expressions. For example, the tree in Figure 4.2.4(a) represents the arithmetic expression $(A+B)\times C$. In general, each internal node of what we call an *operator tree* corresponds to an operator, and the node's children correspond to its operands. Note in the Figure that the left child of the multiplication node is the addition node, which means that the left operand of the product is the sum $A+B$. The operator tree for the arithmetic expression $A+(B\times C)$ is shown in Figure 4.2.4(b).

For the foregoing operator trees, the three canonical traversal sequences are as follows:

	Tree (a)	*Tree* (b)
PREORDER:	$\times + ABC$	$+A\times BC$
POSTORDER:	$AB+C\times$	$ABC\times +$
INORDER:	$A+B\times C$	$A+B\times C$

Preorder and postorder sequences are known as the *prefix* and *postfix*

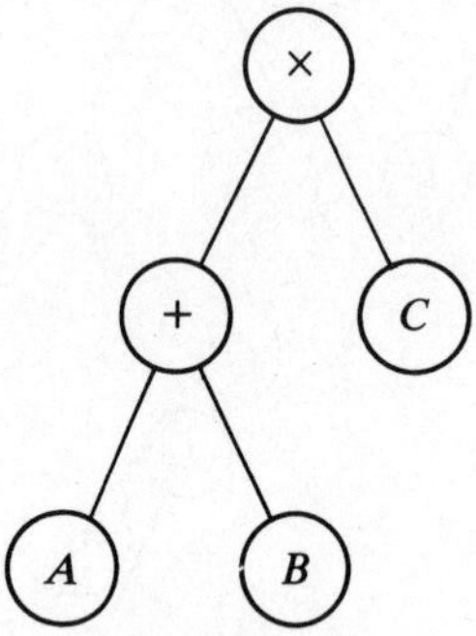

Figure 4.2.4(a)

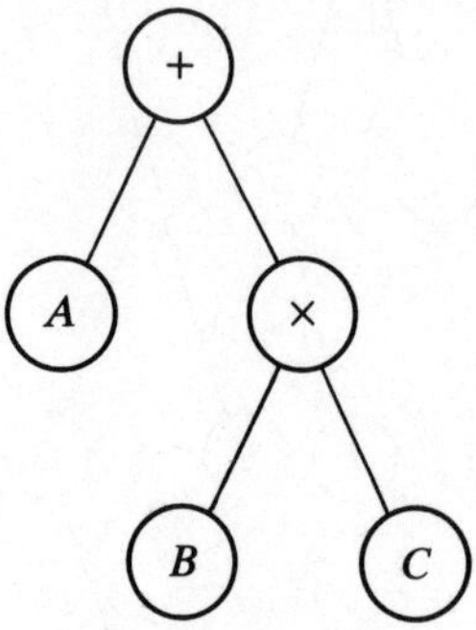

Figure 4.2.4(b)

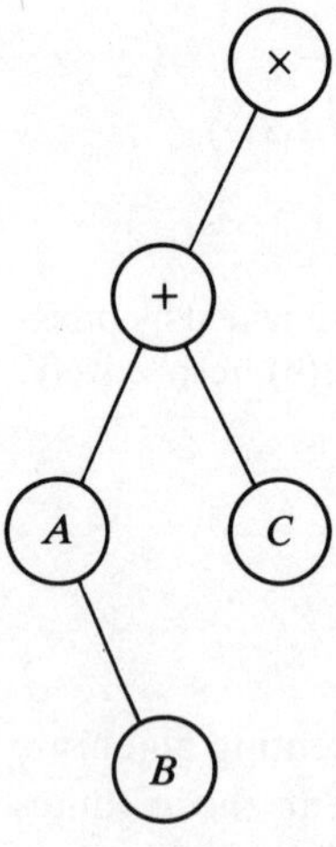

Figure 4.2.4(c)

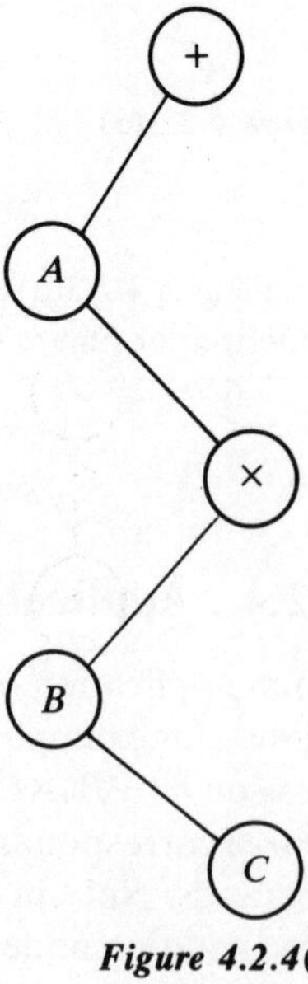

Figure 4.2.4(d)

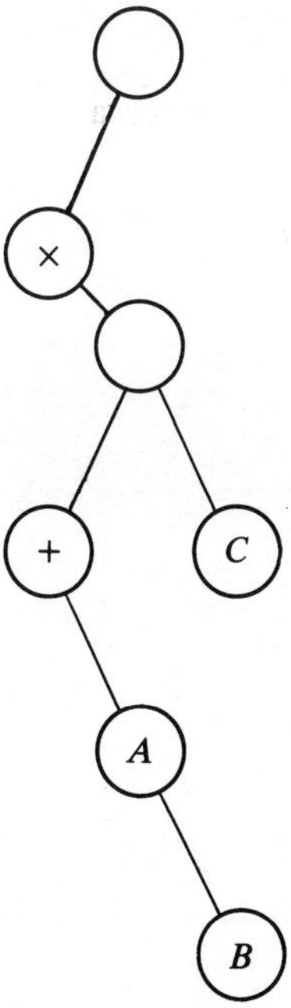

Figure 4.2.4(e)

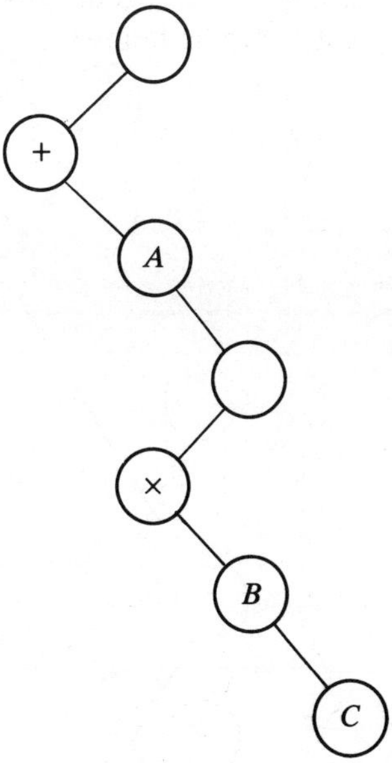

Figure 4.2.4(f)

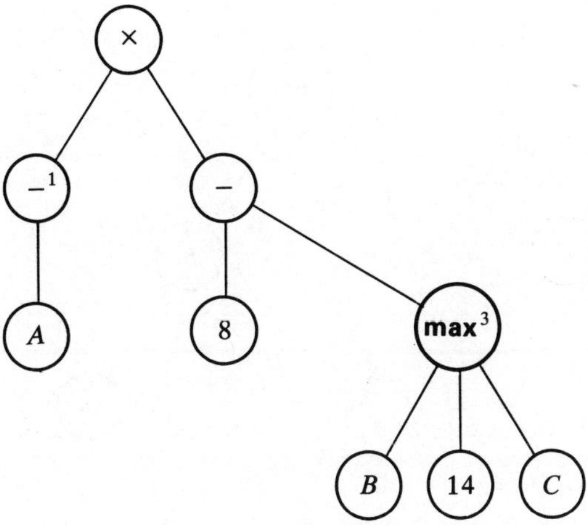

Figure 4.2.4(g)

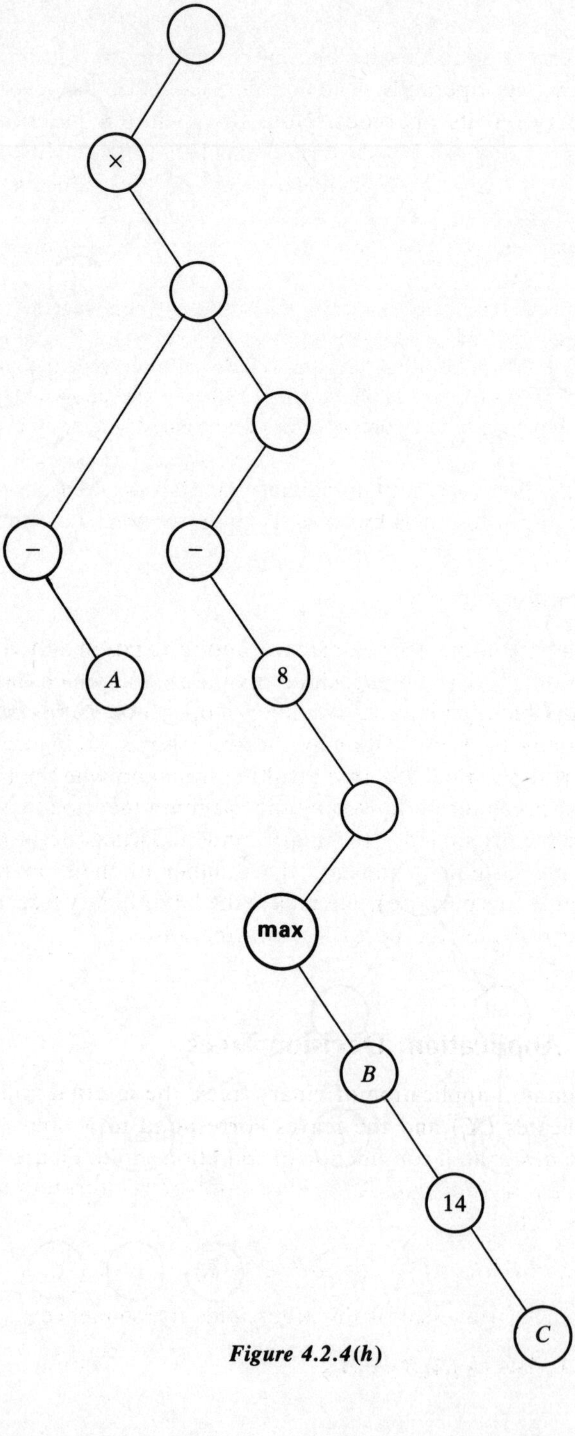

Figure 4.2.4(h)

'Polish-strings' corresponding to algebraic expressions, where in the former case each operator precedes its operands, and in the latter case each operator follows its operands. The inorder sequence has each (binary) operator in between its operands. Note that, without parenthesization, inorder sequences (but not Polish strings) may be ambiguously interpreted; to avoid ambiguity, inorder traversal algorithms may introduce parentheses whether necessary or not (with respect to precedence rules) producing $((A+B)\times C)$ and $(A+(B\times C))$ for trees (a) and (b), respectively.

Example Figures 4.2.4(c) and (d) show the results of applying the binary tree transformation algorithm to the two operator trees. Figures 4.2.4(e) and (f) show a variation, where each operator has been 'moved down' to precede its operands in a linear sublist. The result, which we call a 'LISP tree', is important because it corresponds to how algebraic expressions are represented internally in LISP.

Algebraic expressions containing unary operators and n-ary operators, for $n>2$, may be represented by an n-ary operator tree. For example, the expression

$$-A\times(8-\mathbf{max}(B,14,C))$$

—which has a unary minus operator (representing negation of A, as opposed to subtraction of A from an unspecified operand), and which has an n-ary operator **max** (which can have any number of operands, but in the example has only 3)—may be represented by the tree shown in Figure 4.2.4(g). Figure 4.2.4(h) shows the LISP tree resulting from applying the binary tree transformation procedure followed by the variation mentioned above. We remark that, in the original tree, the unary minus and **max** operators must be marked in some fashion to indicate the number of their operands (the superscripts serve this purpose), whereas in the latter binary tree, the length of each sublist provides the necessary information.

4.2.5 Application: Decision Trees

In another common application of binary trees, the internal nodes correspond to predicates (X_i) and the leaves correspond to actions (A_i) to be performed if a designated conjunction of conditions hold. Figure 4.2.5 is an example of such a *decision tree*. As an illustration of its interpretation, action A_1 is to be executed if

$$(X_3=\boldsymbol{T}\wedge X_1=\boldsymbol{T}\wedge X_2=\boldsymbol{T})\ \vee\ (X_3=\boldsymbol{F}\wedge X_2=\boldsymbol{T})$$

is true. A preorder traversal of this tree yields the sequence

$$X_3X_1X_2A_1A_2A_3X_2A_1X_1A_2A_3;$$

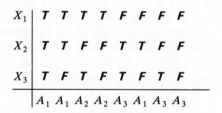

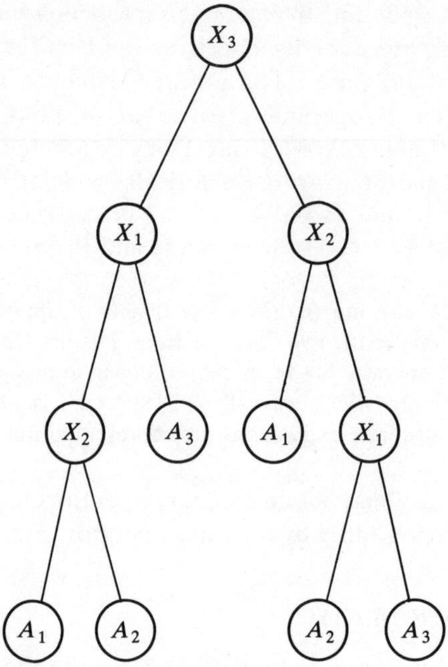

Figure 4.2.5

X_1	T	T	T	T	F	F	F	F
X_2	T	T	F	F	T	T	F	F
X_3	T	F	T	F	T	F	T	F
	A_1	A_1	A_2	A_2	A_3	A_1	A_3	A_3

Table 4.2.5

the traversal algorithm can be modified to add the words **IF**, **THEN**, and **ELSE** appropriately to obtain **IF** X_3 **THEN IF** X_1 **THEN IF** X_2 **THEN** A_1 **ELSE** A_2 **ELSE** A_3 **ELSE IF** X_2 **THEN** A_1 **ELSE IF** X_1 **THEN** A_2 **ELSE** A_3. Decision trees also have a tabular representation called a *decision table,* as shown in Table 4.2.5; we introduced such tables in Section 1.2.3 and will discuss them further in Section 5.6.9. See Harel [4.6] and Moret [4.13] for some related concepts.

4.2.6 Balanced Binary Trees

It is often desirable for ordered binary trees to be 'balanced' in (say) the informal sense that there are as many nodes to the left of the root as to the right. When the tree has an even number of nodes, this kind of balance is impossible. However, there are various less restrictive definitions of 'balance' which are of practical importance. We give several such definitions here.

Definition 1: A binary tree is ***size-balanced*** if the number of nodes in the left and right subtrees of each node differ by no more than 1.

Definition 2: A binary tree is ***depth-balanced*** if the depths of its null leaves differ by no more than 1.

Definition 3: A binary tree is ***height-balanced*** if the heights of the left and right subtrees of each node differ by no more than 1.

Size balance is more restrictive than depth balance, while height balance is less restrictive than depth balance. For example, Figure 4.2.6(a) illustrates a tree which is size-balanced; the label of a node x in the tree is equal to the number of nodes in the subtree rooted at x. This tree is also depth- and height-balanced. Figure 4.2.6(b) illustrates a tree which is depth-balanced, but not size-balanced; the label of a node in the tree is equal to its depth. The depths of the null leaves are either 3 or 4. The tree is also height-balanced. Figure 4.2.6(c) illustrates a tree which is height-balanced, but not depth-balanced (nor size-balanced); the label of a node x in the tree

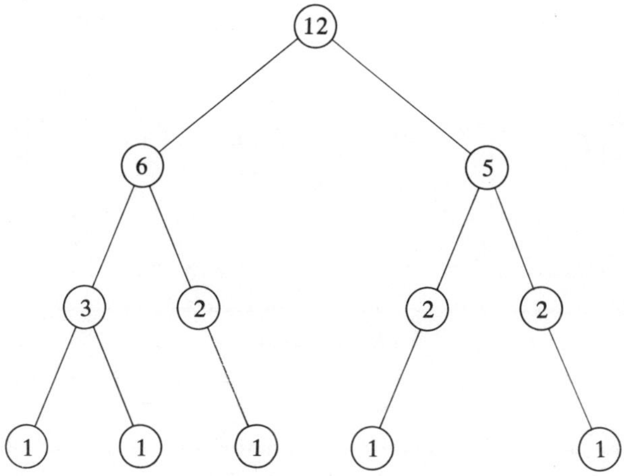

Figure 4.2.6(a)

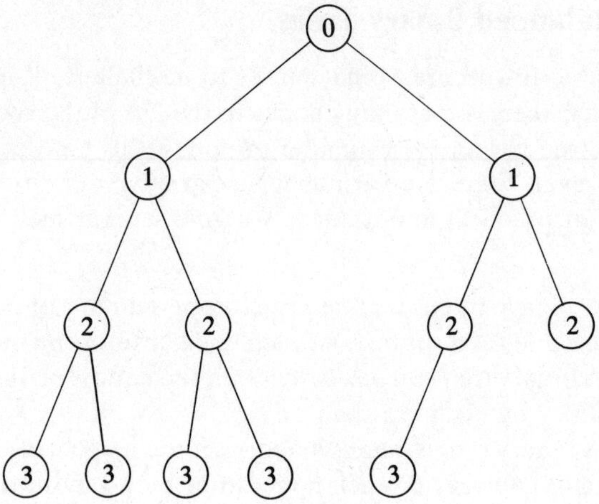

Figure 4.2.6(b)

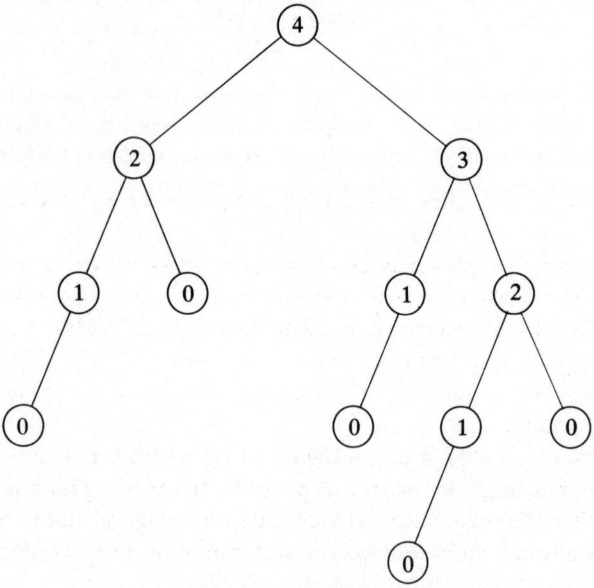

Figure 4.2.6(c)

is equal to the height of the subtree rooted at x. Figure 4.2.6(d) illustrates a tree which is not balanced by any of the foregoing definitions; it also illustrates why null leaves rather than leaves are used in Definition 2, and

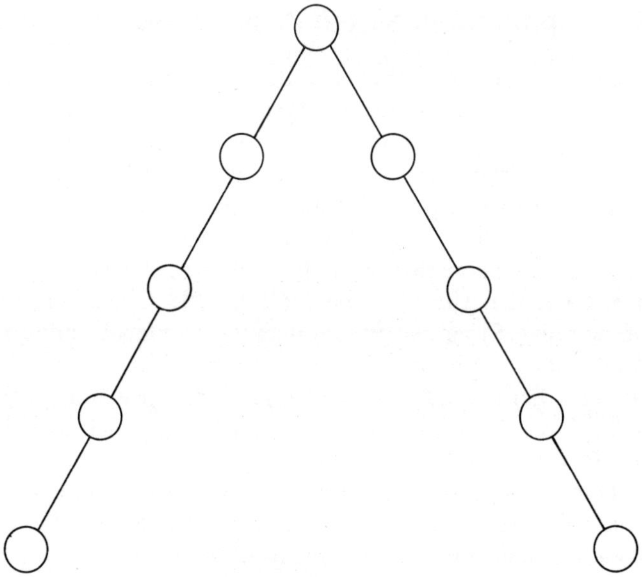

Figure 4.2.6(d)

why subtrees of each node in addition to the root are used in Definitions 1 and 3.

Size balance is generally too restrictive to be of much value. Depth balance is the most common. For example, canonical trees are depth-balanced. Depth-balanced trees need not be canonical, in that the null leaves need not appear only in the rightmost positions. Because depth-balanced trees may be too restrictive for some applications, height-balanced trees are also of importance.

We observe that in a depth-balanced (as well as any canonical) tree, leaves appear as 'high' in the tree as possible, there being no null leaves at a depth less than that of any leaf. Hence, a depth-balanced binary tree consisting of N nodes has minimum height and minimum average depth over all binary trees of N nodes, a very desirable property. In fact, the height H of a depth-balanced binary tree of N nodes is **int** $(\mathbf{log}_2(N))$; H is also equal to, or is one greater than, the depth of each leaf of the tree. (H is also equal to, or one less than, the depth of each null leaf of the tree.) The equality holds (i.e., null leaves are all at the same depth) only in the case that the binary tree is complete; then $N+1=2^{H+1}$, the number of null leaves and twice the number of leaves.

4.2.7 Application: Sorted Binary Trees

Associate a unique label λ_x with each node x in a binary tree. The label λ_x is also called the *key* of x. We say that the binary tree is (*key-*)*sorted* if, for each node x, the left subtree of x contains only nodes whose associated keys are less than λ_x, and the right subtree of x contains only nodes whose associated keys are greater than λ_x. (The less than and greater than relations are with respect to a lexicographical order if the keys are not numeric.) Figure 4.2(b) is a sorted binary tree: note that the node labels in the left subtree of the root [$\{C,J\}$] are all less than the root label [M], that the node labels in the right subtree of the root [$\{S\}$] are all greater than the root label, and furthermore that the node labels in the right subtree of node C are all greater than C [$J>C$]. On the other hand, Figure 4.2.2(a) is not a sorted binary tree since, for example, B is in the left subtree of the root A, which would indicate falsely that B is less than A.

Sorted binary trees (sometimes called 'binary search trees') are used to facilitate the binary searching of a set of data. In essence, pointers to subsets of data to be searched appear in the tree itself, so that the midpoint calculations and boundary adjustments necessary to binary search an array (see Section 5.3.4) are no longer necessary. In Section 5.3, we discuss algorithms for binary searching a key-sorted binary tree, for constructing such a tree in the first place, and for adding and deleting nodes.

4.3 REPRESENTATIONS OF DIGRAPHS

A finite digraph $G=(S,R)$ may be represented in many ways, depending upon how the set S and the relation-set R are represented (see Section 3.2.9); the representation of R is of greater significance since R provides the *structural* information about an arbitrary set of objects S, which is what distinguishes digraphs from general (unstructured) sets. The 'characteristic matrix' representation for R is the most common. Letting the set $S=\{s_1,s_2,\ldots,s_n\}$ be ordered, the *characteristic matrix* of R, denoted $\mathbf{C}_R$, is defined as follows:

$$\mathbf{C}_R(i,j)= \begin{cases} 1 \text{ if } (s_i,s_j)\in R \\ 0 \text{ if } (s_i,s_j)\notin R. \end{cases}$$

Observe that $\mathbf{C}_R$ has one nonzero element for each branch, and has zero elements otherwise.

In the context of digraphs, $\mathbf{C}_R$ (or $\mathbf{C}$, when R is understood) is usually called the (Boolean) *adjacency matrix* of G, or sometimes the *connectivity* matrix. Note that $\mathbf{C}$ is an $(n\times n)$ matrix, where $n=\#(S)$, and depends upon the ordering of the set S. The *incidence matrix* **B** of an irreflexive digraph $G=(S,R)$ is an $(n \times m)$ matrix, where $n=\#(S)$ and $m=\#(R)$, and depends upon the ordering of both S and R $(R=\{r_1,\ldots,r_m\})$:

$$\mathbf{B}(i,j)=\begin{cases} -1 \text{ if branch } r_j \text{ is incident from } s_i \\ \ \ 0 \text{ if branch } r_j \text{ is not incident with } s_i \\ +1 \text{ if branch } r_j \text{ is incident to } s_i. \end{cases}$$

(The choice of signs is arbitrary; by some conventions, they are reversed or omitted.) The restriction on incidence matrices, but not adjacency matrices, to irreflexive digraphs is because $\mathbf{B}(i,j)$ would not be defined uniquely if $r_j=(s_i,s_i)$; in this event, the definition may be generalized so that $\mathbf{B}(i,j)=0$ holds, but the matrix then no longer uniquely represents the digraph.† Adjacency matrices have the advantage that they completely (and uniquely, except for permutations of S) represent the structure of arbitrary digraphs. Furthermore, as Boolean arrays they can be conveniently represented and manipulated by computers. Incidence matrices, on the other hand, are not Boolean. However, they are important because their rows are associated with cuts, as we shall see.

Example For the digraph shown in Figure 4.3(a),

$$\mathbf{C}_R=\begin{matrix} & 1 & 2 & 3 & 4 & 5 \\ 1 & 0 & 1 & 1 & 1 & 0 \\ 2 & 1 & 0 & 1 & 1 & 0 \\ 3 & 0 & 0 & 0 & 1 & 0 \\ 4 & 0 & 0 & 0 & 1 & 0 \\ 5 & 0 & 0 & 0 & 0 & 1 \end{matrix}$$

$$\mathbf{B}=\begin{matrix} & a & b & c & d & e & f & g & h & i \\ 1 & -1 & -1 & -1 & 0 & 0 & 0 & +1 & 0 & 0 \\ 2 & +1 & 0 & 0 & -1 & -1 & 0 & -1 & 0 & 0 \\ 3 & 0 & 0 & +1 & +1 & 0 & -1 & 0 & 0 & 0 \\ 4 & 0 & +1 & 0 & 0 & +1 & +1 & 0 & 0 & 0 \\ 5 & 0 & 0 & 0 & 0 & 0 & 0 & 0 & 0 & 0 \end{matrix}.$$

Here we have assumed that S and R are ordered sets, $S=\{1,2,3,4,5\}$, $R=\{a,b,c,d,e,f,g,h,i\}$. The node and branch labels are shown for convenience; they are not part of the matrices.

One disadvantage of the use of adjacency matrices is that they may be very sparse. Of the n^2 elements of the matrix, only m are nonzero. It makes

†Uniqueness is not lost if $\mathbf{B}(i,j)$ is defined to equal $+1$ (instead of 0 or -1), but this definition leads to other problems; e.g., the matrix equations of Section 4.3.3 would no longer hold.

sense then to list only the m branches, e.g., we may represent the above digraph as follows:

$$\begin{bmatrix} 1 & 2 \\ 1 & 3 \\ 1 & 4 \\ 2 & 1 \\ 2 & 3 \\ 2 & 4 \\ 3 & 4 \\ 4 & 4 \\ 5 & 5 \end{bmatrix}$$

where each row represents one branch, or more compactly as follows:

$$\begin{bmatrix} 2 & 3 & 4 \\ 1 & 3 & 4 \\ 4 & 0 & 0 \\ 4 & 0 & 0 \\ 5 & 0 & 0 \end{bmatrix}$$

where row i is a 'vector' which indicates the successors of node i. The latter is called the 'adjacency vector' representation. Because the number of successors for each node may vary, each vector may be represented as a list instead, yielding what is called the 'adjacency list' representation; these lists may be either linked or sequential.

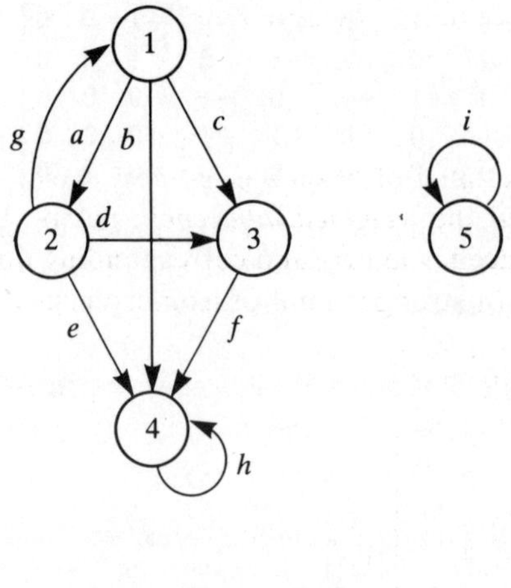

Figure 4.3(a)

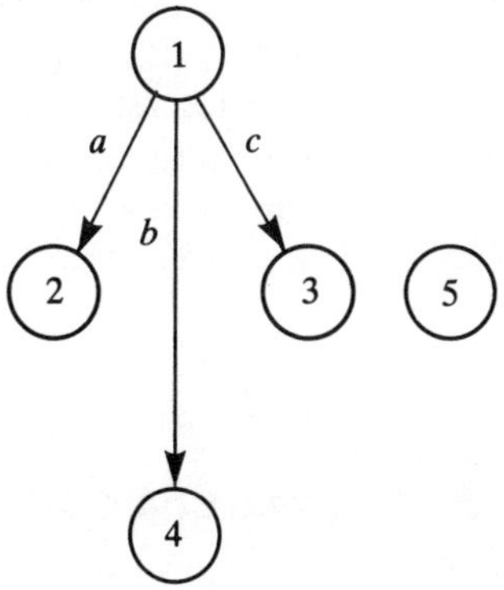

Figure 4.3(b)

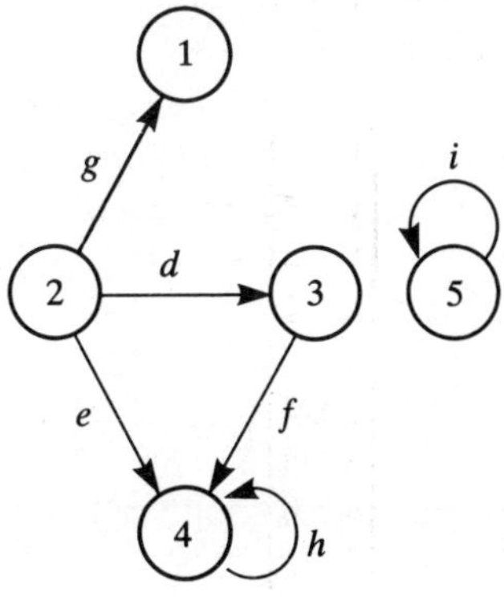

Figure 4.3(c)

4.3.1 Weighted Digraphs

A digraph $G=(S,R)$ is said to be 'node-labeled' if S is a labeled set, and 'branch-labeled' if R is a labeled set. If S is labeled, the label of each node is by convention placed inside (or next to) the circle (or point) in its pictorial representation. (If S is not otherwise explicitly labeled, it is generally considered an ordered set with the ordinate of each node serving as its implicit label.) If R is labeled, the label of each branch is by convention placed somewhere along the line in its pictorial representation.

The term *weighted digraph* refers to a digraph which is branch-labeled, with the label of a branch called its *weight*. A weighted digraph having numerical weights is sometimes called a *network*. Adjacency matrices may be generalized to represent weighted digraphs by defining

$$\mathbf{W}(i,j)= \begin{cases} \lambda(s_i,s_j) & \text{if } (s_i,s_j)\in R \\ \phi & \text{if } (s_i,s_j)\notin R \end{cases}$$

where $\lambda(s_i,s_j)$ is the label of branch (s_i,s_j) and ϕ is a special 'illegal label' symbol; $\mathbf{W}$ is called the *weighted adjacency* matrix. Digraphs having the same weighted adjacency matrix and corresponding node labels are said to 'match identically' (a stronger kind of isomorphism.)

Example For the digraph shown in Figure 4.3(a), the weighted adjacency matrix is

$$\mathbf{W}=\begin{bmatrix} \phi & a & c & b & \phi \\ g & \phi & d & e & \phi \\ \phi & \phi & \phi & f & \phi \\ \phi & \phi & \phi & h & \phi \\ \phi & \phi & \phi & \phi & i \end{bmatrix}.$$

4.3.2 Paths

Often, the complete general structure of a digraph is not of interest, but rather only some specialized (derivable) information is. For example, we may only wish to know whether or not there are paths from one node to another. Let $S=\{s_1,s_2,\ldots,s_n\}$ be an ordered set of nodes in digraph $G=(S,R)$; we define

$$\mathbf{P}_k(i,j)=\begin{cases} 1 & \text{if there is a path of length } k \text{ from } s_i \text{ to } s_j, \\ 0 & \text{otherwise.} \end{cases}$$

Since there is a path of length zero from each node to itself, $\mathbf{P}_0=\mathbf{I}$, the $(n \times n)$ identity matrix. A path of length one is a single branch, so $\mathbf{P}_1=\mathbf{C}$, the adjacency matrix. It turns out that $\mathbf{P}_k=\mathbf{C}^k$, the kth (logical) power of the Boolean matrix $\mathbf{C}$ (see Section 3.2.3). Furthermore, if $\mathbf{P}$ is the *path matrix* defined by

$$\mathbf{P}(i,j)=\begin{cases} 1 & \text{if there is a nontrivial path from } s_i \text{ to } s_j \\ 0 & \text{otherwise.} \end{cases}$$

(where paths may be of any length greater than 0), then

$$\mathbf{P}=\mathbf{P}_1 \circledvee \cdots \circledvee \mathbf{P}_n=\mathbf{C} \circledvee \mathbf{C}^2 \circledvee \cdots \circledvee \mathbf{C}^n$$

(the logical sum of the Boolean matrices $\mathbf{P}_1,\ldots,\mathbf{P}_n$). Paths of lengths greater than n in a digraph having n nodes need not be considered since they are necessarily nonsimple and hence are already represented in $\mathbf{P}$ by simple paths. If trivial (zero-length) paths are also included, we may consider instead

$$\mathbf{P}_0 \circledvee \mathbf{P}_1 \circledvee \cdots \circledvee \mathbf{P}_{n-1}=\mathbf{I} \circledvee \mathbf{C} \circledvee \mathbf{C}^2 \circledvee \cdots \circledvee \mathbf{C}^{n-1}.$$

($\mathbf{P}_n$ is not included since simple paths of length n are necessarily cycles, which are already represented by $\mathbf{P}_0$.) This sum defines the *reachability matrix*, denoted $\mathbf{R}^*$, where

$$\mathbf{R}^*(i,j)=\begin{cases} 1 & \text{if } s_j \text{ is reachable from } s_i \\ 0 & \text{otherwise} \end{cases}$$

(cf. the reachability relation defined in Section 2.2.2).

Path or reachability matrices provide information about the existence of paths, but do not describe the paths themselves in terms of the branches they contain (or the nodes they traverse). To obtain these descriptions, we assume the branches are distinctively labeled. A path $\langle (a_{i_1},b_{i_1}),(a_{i_2},b_{i_2}),\ldots,(a_{i_k},b_{i_k})\rangle$ may then be represented by the string of labels

$$\lambda(a_{i_1},b_{i_1})\lambda(a_{i_2},b_{i_2})\ldots\lambda(a_{i_k},b_{i_k}),$$

which we called a *symbolic path*; symbolic paths can be determined by symbolic matrix calculations on weighted-adjacency matrices whose elements are sets of branch-label strings.

Let **W** be the matrix defined by

$$\mathbf{W}(i,j) = \begin{cases} \{\lambda(s_i,s_j)\} & \text{if } (s_i,s_j) \in R \\ \emptyset & \text{if } (s_i,s_j) \notin R \end{cases}$$

where Ø denotes the empty set, and let $\mathbf{W}^k$ be its k-fold symbolic matrix product computed using the symbolic sum ⊕ and symbolic product ⊗ operations of Section 3.2.4. Then $\mathbf{W}^k(i,j)$ is the set of all symbolic paths of length k from node s_i to node s_j.

Example For the digraph of Figure 4.3(a),

$$\mathbf{P}_1 = \mathbf{C}^1 = \begin{bmatrix} 0 & 1 & 1 & 1 & 0 \\ 1 & 0 & 1 & 1 & 0 \\ 0 & 0 & 0 & 1 & 0 \\ 0 & 0 & 0 & 1 & 0 \\ 0 & 0 & 0 & 0 & 1 \end{bmatrix}$$

$$\mathbf{P}_2 = \mathbf{C}^2 = \begin{bmatrix} 1 & 0 & 1 & 1 & 0 \\ 0 & 1 & 1 & 1 & 0 \\ 0 & 0 & 0 & 1 & 0 \\ 0 & 0 & 0 & 1 & 0 \\ 0 & 0 & 0 & 0 & 1 \end{bmatrix}$$

$$\mathbf{P}_3 = \mathbf{C}^3 = \begin{bmatrix} 0 & 1 & 1 & 1 & 0 \\ 1 & 0 & 1 & 1 & 0 \\ 0 & 0 & 0 & 1 & 0 \\ 0 & 0 & 0 & 1 & 0 \\ 0 & 0 & 0 & 0 & 1 \end{bmatrix}$$

$$\mathbf{P} = \begin{bmatrix} 1 & 1 & 1 & 1 & 0 \\ 1 & 1 & 1 & 1 & 0 \\ 0 & 0 & 0 & 1 & 0 \\ 0 & 0 & 0 & 1 & 0 \\ 0 & 0 & 0 & 0 & 1 \end{bmatrix}$$

$$\mathbf{R}^* = \begin{bmatrix} 1 & 1 & 1 & 1 & 0 \\ 1 & 1 & 1 & 1 & 0 \\ 0 & 0 & 1 & 1 & 0 \\ 0 & 0 & 0 & 1 & 0 \\ 0 & 0 & 0 & 0 & 1 \end{bmatrix}$$

$$\mathbf{W}=\begin{bmatrix} \emptyset & \{a\} & \{c\} & \{b\} & \emptyset \\ \{g\} & \emptyset & \{d\} & \{e\} & \emptyset \\ \emptyset & \emptyset & \emptyset & \{f\} & \emptyset \\ \emptyset & \emptyset & \emptyset & \{h\} & \emptyset \\ \emptyset & \emptyset & \emptyset & \emptyset & \{i\} \end{bmatrix}$$

$$\mathbf{W}^2=\begin{bmatrix} \{ag\} & \emptyset & \{ad\} & \{ae,cf,bh\} & \emptyset \\ \emptyset & \{ga\} & \{gc\} & \{gb,df,eh\} & \emptyset \\ \emptyset & \emptyset & \emptyset & \{fh\} & \emptyset \\ \emptyset & \emptyset & \emptyset & \{hh\} & \emptyset \\ \emptyset & \emptyset & \emptyset & \emptyset & \{ii\} \end{bmatrix}.$$

4.3.3 Fundamental Sets

We say B is a *fundamental* subset of S if all the members of S can be 'generated', in some sense, from the members of B. An example is the node base B of $G=(S,R)$, as defined in Section 4.1.4. Let $B=\{s_1,\ldots,s_m\}$ be a node base, and $S_i=\{s\in S\,|\,(s_i,s)\in R^*\}$, where R^* is the reachability relation. Note that the sets $\{S_1,S_2,\ldots,S_m\}$ cover S, but need not be pairwise disjoint.

A partition of S can be obtained by defining, for each $a\in S$,

$$S_a=\{s\in S\,|\,(s,a)\in R^* \text{ and } (a,s)\in R^*\},$$

the set of all nodes strongly connected to a. 'Strong connectivity' is in fact an equivalence relation, defined by

$$\{(a,b)\in S\times S\,|\,(a,b)\in R^* \text{ and } (b,a)\in R^*\},$$

whose equivalence classes correspond to the MSCRs of G and partition S. Let M be a set containing exactly one member of each MSCR; this set may also be regarded as a fundamental set of nodes. M always contains a node base as a subset, is such that every cycle in G traverses at most one of its nodes, and is such that each of its nodes generates one MSCR.

Let T be a spanning tree of a connected digraph G. It can be shown that each chord of G with respect to T is in a unique simple semicycle of G containing no other chords. The set of simple semicycles of G corresponding to each chord is a fundamental set, called a *loop basis* Λ. All other semicycles or loops are linear combinations of the 'basis loops' (members of Λ), i.e., Λ is an independent set.† (Recall from Section 2.1.5 that a linear combination of sets is a generalized disjoint union of the sets.) The size of Λ is the cyclomatic number of G.

†Independence here is in the vector-space sense; see Section 2.5.1 and Exercise 30 of Chapter 2.

It can also be shown that each branch of a spanning tree T of G is in a unique cut of G containing no other branch of T. The set of cuts of G corresponding to the branches of a given T is a fundamental set, called a *cut basis* K. A *cut-set* is a linear combination of cuts, and every cut-set is a linear combination of the 'basis cuts' (members of K), i.e., K is also an independent set (hence not all disconnecting sets of branches are cut-sets). The size of K is the rank of G.

Let $G=(S,R)$ be a connected digraph, with $S=\{s_1,\ldots,s_n\}$ and $R=\{r_1,\ldots,r_m\}$ being ordered sets, and $\nu=m-n+1$, $\rho=n-1$. Let $\Lambda=\{\ell_1,\ldots,\ell_\nu\}$ be a loop basis and $\mathrm{K}=\{k_1,\ldots,k_\rho\}$ be a cut basis with respect to a given spanning tree T. Then, the *loop matrix* **L** is an $(m\times\nu)$ matrix where

$$\mathbf{L}(i,j)=\begin{cases} +1 \text{ if } r_i \text{ is a collinear branch in loop } \ell_j \\ \ \ 0 \text{ if branch } r_i \text{ is not in loop } \ell_j \\ -1 \text{ if } r_i \text{ is an opposing branch in loop } \ell_j \end{cases}$$

and the *cut matrix* **K** is a $(\rho\times m)$ matrix where

$$\mathbf{K}(i,j)=\begin{cases} +1 \text{ if } r_j \text{ is a collinear branch in cut } k_i \\ \ \ 0 \text{ if branch } r_j \text{ is not in cut } k_i \\ -1 \text{ if } r_j \text{ is an opposing branch in cut } k_i. \end{cases}$$

We adopt the convention that the orientation of a loop or cut corresponds to the direction of that branch contained therein having the lowest ordinate with respect to the given ordering of R. (Alternatively, since there is a unique correspondence between basis loops and chords of T, and between basis cuts and branches of T, the chords and branches of T may serve to define the orientation of the basis loops and cuts.)

It can be shown that

$$\mathbf{B}\circledast\mathbf{L}=0 \qquad \text{and} \qquad \mathbf{K}\circledast\mathbf{L}=0$$

where **B** is the incidence matrix, and 0 is a matrix whose elements are all equal to zero. ($\circledast$ denotes the ordinary matrix product.) These equations enable us to determine relatively easily the loops and cuts of a graph given its incidence matrix. In matrix theory or linear algebra terms, the n rows of **B** are linearly dependent. The $n-1$ rows of **K** are linearly independent and span the same space (i.e., the 'row space' of **B**). The $m-n+1$ columns of **L** are linearly independent and span the 'null space' of **B** or **K**. It should be emphasized that the equations $\mathbf{B}\circledast\mathbf{L}=0$ and $\mathbf{K}\circledast\mathbf{L}=0$ do not have unique solutions since spanning trees are not unique (nor are loop and cut bases). However, any solution of the equations will provide a loop basis and a cut basis, from which all other loops and cut-sets can be derived.

The foregoing can be generalized to handle disconnected digraphs by treating each of the p pieces separately or together. For the latter case, we recall that $\nu=m-n+p$ and $\rho=n-p$. For additional information, see Liu [4.10] and Pearl [4.15].

Example For the digraph of Figure 4.3(a), {1,5} and {2,5} are node bases. The equivalence classes corresponding to the MSCRs are {1,2}, {3}, {4}, and {5}. A spanning tree of the digraph is shown in Figure 4.3(b).

For this tree, the chords are shown in Figure 4.3(c). The 'fundamental' loop (unique simple semicycle) containing chord d and no other chord is the sequence of branches $\langle d,c,a\rangle$. The set of basis loops is tabulated below:

CHORD	BASIS LOOP
d	$\ell_d=dca$
e	$\ell_e=eba$
f	$\ell_f=fbc$
g	$\ell_g=ga$
h	$\ell_h=h$
i	$\ell_i=i$

Note that ℓ_d consists of the ordered set of branches $\{a,c,d\}$ (listed according to the given [alphabetical] branch ordering), the first branch of which can be used to specify the reference direction; branch d is collinear with reference branch a and branch c is opposing. Continuing in this fashion, we obtain the loop matrix

$$\mathbf{L} = \begin{array}{c} \\ a\\ b\\ c\\ d\\ e\\ f\\ g\\ h\\ i \end{array}\begin{array}{c} \begin{array}{cccccc} d & e & f & g & h & i \end{array}\\ \begin{bmatrix} +1 & +1 & 0 & +1 & 0 & 0\\ 0 & -1 & +1 & 0 & 0 & 0\\ -1 & 0 & -1 & 0 & 0 & 0\\ +1 & 0 & 0 & 0 & 0 & 0\\ 0 & +1 & 0 & 0 & 0 & 0\\ 0 & 0 & -1 & 0 & 0 & 0\\ 0 & 0 & 0 & +1 & 0 & 0\\ 0 & 0 & 0 & 0 & +1 & 0\\ 0 & 0 & 0 & 0 & 0 & +1 \end{bmatrix}\end{array}.$$

The set of cuts corresponding to the branches in the tree of Figure 4.3(b) are tabulated below:

BRANCH	CUT
a	$k_a=\{a,d,e,g\}$
b	$k_b=\{b,e,f\}$
c	$k_c=\{c,d,f\}$

k_a is that minimal set of chords whose removal, together with branch a, from the digraph would increase the number of its pieces. Note that removing $\{a,d,e,g\}$ would partition the original digraph into the node-sets {2}, {1,3,4} (and {5}). Note in addition that branch a (which we choose as the reference branch because it is first according to the given [alphabetical] branch ordering) goes from {1,3,4} to {2}, while branches d, e, and g go in the opposite direction. Continuing in this fashion, we obtain the cut matrix

$$\mathbf{K} = \begin{matrix} k_a \\ k_b \\ k_c \end{matrix} \overset{\begin{matrix} a & b & c & d & e & f & g & h & i \end{matrix}}{\begin{bmatrix} +1 & 0 & 0 & -1 & -1 & 0 & -1 & 0 & 0 \\ 0 & +1 & 0 & 0 & +1 & +1 & 0 & 0 & 0 \\ 0 & 0 & +1 & +1 & 0 & -1 & 0 & 0 & 0 \end{bmatrix}}.$$

The reader should verify that $\mathbf{K} \circledast \mathbf{L} = 0$.

It should be noted that k_a, k_b, k_c correspond to the second, fourth and third rows, respectively, of the incidence matrix **B** given earlier. This happenstance results from a fortuitous choice of the spanning tree, and of reference directions. For other choices, all we can say is that rows of **K** are linear combinations of rows of **B**, and vice versa. For example, the first row of **B** is a linear combination (disjoint union) of k_a, k_b and k_c: i.e.,

$$\{a,b,c,g\} = \{a,d,e,g\} \oplus \{b,e,f\} \oplus \{c,d,f\}.$$

The last row of **B** equals (say) $k_a \oplus k_a$.

4.3.4 Data Structure

In the foregoing, we have been concerned primarily with representations of structural information about a set of objects, rather than about the objects themselves. Here, we regard the objects (i.e., the nodes of a digraph) as data items whose values are to be stored within a computer, together of course with information (i.e., the branches) relating the data items. One way of representing digraphs is to store the data-item values in an arbitrary fashion, and to define an array S such that $S(i)$=location of the ith data-item value, together with a (adjacency) matrix **C** such that $\mathbf{C}(i,j)=1$ if the ith data item is related to (i.e., is an immediate predecessor of) the jth data item, else $\mathbf{C}(i,j)=0$. Alternatively, $S(i)$ may contain the ith data-item value itself (rather than its location).

Plexes An alternative to the representation of digraphs utilizing array (mapped) data structures is their representation as linked data structures (recall Section 1.2.4). The values of data items (nodes of the digraph) may be stored in an arbitrary fashion as before, but each is now stored along with explicit pointers to its immediate successors. In other words, each data node contains the value associated with a data item in addition to a number of pointer fields representing branches from the data item to another. Such a linked data structure, with a distinguished node base, is called a *plex* (or sometimes 'network'). (The node base may be represented as a separate set if its size is not 1.) Two special cases (lists and trees) are discussed below.

One difficulty with representing digraphs by plexes is that the number of pointer fields which data nodes contain may vary from zero (for sink nodes) to some maximum M (for nodes having the largest out-degree). It would be wasteful for each data node to have M pointer fields when many or most of them may be null; on the other hand, the alternative is to add to each data node an indicator which specifies how many pointers it has.

Lists A *linear linked list* (or *chain*) is a linked data structure having only one pointer field per data item, so that each item points to one 'next' item (its immediate successor) in the list. The 'last' item in the linear linked list (the *tail*) is distinguished by having a null pointer. The base of a linear linked list is usually a pointer to the 'first' data item (called the *head*); if the linear linked list is empty (i.e., contains no data items), then the base may still be defined, but is a null pointer. A linear linked list may also be regarded as an ordinal data structure in which the data items are ordered by the pointers.

A *circular list* (or 'ring') is similar to a linear linked list except that the pointer in the tail is not null, but instead points to the head. This permits access to all data items in the list starting at any one of them. (While each item may thus be the 'first', it is more common to distinguish one item as the list head.)

A *symmetric* (or 'doubly linked' or '2-way') *list* is similar to a linear linked list except that each data item has a second pointer field which specifies the location of its immediate predecessor in the list. This also permits access to all data items in the list starting at any one of them, but its importance is that it allows rapid access to preceding data items.

Figure 4.3.4(a) illustrates a linear linked list; its head is at address 112 and its tail is at address 130. A null pointer has value -1. Figures 4.3.4(b) and (c) show circular and symmetric lists, respectively.

We remark that the term 'list' is often used to mean any ordinal data structure, not just what we have called a linear linked list. For example, a sequence is sometimes called a 'sequential' list, as opposed to a 'linked' list. Also, lists (sequential or linked) may be characterized by access restrictions to individual data items—i.e., to only the head or tail. A *stack* is a list where only one end (called its 'top') may be accessed; insertions ('pushing') and deletions ('popping') of data items must be made at the top. A *queue* is a list where both ends may be accessed, but insertions are allowed only at one end and deletions at the other. We also say that a stack is a *last-in–first-out* (LIFO) data structure, while a queue is a *first-in–first-out* (FIFO) one.

Trees A *tree* (or 'tree-structure', when we wish to distinguish the data structure from the algebraic system) is a plex whose underlying digraph is acyclic and rooted. It differs from a list in having generally more than one pointer field per data item, two in the most common case of binary tree structures. The base of a tree is usually a pointer to the root.

Figure 4.3.4(d) illustrates a binary tree whose root is at address 106. A convenient pictorial representation is shown in Figure 4.3.4(e).

S-expressions Suppose the data items (or nodes) of a digraph are distinctively labeled: the labels may be symbolic names for the items, the locations of the items, or perhaps the values of the items themselves. A digraph may be

BASE:	112	112	112	106
100				
103	Gu : 106	Gu : 106	121 : Gu : 106	−1 : Gu : −1
106	Ma : 115	Ma : 115	103 : Ma : 115	121 : Ma : 115
109				
112	Ba : 121	Ba : 121	−1 : Ba : 121	−1 : Ba : −1
115	Pa : 130	Pa : 130	106 : Pa : 130	−1 : Pa : 130
118				
121	Co : 103	Co : 103	112 : Co : 103	112 : Co : 103
124				
127				
130	Pi : −1	Pi : 112	115 : Pi : −1	−1 : Pi : −1
133				
136				
	(a)	(b)	(c)	(d)

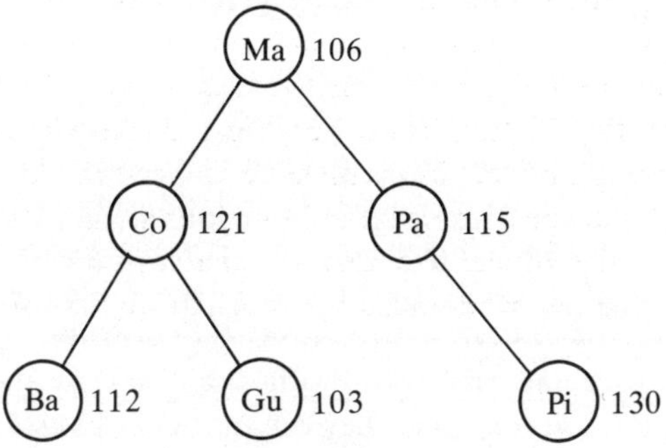

(e)

Figure 4.3.4

represented by a character string as follows. We define an *S-expression* for the subgraph generated by the data item (i.e., node) labeled x, denoted $\mathbf{S}(x)$, as the string $(x,\mathbf{S}(y_1),...,\mathbf{S}(y_n))$, where $y_1,\ldots,y_n$ are the labels of the immediate successors of x; if x has no successor, $\mathbf{S}(x)=x$. This is a recursive definition in that each $\mathbf{S}(y_i)$ is itself an S-expression, which may contain x, and so to avoid circularity, we adopt an expression 'expansion' rule: The expansion of $\mathbf{S}(x)=(x,\mathbf{S}(y_1),...,\mathbf{S}(y_n))$ is obtained by substituting for $\mathbf{S}(y_i)$ the expansion of the string $(y_i,\mathbf{S}(z_{i1}),...,\mathbf{S}(z_{im}))$, in left-to-right order (i.e., for $i=1$, then for $i=2,\ldots$, and finally for $i=n$), with the proviso that if $\mathbf{S}(y)$ has been previously expanded, then it is to be replaced by y. The result is a string of labels within nested parentheses that reflect structural relationships; spaces rather than commas are often used to delimit the elements of each string. (The term 'S-expression' is used because of a similarity with the list structures in LISP; see McCarthy [4.12].)

An analogous *M-expression* may be defined such that a node label is followed by a parenthesized list of its successors, i.e., $\mathbf{M}(x)=x(\mathbf{M}(y_1),...,\mathbf{M}(y_n))$, and $\mathbf{M}(y_i)=y_i$ if y_i has no successor or was previously listed. We remark that the right parentheses in S- or M- expressions may be omitted if each corresponding left parenthesis has an associated *count* equaling the number of terms it parenthesizes, or if each left parenthesis is repeated *count* times.

If, in the above, x is an out-root, then $\mathbf{S}(x)$ represents the entire digraph. Otherwise, we choose a node base $\{x_1,x_2,\ldots,x_r\}$ and represent the digraph by the concatenation $\mathbf{S}(x_1)\mathbf{S}(x_2)...\mathbf{S}(x_r)$, which is to be expanded from left to right, again with the proviso that, for any node-label y (even a successor of two base nodes), $\mathbf{S}(y)$ is to be expanded only once.

4.3.5 Comparisons of Data Structures

The three data structure representations of digraphs (adjacency matrices, plexes, and S-expressions) we have discussed above have different advantages. Adjacency matrices are primarily useful for analyzing structural properties—e.g., to determine connectivity relationships, to compute path lengths, to find SCRs and cliques, and so forth. Plexes are primarily useful for information storage and retrieval, where data-node pointers permit fast access to related pieces of information. S-expressions provide a good method for representing digraphs during input and output operations, at least for 'typographical' reasons; they are particularly useful for representing trees (as we shall see).

Examples Consider the digraph of Figure 4.3(a). Let D_i denote the data-value associated with the ith data item (i.e. with node i). Suppose these data values are stored in an arbitrary fashion as shown in Figure 4.3.5(a). The digraph $G=(S,R)$ can

100	
101	
102	*D*4
103	*D*3
104	
105	*D*2
106	*D*5
107	
108	
109	*D*1
110	

(a)

200	
201	109
202	105
203	103
204	102
205	106
206	
207	
208	
209	
210	

(b)

301	2 : 3 : 4 : −1
302	1 : 3 : 4 : −1
303	4 : −1 : :
304	4 : −1 : :
305	5 : −1 : :

(c)

100	109 : 106 : −1 : : :
101	
102	*D*4 : 102 : −1 : −1 : −1 : −1
103	*D*3 : 102 : −1 : −1 : −1 : −1
104	
105	*D*2 : 109 : 103 : 102 : −1 : −1
106	*D*5 : 106 : −1 : −1 : −1 : −1
107	
108	
109	*D*1 : 105 : 103 : 102 : −1 : −1

(d)

Figure 4.3.5

then be represented by two arrays S and R, where S is a 'dope vector' (cf. Section 2.3.1), as shown in Figure 4.3.5(b), and R is an adjacency matrix

$$\begin{bmatrix} 0 & 1 & 1 & 1 & 0 \\ 1 & 0 & 1 & 1 & 0 \\ 0 & 0 & 0 & 1 & 0 \\ 0 & 0 & 0 & 1 & 0 \\ 0 & 0 & 0 & 0 & 1 \end{bmatrix}$$

which can be represented in conventional matrix fashion (cf. Section 3.2.6), or perhaps using sequential adjacency lists, as shown in Figure 4.3.5(c). A plex representation of the digraph is shown in Figure 4.3.5(d). The S-expression representation of the digraph, for the node base $\{1,5\}$, can be found as follows:

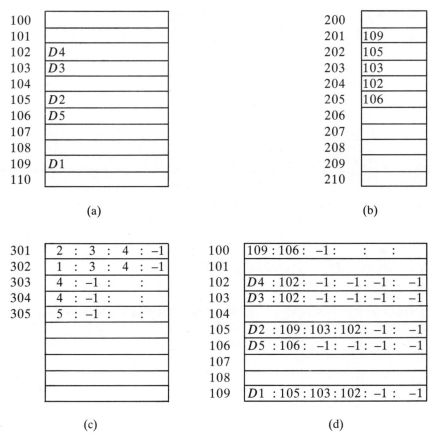

$$\begin{aligned}
\mathbf{S}(1)\mathbf{S}(5) &= (1,\mathbf{S}(2),\mathbf{S}(3),\mathbf{S}(4))\mathbf{S}(5)\\
&= (1,(2,\mathbf{S}(1),\mathbf{S}(3),\mathbf{S}(4)),\mathbf{S}(3),\mathbf{S}(4))\mathbf{S}(5)\\
&= (1,(2,\ 1\ ,\mathbf{S}(3),\mathbf{S}(4)),\mathbf{S}(3),\mathbf{S}(4))\mathbf{S}(5)\\
&= (1,(2,\ 1\ ,(3,\mathbf{S}(4)),\mathbf{S}(4)),\mathbf{S}(3),\mathbf{S}(4))\mathbf{S}(5)\\
&= (1,(2,\ 1\ ,(3,(4,\mathbf{S}(4))),\mathbf{S}(4)),\mathbf{S}(3),\mathbf{S}(4))\mathbf{S}(5)\\
&= (1,(2,\ 1\ ,(3,(4,\ 4\ )),\mathbf{S}(4)),\mathbf{S}(3),\mathbf{S}(4))\mathbf{S}(5)\\
&= (1,(2,\ 1\ ,(3,(4,\ 4\ )),\ 4\ ),\mathbf{S}(3),\mathbf{S}(4))\mathbf{S}(5)\\
&= (1,(2,\ 1\ ,(3,(4,\ 4\ )),\ 4\ ),\ 3\ ,\mathbf{S}(4))\mathbf{S}(5)\\
&= (1,(2,1,(3,(4,4)),4),3,4)\mathbf{S}(5)\\
&= (1,(2,1,(3,(4,4)),4),3,4)(5,\mathbf{S}(5))\\
&= (1,(2,1,(3,(4,4)),4),3,4)(5,5).
\end{aligned}$$

Using spaces rather than commas as delimiters, we have

$$\mathbf{S}(1)\mathbf{S}(5)=(1(2\ 1(3(4\ \ 4))4)3\ 4)(5\ 5).$$

4.4 REPRESENTATIONS OF TREES

Trees, as special cases of digraphs, may be represented in precisely the same ways: e.g., by adjacency or incidence matrices, possibly weighted, or by plex or S-expression data structures. Since trees are much simpler than general digraphs, it is natural for those representations to be much simpler too. For example, since a (out-oriented) tree is acyclic, it follows that for a suitable ordering of its nodes, the adjacency matrix is upper triangular. In fact, the $(n \times n)$ adjacency matrix for a tree containing n nodes contains exactly $n-1$ nonzero elements, hence sparse matrix representations are almost always preferable.

Example Consider the tree shown in Figure 4.4. Its adjacency matrix is

$$\mathbf{C}=\begin{bmatrix} 0 & 1 & 1 & 0 & 0 & 0\\ 0 & 0 & 0 & 1 & 1 & 0\\ 0 & 0 & 0 & 0 & 0 & 1\\ 0 & 0 & 0 & 0 & 0 & 0\\ 0 & 0 & 0 & 0 & 0 & 0\\ 0 & 0 & 0 & 0 & 0 & 0 \end{bmatrix}.$$

4.4.1 Matrix Representation of t-ary Trees

If an n-node tree is t-ary, for a known t (necessarily less than n), then it may be represented by an $(n \times t)$ matrix $\mathbf{M}$ in the following manner. Let $S=\{s_1,s_2,\ldots,s_n\}$ be the ordered set of nodes of an out-oriented tree, with

s_1 being the root, and regard the tree as positional. Then define

$$\mathbf{M}(i,j)=\begin{cases} k & \text{if } s_k \text{ is the } j\text{th child of } s_i \\ 0 & \text{if the } j\text{th child of } s_i \text{ is null} \end{cases}$$

for $1 \le i \le n$, $1 \le j \le t$. (In the above, k ranges from 2 to n.) Row i of **M** identifies the t nodes that are children of s_i. Note that out of $n \cdot t$ elements, again only $n-1$ are nonzero. Although over half the elements are zero, this representation is common, especially for binary trees ($t=2$). (Note that this is a special case of the 'adjacency vector' representation for general directed graphs.)

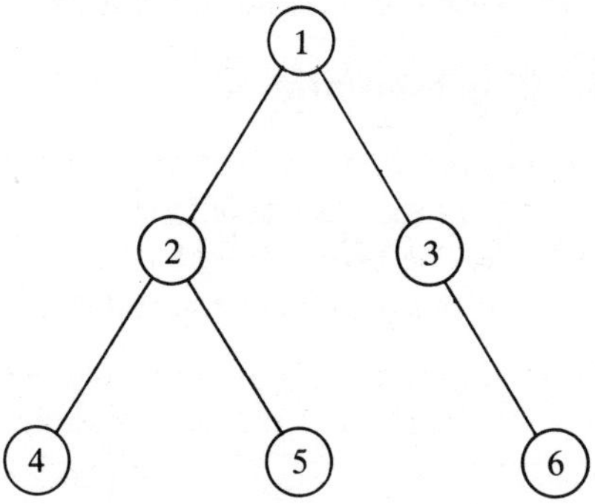

Figure 4.4

Example For the tree of Figure 4.4,

$$\mathbf{M}=\begin{bmatrix} 2 & 3 \\ 4 & 5 \\ 0 & 6 \\ 0 & 0 \\ 0 & 0 \\ 0 & 0 \end{bmatrix}.$$

4.4.2 In-oriented Trees

A variation of the foregoing scheme is worth noting. Let

$$\mathbf{M}'(i)=\begin{cases} j & \text{if } s_j \text{ is the parent of } s_i \\ 0 & \text{if } s_i \text{ has no parent (i.e., is the root)} \end{cases}$$

for $1 \le i \le n$. $\mathbf{M}'$ is an n-vector having only one zero, and so is a particularly compact way of representing a tree. However, it is not as useful for most applications since common tasks (such as traversing the tree, or finding all children of the root, or finding all leaves) require much searching. This is one of many examples of a space vs. time trade-off. We remark that **M**′ is, in essence, an in-oriented tree representation whereas **M** is out-oriented.

Example For the tree of Figure 4.4,

M′=[0 1 1 2 2 3].

Note that positional information is lost; node 6 may be either the left or right child of node 3.

4.4.3 Linked Representations

Representation by the matrix, as above, of an ordered but nonpositional t-ary tree which is not full is rather wasteful, especially when t is large and there are many null leaves. The problem is that each row of **M** always has space reserved for the identities of t children whether or not it has any at all.

100	
101	4 : −1
102	5 : −1
103	2 : 200
104	6 : −1
105	3 : 203
106	1 : 202
200	101 : 201
201	102 : −1
202	103 : 204
203	104 : −1
204	105 : −1

(a)

100	
101	4 : −1 : −1
102	5 : −1 : −1
103	2 : 101 : 102
104	6 : −1 : −1
105	3 : −1 : 104
106	1 : 103 : 105
107	

(b)

100	
101	4 : −1 : 102
102	5 : −1 : −1
103	2 : 101 : 105
104	6 : −1 : −1
105	3 : 104 : −1
106	1 : 103 : −1
107	

(c)

Figure 4.4.3.

However, note that each row (node) in effect contains a list of t sequential 'pointers' to other rows (nodes), and so **M** may be regarded as a linked data structure (or plex) satisfying the requirements of tree structures, as given in Section 4.3.4.

Trees (the digraphs) may be represented by trees (the data structures) in several ways. For example, the aforementioned lists of pointers may be represented as separate linear linked lists which are not physically part of the nodes. For the tree of Figure 4.4, this is illustrated in Figure 4.4.3(a). Of course, the most natural linked representation for a t-ary tree is the data structure where each node includes t pointer fields, each pointing to one of its children. This is illustrated in Figure 4.4.3(b) for $t=2$. Recalling the 'binary tree transformation' discussed in Section 4.2.3, any (nonpositional) t-ary tree can also be represented by a binary tree which has two pointer fields, one pointing to its first child, and the other to its right sibling. This is illustrated in Figure 4.4.3(c). Pointers to null leaves may be eliminated (at the expense of positional information) by associating with each node a number specifying how many 'non-null' pointers it has, followed by these pointers. An alternative is to append to the sequential list of pointers associated with each node a special 'null pointer' to mark the end of the list.

4.4.4 S-expressions

The form of S-expressions (and M-expressions) is simpler for trees than for general digraphs. One simplifying feature is that each node (or node label) appears in an S-expression exactly once, so the complications associated with circularity do not exist. Furthermore, the S-expression σ for a tree T is such that each subtree of T is associated with a parenthesized substring of σ. For the tree of Figure 4.4, the S-expression is

(1 (2 4 5)(3 6)).

An S-expression provides a linear (one-dimensional) representation for a two-dimensional tree. This is of special importance when a tree is to be input to or output from a computer in what is necessarily a linear (character-by-character) form. Fortuitously, simple procedures exist for translating from (to) S-expression strings to (from) linked tree structures, as may be found in LISP compiler/interpreters. We describe such procedures in Section 5.3.2.

EXERCISES

1. For the digraph G shown in Figure 4A, specify whether it is or is not

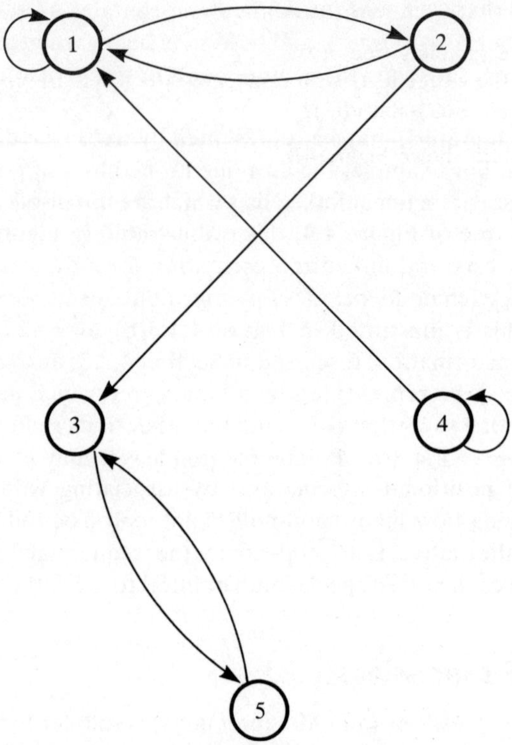

Figure 4A

(a) connected, strongly connected, or disconnected
(b) symmetric, asymmetric, or antisymmetric
(c) transitive
(d) reflexive or irreflexive
(e) cyclic or acyclic.

2. For the given digraph G shown in Figure 4A,

(a) show the symmetrized digraph of G
(b) show its complement $-G$
(c) show a diametric path of G
(d) find its strong components
(e) determine the condensed graph of G
(f) determine the cliques of G and $-G$.

3. Show that a connected digraph contains an Eulerian cycle iff each node has in-degree equal to its out-degree.

4. Show that completeness is a sufficient, but not a necessary, condition for a digraph to contain a Hamiltonian cycle.

5. Give an example of a digraph which is unilaterally (but not strongly) connected.

6. For the graph shown in Figure 4A,

(a) show a spanning tree of G
(b) show the cotree of G with respect to your spanning tree
(c) what are $\rho(G)$ and $\nu(G)$?

7. For the tree in shown Figure 4B,

(a) show the preorder and postorder traversal sequences
(b) show the binary tree resulting from the transformation of Section 4.2.3.

8. Show that the height of a depth-balanced binary tree having N nodes is $\mathbf{int}(\mathbf{log}_2(N))$.

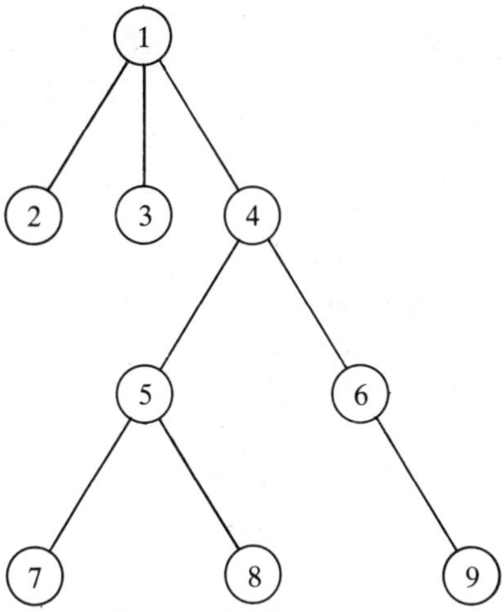

Figure 4B

9. For the digraph of Figure 4C(a), give its

(a) (Boolean) adjacency matrix **C**
(b) incidence matrix **B**
(c) weighted adjacency matrix **W**
(d) path matrix **P**
(e) reachability matrix $\mathbf{R}^*$
(f) weighted path matrix $\mathbf{W}^*$
(g) loop matrix **L** (with respect to the spanning tree of Figure 4C(b))
(h) cut matrix **K** (with respect to the spanning tree of Figure 4C(b))
(i) plex representation
(j) S-expression representation
(k) condensed graph.

10. (a) How are the data items of Figure 4.3.4 logically ordered?
(b) Show the result of inserting 'LYCHEE' into the three lists of Figure 4.3.4(a), (b) and (c) while preserving the logical ordering.

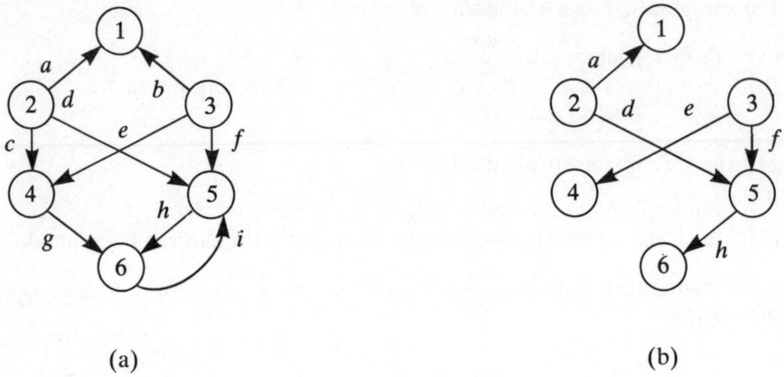

Figure 4C

11. For the tree of Figure 4D, give its

(a) matrix representation **M**
(b) *n*-vector representation **M**′
(c) linear linked list representation
(d) S-expression representation.

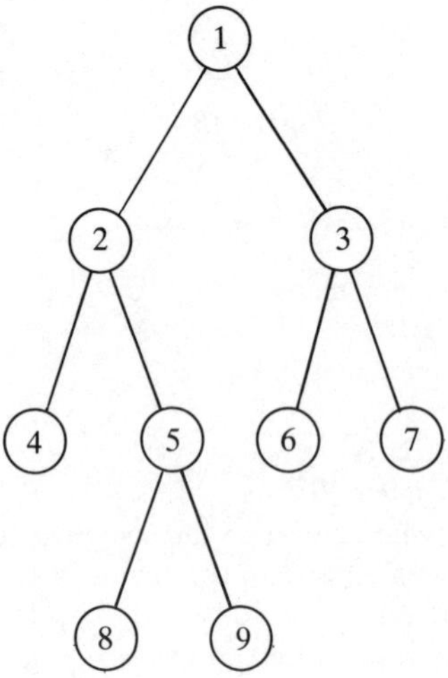

Figure 4D

PROGRAMMING ASSIGNMENTS

1. Write a program which, given an adjacency matrix **C** representation of a digraph, computes

(a) **P** (b) $\mathbf{R}^*$ (c) **L** (d) **K**.

2. Write a program which, given a weighted adjacency matrix **W** representation of a digraph, determines

(a) $\mathbf{W}^*$
(b) its S-expression representation.

3. Write a program which, given an adjacency matrix **C** representation of a digraph, determines

(a) whether it is symmetric, antisymmetric, asymmetric, reflexive, irreflexive, and/or transitive
(b) its MSCRs.

4. Create the three lists of Figure 4.3.4 by using
(a) LISP
(b) PL/I pointer facilities
(c) the array facilities of any language.

5. Create the trees of Figure 4.4.3 by using
(a) PL/I pointer facilities
(b) the array facilities of any language.

5 APPLICATIONS—ALGORITHMS AND FORMAL COMPUTATION

5.1 A SURVEY OF APPLICATIONS

We introduce in this section various applications of digraphs and trees to the modeling of software systems. These applications will be described in greater detail in later sections and chapters, but are mentioned here briefly to provide motivation.

5.1.1 Flowgraphs and Programs

A *flowgraph* is defined as a digraph which models a program flowchart, as follows. Each box of the flowchart is a node of the flowgraph, and each line connecting boxes of the flowchart is a branch connecting corresponding nodes. Some flowcharts are drawn with a branch intersecting and terminating at another branch, as shown in Figure 5.1.1. In a flowgraph, these branches are redrawn to terminate at a node. While there are no inherent differences between the 'chart' and the 'graph', from the latter it is clear that any program structure can be represented as a matrix (specifically, an adjacency matrix). Matrix algorithms can then be used to analyze structural properties of programs. For example, structural errors such as unreachable code or dead-end code can be easily found by inspection of the reachability matrix. We remark that cross-reference tables produced by compilers are just adjacency lists (cf. Section 4.3). Furthermore, optimizing compilers generally attempt to move loop-invariant code from inside a loop to the

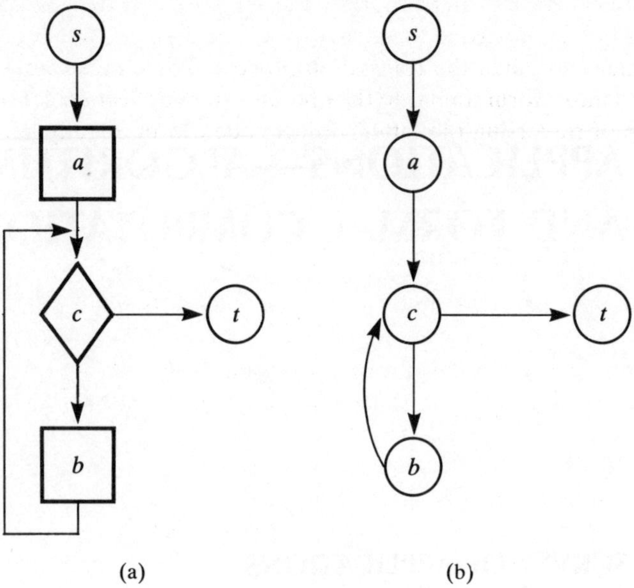

Figure 5.1.1

outside; loop or SCR detection algorithms are employed as a part of this process.

A computer program may be thought of as a set of data, portions of which are to be interpreted (executed) by a computer as instructions which operate on other portions. A *computer* is a machine (or *automaton*) which moves from configuration to configuration according to a set of transition rules. Automata can be represented as directed graphs, allowable transitions being associated with branches connecting configuration nodes.

5.1.2 Languages and Compilers

A flowgraph may be thought of as a pictorial representation of an algorithm. An algorithm may be represented in a variety of other ways, the way most suitable for processing (interpretation) by a computer being that of a program expressed in some programming language—i.e., as a string of characters from some alphabet, where the string must satisfy given structural rules (to be 'syntactically' correct). A string is syntactically correct if it can be 'parsed' according to a set of syntax (string substitution) rules, called a *grammar*. Certain grammars (of the 'context-free' class) can be represented as directed graphs, called 'syntactic charts'. A parse of a string with respect to a context-free grammar can be represented as an ordered tree. A 'compiler' which translates input strings to parse trees can be designed to operate using syntactic charts to guide the translation process. These parse trees can be transformed into a form similar to the operator trees of Section 4.2.4, and in the process of traversing the latter, object code can be readily generated.

5.1.3 Data Structures

We have previously shown how digraphs can be represented using various data structures. Taking the opposite point of view, we observe that certain data structures can be modeled formally by a digraph. Consider the case of plexes (networks) or trees which arise naturally as data base structures, or linear linked lists which are used in list processing applications. In the former case, digraph algorithms may be used to determine access paths to specified data nodes, to traverse and search all nodes, to retrieve related nodes (e.g., cliques), or to find matching (isomorphic) graphs or subgraphs. Analysis and optimization techniques for general digraphs are also applicable to the evaluation and comparison of the performances of given data structures, and to the determination of the best way to organize data in a given structure.

5.1.4 Operating Systems

One of the main functions of operating systems is to provide the environment necessary for parallel processing. We use the term 'parallel processing' to mean the concurrent execution of several processes—whether with one processor (by 'interleaving' operations), or with several processors (whereby the operations of different processes overlap in time). Of primary importance is the resource allocation task: the operating system must decide when to allocate resources, and how much to allocate to which processes; examples of resources are processors, storage, and I/O units. These decisions must be made according to a number of different and often competing criteria (such as minimizing turnaround or response times, or maximizing throughput), subject to constraints of various sorts (e.g., priorities, functional determinacy, and synchronization). Many problems of the sort suggested here can be treated by formalizing the 'state' of the operating system as essentially the composition of individual process states, and defining state transitions according to allocation decisions. Relationships between processes can, of course, be modeled by a graph whose nodes are the processes and whose (labeled) branches correspond to possible interactions. A graph model in which some nodes represent events is also useful.

5.2 DIGRAPH ALGORITHMS

5.2.1 Reachability Matrix

The reachability matrix of $G=(S,R)$ was defined in Section 4.3.2 as

$$\mathbf{R}^* = \mathbf{I} \circledvee \mathbf{C} \circledvee \mathbf{C}^2 \circledvee \cdots \circledvee \mathbf{C}^{n-1}$$

where $\mathbf{C}$ is the (Boolean) adjacency matrix of G. Evidently, $\mathbf{R}^*$ can be computed via a sequence of logical matrix multiplications and additions. However, more efficient algorithms exist, one of which is that of Warshall [5.46]. The matrix $\mathbf{R}^*$ may be computed as follows. Let $\mathbf{R}(i,j)$ denote the row i, column j element of an array $\mathbf{R}$, which is initialized to $\mathbf{C}$. Then scanning $\mathbf{R}$ columnwise (i.e., for $j=1, \ldots, n$), we perform the following:

for $i=1,\ldots,n$
  if $\mathbf{R}(i,j)=1$ *then*
  for $k=1,\ldots, n$, *set* $\mathbf{R}(i,k)=\mathbf{R}(i,k) \vee \mathbf{R}(j,k)$.

On completion of the j loop, $\mathbf{R}^*$ is given by $\mathbf{I} \circledvee \mathbf{R}$. The foregoing is based on the observation that a path must exist from i to k in the event that we already have so concluded, or in the event that there are paths from i to j and from j to k.

As suggested in Section 5.1.1, reachability information may be used in compilers to detect structural errors in programs. See Karp [5.27], for example. Related to this are methods of determining 'representative' sets of data to test programs. Ideally, each path from entry to exit should be traversed; e.g., see Howden [5.25].

5.2.2 Traversals

A traversal of a digraph is a means of accessing (or 'visiting') every node of the graph. A desirable traversal sequence would follow a Hamiltonian path, but unfortunately such a path does not exist for all graphs. In general, some nodes must be revisited, hence traversal algorithms require 'backtracking'.

Depth-first traversal starting at a designated node s_0 is best described as a recursive algorithm. First, the starting node s_0 is visited. Next, an unvisited node adjacent to node s_0 is designated as the starting node for another depth-first traversal. When all the nodes adjacent to the last starting node have been visited, then the most recently visited starting node having an unvisited node adjacent to it is revisited, and the unvisited node is designated as the starting node for another depth-first traversal. The algorithm terminates when no unvisited nodes are adjacent to any of the visited ones. (If the nodes are numerically labeled, then the canonical order in which to visit the unvisited nodes adjacent to a starting node is by increasing label.)

Breadth-first traversal starting at a designated node s_0 is best described as an iterative algorithm. First, the starting node s_0 is visited. Next, all nodes adjacent to it are visited in some order (canonically, by increasing node label). All nodes adjacent to the latter set of nodes are then visited in turn, and so forth, until no more unvisited nodes can be reached. Note that nodes are traversed in order of increasing (minimal) distance from the start node.

Example Consider the symmetric connected digraph shown in Figure 5.2.2. A depth-first traversal starting at node 1 will visit nodes (canonically) in the following sequence:

$$\langle 1,4,2,5,3,6\rangle .$$

A breadth-first traversal starting at node 1 will visit nodes in the following sequence:

$$\langle 1,4,5,6,2,3\rangle .$$

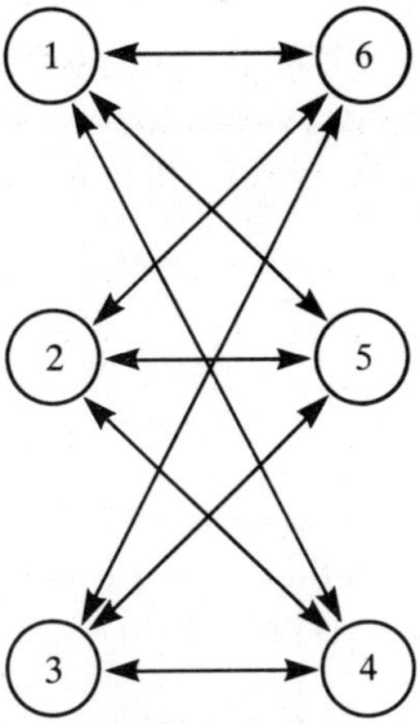

Figure 5.2.2

5.2.3 MSCRs, Cycles, and Cliques

Maximal strongly connected regions in a digraph can be found by examining its reachability matrix for symmetries. For nodes i and j to be in the same MSCR, elements (i,j) and (j,i) of the matrix must both be 1. If corresponding rows and columns of the reachability matrix are permuted so that nodes in the same MSCR are associated with adjacent rows and columns, then the permutation will have almost a block-diagonal form, with each 'block' corresponding to an MSCR. An efficient algorithm for finding MSCRs is that of Tarjan [5.42].

Example The matrix

$$
\begin{bmatrix}
1 & 1 & 1 & 1 & 0 & 0 & 0 & 0 & 0 & 0 \\
1 & 1 & 1 & 1 & 0 & 0 & 0 & 0 & 0 & 0 \\
1 & 1 & 1 & 1 & 0 & 1 & 1 & 0 & 0 & 0 \\
1 & 1 & 1 & 1 & 0 & 0 & 0 & 0 & 1 & 0 \\
0 & 0 & 0 & 0 & 1 & 1 & 0 & 0 & 0 & 0 \\
0 & 0 & 0 & 0 & 1 & 1 & 0 & 0 & 0 & 0 \\
0 & 0 & 0 & 0 & 0 & 1 & 1 & 1 & 1 & 0 \\
0 & 0 & 0 & 0 & 0 & 0 & 1 & 1 & 1 & 0 \\
0 & 0 & 0 & 0 & 0 & 0 & 1 & 1 & 1 & 0 \\
0 & 0 & 0 & 0 & 0 & 0 & 0 & 0 & 0 & 1
\end{bmatrix}
$$

has four MSCRs, containing 4, 2, 3 and 1 nodes, respectively; matrix elements (3,6), (3,7), (4,9) and (7,6) are the 'links' connecting the MSCRs.

Each cycle in a digraph must be wholly contained within a single MSCR. Furthermore, each cycle can be expressed as a combination of simple cycles from a 'loop basis' (see Section 4.3.3). All the simple cycles in a digraph can be found by taking one MSCR at a time, arbitrarily choosing a spanning tree for that MSCR, finding the associated simple cycle for each chord of the spanning tree, and then taking various combinations of these cycles. A more efficient algorithm for enumerating all of the simple cycles of a digraph is that of Tarjan [5.43], which is based on a depth-first traversal.

A clique of a digraph is an SCR, each (nonempty) subset of which is also an SCR. In principle, it is possible to start with each MSCR, and examine all of its subsets recursively for the desired property, in order to find the cliques. An alternative procedure is based on the observation that suitable permutations of the adjacency matrix of a (reflexive) digraph will have blocks along the diagonals corresponding to the cliques. If the digraph is symmetric (i.e., equivalent to a nondirected graph), then a more efficient algorithm for enumerating all of its cliques is given by Paull and Unger [5.47].

5.2.4 Isomorphisms

In Section 4.1, two digraphs were said to be (structurally) isomorphic if they are identical except possibly for the labeling of their nodes and branches—in other words, if their adjacency matrices are equal for some permutation of corresponding rows and columns. Two digraphs 'match identically' if they are isomorphic and have the same respective node and branch labels. Clearly, one means of determining whether two given digraphs are isomorphic is to test whether an adjacency matrix of one can be permuted to obtain an adjacency matrix of the other. Unfortunately, there are $n!$ permutations of an $(n \times n)$ matrix, and, in general, no nonheuristic algorithm of less complexity appears possible. However, various heuristic means to eliminate the need to test certain permutations have been suggested. For example, if node a in digraph G_1 and node b in digraph G_2 are to correspond, then they must have the same structural properties (in-degree, out-degree, distance from nodes already known to correspond, memberships in cycles, etc.). Node and branch labels help further to reduce the number of permutations that must be considered. We discuss graph matching further in Section 7.3.2.

5.2.5 Topological Sort

Let $G=(S,R)$ be a digraph, where $S=\{s_1,s_2,\ldots,s_N\}$, $R\subseteq S\times S$. We wish to find an ordering of the nodes, i.e., a sequence $S'=\langle s_{i_1}, s_{i_2},\ldots,s_{i_N}\rangle$, such that there is no branch from s_{i_j} to any of the nodes which appear earlier in the sequence; in general, many such sequences may be possible, i.e., S' may not be unique. Recalling Section 2.2.4, we then say that S' is a ***topologically sorted*** set with respect to R. The process of obtaining S' or the reverse of S', given $G=(S,R)$, will be called 'topologically sorting the graph'. Pictorially, if a topologically sorted graph is drawn with the nodes on a straight line according to the sequence S', then the arrows can go only in one direction. Clearly, not all graphs can be topologically sorted, e.g., the graph having a symmetric pair of branches (a,b) and (b,a). It is the absence of (nontrivial) loops that is crucial: an irreflexive digraph can be topologically sorted iff it is acyclic. (A digraph with reflexive branches can be topologically sorted iff deletion of its reflexive branches yields an acyclic digraph.) The acyclic property is equivalent to the condition given in Section 2.2.4, namely, that the transitive closure of R, R^+, be antisymmetric and hence that R^+ be a partial order. If R^+ is a total order, S' is unique.

An iterative algorithm for topological sorting, which produces S' given the adjacency matrix **C** representation of an acyclic graph $G=(S,R)$, is the following:

Repeat the following until all nodes have been placed in S':
find a minimal member $\hat{s}$ of S by looking for a column of **C** that has all zeros, delete $\hat{s}$ from S and **C** by deleting its corresponding rows and columns from **C**, and place $\hat{s}$ in sequence in S'.

If the graph has reflexive branches, but is otherwise acyclic, this algorithm may still be used if the reflexive branches are first deleted by setting the diagonal elements of **C** to zero. If the graph is not acyclic, the repetitive algorithm will 'block' when there is no all-zero column.

Example For the first graph (S,R_1) of Figure 2.2.4, $S=\{1,2,3,4\}$ and the adjacency matrix is

$$\mathbf{C}=\begin{bmatrix}0&0&0&0\\1&0&0&0\\0&0&0&0\\1&1&1&0\end{bmatrix}.$$

The only minimal member of S is $\hat{s}=4$ and $S'=\langle 4\rangle$, hence node 4 should be deleted from S and **C**, resulting in $S=\{1,2,3\}$ and

$$\mathbf{C}=\begin{bmatrix}0&0&0\\1&0&0\\0&0&0\end{bmatrix}.$$

Nodes 2 and 3 are now both minimal ones, so if the former is deleted, $S'=\langle 4,2\rangle$, $S=\{1,3\}$, and **C** contains only zero elements. Continuing, we find that $S'=\langle 4,2,1,3\rangle$ is one (of three) topological sorts.

5.3 TREE ALGORITHMS

5.3.1 Tree Traversals

In Section 4.2.2, a *canonical* tree traversal algorithm was given. For binary trees, a particularly simple implementation using a stack† is the following recursive 'visiting' algorithm:

> On the first visit to a node, push a pointer (PTR) to its right child onto the stack, and then visit its left child; if it has no left child, pop the stack into PTR (repeatedly until not null) and then visit the node indicated by PTR.

The algorithm starts by visiting the root, and terminates when the stack is empty when popped. [Null pointers (not to be confused with an empty stack indicator) are actually pushed onto the stack, but in the algorithm above they are ignored when popped; they are not ignored in subsequent algorithms.]

To 'process' (access, print, etc.) the nodes of a binary tree in *preorder*, each node should be processed when first visited—i.e., just prior to pushing its right-child pointer onto the stack. To process the nodes in *inorder*, each node should be processed when visited on return from its left child—i.e., in the order that the right-child pointers are popped; this is simple to implement if the identity of a node is stacked along with its (possibly null)

†Recall that a stack is a list where items are 'pushed' (inserted) and 'popped' (deleted) at the same end; when the stack is popped, the item that is removed is the last (most recent) one that was pushed.

right-child pointers. To process the nodes in *postorder,* each node should be processed when visited on return from its right-child (hence just before returning to its parent). To implement this, the identity of a node should be stacked along with its possibly null right-child pointer, as in inorder processing. However, in this case, null pointers should be distinguished from non-null pointers when popped. When a sequence of null pointers is popped, the associated nodes should be processed in the same sequence, until a non-null pointer is popped; at that time, if the non-null pointer corresponds to the node just processed (true only on return from a right-child), then its associated node should be processed and the stack should be popped again. Otherwise, when the stack is popped, the entry should be immediately restacked in order to defer processing until return from the right-child, which is to be visited next.

A simpler *postorder* traversal algorithm, which requires that the top of the stack be accessed without popping it,† follows:

> On the first visit to a node, push a pointer to its right child (if not null) onto the stack, followed by a pointer to its left child (if not null), and then visit the node identified by the top entry of the stack, marking it as visited; if the top entry on the stack has been marked as visited, then pop and process it, otherwise visit it.

The stack is initialized with a pointer to the root, which is then marked and initially visited, and the algorithm terminates when the root has been processed. Applied to the operator tree of Figure 4.2.4(a), the output of the algorithm is the postfix expression $AB+C\times$.

5.3.2 Linear Representation Conversion

A tree is a two-dimensional object, but for computer input and output, conversion from and to a one-dimensional (list) representation is desirable. Conversion algorithms mapping the linear S-expression representation to the two-dimensional linked binary tree representation for general trees are particularly simple. For example, in scanning an S-expression one symbol at a time from left to right, the following tree construction algorithm can be used:

Do repeatedly:
(i) if the scanned symbol is a '(', then
    (a) obtain a new-node

†This can be implemented by popping the top entry, and then immediately restacking it.

(b) if old-node pointer=ϕ
then put pointer to new-node in left-half of stack node
else put pointer to new-node in right-half of old-node
(c) push pointer to new-node onto stack
(d) let old-node pointer=ϕ

(ii) if the scanned symbol is a data-item α, then
(a) obtain a new-node
(b) if old-node pointer=ϕ
then put pointer to new-node in left-half of stack node
else put pointer to new-node in right-half of old-node
(c) put α in left-half of new-node
(d) let old-node pointer=pointer to new-node

(iii) if the scanned symbol is a ')', then
(a) if old-node pointer=ϕ
then put null pointer in left-half of stack node
else put null pointer in right-half of old-node
(b) pop stack to old-node pointer

(iv) if the last character has been scanned, then
(a) put null pointer in right-half of old-node
(b) stop.

This iterative algorithm starts with the stack initially empty and with old-node pointer set to the base address (or 'list head'), and terminates when the final symbol in the S-expression string has been scanned; if the string is well-formed (has balanced nested parentheses), the procedure will terminate with an empty stack. Note that step (iv) would place a null pointer, denoted ϕ, in the list head for an empty S-expression.

An algorithm which would print (linearly) the S-expression given its linked binary tree representation is as follows:

Do repeatedly:

(i) if the node pointed to by PTR contains a left pointer (to a sublist), then
(a) print a left parenthesis
(b) push its right pointer onto the stack
(c) repeat with PTR set to the left pointer

(ii) if the node pointed to by PTR contains a data item in its left-half, then
(a) print the data item
(b) repeat with PTR set to its right pointer

(iii) if PTR is the null pointer, then
(a) pop stack to PTR
(b) if the stack was not empty
then print a right parenthesis
else stop.

This algorithm should be initially called with PTR set to the value contained in the list head, and with the stack empty; it terminates when an attempt to pop an empty stack is made. (Note that it traverses the binary tree canonically.)

Example Application of the first algorithm to the S-expression (1 (2 4 5) (3 6)) yields the tree structure shown in Figure 5.3.2. Application of the second algorithm to this tree yields the original S-expression.

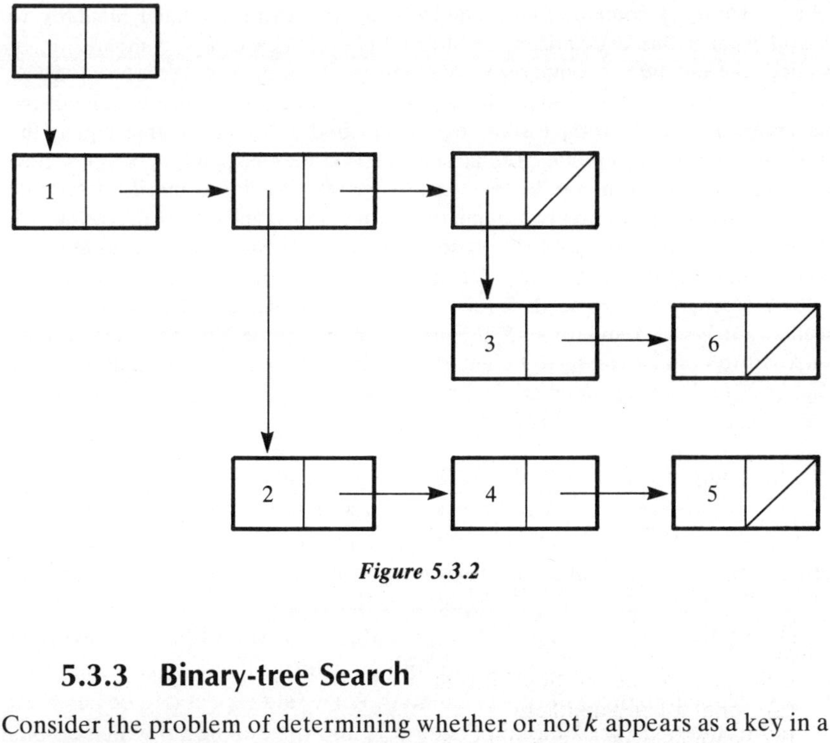

Figure 5.3.2

5.3.3 Binary-tree Search

Consider the problem of determining whether or not k appears as a key in a sorted binary tree (cf. Section 4.2.7). We may proceed as follows.

Do repeatedly:

Compare k with the key λ_x of a given node x:

(a) if $k<\lambda_x$, repeat with x replaced by the left-child of x
(b) if $k>\lambda_x$, repeat with x replaced by the right-child of x
(c) if $k=\lambda_x$, stop (the key of node x is k)
(d) if x is a null leaf, stop (k is not in the tree).

This procedure starts with x set equal to the root of the tree. It differs from the procedure for binary searching a sequential array in that we may simply follow (precalculated) pointers to 'middle' keys rather than calculate where these middle keys are each time a new key is sought. This precalculation may be done using the procedure for constructing the tree that is presented in the next section. This is another example where processing time (for the search) can be reduced at the expense of space (for pointers) and preprocessing overhead.

Example Refer to Figure 4.2(b). For $k=J$, it takes three comparisons to search for the key. First, k is compared with the root node M, and since $J<M$ we proceed to M's left subtree. Second, k is compared with C, and since $J>C$ we proceed to C's right subtree. Third, k is compared with J, and we stop since there is a match. Similarly, to find M it takes one comparison, and to find C or S it takes two comparisons. In general, the number of comparisons required to find a node in the tree equals its depth in the tree plus one (where the depth of the root equals zero, etc.). Hence the maximum number of comparisons required to find a key in the tree equals the maximum depth plus one, which equals the height of the tree plus one. Suppose now that the desired key is not in the tree, e.g., that $k=B$ in the foregoing. To determine that this k is missing requires two comparisons (of k with M and then with C), at which point we encounter the null left pointer at node C. Note that the number of comparisons required to find that k is missing depends on where k should be: for $k<C$, 2 comparisons are needed; for $C<k<J$ and for $J<k<M$, 3 comparisons are needed; for $M<k<S$ and for $k>S$, 2 comparisons are needed. In fact, each of these ranges of values of k can be represented by a square node in the augmented tree of Figure 4.2(c), and the number of comparisons required equals the depth of the square node.

It should be emphasized that the foregoing search procedure requires that the binary tree be key-sorted. If the tree is not sorted, then the key of each node in the tree may need to be examined, using (say) any of the general tree traversal algorithms presented earlier, in order to determine whether or not a given key k appears in the tree.

In Section 7.2.4, under certain probabilistic assumptions, the average number of comparisons required to find an arbitrarily given key using the binary-tree search procedure is calculated. We would expect it to be between the minimum of 1 (for k equal to the root) and the maximum which equals the height of the tree plus one. Furthermore, we would expect binary-tree searching to be most efficient when the height of the tree is minimal, i.e., when the tree is depth-balanced.

5.3.4 Construction of Key-sorted Binary Trees

Suppose an array, K, of keys is provided in increasing order—i.e., $K=\{k_1, k_2, \ldots, k_N\}$ is an ordered set such that $k_i<k_{i+1}$ for $i=1,\ldots,N-1$. A sorted binary tree $T(K)$ may be constructed inductively as follows, where initially $\ell=1$ and $u=N$:

Find the 'middle' key k_i in the sequence $K=\langle k_\ell,\ldots,k_u\rangle$ by calculating $i=\textbf{int}((u-\ell)/2)+\ell$, and define two new sequences $K_L=\langle k_\ell,\ldots,k_{i-1}\rangle$ and $K_R=\langle k_{i+1},\ldots,k_u\rangle$. Let k_i be the key of the root of the desired tree $T(K)$, let its left subtree be the sorted binary tree associated with K_L, $T(K_L)$, and similarly, let its right subtree be $T(K_R)$. If K_L or K_R is an empty sequence, which occurs when $u-\ell<2$, then let $T(K_L)$ or $T(K_R)$, respectively, be a null leaf. (If $u=\ell$, then the single key is chosen as a root with two null subtrees.)

This procedure constructs the tree by an adaptation of the method for binary searching a sequential array; it requires that a complete ordered set of keys be available from the start. Each key in K is guaranteed to be chosen as the root of a subtree at some point.

A procedure which inserts one key at a time into an evolving tree, and does not require that the keys $K=\{k_1,k_2,\ldots,k_N\}$ be provided in order, operates by searching a sorted binary tree (using the algorithm discussed in Section 5.3.3). It proceeds as follows:

Do repeatedly for $i=2,3,\ldots,N$:

To insert a node with key k_i into a sorted binary tree T (containing just k_1, initially), binary search T for the occurrence of k_i. If the key is found, then there is a duplication (which we regard here as an error). Otherwise, the binary-tree search algorithm will terminate at a null leaf x. Key k_i should then be made the key of a newly inserted leaf node in place of x (which in turn has two new null leaves). Let T be this new tree.

Note that this insertion algorithm starts with T consisting of a single node with key k_1, and with $i=2$; a node with key k_2 is then inserted as the left or right child of T, depending upon whether $k_2<k_1$ or $k_2>k_1$, respectively. (If $N=1$, the tree contains just k_1.)

This second algorithm is easy to implement, and may be used as a general *insertion* algorithm to add new keyed nodes to a sorted binary tree, but unfortunately it may produce a very off-balanced tree. In fact, if the nodes are inserted in increasing key order, then the worst case situation in which every node has a null left pointer will arise. The first algorithm, which employs binary searching of a sequential array, will produce a size and depth balanced binary tree.

It should be observed that an inorder traversal of any sorted binary tree will traverse the nodes in increasing key order. Hence one way to transform an off-balanced sorted binary tree into a balanced one is to use an inorder traversal algorithm to produce a sequential array of keys, and then to apply the first algorithm to this array. However, a general insertion algorithm that maintains balance as the tree evolves is highly desirable. Such an algorithm for height-balanced (also called 'AVL') trees was devised by Adel'son-Vel'skii and Landis; analyses of this algorithm and many others appear in Knuth [5.28].

5.4 APPLICATIONS TO PROGRAMMING

5.4.1 Flowcharts

Flowcharts are pictorial representations of computer programs, and are popular because 'a picture is worth a thousand words'. Flowcharts consist of 'processing-blocks' or nodes of various shapes and sizes, and of 'control-flow' branches which interconnect the blocks. It is natural then to use directed graphs to model computer programs or algorithms. We restrict ourselves to deterministic algorithms here.

A flowchart, as a digraph, has three kinds of nodes: *condition* nodes (commonly diamond-shaped), *action* nodes (commonly rectangular), and *junction* nodes (commonly circular and small). Condition nodes have in-degree=1 and out-degree>1; condition nodes having out-degree=2 are called *predicate* nodes. Action nodes have in-degree=1 and out-degree=1. Junction nodes have in-degree>1 and out-degree=1, except for two special cases: the *entry* node has in-degree=0 and out-degree=1, and the *exit* node has in-degree>0 and out-degree=0. (Each flowchart must have at least two junction nodes, one entry and one exit node, these being distinct.) A branch from one node to another means that the latter 'may' be processed after the former: if a node has only one successor node, then the latter *must* be processed after the former; if a node has more than one successor, true only for condition nodes, then one of the successors must be processed, the choice depending upon the specified condition. Processing begins with the entry node and ends with the exit node (provided the underlying algorithm is correct).

The simplest nontrivial flowchart consists of a single action node having one predecessor node (the entry or 'begin' node) and one successor node (the exit or 'end' node), as shown in Figure 5.4.1(a); we call this a *basic* flowchart. More complex flowchart structures can be constructed by combining simpler flowcharts in one of three ways. Flowcharts can be:

(i) combined 'sequentially' by letting the exit node of one coincide with the entry node of another, resulting in a *composite* structure

(ii) combined 'in parallel' by letting two (or more) be successors of a common predicate (or condition) node and also letting them have a common exit node, resulting in an *alternation* (or *case*) structure

(iii) placed within a 'loop' with a predicate node at its beginning or end to govern its termination, resulting in an *iterative* structure.

Figure 5.4.1(b) illustrates the composition of flowcharts $\mathbf{F}_1$ and $\mathbf{F}_2$. Figure 5.4.1(c) illustrates the alternation of flowcharts $\mathbf{F}_1$ and $\mathbf{F}_2$; it is also called an

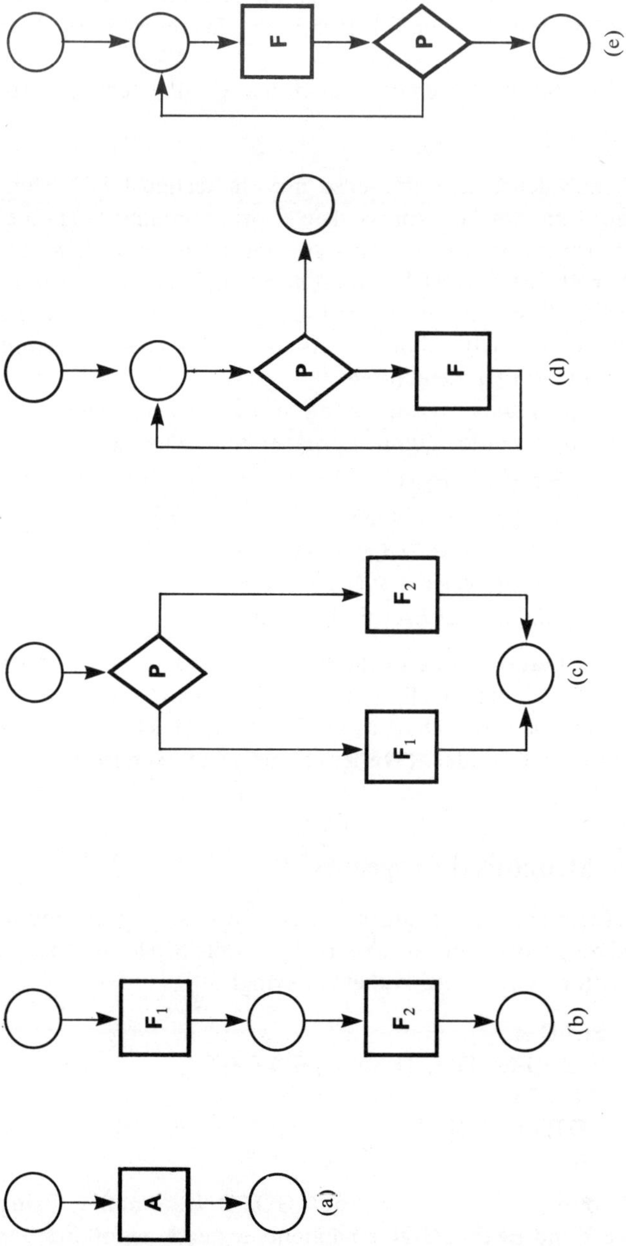

Figure 5.4.1

IF–THEN–ELSE structure. Figure 5.4.1(d) and (e) illustrate the iteration of flowchart **F** in WHILE–DO and REPEAT–UNTIL structures, respectively.

We say that a flowchart is *structured* if it can be constructed from basic flowcharts using only these composition, alternation, and iteration structures. A fundamental theorem (cf. Böhm and Jacopini [5.10]) is that:

> Every (unstructured) flowchart is functionally equivalent to a structured flowchart.

('Functional equivalence' is in the sense given in Section 1.2.3). This implies that any algorithm can be expressed in a programming language whose 'control-flow' (branching) statements are limited to IF–THEN–ELSE, and WHILE–DO or REPEAT–UNTIL (assuming normal instruction sequencing). Specifically, GOTO statements are not necessary, and their indiscriminate use may result in unstructured flowcharts. We say that a *structured program* is one whose flowchart is structured.

Structured programs corresponding to the flowcharts shown in Figure 5.4.1 may be expressed in a typical programming language as follows:

(a) BEGIN A END
(b) BEGIN $\mathbf{F}_1$; $\mathbf{F}_2$ END
(c) BEGIN IF **P** THEN $\mathbf{F}_1$ ELSE $\mathbf{F}_2$ END
(d) BEGIN WHILE **P** DO **F** END
(e) BEGIN REPEAT **F** UNTIL **P** END

We remark that only one of the two iteration structures (WHILE and REPEAT) is sufficient in that a flowchart using one can be transformed into a functionally equivalent flowchart using the other. This transformation may require duplication of nodes. (We leave this as an exercise.)

5.4.2 Structured Programs

Figure 5.4.2(a) illustrates an unstructured flowchart corresponding to an unstructured program having an 'exit' in the middle of a loop. An example of a program with this structure is the following:

```
BEGIN: READ X
       IF END-OF-FILE THEN GOTO EXIT
       WRITE F(X)
       GOTO BEGIN
EXIT:  END
```

This can also be expressed without GOTO statements, using (say) DO–FOREVER and EXIT–LOOP statements instead, as follows:

```
DO-FOREVER
  READ X
  IF END-OF-FILE THEN EXIT-LOOP
  WRITE F(X)
END-DO
```

While in this form the program has some attraction, it still is unstructured by our definition. A common way to restructure such a program is to place the **READ** outside (prior to) the loop and also at the end of the loop, as follows:

```
BEGIN
  READ X
  WHILE ¬END-OF-FILE DO
    BEGIN WRITE F(X); READ X END
END
```

The flowchart for this program is shown in Figure 5.4.2(b); note that it is structured, consisting of the composition of $\mathbf{F}_2$ and $\mathbf{F}_1$ within a **WHILE–DO** loop, which in turn is composed with $\mathbf{F}_1$. The 'cost' of this restructuring is duplication of the code associated with $\mathbf{F}_1$, which in general may be much larger than the single **READ** of our example.

Figure 5.4.2(c) illustrates another unstructured flowchart corresponding to an unstructured program having a loop with two exits—e.g., a 'normal' exit, and an 'error' exit. An example of a program with this structure is the following:

```
BEGIN
  READ X
  WHILE ¬END-OF-FILE DO
    BEGIN
      Y:=F(X)
      IF Y<0 THEN ERROR-EXIT
      WRITE SQR(Y); READ X
    END
END
```

The error exit can be eliminated by setting an error flag instead, which is then tested along with the condition for a normal exit. This error flag must of course be 'cleared' initially (outside the loop), as illustrated in the following:

```
BEGIN
  READ X
  ERROR:="false"
  WHILE ¬END-OF-FILE & ¬ERROR DO
    BEGIN
```

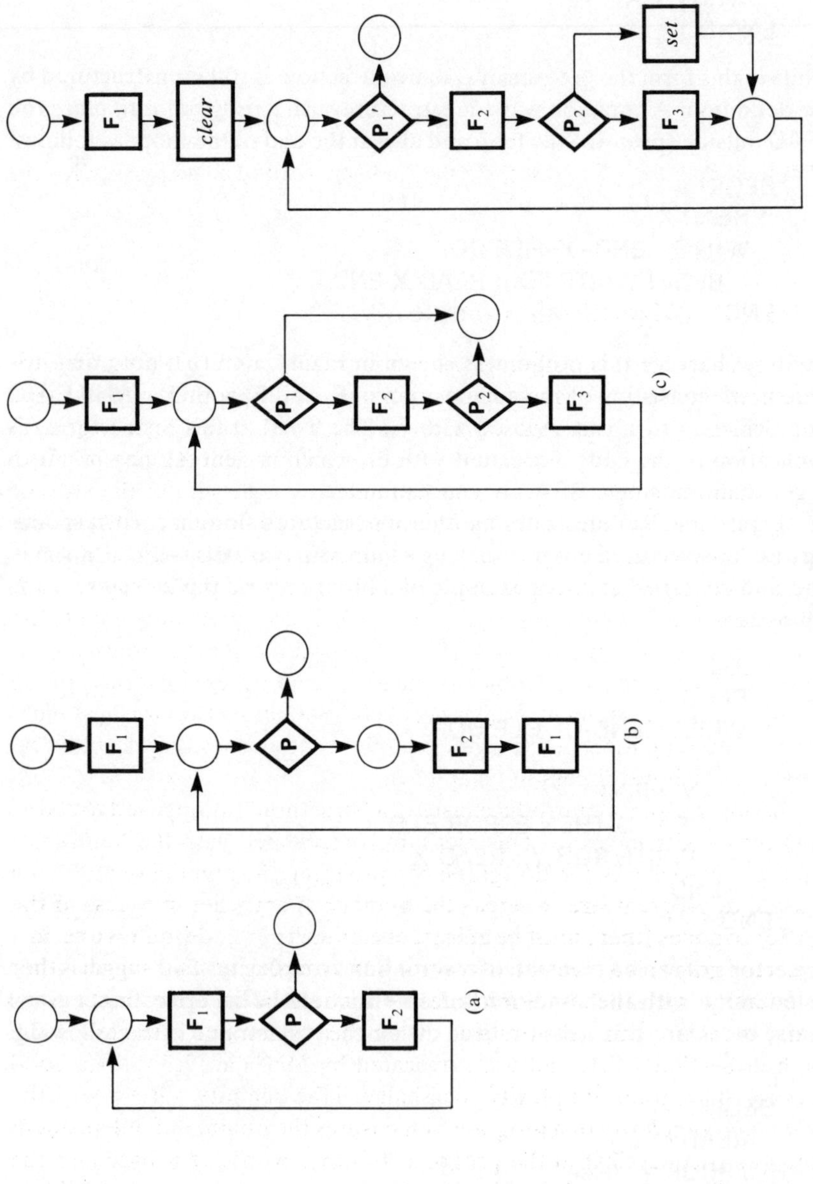

Figure 5.4.2

```
        Y:=F(X)
        IF Y<0 THEN ERROR:="true"
        ELSE BEGIN WRITE SQR(Y); READ X END
      END
  END
```

The flowchart for this program is shown in Figure 5.4.2(d); as before, this flowchart is structured. The cost of this restructuring is the addition of new action nodes to clear and set the error flag, as well as the modification of the **WHILE** predicate ($\mathbf{P}_1'$) to test the flag. For some related concepts, see Knuth [5.29], Peterson [5.39], and Lew [5.31].

5.4.3 Algorithmic Complexity

The two most useful measures of the complexity of computer programs are execution time (usually averaged over possible sets of input data) and space (i.e., the amount of main and auxiliary storage required). Space requirements are relatively easy to calculate. Given the flowgraph of a program, the space required for the program instructions equals the sum of the sizes of each node; the size of each node depends upon computer hardware *and* software, but compilers can easily provide such information. The space required for static data structures is generally specified by the programmer. For dynamic data structures, e.g., those which are input dependent, the maximum space required is also generally specified by the programmer. However, it is often preferable to measure complexity in terms of the average space requirement, which necessitates analysis by statistical techniques. Determination of average execution times also requires statistical analyses, which we discuss in Chapter 6.

In this section, we introduce instead a 'structural' (as opposed to statistical) complexity measure. This measure is associated with the number of nodes and branches in the flowgraph of a program: the number of nodes is a measure of program size, whereas the number of branches in excess of the number of nodes (there must be at least one branch per node, minus one, in a connected graph) is a measure of control flow complexity. This suggests that the quantity $\#B$(ranches) $-\#N$(odes) $+1$ would be an appropriate complexity measure; this quantity was called the 'cyclomatic number' of the graph in Section 4.2.1, and was advocated by McCabe [5.33]† as a good measure of program complexity. We shall call the quantity $\#B-\#N+1$ the *cyclomatic complexity* of a program: it measures the number of independent loops (i.e., semicycles) in the program. In other words, it is based on the branching (both forwards and backwards) structure of the program. While

†McCabe actually suggested the use of $\#B-\#N+2$, in effect adding a branch from the exit node to the entry node.

the cyclomatic complexity may not in itself be very significant, it does provide one way to compare alternative programmed solutions for the same problem.

Kirchhoff Analysis In a program flowgraph without loops, each node and each branch is executed exactly once. When there are loops, some nodes and branches may be executed more than once or not at all, generally depending upon input data. Our objective here is to determine the node and branch execution frequencies as a function of input parameters. Let N_a denote the execution frequency of branch a, and N_x denote the execution frequency of node x; these frequencies may be fractional, representing relative frequencies (or 'probabilities') of execution. We proceed by noting that a conservation law holds for each node:

> ***Kirchhoff's Law***† For any node x (other than a source or sink), the sum(-in) of the execution frequencies of the branches leading into node x equals the sum(-out) of the execution frequencies leading out of node x, which also equals the execution frequency of node x.

For example, if node x has two 'input' branches, a and b, and two 'output' branches, c and d, then $N_a+N_b=N_c+N_d=N_x$. For a source (entry) node s, the sum-out equals N_s; for a sink (exit) node t, the sum-in equals N_t. For single-entry, single-exit flowgraphs, $N_s=N_t=1$. Since, given all the branch execution frequencies, we can compute all the node execution frequencies, we regard the branch execution frequencies as our unknowns. For each node, we can write a conservation equation relating these unknowns. So far, we have $\#N$ (linear) equations in $\#B$ unknowns, where $\#B>\#N$. Thus, these equations are not linearly independent.

Example 1 Consider the strictly sequential flowgraph of Figure 5.4.3(a): its node equations are

$N_a=1$ (for node s)
$N_a=N_b$ (for node x)
$N_b=1$ (for node t).

Note that only two of the foregoing three equations are independent.

In general, only $(\#N-1)$ of the $\#N$ node equations in $\#B$ unknowns are independent, hence we need $\#B-(\#N-1)$ additional equations in order to solve uniquely for the unknowns. This number is precisely the cyclomatic complexity of the program flowgraph. Recalling that this number is also the number of independent (acyclic) loops in the flowgraph, it is clear that what remains to be found is the

†This law is analogous to the physical law named after Kirchhoff which states that current must be conserved in an electrical circuit. Further analogies are possible (see Davies [5.15]), but they are of limited value.

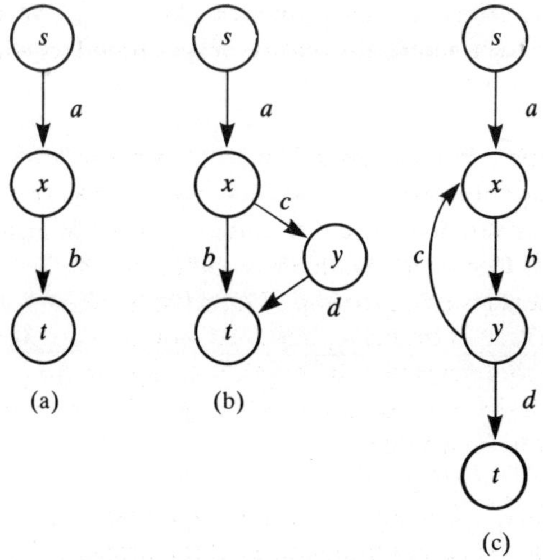

Figure 5.4.3

execution frequency for at least one branch in each loop, and that these unknown frequencies may be data dependent.

Example 2 Consider the flowgraph of Figure 5.4.3(b): its node equations are

$N_a=1$ (for node s)
$N_a=N_b+N_c$ (for node x)
$N_c=N_d$ (for node y)
$N_b+N_d=1$ (for node t)

only three of which are independent; to solve for the four unknowns, we need one additional equation, e.g., that $N_b=1/2$ or $N_b=N_c$ (assuming 'equilikelihood').

Example 3 Consider the flowgraph of Figure 5.4.3(c): its node equations are:

$N_a=1$ (for node s)
$N_a+N_c=N_b$ (for node x)
$N_b=N_c+N_d$ (for node y)
$N_d=1$ (for node t)

only three of which are independent; to solve for the four unknowns, we need one additional equation, e.g., that the loop is iterated L times ($N_c=L$).

We emphasize that branching probabilities and loop iterations are generally data dependent and not determinable from an analysis of a program's graphical structure; we must use 'semantic' information instead to determine (or estimate) one branch execution frequency for each independent loop.

Given the branch execution frequencies of a flowgraph, the node execution frequencies are easily computed: the execution frequency of a node x equals the sum-in of the execution frequencies of the branches leading into x (which also equals the sum-out). The *execution time* $T_{\mathbf{P}}$ of a program **P** equals $\Sigma_x N_x \cdot \tau_x$, where τ_x is the execution time of node x, for each node x in the program flowgraph. If the times τ_x are data dependent, say, a function of I, then $T_{\mathbf{P}}$ is also a function of I. If I is an integer representing (say) the amount of data processed by a program, or the number of iterations made, then the asymptotic complexity (cf. Section 3.1.6) of the program with respect to I is of interest.

5.4.4 Computability

Any computer program **P** may be thought of as an 'implementation' of a function $\mathbf{f}_{\mathbf{P}}$: $\mathcal{I} \rightarrow \mathcal{I}$, where $\mathcal{I}$ is the set of natural numbers. The domain of $\mathbf{f}_{\mathbf{P}}$ is the set of 'input' data for which the function is defined, and the range of $\mathbf{f}_{\mathbf{P}}$ is the set of 'output' data computed for the data in the domain. We say **P** 'implements' $\mathbf{f}_{\mathbf{P}}$ if they are functionally equivalent, i.e., if **P** computes and outputs $\mathbf{f}_{\mathbf{P}}(i)$ for any input i. Both the input and output can be expressed (say) as strings of binary-coded characters, which have unique interpretations as numbers in $\mathcal{I}$; the Gödel numbering scheme (discussed in Section 1.1.6) can also be used to ensure that each program has a unique integer encoding. The program itself may also be regarded as a single string of characters, from a finite character set A, where the string must satisfy a finite set of syntactic rules (grammar) for a given programming language $L \subseteq A^*$. For convenience, we may assume a binary encoding, where $A = \{0, 1\}$, so that inputs, outputs, and programs are each represented by single binary strings (or numbers). We emphasize that any 'real' computer program **P** must be a *finite* string, and so must its input i and output $o = \mathbf{f}_{\mathbf{P}}(i)$. A compiler, used to check whether or not $\mathbf{P} \in L$, would never reach its end if **P** were not finite. Furthermore, we require the output to be computed within a finite amount of time (and that input/output times be finite). We say that a program is ***finitary*** if all these finiteness assumptions hold; recall that, although a number is necessarily finite, there is no upper limit on what the number can be.

Given an arbitrary (numerical) function $\mathbf{f}$: $\mathcal{I} \rightarrow \mathcal{I}$, the programmer's task is to write a computer program $\mathbf{P} \in L$ which implements $\mathbf{f}$, provided $\mathbf{f}$ is computable. We pose the question:

Is every function $\mathbf{f}$: $\mathcal{I} \rightarrow \mathcal{I}$ computable?

The answer is ***no***! We can prove this by observing that (a) there are uncountably many functions $\mathbf{f}$: $\mathcal{I} \rightarrow \mathcal{I}$, but (b) there are only countably many different programs $\mathbf{P} \in L$. (Recall our discussion of countable sets in Section 2.3.2.)

The latter follows from the fact that A^* is countable (for any finite set A), hence $L \subseteq A^*$ is also countable: A^*, the set of all strings over A, is countable since these strings have a natural ordering, the lexicographical one (with respect to any ordering of A); each member of A^* is uniquely determined by its position in such a lexicographical ordering, so a 1–1 correspondence between $\mathcal{I}$ and A^* exists. That there are uncountably many functions from $\mathcal{I}$ to $\mathcal{I}$ follows from a proof by contradiction: assuming there are a countable number of functions **f**: $\mathcal{I} \to \mathcal{I}$, they can be denoted $\mathbf{F}_0$, $\mathbf{F}_1$, $\mathbf{F}_2$, ...; but if **g**: $\mathcal{I} \to \mathcal{I}$ is defined such that $\mathbf{g}(i)=\mathbf{F}_i(i)+1$, then $\mathbf{g} \neq \mathbf{F}_i$ for any i, which is a contradiction. We conclude that there exist functions which no computer program can implement.

We emphasize that the foregoing 'theorem of noncomputability' is based on the fact that computer programs are necessarily finite in length. If infinite-length programs were permitted, a noncomputable function $\mathbf{f}^*$ could be programmed 'by cases'—e.g., by a program of the form:

if $i=0$ **then** $\mathbf{f}^*$ $(i):=j$
if $i=2$ **then** $\mathbf{f}^*$ $(i):=k$
...

where the ellipsis denotes an infinite set of additional cases for which no closed formula specification exists. Furthermore, the domain of $\mathbf{f}^*$ may not be known in closed (finite) form; for example, $\mathbf{f}^*$ (i) may be undefined for an infinite number of values in addition to $i=1$. This implies that a program which 'enumerates' (lists each and only the members of) the range of a noncomputable function may not exist.

The Halting Problem As every computer programmer knows, there are programs for which the function $\mathbf{f_P}$: $\mathcal{I} \to \mathcal{I}$ is not total, i.e., for which $\mathbf{f_P}(i)$ is undefined for some input i. For example, the logarithm function is undefined for an input of zero. In practice, the fact that $\mathbf{f_P}(i)$ is undefined may be manifested by the occurrence of a system-detectable error (e.g., a 'divide-by-zero' interrupt), or by the occurrence of an 'infinite loop' where the program does not output in a finite amount of time. In the latter event, we say that the program does not 'halt'.

The *halting problem* is that of deciding whether an arbitrary program **P** will halt (and output) for an arbitrary input i. For a particular **P** and i (e.g., $\mathbf{P}_1$: **input**(i); $o:=i$; **output** (o); **end**, or $\mathbf{P}_2$: ℓ: **goto** ℓ; **end**), the problem may be solvable; however, we are not always so fortunate.

> The halting problem is unsolvable in general; i.e., it is not possible to tell (in a finite amount of time) whether or not an arbitrary program **P** will halt for an arbitrary input i.

The finite-time requirement is necessary because, otherwise, we could just run the program and wait forever. We distinguish this from the case of waiting indefinitely, but for a necessarily finite amount of time.

Example Consider the following program:

```
P3: input (i);
    if i is prime
    then output(1)
    else output(0); end
```

where the test for primeness is made by dividing i by each number less than i. For any 'legal' (positive integer) input i, this program will always halt, although no ***a priori*** time limit may be known; in other words, we may have to wait indefinitely for output, but not forever.

Software Implications In more practical terms, the unsolvability of the halting problem means that a compiler cannot guarantee that an arbitrary computer program will be free of the possibility of infinite loops. We emphasize that we are speaking in general; for specific programs, or specific inputs, the halting problem may be solvable, but not for *arbitrary* programs and inputs.

A related consequence of the unsolvability of the halting program is the unsolvability of the *equivalence problem*. It is not possible to tell (in a finite amount of time) whether or not two arbitrary programs are functionally equivalent. It also is not possible to tell whether an arbitrary instruction within a computer program will ever be executed. For further discussion, see Hoare [5.23].

5.5 APPLICATIONS TO LANGUAGES

5.5.1 Context-free Languages

A *context-free language* (CFL) is a set of terminal strings (cf. Section 3.3.2) which can be generated by a context-free grammar (CFG). Any given CFL can be generated by more than one CFG. By convention, for a context-free grammar (A,T,P,σ_0), it is sufficient to specify only the set of productions P and have A, T and σ_0 defined implicitly as follows: $A-T$ is the set of symbols appearing on the left-hand side of a production; T is the set of symbols appearing *only* on the right-hand side of a production; and σ_0 is that member of $A-T$ which does not appear on the right-hand side of any production whose left-hand side is not also σ_0. Recall that only nonterminals can appear on the left-hand side by definition. Furthermore, if a nonterminal σ' does not appear on the left-hand side, it cannot ever be rewritten, so no terminal string can be derived from a sentential form containing σ'; hence we do not permit nonterminals to appear only on the right-hand side of a production.

Example Given the set of productions

$$
\begin{aligned}
P_0=\{&\sigma_0\rightarrow\sigma_0+\sigma_1\\
&\sigma_0\rightarrow\sigma_1\\
&\sigma_1\rightarrow\sigma_1\times\sigma_2\\
&\sigma_1\rightarrow\sigma_2\\
&\sigma_2\rightarrow\lambda\}
\end{aligned}
$$

we deduce that $G_0=(A,T,P_0,\sigma_0)$, where $A=\{\sigma_0, \sigma_1, \sigma_2\}\cup T$, and $T=\{+, \times, \lambda\}$.

Context-free languages are of great importance since a standard way of defining the 'syntax' of computer programming languages is by means of context-free grammars. The productions of a CFG (also called BNF rules) serve to define the legal (i.e., syntactically correct) statements of programming languages: strings of terminal symbols are legal (correct) statements if they can be derived from the initial symbol of the grammar, and are illegal (erroneous) otherwise. More specifically, the statement types of a programming language are defined by separate grammars, and a program is defined as the concatenation of such statements. While individual statements can usually be defined by context-free grammars, programs as a whole cannot, since the legality of one statement often depends upon other statements. For example, consider a language which has the requirement that a variable V must be declared before it is used, i.e., which contains strings of the form $aVbVc$ *where a, b,* and c are arbitrary; in Chapter 3 (Exercise 22), we stated that this language was not context-free. To handle such 'context-sensitive' problems, it is common to define programming languages by a combination of 'syntactic' rules (Type 2 productions) and 'semantic' rules (to resolve any problems).

5.5.2 Parse Trees

For any CFG (A,T,P,σ_0), a *leftmost* [*rightmost*] derivation of a sentential form w from σ_0, $\sigma_0=w_0\Rightarrow\cdots\Rightarrow w_r=w$, is one where each production applied to derive w_{i+1} from w_i involves a substitution for the leftmost [rightmost] nonterminal in w_i. Each derivation can be represented by an ordered (out-rooted) tree, with σ_0 as the root node, where application of the production $\sigma\rightarrow\alpha_1\alpha_2\ldots\alpha_k$ is represented by having each α_i (in order) be a child of σ; such derivation trees are called *parse* trees.

Example With respect to the grammar G_0 defined above, the string $\lambda+\lambda\times\lambda$ has as a leftmost derivation the sequence

$$\begin{aligned}\sigma_0 &\Rightarrow \sigma_0+\sigma_1\\ &\Rightarrow \sigma_1+\sigma_1\\ &\Rightarrow \sigma_2+\sigma_1\\ &\Rightarrow \lambda+\sigma_1\\ &\Rightarrow \lambda+\sigma_1\times\sigma_2\\ &\Rightarrow \lambda+\sigma_2\times\sigma_2\\ &\Rightarrow \lambda+\lambda\times\sigma_2\\ &\Rightarrow \lambda+\lambda\times\lambda\end{aligned}$$

and as a rightmost derivation the sequence

$$\begin{aligned}\sigma_0 &\Rightarrow \sigma_0+\sigma_1\\ &\Rightarrow \sigma_0+\sigma_1\times\sigma_2\\ &\Rightarrow \sigma_0+\sigma_1\times\lambda\\ &\Rightarrow \sigma_0+\sigma_2\times\lambda\\ &\Rightarrow \sigma_0+\lambda\times\lambda\\ &\Rightarrow \sigma_1+\lambda\times\lambda\\ &\Rightarrow \sigma_2+\lambda\times\lambda\\ &\Rightarrow \lambda+\lambda\times\lambda.\end{aligned}$$

Both derivations can be represented by the parse tree shown in Figure 5.5.2(a).

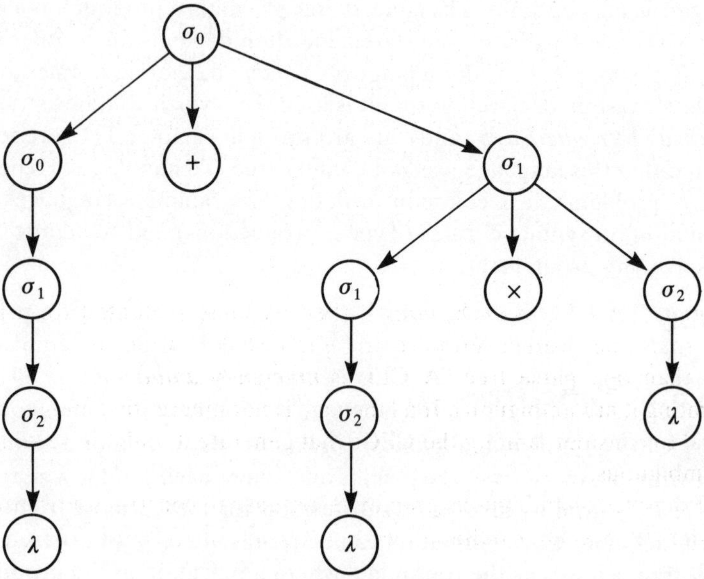

Figure 5.5.2 (a)

Observe that although several derivations are possible for the above example, there is only one parse tree. The two derivations correspond to different orders in which this parse tree can be constructed. Very significantly, the precedence of $\times$ over $+$ is reflected in the tree: $\times$ is combined in a subtree before $+$.

To the preceding arithmetic expression grammar G_0, we add a parenthesization production, change the variable symbols, and add the 'end-marker' symbol $\perp$:

$$\begin{aligned}\{G_{\text{pae}} &\to \perp E \perp \\ E &\to E+T \\ E &\to T \\ T &\to T\times F \\ T &\to F \\ F &\to \lambda \\ F &\to (E)\}.\end{aligned}$$

The end-marker symbol serves to delimit the beginning and end of the arithmetic expressions; in practice, string processing programs may add the delimiters on input. It should be emphasized that *by design* this grammar, G_{pae}, incorporates the three arithmetic expression evaluation rules given in Section 2.4.6. Nesting of parentheses is handled by treating a parenthesized expression as just a simple variable. The precedence of $\times$ before $+$ follows since the T in $E+T$ may be rewritten as $T\times F$, but T and F in $T\times F$ cannot be rewritten by a string containing a $+$ (unless parenthesized).

The left-to-right rule for $+$ follows from the production $E\to E+T$, as opposed to $E\to T+E$ or $E\to E+E$; an analogous statement applies for $\times$. We also say that $+$ [and $\times$] are 'left-associative' in that the structure of the parse tree for $\lambda+\lambda+\lambda$ is like that for $(\lambda+\lambda)+\lambda$, as opposed to $\lambda+(\lambda+\lambda)$. The parse tree for the expression $\perp\lambda+(\lambda)\times\lambda\perp$ is shown in Figure 5.5.2(b).

Ambiguity A CFG is *ambiguous* if there is some sentential form having more than one distinct leftmost or rightmost derivation, or equivalently more than one parse tree. A CFL is *inherently ambiguous* if all CFGs generating it are ambiguous. If a language is not inherently ambiguous, it is natural to consider, among the CFGs that generate it, only those which are not ambiguous.

Examples of ambiguous grammars are easy to construct. For example, consider the grammar defined by

$$\begin{aligned}P=\{&\sigma_0\to\sigma_0+\sigma_0 \\ &\sigma_0\to\lambda\}.\end{aligned}$$

The language generated by this grammar is $\lambda(+\lambda)^*$. The sentential form $\lambda+\lambda+\lambda$ has the two leftmost derivations

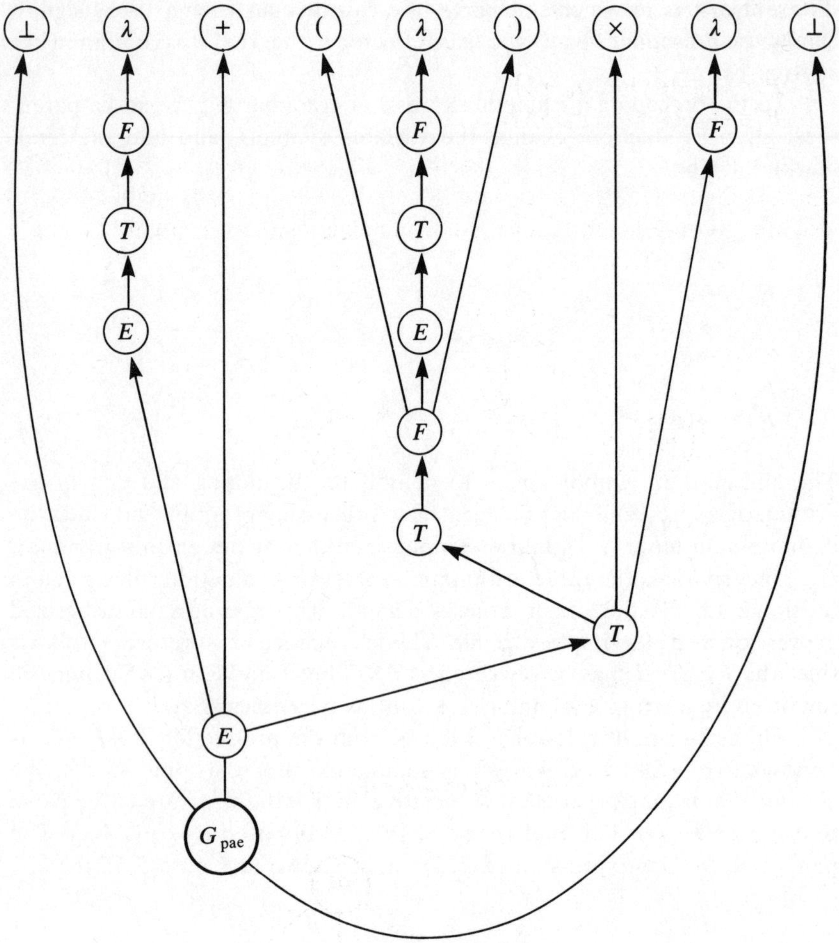

Figure 5.5.2 (b)

$$\sigma_0 \Rightarrow \sigma_0 + \sigma_0 \Rightarrow \lambda + \sigma_0 \Rightarrow \lambda + \sigma_0 + \sigma_0$$
$$\Rightarrow \lambda + \lambda + \sigma_0 \Rightarrow \lambda + \lambda + \lambda$$

and

$$\sigma_0 \Rightarrow \sigma_0 + \sigma_0 \Rightarrow \sigma_0 + \sigma_0 + \sigma_0 \Rightarrow \lambda + \sigma_0 + \sigma_0$$
$$\Rightarrow \lambda + \lambda + \sigma_0 \Rightarrow \lambda + \lambda + \lambda.$$

Their corresponding parse trees are shown in Figures 5.5.2(c) and (d), respectively. Note that in the former case sums are 'evaluated' from right to left, while in the latter case sums are evaluated from left to right. Unambiguous grammars which are equivalent (i.e., which generate the same languages) are defined by

$$P' = \{\sigma_0 \to \sigma_0 + \lambda,\ \sigma_0 \to \lambda\}$$
$$P'' = \{\sigma_0 \to \lambda + \sigma_0,\ \sigma_0 \to \lambda\}.$$

These enforce left-to-right and right-to-left evaluation rules, respectively. Fortunately, useful languages are generally *not* inherently ambiguous; an example of an inherently ambiguous language is $\{a^i b^i c^j\} \cup \{a^i b^j c^j\}$.

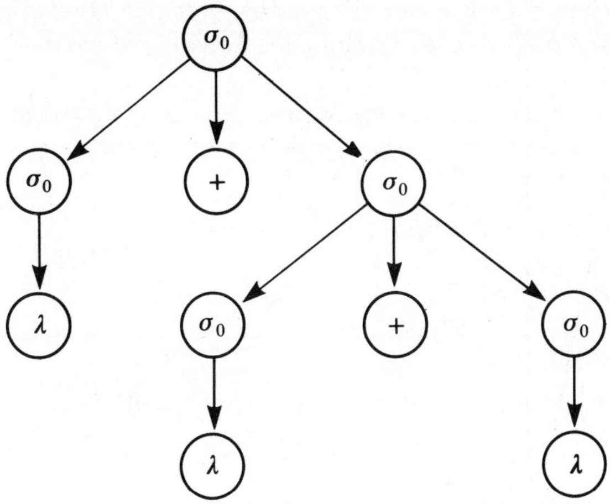

Figure 5.5.2 (c)

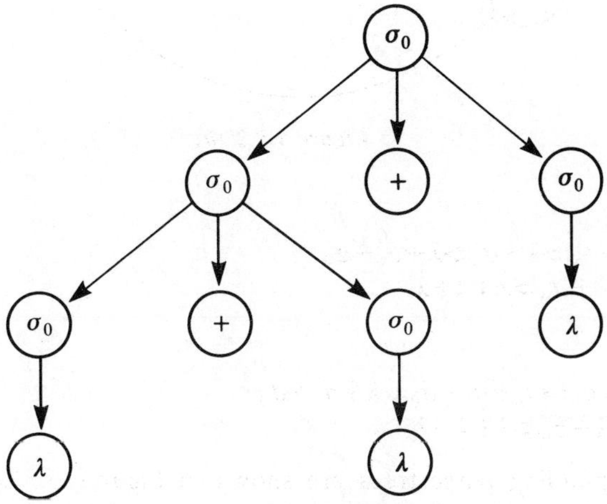

Figure 5.5.2 (d)

5.5.3 Digraph Representation

Any context-free grammar $G=(A, T, P, \sigma_0)$ can be represented by a directed graph (S, R) in the following fashion. Define a distinct node in S for each member of A and P; we may, in other words, let $S=A \cup P$. If production $p \in P$ is of the form $\sigma \rightarrow \alpha$, for $\alpha \in A \cup \{\varepsilon\}$, then let (σ, p) and (p, α) be branches in R [or just (σ, p) if $\alpha=\varepsilon$]. If production $p \in P$ is of the form $\sigma \rightarrow \alpha_1 \alpha_2 \ldots \alpha_k$, for $k>1$, each $\alpha_i \in A$, then let (σ, p) and (p, α_i) for each α_i be branches in R; in this case, the branches (p, α_i) are considered ordered by increasing i. The resulting ordered digraph is called the *syntax graph* of G; it and slight variations have also been called 'syntactic charts'.

Example Consider the right linear grammar generating the language $a^* bc^*$, as defined in Section 3.3.4. $P_1=\sigma_0 \rightarrow a\sigma_0$, $P_2=\sigma_0 \rightarrow b\sigma_1$, $P_3=\sigma_1 \rightarrow c\sigma_1$ and $P_4=\sigma_1 \rightarrow \varepsilon$. Then $G=(S,R)$, where

$$
\begin{aligned}
S &= \{\sigma_0, \sigma_1, a, b, c, P_1, P_2, P_3, P_4\} \\
R &= \{(\sigma_0, P_1), (P_1, a), (P_1, \sigma_0), \\
&\quad (\sigma_0, P_2), (P_2, b), (P_2, \sigma_1), \\
&\quad (\sigma_1, P_3), (P_3, c), (P_3, \sigma_1), \\
&\quad (\sigma_1, P_4)\}.
\end{aligned}
$$

This is depicted in Figure 5.5.3(a). Figure 5.5.3(b) shows the digraph representation for the arithmetic expression grammar G_0 defined previously.

Because digraphs have convenient internal computer representations, syntax graphs are quite useful for grammar-processing programs. Specifi-

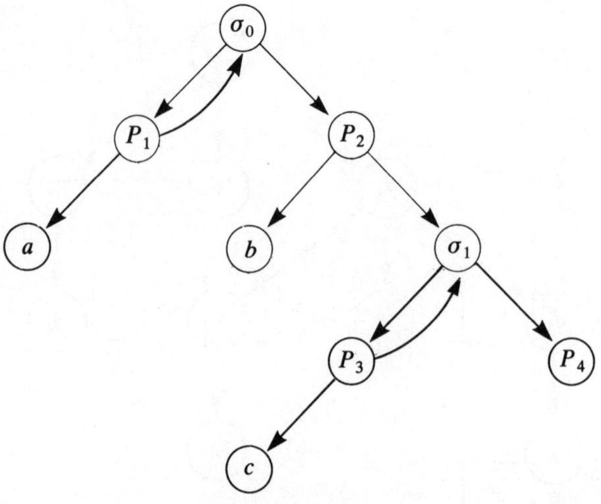

Figure 5.5.3 (a)

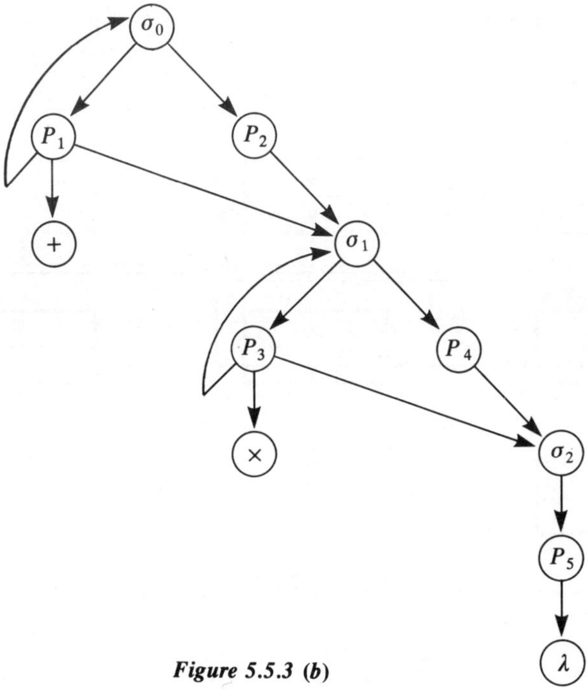

Figure 5.5.3 (b)

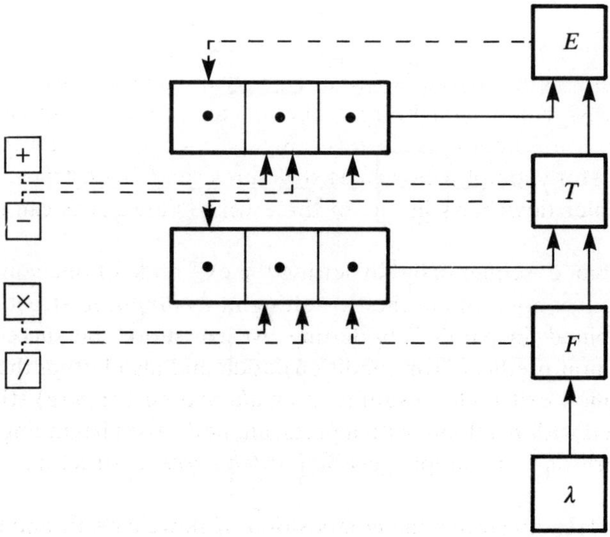

Figure 5.5.3 (c)

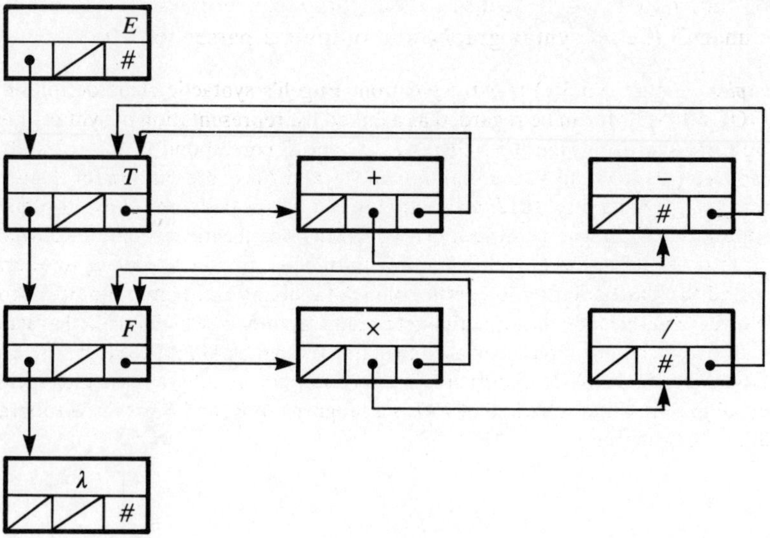

Figure 5.5.3 (d)

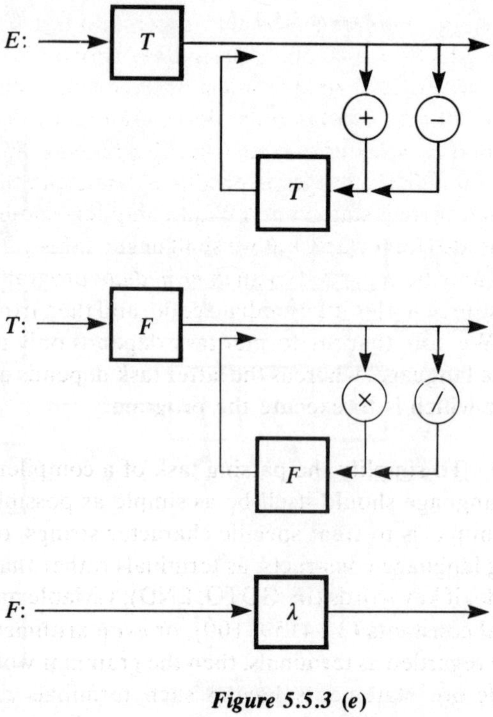

Figure 5.5.3 (e)

cally, they may be used by a so-called 'compiler compiler', which inputs a grammar G (i.e., a syntax graph) and outputs a parser for $\mathbf{L}(G)$.

Examples Figure 5.5.3(c) is extracted from Engeli's syntactic chart definition of ALGOL 60 [5.5]. It can be regarded as a linked list representation of syntax graphs of the form shown in Figure 5.5.3(b). (E, T, and F correspond to σ_0, σ_1, and σ_2, respectively; P_2, P_4, and P_5 are omitted; and $-$ and $/$ are alternatives for $+$ and $\times$, respectively.) Figure 5.5.3(d) is a variation which is especially useful for 'top-down' compiling (see Cohen and Gottlieb [5.13]). Each node has three pointer fields, one pointing to its definition (which is null for terminals), another pointing to alternatives, and the third pointing to continuations of a definition. Figure 5.5.3(e) shows yet another graphical means of defining the same grammar, which is called a 'syntax diagram'; for example, it has been used to define the syntax of PASCAL [5.26]. Each of the nonterminals (E, T, and F) are, in effect, defined as subgraphs, which include other subgraphs; F is a subgraph of T, T is a subgraph of E, and E may be a subgraph of still other subgraphs.

5.5.4 Compilers

A compiler may be thought of as a translator from any language to another. More usually, we say that a compiler translates *source* programs written in a 'high-level' language to *object* programs written in a 'low-level' language. For a general introduction to compiler design, see Brown [5.12]. Regardless of the languages involved, a compiler must first recognize statements in the source program, which is commonly done grammatically. The source language is defined by a formal (context-free) grammar, and the compiler parses each statement in the program, producing as output the derivation or parse tree for each correct statement. (What compilers should do for incorrect statements is also important, but we shall neglect this.) The sequence of such parse trees may be regarded as an *intermediate* program: the compiler translates from source to this intermediate code, and then from intermediate to object code. We note that the former task depends only upon the grammar of the source language, whereas the latter task depends upon the nature of the computer which is to execute the program.

Lexical Analysis To simplify the parsing task of a compiler, the grammar for the source language should itself be as simple as possible. One way to simplify the grammar is to treat specific character strings, or certain commonly occurring language constructs, as terminals rather than as nonterminals. For example, if key words (IF, GOTO, END), variable names (X, SUM, AVG), numerical constants (3.14159, 100), or even arithmetic expressions ($A+B+C$) were regarded as terminals, then the grammar would not have to define them. Before statements having such terminals can be parsed, however, the terminals must be recognized by a pre-processor, called a *lexical analyzer,* and the terminals formed are called *tokens*. A trade-off is clearly involved: parsing may be simplified at the expense of complicating the lexical analyzer.

Lexical analyzers may translate source programs, one statement at a time, into strings of tokens; the string of tokens corresponding to each statement is then parsed. However, the lexical analysis task is usually performed in conjunction with the parsing task, often as a subroutine which finds the next token, or which just checks whether or not a particular token may be next. It should be noted that a statement may have to be scanned well past a token before it can be recognized. For example, in the FORTRAN statement

```
DO123I=1,N
```

whether the first token is (say) the keyword "DO" or the variable name "DO123I" cannot be ascertained until the comma is reached, whereupon the possibility of an assignment statement can be ruled out for grammatical reasons. In other words, when terminals may have common prefixes, lexical analysis cannot be separated from grammatical analysis.

Phrase Reduction An algorithm which accepts any input $w \in \mathbf{L}(G)$ and produces as output some representation of a derivation sequence for w, is called a *parser* for G; the parser should block for $w \notin \mathbf{L}(G)$. A parser that always outputs leftmost derivations is called a 'left-parser'; one that outputs rightmost derivations is called a 'right-parser.' So that the leftmost and rightmost derivations are unique, we require that G be unambiguous.

Let T_w denote the derivation or parse tree of a sentential form w of a context-free grammar $G=(A, T, P, \sigma_0)$. A *phrase* of w is any subtree of T_w; phrases are denoted by the leaves of the (ordered) subtree in left-to-right order, hence by a substring of w. (But not every substring w' of w denotes a phrase, since w' may not have an associated subtree.) A phrase is *simple* if it contains no other phrase. The *handle* of a sentential form is its leftmost simple phrase. A phrase $v_1v_2...v_k$ is *reduced* (or the tree containing it is 'pruned') by replacing it by u, if $u \to v_1v_2...v_k$ is a production.

It can be shown that the successive reduction of handles of a sentential form w corresponds to its rightmost derivation in the sense that the sequence of reductions is the *reverse* of the sequence of productions used in the rightmost derivation of w from σ_0. A deterministic rewriting system that performs this sequence of handle reductions (with an associated output) is a right-parser. We shall say that a *canonical* parser recognizes the handle of a given sentential form, after which the handle is reduced and a record of this action is made; the canonical parser is then restarted with the rewritten sentential form as a new input. The design and operation of such a parser, and the class of grammars which can be so parsed, is discussed in detail in textbooks on compiler design. For example, see Gries [5.21].

Code Generation As stated above, the parsing task of a compiler is that of constructing a parse tree for grammatically correct statements. A matrix or linked representation (see Section 4.4) of the parse tree may be easily constructed using a canonical parser. As each handle is recognized, a new node is added to the tree. The final node added will be the root of the parse tree. In the resulting tree, the leaves identify tokens while the internal nodes identify productions.

Given the parse tree, the reverse of the rightmost derivation sequence can be found by canonically traversing the tree. When a node is traversed for the last time (on return from its rightmost child), it is the root of the handle-subtree, which may then be pruned. As each handle is pruned, some appropriate action may be taken, such as output of intermediate code or object code generation. The intermediate code can be in the form of an 'operator tree'; for example, the parse tree of Figure 5.5.2(a) can be easily transformed into the operator tree of Figure 4.2.4(b). Machine or assembly language object code can in turn be easily generated by traversing the operator tree. For details, including code optimization techniques utilizing graph and tree concepts, see Allen [5.4] and Nakata [5.35] for some pioneering work; Aho and Ullman [5.3] describe more recent developments. Of related interest are the 'debugging compiler' concepts mentioned in Section 5.2.1; see Ramamoorthy [5.40] and Lew [5.30] for a further discussion.

5.6 AUTOMATA

An algebraic system consisting of a set S and a relation R (i.e., a digraph) may be thought of as a set of ***states*** and allowable ***transitions*** between states: $(a,b)\in R$ means that it is possible to 'move' from state a to state b. Any algebraic system so interpreted is called a 'state-transition system' or 'machine'. An ***automaton*** is a machine where a state transition can be made only in response to an input, and then only according to a fixed set of rules. An input must be a symbol in some alphabet A, and we say that a sequence of inputs (i.e., a string over A) moves the automaton through a sequence of transitions from an 'initial' state into a 'final' state, by 'reading' one symbol at a time (in left-to-right order) until the input is exhausted. Transition rules specify the (next) state into which the automaton should move as a function of its (current) ***configuration***, which consists of its current state and the input substring remaining to be read. We also say that the automaton moves from configuration to configuration. Suppose now that we distinguish a subset of states $F \subseteq S$. We can then associate with the automaton the language consisting of those strings which move it from some given intial state into a final state that is in F. (A 'final' state may be defined as that of a configuration having no remaining input.)

Formally, a *finite-state acceptor* (FSA) is an algebraic system $M=(S,A,\delta,s_0,F)$ where S is a finite nonempty set (of states), A is a finite nonempty set (of input symbols), $\delta: S \times A \rightarrow S$ (the state transition function), $s_0 \in S$ (the initial state), and $F \subseteq S$ (the set of *accepting* states). If $\delta(s,a)=s'$, we sometimes write $s \underset{a}{\rightarrow} s'$. We extend the definition of δ so that $\delta(s,\varepsilon)=s$, and

$$\delta(s,a_1a_2\ldots a_n)=\delta(\delta(\ldots\delta(\delta(s,a_1),a_2),\ldots),a_n);$$

δ so extended is a function from $S \times A^*$ to S. The language *accepted* by an FSA M, denoted $\Lambda(M)$, is the set of input strings $\{w \in A^* \mid \delta(s_0,w) \in F\}$.

If δ is a total function, then the set of strings *rejected* by the FSA, $\{w \in A^* \mid \delta(s_0,w) \notin F\}$, is equal to $A^* - \Lambda(M)$. However, if δ is a partial function, i.e., if there are pairs $(s,a) \in S \times A$ for which δ is undefined, then the FSA may *block* (at other than an accepting state) for some input strings; we say these strings are not 'recognized' (neither accepted nor explicitly rejected) by the FSA. If δ is not a (single-valued) function from $S \times A$ to S, but instead maps pairs (s,a) to subsets of states (i.e., $\delta: S \times A \rightarrow 2^S$), we say that the FSA is *nondeterministic*; the FSA then generally has a choice among a number of states into which it may move from any configuration. In some cases, 'autonomous' moves, $\delta(s,\varepsilon)$, to other states are permitted; in these cases, $\delta: S \times (A \cup \{\varepsilon\}) \rightarrow 2^S$. The language accepted by a nondeterministic FSA is then given by

$$\{w \in A^* \mid \delta(s_0,w) \cap F \neq \emptyset\};$$

the set of final states $\delta(s_0,w)$ need not be a subset of F. It can be shown that, for any nondeterministic FSA, there exists a deterministic FSA which accepts the same language; consequently, we may restrict ourselves to the simpler FSAs with no loss of generality. See Ginsburg [5.20] and Minsky [5.34].

For later convenience, we also define a 'move' relation $\vdash$ on configurations $S \times A^*$. We write $[s,w] \vdash [s',w']$ if $w=aw'$ and $s \underset{a}{\rightarrow} s'$, i.e., the FSA moves from state s to s' by reading the leading input symbol a. We define $\vdash^*$ as the star closure of $\vdash$. In terms of $\vdash$,

$$\Lambda(M)=\{w \in A^* \mid [s_0,w] \vdash^* [s,\varepsilon], \text{ for some } s \in F\}.$$

Example Consider the FSA (S,A,δ,s_0,F) where $S=\{s_0,s_1,s_2\}$, $A=\{\lambda,+,\times\}$, $F=\{s_0,s_1\}$, and δ is defined by cases as follows:

$$\delta(s_0,\lambda)=s_1,\ \delta(s_1,+)=s_2,\ \delta(s_1,\times)=s_2,\ \delta(s_2,\lambda)=s_1.$$

Note δ: $S \times A \to S$ is a partial function. It is also common to represent (or define) δ tabularly as shown in Table 5.6, where blank entries represent arguments for which δ is not defined. The sequence of configuration moves as the FSA processes the string $\lambda + \lambda \times \lambda$ is as shown:

$$[s_0, \lambda + \lambda \times \lambda] \vdash [s_1, +\lambda \times \lambda] \vdash [s_2, \lambda \times \lambda] \vdash [s_1, \times \lambda] \vdash [s_2, \lambda] \vdash [s_1, \varepsilon].$$

Since the final state is in F, the string is accepted. In terms of state transitions, the foregoing is equivalent to the sequence

$$s_0 \underset{\lambda}{\to} s_1 \underset{+}{\to} s_2 \underset{\lambda}{\to} s_1 \underset{\times}{\to} s_2 \underset{\lambda}{\to} s_1.$$

S \ A	λ	$+$	$\times$
s_0	s_1		
s_1		s_2	s_2
s_2	s_1		

Table 5.6

5.6.1 State-transition Diagrams

As has been noted above, a digraph (S,R) may represent possible transitions between states of an FSA (S,A,δ,s_0,F), where $(s,s') \in R$ iff $\delta(s,a)=s'$ for some a. We may then give the branch (s,s') the label a, and write $s \underset{a}{\to} s'$ to denote the transition. Since any branch (s,s') may have more than one such label, we may instead label it with the set $\{a \in A \mid \delta(s,a)=s'\}$. Such a branch-labeled digraph is called a *state(-transition) diagram*. In its pictorial representation, it is often convenient to draw a separate branch for each distinct label from node s to s', so the diagram is actually a multigraph.

Examples In the state-diagram representations of FSAs given in Figure 5.6.1, initial states are shown as the leftmost nodes, and accepting states are shown as hatched nodes. The language accepted by the FSA in Figure 5.6.1(a) is $L_1 = \lambda(+\lambda)^*$. The FSA in Figure 5.6.1(b) accepts, in addition to L_1, the empty string (since the initial state is also an accepting state), and allows $\times$ in place of $+$. The FSA in Figure 5.6.1(c) accepts $a^* bc^*$, the regular set introduced in Section 3.3.3. The languages accepted by the FSAs in Figures 5.6.1(d) and (e) are $\{ab,c\}^* a$ and $(ca)^*$, respectively; their union $\{ab,c\}^* a \mid (ca)^*$ is accepted by the FSA in Figure 5.6.1(g), which is much more complicated than might be expected. The simpler FSAs

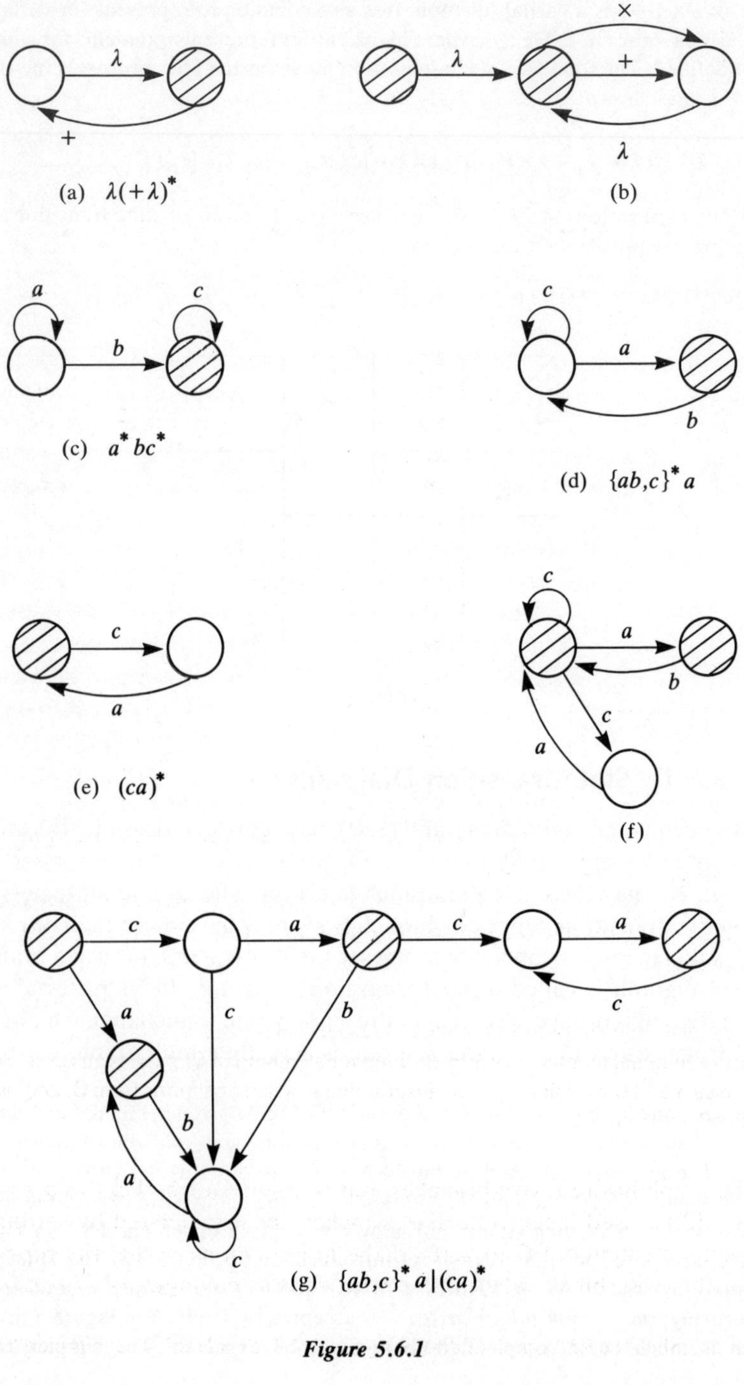

(a) $\lambda(+\lambda)^*$

(b)

(c) $a^* bc^*$

(d) $\{ab,c\}^* a$

(e) $(ca)^*$

(f)

(g) $\{ab,c\}^* a \mid (ca)^*$

Figure 5.6.1

in Figures 5.6.1(d) and (e) cannot simply be combined to yield that in Figure 5.6.1(f) because acceptance of additional strings not in the desired language would result; e.g., the latter FSA accepts *ccaa*. The FSA in Figure 5.6.1(f) is also nondeterministic, whereas all the preceding ones are deterministic.

5.6.2 Transducers

A *transducer* is an automaton which outputs a symbol or a string of symbols each time it moves from one state or configuration into another. A *finite-state transducer* (FST) is an FSA, $M=(S,A,\delta,s_0,F)$ to which is added an *output function* Θ: $S\times A\to B$; if the FST moves from state s to $\delta(s,a)$ on reading input symbol a in A, it also outputs the symbol $\Theta(s,a)$ in B. (The output alphabet B may, but need not, be the same as the input alphabet A.) A state diagram for an FST has a second set of labels $\{\Theta(s,a)\mid\delta(s,a)=s'\}$ for branch (s,s') in the associated digraph. The output function Θ may be extended to map from $S\times A^*$ to B^* by defining

$$\Theta(s,\varepsilon)=\varepsilon \qquad \text{and} \qquad \Theta(s,a_1a_2\ldots a_n)=\Theta(\Theta(\ldots\Theta(\Theta(s,a_1),a_2),\ldots),a_n).$$

A (deterministic) FST defines a translation τ from A^* to B^* by letting $\tau(w)=\Theta(s_0,w)$ if $\delta(s_0,w)\in F$. FSTs 'preserve' languages in the sense that, for any FST, $\{\tau(w)\mid w\in L\}$ is a Type 3 (or Type 2) language if L is also a Type 3 (or Type 2, respectively) language. We remark that the state transition and output functions may be combined, so that δ': $S\times A\to S\times B$, where $\delta'(s,a)=(s',b)$ iff $\delta(s,a)=s'$ and $\Theta(s,a)=b$.

Automata with more general output capabilities may also be defined. For example, B may itself be a set of strings, including ε. There may also be more than one output function, in which case we say that the automaton has multiple 'write heads'. Of greater significance are automata having the capability of reading what they write, generally in some restricted fashion; such automata have in essence an auxiliary memory. We discuss three important classes below.

5.6.3 Push-down Acceptors

An important class of automata having a restricted capability of reading what they write is that in which each automaton has a 'push-down' tape, in addition to input and output tapes. A 'push-down' tape is one that operates as a *stack*, where what can be read (and unstacked) at any time is restricted to the last or 'top' symbol written (stacked) onto the tape. A *push-down acceptor* (PDA) is a nondeterministic automaton (S,A,Z,δ,z_0,s_0,F), where

S is a finite nonempty set (of states)
A is a finite nonempty set (of input tape symbols)
Z is a finite nonempty set (of push-down tape symbols)
δ: $S\times(A\cup\{\varepsilon\})\times Z\to 2^{S\times Z^*}$
$z_0\in Z$ is the initial symbol on the push-down tape
$s_0\in S$ is the initial state
$F\subseteq S$ is the set of accepting states.

Note that S,A,s_0, and F are as for FSAs, but now the transition function δ is a function of the current state, the leading input symbol (to be read), and the top symbol on the push-down tape (to be unstacked), and specifies a new state and a string of output symbols to be written onto the push-down tape one at a time (so that the trailing symbol of the string ends up on the top). If $(s',\varepsilon)\in\delta(s,a,z)$, then the top symbol of the push-down tape (z) may be simply unstacked, with nothing being restacked, in the transition from state s to s'.

The configuration of a PDA, including the contents of the push-down tape, is a triple $[s,\alpha,\zeta]$, where $s\in S$ is the current state, $\alpha\in A^*$ is the remaining input string, and $\zeta\in Z^*$ is the string on the push-down tape. By convention, the leading symbol of α is the current input, and the trailing symbol of ζ is the top of the push-down tape. The initial configuration is then $[s_0,w,z_0]$, where $w\in A^*$ is an input string which is to move the PDA into a final configuration $[s_1,\varepsilon,\zeta_1]$. We say that the PDA *accepts* w if $s_1\in F$. (An alternative definition of acceptance requires that $\zeta_1=\varepsilon$, with s_1 unspecified; note that $\delta(s,a,\varepsilon)$ is undefined for any s and a.)

Formally, we write $[s,aw,\beta z]\vdash[s',w,\beta\gamma]$ if $\delta(s,a,z)$ contains (s',γ); $\vdash$ is the 'move' relation on configurations, and we define $\vdash^*$ to be its star closure. The language accepted by a PDA $P=(S,A,Z,\delta,z_0,s_0,F)$ is the set of input strings

$$\Lambda(P)=\{w\in A^* \mid [s_0,w,z_0]\vdash^* [s,\varepsilon,\zeta] \text{ for some } s \text{ in } F,\ \zeta \text{ in } Z^*\}.$$

We remark that, unlike the case for FSAs, a deterministic PDA (i.e., one for which each set $\delta(s,a,z)$ may contain at most one member for $a\in A\cup\{\varepsilon\}$, and contains none if $\delta(s,\varepsilon,z)\neq\emptyset$) is less general, in that there are languages that can be accepted by a nondeterministic PDA but cannot be accepted by a deterministic one. Furthermore, a PDA which also has a separate output tape, called a push-down transducer (PDT), does not 'preserve' languages (as does an FST).

Example We shall define a PDA which accepts $\{({}^n\lambda)^n \mid n\geq 0\}$. Let

$$S=\{s_0,s_1,s_2,s_f\},\ A=\{(,\lambda,)\},\ Z=\{z_0,(\},\ F=\{s_f\},$$

and δ be defined by cases as follows:

(1) $\delta(s_0, \lambda, z_0) = \{(s_f, \varepsilon)\}$
(2) $\delta(s_0, (, z_0) = \{(s_1, z_0()\}$
(3) $\delta(s_1, (, () = \{(s_1, (()\}$
(4) $\delta(s_1, \lambda, () = \{(s_2, \varepsilon)\}$
(5) $\delta(s_2,), () = \{(s_2, \varepsilon)\}$
(6) $\delta(s_2,), z_0) = \{(s_f, \varepsilon)\}$.

(1) results in acceptance of λ (i.e., when $n=0$). (2) and (3) stack consecutive input occurrences of '(', until an occurrence of a single λ is encountered, at which time one '(' is unstacked using (4). (5) results in the unstacking of a '(' for each input occurrence of a ')'. If there are n occurrences of '(', then each is stacked: one is unstacked when λ is encountered, and the other $n-1$ are unstacked for $n-1$ occurrences of ')'; at the nth occurrence of a ')', only z_0 remains on the stack, so (6) will result in acceptance (while also emptying the stack). Too many ')'s would lead to blocking since $\delta(s_f,), z)$ is undefined, while too few ')'s would leave the PDA in a nonaccepting state s_2 [with more '('s on the stack]. The operation of the PDA as it processes (a) ((λ)), (b) ((λ))), and (c) ((λ) is traced below:

(a) $[s_0, ((\lambda)), z_0] \vdash [s_1, (\lambda)), z_0(]$
$\vdash [s_1, \lambda)), z_0((]$
$\vdash [s_2,)), z_0(]$
$\vdash [s_2,), z_0]$
$\vdash [s_f, \varepsilon, \varepsilon]$

(b) $[s_0, ((\lambda))), z_0] \vdash [s_1, (\lambda))), z_0(]$
$\vdash [s_1, \lambda))), z_0((]$
$\vdash [s_2,))), z_0(]$
$\vdash [s_2,)), z_0]$
$\vdash [s_f,), \varepsilon]$

(c) $[s_0, ((\lambda), z_0] \vdash [s_1, (\lambda), z_0(]$
$\vdash [s_1, \lambda), z_0((]$
$\vdash [s_2,), z_0(]$
$\vdash [s_2, \varepsilon, z_0]$.

We remark that this PDA is deterministic.

5.6.4 Linear-bounded Acceptors

The auxiliary push-down tape of a PDA gives it a memory of unbounded size, but the 'top-of-stack' access restriction (and the inability to re-read [by backspacing or rewinding] the input tape) significantly limits the class of languages which it can accept. We next define an automaton which has a tape that can be accessed (read or written) at either end or anywhere in the middle; this tape will in fact be the input tape, which we permit to move forwards or backwards, one symbol at a time, relative to a stationary read/write head. Preferably, for notational reasons, we may regard the tape as stationary and permit the read/write head to move across the tape. The symbol 'under' the read/write head is called the *current* symbol and only this symbol may be read or rewritten at any given time. While this new automaton is permitted to write on its input tape, it is not permitted to write before the leading input symbol or after the trailing input symbol. Thus, with this restriction, if the automaton writes at all, it must overwrite the input string. The contents of the tape then constitutes the auxiliary memory of the automaton, and the size of this memory is bounded by the length of the input string.

A *linear-bounded acceptor* (LBA) is a nondeterministic automaton $B=(S,A,\delta,s_0,F)$ where S, A, s_0 are as for an FSA, $F\subseteq S-\{s_0\}$, and

$$\delta\colon S\times A\to 2^{S\times A\times\{-1,0,1\}}.$$

If $(s',a',k)\in\delta(s,a)$, the LBA may make a transition from state s to s' if the current symbol (being read) is a; in the transition, the symbol a is to be overwritten by the symbol a', and, in addition, the tape is to be moved k symbols to the left (relative to the head), or equivalently, the read/write head is moved k symbols to the right (relative to the tape). If k is negative, the movement is in the opposite direction, In other words,

$k=-1$	means head moves one symbol to left
$k=0$	means head does not move
$k=1$	means head moves one symbol to right.

The boundedness restriction stated earlier is reflected in this definition by the fact that the empty symbol ε cannot be read or written.

The configuration of an LBA is the triple consisting of the current state s, the string $a_1a_2\ldots a_n$ on the tape, and the subscript i indicating that the head is currently positioned under symbol a_i; for notational convenience, we denote this configuration by writing $[a_1a_2\ldots a_{i-1}\underline{s}a_i\ldots a_n]$. Then for u,v in A^*, b in A, we write

$[ub\underline{s}av]\vdash[u\underline{s}'ba'v]$	if $\delta(s,a)$ contains $(s',a',-1)$
$[u\underline{s}av]\vdash[u\underline{s}'a'v]$	if $\delta(s,a)$ contains $(s',a',0)$
$[u\underline{s}av]\vdash[ua'\underline{s}'v]$	if $\delta(s,a)$ contains $(s',a',1)$

and define $\vdash^*$ as the star closure of $\vdash$. Note that since $\delta(s,\varepsilon)$ is not defined, the LBA can no longer move if it enters a configuration $[u\underline{s}]$; we say then that the LBA goes off the tape on the right. (If we let b in the foregoing be ε, it may also go off on the left.) An alternative to going off the tape is to assume that the input string is enclosed between distinguishable end-markers which cannot be rewritten. The language *accepted* by an LBA B is defined as the set of strings

$$\Lambda(B)=\{w\in A^*\mid[\underline{s}_0w]\vdash^*[u\underline{s}]\text{ for some } s \text{ in } F,\ u \text{ in } A^*\}.$$

5.6.5 Turing Machines

The most general class of automata is that of Turing machines, which may be defined in numerous equivalent ways, but whose distinguishing feature is that their input tape may be written upon, as for LBAs, but without the boundedness restriction (nor the nondeterminism). This means, in particular, that a symbol may be written beyond either end of the input tape, and furthermore, that its memory is not bounded by the length of the input string. Unlike a PDA, which also has an unbounded memory, a Turing machine can freely access any portion of its tape.

We define a *Turing machine* (TM) as a deterministic automaton $T=(S,A,\delta,s_0,F)$ where S, A, s_0, and F are as for an FSA, and

$$\delta: S\times A\to S\times A\times\{-1,0,1\}.$$

$\delta(s,a)=(s',a',k)$ has the same meaning as for an LBA (whose transition function δ specifies single-element sets), as does the move relation $\vdash$ on configurations of the form $[u\underline{s}av]$. We must, however, distinguish one element of A as the ***blank*** character, denoted β, and assume that the input string is preceded and followed by an indefinite number of βs. We say then that a TM *halts* for an input string w in A^* if

$$[\beta^*\,\underline{s}_0 w\beta^*\,]\vdash^*\,[\beta^*\,u\underline{s}av\beta^*\,]$$

such that $\delta(s,a)$ is undefined; in addition, we say w is *accepted* if s is in F, and *rejected* otherwise. Furthermore, we say that the TM *computes* the partial function $\mathbf{T}(w)=uav$. If the TM does not halt for w in a finite number of moves, then we say that it *loops* for w.

Example The TM shown in Table 5.6.5 (where we have used the same tabular representation as was introduced earlier for FSAs) computes the greatest common divisor (cf. Section 1.1.7) of two numbers m and n using Trakhtenbrot's algorithm [5.45]. These numbers are represented on the input tape by strings of m and n copies of the symbol '|', respectively, without a delimiting symbol separating them; instead of a delimiter, the numbers are distinguished by assuming that the tape head is initially positioned under the leftmost '|' of the right-hand number. For (say) $m=2$ and $n=4$, the initial configuration is

$$[\beta^*\,|\,|\underline{s}_0|\,|\,|\,|\beta^*\,].$$

The TM moves are as follows:

$$\begin{aligned}
[\beta^*\,|\,|\underline{s}_0|\,|\,|\,|\beta^*\,] &\vdash [\beta^*\,|\,|\underline{s}_1a|\,|\,|\beta^*\,]\\
&\vdash [\beta^*\,|\underline{s}_1|a|\,|\,|\beta^*\,]\\
&\vdash [\beta^*\,|\underline{s}_0ba|\,|\,|\beta^*\,]\\
&\vdash [\beta^*\,|b\underline{s}_0a|\,|\,|\beta^*\,]\\
&\vdash [\beta^*\,|ba\underline{s}_0|\,|\,|\beta^*\,]\\
&\vdash [\beta^*\,|ba\underline{s}_1a|\,|\beta^*\,]\\
&\vdash [\beta^*\,|b\underline{s}_1aa|\,|\beta^*\,]
\end{aligned}$$

$$
\begin{aligned}
&\vdash [\beta^* \,|\underline{s}_1 baa\,|\,|\beta^*\,] \\
&\vdash [\beta^* \underline{s}_1 |baa\,|\,|\beta^*\,] \\
&\vdash [\beta^* \underline{s}_0 bbaa\,|\,|\beta^*\,] \\
&\vdash \ldots
\end{aligned}
$$

We leave the generation of the remainder of this 'trace' as an exercise.

Generalizing, we make the following observation about the states of this TM:

(s_0) when in s_0, the TM scans toward the right, looking for the nearest '|': if found, it changes it to a and switches to state s_1; if not found (i.e., if β is encountered), it backs up one character and switches to state s_3.

(s_1) when in s_1, the TM scans toward the left, looking for the nearest '|': if found, it changes it to b and switches to state s_0; if not found (i.e., if β is encountered), it backs up one character and switches to state s_2.

(s_2) when in s_2, the TM scans toward the right, erasing each b as it goes, and then changes each a to '|' and, after backing up one character, it returns to state s_0.

(s_3) when in s_3, the TM scans toward the left, erasing each a as it goes, and then changes each b to '|', after which it 'stops' (by going to state s_4) or returns to state s_0, depending upon whether the left end was reached.

(s_4) this is the 'accepting' state at which the TM halts.

The operation of the TM, starting from configuration $[\beta^* \,|^m \underline{s}_0 |^n \beta^*\,]$, can be traced as follows.

Case A ($m<n$):

$$
\begin{aligned}
[\beta^* \,|^m \underline{s}_0 |^n \beta^*\,] &\vdash^* [\beta^* \underline{s}_0 b^m a^m |^{n-m} \beta^*\,] \\
&\vdash^* [\beta^* b^m a^m \underline{s}_0 |^{n-m} \beta^*\,] && \text{(scan to right)} \\
&\vdash [\beta^* b^m a^m \underline{s}_1 a\,|^{n-m-1} \beta^*\,] && \text{(change extra | to } a) \\
&\vdash^* [\beta^* \underline{s}_1 b^m a^m a\,|^{n-m-1} \beta^*\,] && \text{(scan to left)} \\
&\vdash [\beta^* \underline{s}_1 \beta b^m a^m a\,|^{n-m-1} \beta^*\,] && \text{(end reached)} \\
&\vdash [\beta^* \underline{s}_2 b^m a^m a\,|^{n-m-1} \beta^*\,] && \text{(back up)} \\
&\vdash^* [\beta^* \beta^m \underline{s}_2 a^m a\,|^{n-m-1} \beta^*\,] && \text{(erase } b\text{s)} \\
&\vdash^* [\beta^* \,|^m \underline{s}_2 a\,|^{n-m-1} \beta^*\,] && \text{(change } a\text{s to |s)} \\
&\vdash [\beta^* \,|^m |\underline{s}_2 |^{n-m-1} \beta^*\,] && \text{(change extra } a \text{ to |)} \\
&\vdash [\beta^* \,|^m \underline{s}_0 |\,|^{n-m-1} \beta^*\,] && \text{(back up)} \\
&= [\beta^* \,|^m \underline{s}_0 |^{n-m} \beta^*\,]
\end{aligned}
$$

Case B ($m=n$):

$$
\begin{aligned}
[\beta^* \,|^m \underline{s}_0 |^n \beta^*\,] &\vdash^* [\beta^* \underline{s}_0 b^m a^n \beta^*\,] \\
&\vdash^* [\beta^* b^m a^m \underline{s}_0 \beta^*\,] && \text{(scan to right)} \\
&\vdash [\beta^* b^m a^{n-1} \underline{s}_3 a \beta^*\,] && \text{(back up)} \\
&\vdash^* [\beta^* b^{m-1} \underline{s}_3 b \beta^n \beta^*\,] && \text{(erase } a\text{s)} \\
&\vdash^* [\beta^* \underline{s}_3 \beta\,|^m \beta^*\,] && \text{(change } b\text{s to |s)} \\
&\vdash [\beta^* \underline{s}_4 \beta\,|^m \beta^*\,] && \text{(stop)}
\end{aligned}
$$

Case C ($m>n$):

$$\begin{aligned}
[\beta^* |^m \underline{s}_0 |^n \beta^*] &\vdash^* [\beta^* |^{m-n} \underline{s}_0 b^n a^n \beta^*] \\
&\vdash^* [\beta^* |^{m-n} b^n a^n \underline{s}_0 \beta^*] && \text{(scan to right)} \\
&\vdash [\beta^* |^{m-n} b^n a^{n-1} \underline{s}_3 a \beta^*] && \text{(back up)} \\
&\vdash^* [\beta^* |^{m-n} b^{n-1} \underline{s}_3 b \beta^n \beta^*] && \text{(erase } a\text{s)} \\
&\vdash^* [\beta^* |^{m-n-1} \underline{s}_3 | |^n \beta^*] && \text{(change } b\text{s to } |\text{s)} \\
&\vdash [\beta^* |^{m-n-1} | \underline{s}_0 |^n \beta^*] && \text{(back up)} \\
&= [\beta^* |^{m-n} \underline{s}_0 |^n \beta^*]
\end{aligned}$$

Hence, for $m=2$ and $n=4$, we have

$$[\beta^* |^2 \underline{s}_0 |^4 \beta^*] \vdash^* [\beta^* \underline{s}_0 b^2 a^2 |^2 \beta^*] \vdash^* [\beta^* |^2 \underline{s}_0 |^2 \beta^*] \vdash^* [\beta^* \underline{s}_4 \beta |^2 \beta^*].$$

S \ A	β	$\vert$	a	b
s_0	$(s_3, \beta, -1)$	$(s_1, a, 0)$	$(s_0, a, +1)$	$(s_0, b, +1)$
s_1	$(s_2, \beta, +1)$	$(s_0, b, 0)$	$(s_1, a, -1)$	$(s_1, b, -1)$
s_2	$(s_0, \beta, -1)$	$(s_0, \vert, -1)$	$(s_2, \vert, +1)$	$(s_2, \beta, +1)$
s_3	$(s_4, \beta, 0)$	$(s_0, \vert, +1)$	$(s_3, \beta, -1)$	$(s_3, \vert, -1)$
s_4				

Table 5.6.5

5.6.6 Relationships to Grammars

A set of strings $L \subseteq A^*$ is called a *regular* language if there exists an FSA, M, such that $L = \Lambda(M)$. We shall show that a language is regular iff there exists a right-linear grammar that generates it. (Regular languages then are also regular sets.)

Theorem 1 $L = \Lambda(M)$ for an FSA M iff $L = \mathbf{L}(G)$ for an RLG G.

Proof (by construction): Let $M = (S, A, \delta, s_0, F)$ be an arbitrary FSA. Define an RLG $G = (S \cup A, A, P, s_0)$ where

$$P = \{s \to a\,\delta(s,a) \mid a \in A,\ s \in S\} \cup \{s \to \varepsilon \mid s \in F\}.$$

We have assumed that $S \cap A = \emptyset$, i.e., the states and inputs of M are different, and so the nonterminal symbols of G correspond to the states of M. Now for any $w \in A^*$, $s_0 \Rightarrow^* w$ iff $s_0 \Rightarrow ws$, where $\delta(s_0, w) = s$, and $s \rightarrow \varepsilon$ is in P. Therefore, $\delta(s_0, w)$ is in F, hence $w \in \Lambda(M)$.

On the other hand, let $G = (V, A, P, \sigma_0)$ be an arbitrary RLG. Define a (nondeterministic) FSA $M = (S', A, \delta, \sigma_0, F)$ as follows. First, replace each production in P, say $\sigma_1 \rightarrow a_{11} \ldots a_{1n} s_1$, where $a_{1i} \in A$ and $s_1 \in (V - A) \cup \{\varepsilon\}$, by the set of productions

$$\{\sigma_1 \rightarrow a_{11}\sigma_{11},\ \sigma_{11} \rightarrow a_{12}\sigma_{12},\ \ldots,\ \sigma_{1(n-1)} \rightarrow a_{1n}s_1\}.$$

Let $G' = (V', A, P', \sigma_0)$ denote this augmented grammar. Now define δ by

$$\delta(\sigma_1, a_{11}) = \sigma_{11},\ \ \delta(\sigma_{11}, a_{12}) = \sigma_{12},\ \ldots,\ \ \delta(\sigma_{1(n-1)}, a_{1n}) = s_1.$$

If $s_1 = \varepsilon$, define $\delta(\sigma_{1(n-1)}, a_{1n}) = \hat{s}$ instead, where $\hat{s}$ is a new symbol. Thus $\delta(\sigma_1, a_{11} \ldots a_{1n}) = s_1$, or $\hat{s}$ (if $s_1 = \varepsilon$). The set of states S' of the FSA is then $(V' - A) \cup \{\hat{s}\}$, and $F = \{\hat{s}\}$. Note that the final production used in the derivation $\sigma_0 \Rightarrow^* w$ must rewrite a variable by a string of terminals, hence $\delta(\sigma_0, w) = \hat{s}$.

We leave it to the reader to verify that, for both constructions, $w \in \Lambda(M)$ iff $w \in \mathbf{L}(G)$. *Q.E.D.*

The following relationships between the languages accepted by automata and the languages generated by grammars also hold.

Theorem 2 $L = \Lambda(P)$ for a PDA P iff $L = \mathbf{L}(G)$ for a CFG G.
Theorem 3 $L = \Lambda(B)$ for a LBA B iff $L = \mathbf{L}(G)$ for a CSG G.
Theorem 4 $L = \Lambda(T)$ for a TM T iff $L = \mathbf{L}(G)$ for a PSG G.

Proofs of these theorems, which we omit, can also be obtained by construction; see Ginsburg [5.19] or Harrison [5.22].

Application to Compilers The foregoing theorems imply that an automaton can be designed to recognize any given computer language. However, unless the language or one of its defining grammars, has special properties, the automaton will not be efficient enough for use in practice.

For a context-free grammar $G = (A, T, P, \sigma_0)$, consider only those sentential forms that can be generated from σ_0 by a rightmost derivation. If the handle of any such sentential form is uniquely (unambiguously) determined by the string to its left and at most k (necessarily terminal) symbols to its right, then the grammar is said to be $LR(k)$. For this class of grammars, a canonical parser as described in Section 5.5.4 exists: it incorporates a deterministic finite-state automaton that scans an input sentential form from left-to-right, moving from state to state, until k symbols past the handle are read, or until the end of the string is reached, whereupon (since the $LR(k)$

property guarantees that the handle be recognizable and uniquely reducible) an accepting state is entered; the handle is then reduced, using a production associated with the state, an appropriate output is made, and the automaton is restarted with the rewritten string as the input. This procedure can be reiterated until the resulting string consists solely of the initial symbol σ_0 of the grammar.

Algorithms exist for constructing canonical parsers for arbitrary $LR(k)$ grammars. In practice, it is desirable to have as few states in a parser as possible, and much research has been devoted to ways of merging states together in the construction process (e.g., see Pager [5.36]). Furthermore, so that a rewritten string need not be read from its beginning unnecessarily, $LR(k)$ parsers generally incorporate a stack so that, following a handle reduction, some earlier configuration may be re-entered. For details, see Backhouse [5.6].

In addition to their application to parsing, which is associated with the syntax of programming languages, automata have also been applied to the modeling of program semantics; see, for example, the 'Vienna definition' of PL/I [5.32]. Other discussions of semantics appear in Blum [5.9], Stoy [5.41], and Tennent [5.44].

5.6.7 Relationship to Computer Architecture

A digital computer can be modeled abstractly as an automaton, which moves from state to state according to a fixed set of rules. There is one rule for each of the *machine language instructions* in the computer's repertoire (or 'instruction set'); examples of instructions are LOAD, STORE, ADD, GOTO. The state of a computer consists of the data or instructions stored in its memory cells, including any special registers such as arithmetic accumulators (see Section 1.1.4), condition flags, and a 'program location counter' (PLC) which keeps track of which instruction in a program is to be executed next.

As a simple illustration, suppose that a computer has N memory cells, denoted C_1 to C_N, in addition to an accumulator (ACC), a flag (FLG), and the PLC. Let the state of this computer be denoted by

$$\langle \text{PLC, ACC, FLG}; C_1, C_2, \ldots, C_n \rangle.$$

We assume each memory cell can have as its (binary) value either a number or an instruction which has two parts. We denote the left-part of cell C_i by OPCODE_i, and the right-part by ADDRESS_i; these notations have meaning only if the cell contains an instruction. Whether a cell is to be interpreted as an instruction or data is determined by whether its address ever becomes the value of PLC! When the computer is started, the initial value of PLC is used to specify the cell address of the first instruction to be executed; PLC is then changed to point to another cell (which hopefully contains an instruction), usually by letting PLC=PLC+1. To be more specific, the state transition rules for our simple computer, whose instruction set is {LOAD, STORE, ADD, GOTO, COMPARE, IFZERO, STOP}, are as follows:

$$\langle p,a,f;C_1,C_2,\ldots,C_N\rangle \vdash \langle p',a',f';C_1',C_2',\ldots,C_N'\rangle$$

where C_p=[OPCODE$_p$:ADDRESS$_p$] and

(a) if OPCODE$_p$=LOAD,
then $p':=p+1$
$a':=C_{\text{ADDRESS}_p}$
$f':=f$
$C_i':=C_i$ each i

(b) if OPCODE$_p$=STORE,
then $p':=p+1$
$a':=a$
$f':=f$
$C_i':=C_i$ for each i except ADDRESS$_p$
$C_{\text{ADDRESS}_p}':=a$

(c) if OPCODE$_p$=ADD,
then $p':=p+1$
$a':=a+C_{\text{ADDRESS}_p}$
$f':=\begin{cases} 1 \text{ if overflow} \\ 0 \text{ otherwise} \end{cases}$
$C_i':=C_i$ each i

(d) if OPCODE$_p$=GOTO,
then $p':=\text{ADDRESS}_p$
$a':=a$
$f':=f$
$C_i':=C_i$ each i

(e) if OPCODE$_p$=COMPARE,
then $p':=p+1$
$a':=a$
$f':=\textbf{sgn}\ (a-C_{\text{ADDRESS}_p})$
$C_i':=C_i$ each i

(f) if OPCODE$_p$=IFZERO
then $p':=\begin{cases} \text{ADDRESS}_p & \text{when } f=0 \\ p+1 & \text{when } f\neq 0 \end{cases}$
$a':=a$
$f':=f$
$C_i':=C_i$ each i

(g) if OPCODE$_p$=STOP,
then the computer 'halts'.

Note that if $OPCODE_p$ is ever undefined, which is possible should the computer execute (say) an instruction [GOTO: i] where by mistake i is the cell address of data whose bit configuration does not correspond to any legal instruction, then the computer 'blocks'. On the other hand, if the data should by coincidence have, as its left part, a bit configuration corresponding to an instruction in the computer's repertoire, then it will be executed just as if it were a program instruction. (This is worse than blocking since erroneous results may go undetected.)

For example, suppose that the initial state of our simple computer is as follows:

$$\langle 4, 0, 0; X, Y, 0, [\text{LOAD:}1], [\text{ADD:}2], [\text{IFZERO:}5], \; [\text{STOP:}0], \ldots \rangle.$$

A trace of the state transitions would reveal that the machine language program (starting at cell 4) adds Y to X successively until an overflow results. Note that it is possible for infinite looping to occur, e.g., when $Y=0$.

The foregoing model of a computer is oversimplified, but does illustrate the most important features. Missing features include how a program and data are loaded in the first place, to yield the initial state, and how input and output operations take place. We re-emphasize, in summary, that programs are stored in the main memory of a computer (and so can be treated as data by other programs), and are distinguishable from ordinary data only by 'context'—i.e., by the values which the PLC can assume. This is commonly known as the 'stored-program' concept (often attributed to von Neumann), perhaps the most important principle governing the original design of general-purpose computers. (For some modern design concepts, see the next section.)

It should also be noted that, at a lower level, computers execute 'microprograms' which specify state transition rules for each OPCODE; for example, one instruction of such a microprogram would increment the PLC. These microprograms are commonly an irreplaceable part of a computer's hardware control unit. However, there exist computers, said to be 'micro-programmable', where it is possible to write new microprograms or alter existing ones. If microprograms are replaceable or changeable, they are said to constitute the *firmware* of a computer. It is also possible (albeit usually impractical) to 'emulate' any computer on any other computer using software, e.g., by writing a higher-level language program which implements a given set of state transition rules, and in which, for example, ordinary program variables are used to represent the special registers and an array is used for the memory cells.

5.6.8 Petri Nets

Automata of the kind discussed up to this point are useful models for

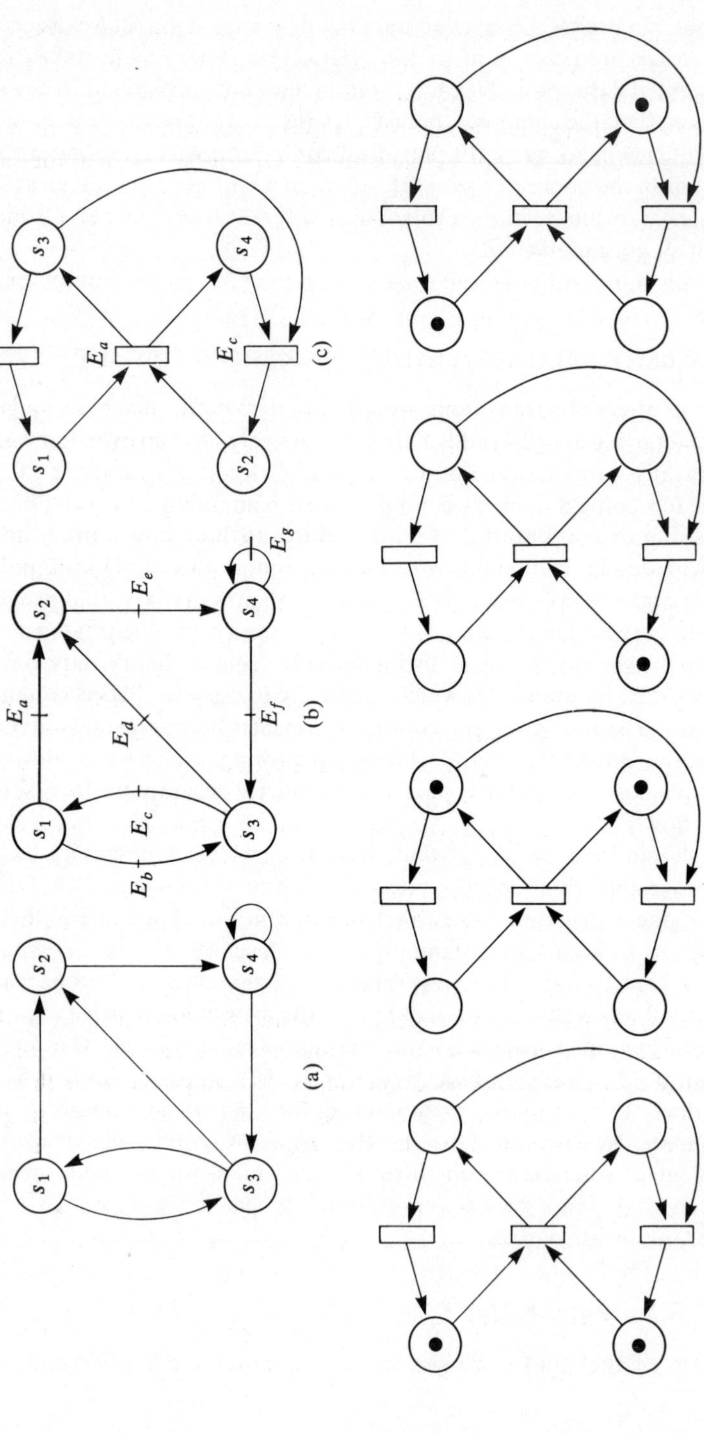

Figure 5.6.8

'uniprocessing' systems but are not well-suited for modeling multiprocessing systems where the state of the system depends upon the states of independently and concurrently executing processes. Suppose that we generalize our concept of a nondeterministic automaton to the case where the system can be in several individual states concurrently. Of course, we may define a composite state as a set of concurrent states, so that the system can be in only one composite state at a time, but it is useful to treat the individual states as independent entities. As an example, in the graph shown in Figure 5.6.8(a), where nodes represent states, the system may be in states s_1 and s_2 concurrently, or in a composite state $\underline{s}_1\underline{s}_2$. It may then make a transition to composite states $\underline{s}_3\underline{s}_2$, $\underline{s}_1\underline{s}_4$, $\underline{s}_3\underline{s}_4$, $\underline{s}_2\underline{s}_4$, or $\overline{\underline{s}}_2$,† depending upon whether transitions are made from s_1 to s_2 or s_3, or from s_2 to s_4. We say these individual transitions correspond to the occurrences of 'events', depicted by labeled branches in Figure 5.6.8(b): for example, the transition from $\underline{s}_1\underline{s}_2$ to $\underline{s}_2\underline{s}_4$ is made when both events E_a and E_e occur, and the transition from $\underline{s}_1\underline{s}_2$ to $\overline{\underline{s}}_2$ is made when only event E_a occurs.

Suppose now that certain events can occur only if the system is in two (or more) given states concurrently, and that an occurrence of an event can cause a transition to several states concurrently. To depict this situation, events can no longer be represented by labeled branches, but can be represented by a distinguished (say, rectangular) class of nodes. For example, in Figure 5.6.8(c), event E_a can occur only if the system is in states s_1 and s_2 concurrently; after occurrence of event E_a, the system enters both states s_3 and s_4. A general system of this sort is called a Petri net.

Formally, a *Petri net* is an algebraic system $(S, E, R_1, R_2, \sigma_0)$, where S is a set of states, E is a set of events, $R_1 \subseteq S \times E$ and $R_2 \subseteq E \times S$ are relations whose members (branches) connect state nodes with event nodes and vice versa, and $\sigma_0 \subseteq S$ (actually, a multiset of S) specifies the initial composite state of the system. Composite states can be represented by 'marking' the individual states, for example, by dots called 'tokens', as shown in Figure 5.6.8(d); this marking represents $\sigma_0 = \underline{s}_1\underline{s}_2$. An event can occur only if all of its predecessor state nodes are marked; occurrence of an event, called its 'firing', results in the removal of a token from each of its predecessors and the placing of a token onto each of its successors. Some state nodes may have more than one token (cf. $\overline{\underline{s}}_2$, in the above). Figure 5.6.8(e) shows the result of firing E_a, when initially in composite state $\sigma_0 = \underline{s}_1\underline{s}_2$. Figure 5.6.8(f) and (g) show the two possible next states; in both cases, no subsequent firings are possible, so the Petri net is said to be 'deadlocked'.

The main problems addressed in the theory of Petri nets include derivations of conditions which ensure 'liveness' (where deadlock is not possible)

†The overbar is used to distinguish this composite state from the case $\underline{s}_2$, where the system is (say, initially) in the single state s_2.

and 'safeness' (having at most one token per state node). See Peterson [5.38] for theoretical details. Related operating systems concepts are discussed in Chapter 9.

We remark, in conclusion, that Petri nets also serve as useful models for *data flow* computer architectures, where instructions are not executed in a prescribed sequence, but rather are executed upon the occurrence of specified events. An instruction can execute whenever its operands (input data) become available. See Dennis [5.17] for further information.

5.6.9 Relationship to Computability

Turing machines are the most general kind of automata in the sense that they can implement any computable function. No other kind of automaton can compute a function which cannot also be implemented by a TM. In fact, there are several ways to implement computable functions:

1. using Markov algorithms (cf. SNOBOL);
2. using λ-calculus (a generalization of LISP S-expressions);
3. using Turing machines;
4. using computer programs (or flowcharts).

That the first three are equivalent in their computational power is one of the main results of 'computability theory' (cf. Church's thesis). That the fourth is also equivalent follows from the fact that computer programs can be written to implement the first three.

It has been shown that a *structured* program (in the sense of Section 5.4.1) is sufficient to implement any computable function, so unstructured programming is never necessary. It can also be shown that a decision table program can implement any computable function, since a decision table can emulate any flowchart or Turing machine (see Lew [5.31]). For example, the Turing machine of Section 5.6.5 can be emulated by the decision table shown in Table 5.6.9, where e denotes the blank character and i=0 initially.

i=?	0	0	0	0	1	1	1	1	2	2	2	2	3	3	3	3
c=?	e	1	a	b	e	1	a	b	e	1	a	b	e	1	a	b
WRITE	e	a	a	b	e	b	a	b	e	1	1	e	–	1	e	1
MOVE HEAD	R	–	L	L	L	–	R	R	R	R	L	L	–	L	R	R
SET i=	3	1	0	0	2	0	1	1	0	0	2	2	–	0	3	3
EXIT	–	–	–	–	–	–	–	–	–	–	–	–	X	–	–	–

Table 5.6.9

EXERCISES

1. Apply the procedure of Section 5.2.1 to find the reachability matrix for the digraph of Figure 5A.
2. Show the depth-first and breadth-first traversals of the digraph of Figure 5A, starting at node 1.
3. For the digraph of Figure 5A,

   (a) find its MSCRs;
   (b) enumerate its simple cycles.

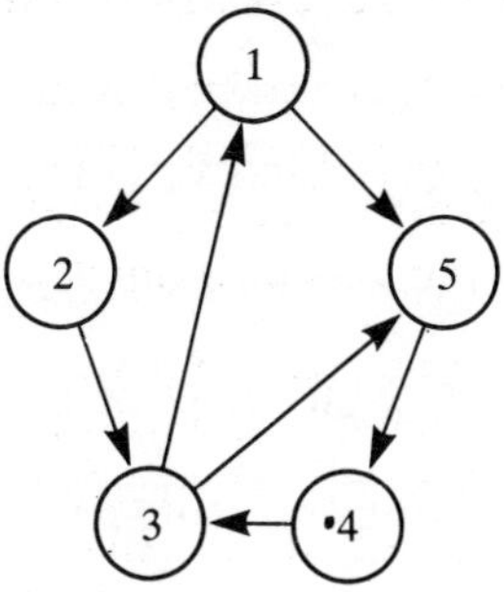

Figure 5A

4. Find the cliques of the graph of Figure 5B, and of the complement of the graph. Show that each clique can be represented by a block in some permutation of the reflexive adjacency matrix **C** ∪ **I**.

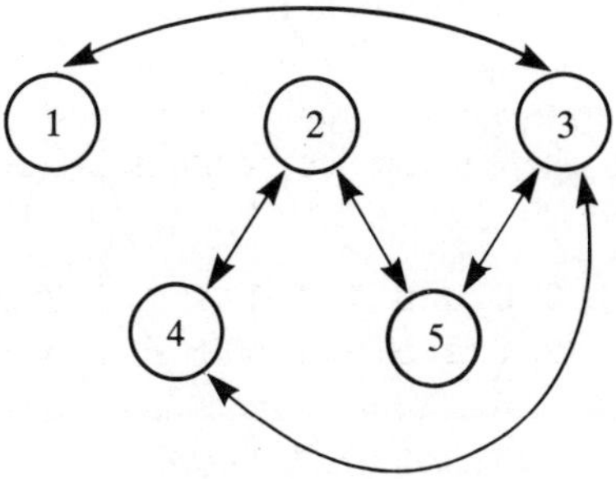

Figure 5B

5. Find nontrivial subgraphs of the digraph of Figure 5C that are isomorphic to each other.
6. Apply the topological sort procedure of Section 5.2.5 on the digraph of Figure 5C.

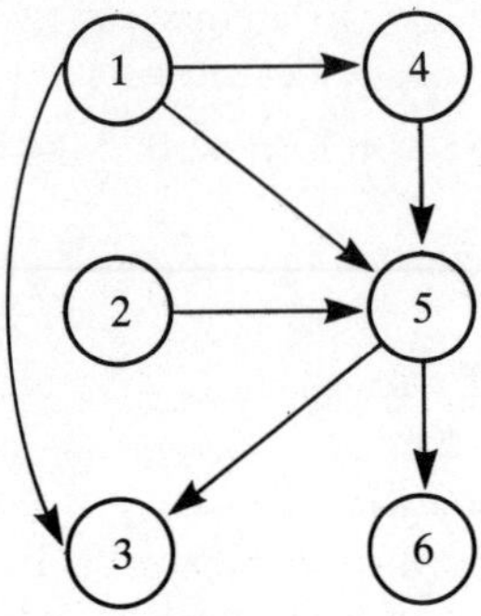

Figure 5C

7. Trace the canonical traversal ('visiting') algorithm of Section 5.3.1 applied to the tree of Figure 5D.
8. Trace the postorder traversal algorithm of Section 5.3.1 applied to the tree of Figure 5D.
9. Trace the tree construction algorithm of Section 5.3.2 applied to the S-expression (1(2(3 4)5)6).
10. Trace the S-expression print algorithm of Section 5.3.2 applied to the tree of Figure 5D.
11. Show that a binary tree search is most efficient when the tree is depth-balanced.
*12. To maintain a balanced tree as nodes are inserted, it is occasionally necessary to change the root. Develop a criterion for choosing the root. (*Hint*: see Knuth [5.28].)

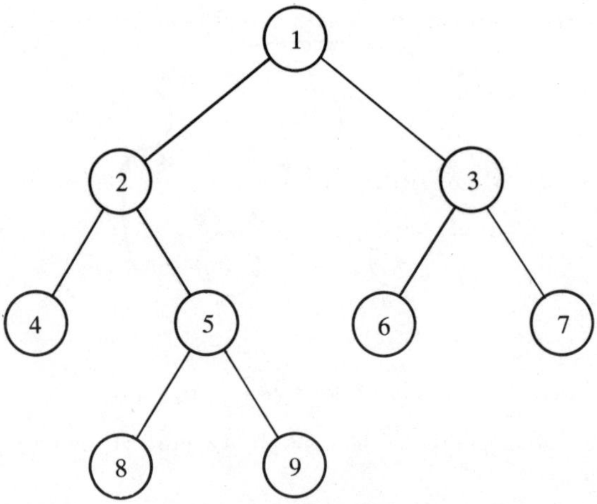

Figure 5D

13. Flowchart a sorting algorithm. (a) Determine its cyclomatic complexity. (b) From computer manufacturer data, determine the execution time of each node in the flowchart.

14. Flowchart a sorting algorithm. Determine sets of test data so that each path from entry to exit is traversed.

15 Using Kirchhoff analysis, estimate the execution time of a sorting algorithm as a function of the size of the array to be sorted.

16. Show that the unsolvability of the 'halting problem' implies the unsolvability of the 'equivalence problem'.

17. Show that it is not possible to tell whether an arbitrary instruction within an arbitrary computer program will ever be executed.

18. Show that every flowchart with **WHILE** loops can be transformed to a functionally equivalent flowchart with **REPEAT** loops, and vice versa.

19. Give the parse tree for $\lambda+\lambda\times\lambda+\lambda$ with respect to G_{pae}.

20. Extend the Type 2 arithmetic expression grammar G_{pae} to include (a) exponentiation ($\uparrow$), which should have greater precedence than $\times$, and should be 'right-associative'; (b) unary minus ($\ominus$), which should have greater precedence than $\uparrow$; (c) subscripted variables.

21. Give the syntax graph of G_{pae}.

22. Verify that $w\in\Lambda(M)$ iff $w\in\mathbf{L}(G)$ for the FSA M defined given the RLG G in the proof of Theorem 1.

23. What language is accepted by the nondeterministic FSA shown in Figure 5.6.1(f)?

24. Show the sequence of configuration moves for the FSA of Figure 5.6.1(g) as it processes the following input strings:

(a) *caca* (b) *caba* (c) *ccaa*

25. Define an FSA which accepts the language $a\{ba,c\}^* \mid (ab)^*$.

26. Show the sequence of configuration moves for the TM of Table 5.6.5 for the inputs $m=2$ and $n=4$.

***29.** Show that a PDA which has two stacks, rather than just one, is equivalent to a TM.

PROGRAMMING ASSIGNMENTS

1. Implement the depth-first traversal procedure of Section 5.2.2.
2. Implement the breadth-first traversal procedure of Section 5.2.2.
3. Implement the topological sort procedure of Section 5.2.5.
4. Implement the canonical traversal algorithm of Section 5.3.1.
5. Implement the postorder traversal algorithm of Section 5.3.1.
6. Implement the tree construction algorithm of Section 5.3.2.
7. Implement the S-expression print algorithm of Section 5.3.2.
8. Implement the binary-tree construction algorithm of Section 5.3.3.
9. Implement the binary-tree search algorithm of Section 5.3.3.
10. Implement the sorted binary-tree insertion procedure of Section 5.3.4.

11. Write a program which simulates a finite-state acceptor.

12. Write a program which simulates a push-down acceptor.

13. Write a program which simulates a Turing machine.

***14.** Write a program which translates decision table programs into a conventional language.

***15.** Write a compiler for a simple assignment statement language, where each statement is of the form ⟨variable⟩:=⟨arithmetic expression⟩.

6 ELEMENTS OF PROBABILITY AND STATISTICS

6.1 STATISTICAL CONCEPTS

Statistics is that branch of mathematics concerned primarily with quantitative descriptions of information. In particular, it concerns the characterization of large amounts of data by a few numbers. The main objectives of such data 'compression' are practicality (it is easy to deal with 'averages' but infeasible to deal with each of a billion different data items) and 'inference' (the appearance of certain patterns in a given set of data provides information about missing or future data). The use of statistics for descriptive purposes is in itself of limited value. However, statistics can also be used to draw conclusions about the source of data, i.e., about the nature of the process from which data is derived. For example, the average of a set of examination scores made by a student in a class is commonly used to infer how well the student knows a subject; also, the average over several classes may also be used to predict performance in subsequent classes. While statistical inferences or predictions may well prove wrong after the fact, they are often one's only recourse before the fact.

It is difficult to overstate the importance of statistics in the computing world. With the exception of ordinary arithmetic, and perhaps the deductive logic that is inherent in rational thought, statistics is the branch of mathematics which is most universally applied.[†] To meet this need, several major statistical programming packages are available—e.g., SPSS and BMD. For a survey, see Schucany [6.29] and Moore [6.25]. Computer scientists

[†]This statement is supported by the number of departments in universities which offer or require a course in statistics for their majors.

should be aware of what is offered in these packages, and, more importantly, their deficiencies. As with all packaged programs, they cannot and should not be used for all applications, and only a good knowledge of statistics permits one to know when and how to use alternatives. Unfortunately, we cannot treat this subject in this book. Only a few basic concepts will be discussed below, primarily those we shall later apply to computer science problems. See Hoel [6.15] for a more extensive elementary treatment; computational aspects are emphasized in Afifi [6.1].

6.1.1 Averages

The term 'average' is used in many ways. The most common usage refers to the (*arithmetic*) ***mean*** of a multiset of numbers, which is the sum of the numbers divided by the size of the multiset. A ***mode*** of a multiset is a member of the multiset with largest multiplicity. (A multiset may have more than one mode; each member of a set is a mode.) The ***median*** of a multiset is the 'middle' member of the 'sorted' multiset, obtained by listing the members of the multiset in increasing (numerical) order: if the size N of the sorted multiset is odd, the $(N+1)/2$th member is the median; if N is even, the median is 'between' the two middle members, usually taken to be their arithmetic mean. For example, consider the multiset $\{5, 2, 3, 2, 5, 1\}$. The mean is $(5+2+3+2+5+1)/6=3$, the modes are 2 and 5 (both have a multiplicity of 2), and the median is the mean of the two middle (3rd and 4th) members of the sorted multiset $\{1, 2, \underline{2}, \underline{3}, 5, 5\}$, i.e., $(2+3)/2=2.5$.

We remark that the mean of a multiset is also equal to the sum of the products of each distinct member of the multiset times its multiplicity, all divided by the size of the multiset. For example, the mean of

$$\{5,2,3,2,5,1\}=(1\times1+2\times2+3\times1+5\times2)/6=3.$$

This leads us to introduce the concept of the ***weighted average*** of a set, where each member of the set has a (generally nonnegative) ***weight***, and the ***weighted size*** of the set is considered to be the sum of the weights of its members: if 1 has weight 3, 2 has weight 1.5, 3 has weight 0, and 5 has weight 1, then the sum of the weights is 5.5 and the weighted average of $\{1,2,3,5\}$ [weighted by $\{3,1.5,0,1\}$] is

$$(1\times3+2\times1.5+3\times0+5\times1)/5.5=2.$$

Formally, we define

$$\text{weighted average of } \{a_1,a_2,\ldots,a_N\} \text{ [weighted by } \{w_1,w_2,\ldots,w_N\}]$$
$$=\frac{a_1\times w_1+a_2\times w_2+\cdots+a_N\times w_N}{w_1+w_2+\cdots+w_N}. \tag{6.1.1}$$

If all of the weights are unity, the weighted average reduces to the arithmetic mean.

While arithmetic means are defined only for multisets of numbers, modes and medians may be defined for nonnumerical data. A mode of a multiset of words, for example, is a (or 'the' if unique) word that appears most frequently. A median for a multiset of words requires that the words be ordered (e.g., alphabetically); it is then the middle word (or middle two words, if the multiset is of even size) in the ordering. Nevertheless, averages are generally defined only for numerical data, to which we now restrict ourselves.

Averages are often called measures of *central tendency* and provide an indication of what a typical member of a multiset may be. However, an average provides no indication of *dispersion* or *uniformity* (i.e., how different members may be from the averages, or from each other) or of *symmetry* (i.e., whether the distribution of data is skewed). The minimal and maximal members of a set specify the *range* of the multiset, one measure of dispersion. The *modality* (number of modes) provides one measure of uniformity.

6.1.2 Deviations

Another measure of dispersion is based upon the amounts by which the members of a multiset differ from its mean. Let m denote the mean of the multiset $X=\{x_1, x_2,\ldots, x_N\}$,

$$m=\frac{1}{N}\sum_{i=1}^{N} x_i;$$

we say x_i'deviates' from the mean by the absolute value $|x_i-m|$. The ***mean deviation*** is then defined as the quantity

$$\frac{1}{N}\sum_{i=1}^{N} |x_i-m|.$$

We note that absolute values are taken for the deviations in order to avoid cancellation of positive and negative amounts.

Another way to avoid cancellations is to square the deviations, and then take the mean of the squared deviations; this quantity, denoted by s^2, is given by

$$s^2=\frac{1}{N}\sum_{i=1}^{N}(x_i-m)^2 \qquad (6.1.2(a))$$

and is called the (biased) *variance*. [The quantity

$$\frac{1}{N-1}\sum_{i=1}^{N}(x_i-m)^2$$

is called the 'unbiased' variance, and will be discussed further in Section 6.2.6; the two quantities are nearly equal for large N, and neither is very meaningful for small N.] The positive square root of the variance, i.e., s, is called the *standard deviation*. It is often more convenient to compute s by means of the formula

$$s=\mathbf{sqr}[(\frac{1}{N}\sum_{i=1}^{N} x_i^2)-m^2] \tag{6.1.2(b)}$$

since m and the parenthesized term can be computed in the same pass through the set of data X. That Eqs. (6.1.2(a)) and (6.1.2(b)) are equivalent is established later. We emphasize here that numerical errors may lead to trouble (in this case, attempting to take the square root of a negative number) if Eq.(6.1.2(b)) is used without care. (See Exercise 10, Chapter 1.)

6.1.3 Histograms

One popular means of characterizing large amounts of data by a few numbers is to partition the data into a few groups, and to characterize each group by the number of its members. For example, a set of numbers between -100 and $+100$, such as

$$S=\{5,\ 2,\ 3,\ 2,\ 5,\ 1,\ 10,\ -7,\ 8,\ -10,\ 73,\ 20\},$$

can be partitioned in three groups

$$G_1=\{i\,|\,i<0\}$$
$$G_2=\{i\,|\,0\le i\le 10\}$$
$$G_3=\{i\,|\,i>10\}$$

and characterized by the group 'frequencies'

$$f_1=\#(G_1)=2$$
$$f_2=\#(G_2)=8$$
$$f_3=\#(G_3)=2.$$

A 'frequency diagram' depicts this information, as shown in Figure 6.1.3(a). If the areas (rather than heights) of each group are proportional to its frequency, as shown in Figure 6.1.3(b), then the diagram is called a *histogram*.

6.1.4 Other Statistical Measures

In addition to the basic statistical measures defined above, other common measures are defined below. Let $X=\{x_1,\ x_2,\ldots,\ x_N\}$ be a set (actually a

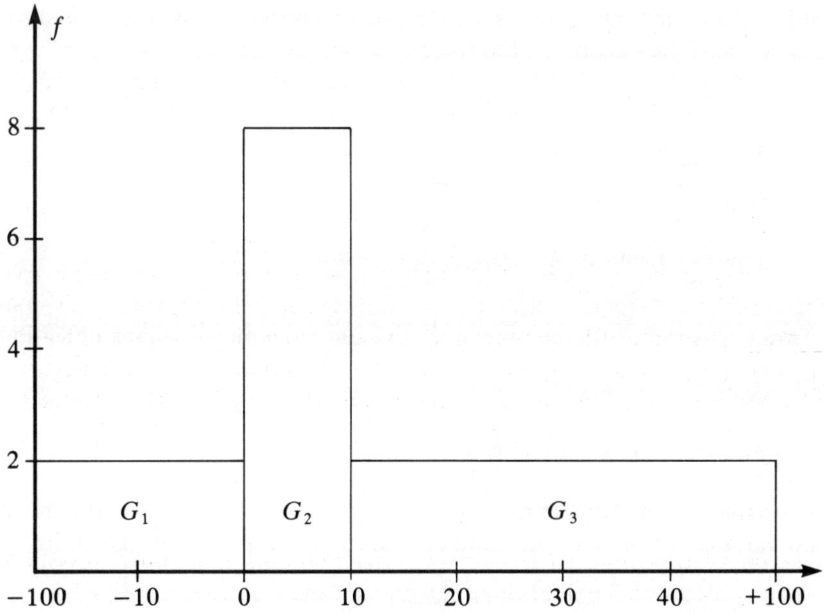

Figure 6.1.3(a)

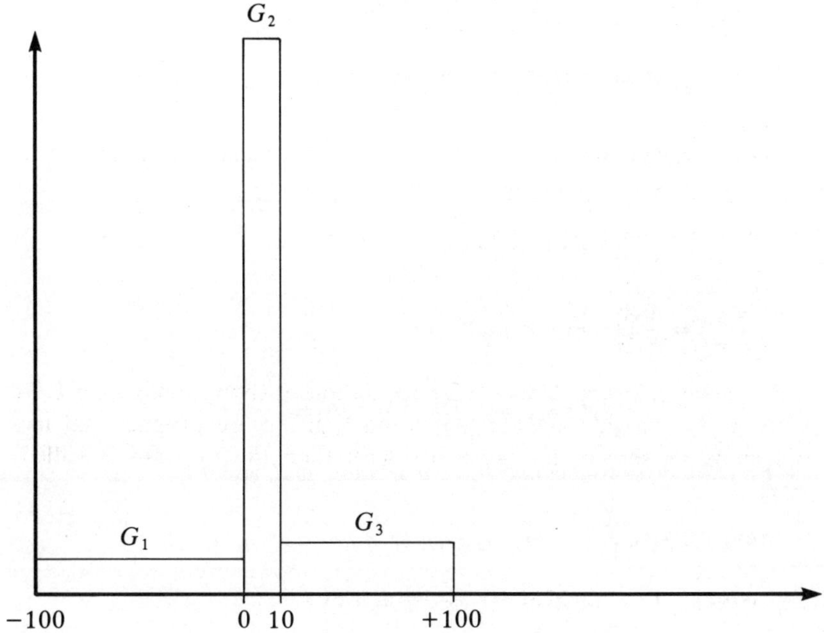

Figure 6.1.3(b)

multiset, but the term 'set' is used in this section instead of 'multiset' because it is standard in statistical literature) of numbers. Then

$$\text{geometric mean of } X = (x_1 x_2 \cdots x_N)^{1/N}$$

$$\text{harmonic mean of } X = \frac{N}{\Sigma_{i=1}^{N} 1/x_i}$$

$$\text{quadratic mean of } X = \textbf{sqr}\left(\frac{1}{N}\sum_{i=1}^{N} x_i^2\right) = \textbf{sqr}(m_X^2 + s_X^2),$$

where m_X is the arithmetic mean $\frac{1}{N}\Sigma x_i$, and s_X is the standard deviation **sqr** $[(\frac{1}{N}\Sigma x_i^2) - m_X^2]$. Furthermore,

$$\text{coefficient of variation} = \frac{s_X}{m_X} .$$

Given two sets of numbers, $X = \{x_1, \ldots, x_{N_X}\}$ and $Y = \{y_1, \ldots, y_{N_Y}\}$, the mean and variance of the (set of) numbers considered together are given by:

$$m_{X,Y} = \frac{\Sigma x_i + \Sigma y_i}{N_X + N_Y} = \frac{N_X m_X + N_Y m_Y}{N_X + N_Y}$$

$$s_{X,Y}^2 = \frac{\Sigma(x_i - m_{X,Y})^2 + \Sigma(y_i - m_{X,Y})^2}{N_X + N_Y}$$

$$= \frac{N_X s_X^2 + N_Y s_Y^2}{N_X + N_Y} + \frac{N_X N_Y (m_X - m_Y)^2}{(N_X + N_Y)^2} .$$

If X and Y are of equal size (i.e., $N_X = N_Y = N$), the mean and variance of the set of sums of corresponding numbers are given by:

$$m_{X+Y} = \frac{1}{N}\Sigma(x_i + y_i) = m_X + m_Y$$

$$s_{X+Y}^2 = \frac{1}{N}\Sigma(x_i + y_i - m_{X+Y})^2$$

$$= s_X^2 + s_Y^2 + 2\left[\frac{1}{N}\Sigma(x_i - m_X)\ (y_i - m_Y)\right].$$

The bracketed term is called the *covariance* of X and Y:

$$\textbf{cov}(X,Y) = \frac{1}{N}\sum_{i=1}^{N} (x_i - m_X)\ (y_i - m_Y).$$

The *correlation* r_{XY} of X and Y is defined by

$$r_{XY} = \frac{\textbf{cov}(X,Y)}{s_X s_Y} .$$

If $r_{XY}=0$, then the sets X and Y are said to be *uncorrelated*. Otherwise each x_i is related to y_i to some degree.

6.1.5 Regression

If sets $X=\{x_1,\ldots,x_N\}$ and $Y=\{y_1,\ldots,y_N\}$ are correlated, it is often desirable to find a function $\mathbf{F}$: $X \rightarrow Y$ such that $\mathbf{F}(x_i)=y_i$. Suppose we plot each pair (x_i,y_i) as a point in a Cartesian graph or space, where x_i is the distance from the origin along a horizontal axis, and y_i is the distance along the vertical axis of the graph. The desired function then is the equation of a curve which passes through each of the N points. For example, if $X=\{1,2,3\}$ and $Y=\{2,5,10\}$, then

$$\mathbf{F}(x)=x^2+1=y$$

is one such curve: this particular curve is a parabola (or polynomial of degree 2). In general, it is possible to find a polynomial of degree $N-1$ which passes through N distinct points. If the degree of the polynomial is restricted to be less than $N-1$, then it is no longer possible for it to pass through N arbitrary points. What may be done instead is to fit a curve which passes as close to these points as possible, where closeness may be measured in various ways. Formally, a distance function between an approximating curve $\mathbf{f}(x)$ and the given points is defined, and the object is to find that curve which minimizes the distance. One measure of distance is the maximum deviation,

$$\mathbf{max}_i\{|\mathbf{f}(x_i)-y_i|\},$$

minimization of which is called *Chebyshev* approximation. Another measure of distance is the average deviation,

$$\frac{1}{N}\sum_{i=1}^{N}(\mathbf{f}(x_i)-y_i)^2,$$

minimization of which is called *least-squares* approximation. A special case of least-squares approximation is where the curve $\mathbf{f}(x)$ is restricted to be a straight line

$$\mathbf{f}(x)=y_0+b\cdot x$$

which is called the (linear) *regression* line. Its slope b and y-intercept y_0 are given by:

$$b=\frac{\mathbf{cov}(X,Y)}{s_X^2}=\frac{r_{XY}s_Y}{s_X}$$

$$y_0=m_Y-b\cdot m_X.$$

In addition to characterizing a set of data points by two numbers b and

y_0, the regression line permits us to predict what other sample points might be. For example, we would expect (x',y') to be a sample point if $y'=y_0+b\cdot x'$. Much of the field of statistical data analysis is concerned with testing how well the regression line performs as a predictor. Extensions include fitting nonlinear (but usually polynomial) curves to a given set of data, and considering dependence of y on more than one variable $x_1,x_2,\ldots,x_K$ [i.e., finding a multidimensional curve $\mathbf{f}(x_1,x_2,\ldots,x_K)=y$]. See Freiberger [6.9] and Ferrari [6.8] for applications to the performance evaluation of computers. One other application is to the estimation of y_0 and b for an $\mathbf{O}(N)$ algorithm (i.e., one with complexity y_0+bN) given measurements of its execution time. (Two measurements are sufficient only if they are exact; if measurements are approximate, regression is necessary.)

6.2 DISCRETE PROBABILITY CONCEPTS

Probability theory differs from statistics in that the latter is concerned with data whereas the former is concerned with conceptual processes (which may generate data in an uncertain fashion). Suppose a process can at any given time enter only one of a specified number of states. The set of all possible states is called the *sample space* S. Unless stated otherwise, we restrict ourselves to *discrete* processes, which we define as those in which S is a finite or countable set. Suppose that the process is more likely to enter some states than others. We formalize the intuitive concept of likelihood by associating with each state s a nonnegative number, denoted $\mathbf{Pr}(s)$, and called its *probability*, which satisfies certain properties (including $\mathbf{Pr}(s)\geq 0$).

First, we wish to define the probability function $\mathbf{Pr}(s)$ so that it is high if the state s is likely, and low if the state is not likely; furthermore, we wish $\mathbf{Pr}(s_1)>\mathbf{Pr}(s_2)$ if state s_1 is more likely than s_2, and $\mathbf{Pr}(s_1)=\mathbf{Pr}(s_2)$ if the states are equally likely. We also wish to choose (*normalize*) the unit for probabilities so that, if a state is absolutely certain (to the exclusion of all other states) then it has a probability equal to 1 (100%). On the other hand, if a state is impossible, then we wish its probability to equal 0 (0%); such states are commonly omitted from the sample space. Finally, we want states into which there is a '50–50' chance of entering to have probability 1/2 (50%). These informal notions are consistent with the following axiom, which we use to *define* probabilities of states: if $S=\{s_1,s_2,\ldots\}$, then

$$\mathbf{Pr}(s_1)+\mathbf{Pr}(s_2)+\cdots=1.$$

Probability may be developed as an axiomatic theory, in the same sense as Euclidean geometry; from the foregoing definitions or axioms, all of the formal theory can be derived.[†] In order to apply the theory to actual processes, we must define a sample space and find a set of numbers to assign to the states as their probabilities. One way to proceed is to utilize 'experiments', either real, simulated, or conceptualized (imaginary): an *experiment* is a sampling (or observation) of the state of a process; we also say that the *outcome* of an experiment is the state that the process enters.

Suppose we perform an experiment N times, and count the number of times that the state of the process turns out to be s; denote this number N_s. Then the *relative frequency* of s is defined as the ratio N_s/N. This ratio may be regarded as a statistical average, in the following sense. If the outcome of the ith performance of the experiment is state s, we let $a_i=1$; otherwise, we let $a_i=0$. Then the arithmetic mean of the multiset of numbers $\{a_1,a_2,\ldots\}$ is equal to N_s/N.

We say (informally) that N_s/N is an approximation to the probability of s, which gets closer and closer to $\mathbf{Pr}(s)$ as N gets larger and larger. We may now take one of two viewpoints. We may 'define' the probability of s as the mathematical limit of N_s/N as N approaches infinity. On the other hand, we may say that probabilities are inherent (conceptual) properties of states, and that N sampling experiments will 'almost certainly' result in state s being observed $N\times\mathbf{Pr}(s)$ times, for sufficiently large N. ('Almost certainly' is in the probabilistic sense that the probability of $(|N_s/N-\mathbf{Pr}(s)|<\epsilon)$ approaches 1, as N approaches infinity, for any given $\epsilon>0$: this is known as the [weak] *law of large numbers*.[‡]) Since both viewpoints formally require an infinite number of experiments, we shall in practice equate probabilities with relative frequencies of a representative number of experiments, either conceptual or real.

We remark that we often make an 'equilikelihood' assumption: namely, that the probability of each of n distinct states is equal to $1/n$. This assumption is generally and *of necessity* made whenever we have no reason to suppose otherwise. Of course, any conclusions deriving from this should always be interpreted with caution.

[†]It should be emphasized that mathematical theories serve only as abstract models for real world processes, and the notion of a probability is just one useful way to represent *uncertainty* in nature. This does not mean that nature always behaves probabilistically, as 'unlucky' gamblers will attest.

[‡]See Feller [6.7], Vol. I, pp. 141, 228.

6.2.1 Expected Values

Probabilities are often used (as the 'weights') to define weighted averages of sets of numbers. This is a natural extension of our use of multiplicities (frequencies) to define the mean of a multiset in Section 6.1.1. Formally, let $S=\{s_1,s_2,\ldots\}$ be the sample space for a process, let $\{a_1,a_2,\ldots\}$ be a set of numbers quantifying (measuring) some common property A of the states, and let $\{p_1,p_2,\ldots\}$ be the respective probabilities of the states. In other words, $p_i=\mathbf{Pr}(s_i)$. We say then that the *expected value* of property A is the weighted average of $\{a_1,a_2,\ldots\}$ [weighted by $\{p_1,p_2,\ldots\}$] and is given by

$$\mathbf{Exp}(A)=a_1\times p_1+a_2\times p_2+\cdots \tag{6.2.1}$$

where the denominator of Eq.(6.1.1) is $(p_1+p_2+\cdots)=1$. **Exp**† is called the *expectation* operator, and Eq.(6.2.1) is called the 'expectation formula'.

Example*: *Average distance We will show here that the average distance between two distinct points in a line is one-third the length of the line. We assume that the line has N equally spaced points (denoted 1, 2, ..., N), and that they have equal probability. Note that there are $N-1$ ways for the distance to be 1, $N-2$ ways for the distance to be 2, 1 way for the distance to be $N-1$, or in general $N-d$ ways for the distance to be d. Since $\Sigma_{d=1}^{N-1}(N-d)=(N-1)N/2$, the expected value for the distance between two distinct points is

$$\mathbf{Exp}(d)=\sum_{d=1}^{N-1} d\times\mathbf{Pr}(d)=\sum_{d=1}^{N-1} d\times\frac{(N-d)}{(N-1)N/2}$$

$$=\frac{2}{N-1}\sum_{d=1}^{N-1} d-\frac{2}{N(N-1)}\sum_{d=1}^{N-1} d^2$$

$$=\frac{2}{N-1}\,\frac{N(N-1)}{2}-\frac{2}{N(N-1)}\,\frac{N(N-1)(2N-1)}{6}=\frac{N+1}{3}.$$

In the limit (for large N), the average distance is $N/3$, as predicted. An immediate application of this result is to the estimation of the average seek distance for a disk unit or file consisting of N cylinders.

Application*: *Expected seek time If seek time t depends linearly on distance n, then, from the foregoing example, we may conclude that

$$\mathbf{Exp}(t)=a\cdot\mathbf{Exp}(n)+b$$

for constants a and b. But in reality, seek time varies nonlinearly with distance, for example, according to curves typically found in disk specification manuals. From such a curve, or from a table of values of the time required to move i cylinders, t_i, the average seek time is given by

†**Exp** should not be confused with the exponential function of Section 3.1.4.

$$\mathbf{Exp}(t)=\sum_{i=1}^{W-1} t_i \cdot p_i$$

where p_i is the probability of requiring a move of i cylinders (given that there is a move), and W is the number of cylinders (access arm positions) for a given disk unit. To determine p_i, we proceed as follows. Suppose $W=4$. Then the possible moves are listed below:

1 to 2,	1 to 3,	1 to 4
2 to 1,	2 to 3,	2 to 4
3 to 1,	3 to 2,	3 to 4
4 to 1,	4 to 2,	4 to 3.

Note that this listing has W rows and $W-1$ columns. In general, there are $W(W-1)$ possible moves; if we assume the moves are equally likely, then (not counting 'non-moves' from a cylinder to itself) the probability of each move is

$$\frac{1}{W(W-1)}.$$

Now for $i>0$, a move of i cylinders toward cylinder number 1 is possible only if we start at cylinder number $i+1$ or above, and a move of i cylinders toward cylinder number W is possible only if we start at cylinder number $W-i$ or below. Hence, there are $\sum_{j=i+1}^{W} 1$ ways to move i cylinders toward cylinder number 1, and $\sum_{j=1}^{W-i} 1$ ways to move i cylinders toward cylinder number W. We conclude that

$$p_i=\frac{\sum_{j=i+1}^{W} 1+\sum_{j=1}^{W-i} 1}{W(W-1)} = \frac{2(W-i)}{W(W-1)}$$

[where we have used Exercise 27 of Chapter 1]. Note that $p_W=0$. We leave as an exercise the proof that, if $t_i=i$ for each i, then $\mathbf{Exp}(t)=(W+1)/3$.

6.2.2 Probability of Events

Earlier, we introduced the concept of probabilities associated with the set of states (sample space) $S=\{s_1,s_2,\ldots\}$ of a discrete process. Briefly, $\mathbf{Pr}(s)$, the probability that the outcome of an experiment results in the process entering into a state s in S, is defined such that $0\le\mathbf{Pr}(s)\le 1$ and

$$\mathbf{Pr}(s_1)+\mathbf{Pr}(s_2)+\cdots=1. \qquad (6.2.2(a))$$

Note that we may regard **Pr** as a function from S to the interval of reals $[0,1]$; *any* function $\mathbf{Pr}:S\to[0,1]$ satisfying Eq.(6.2.2(a)) may be regarded as a 'probability function' on S. We next generalize the concept of probabilities to sets of states.

We say that an *event* E is a subset of states (i.e., $E \subseteq S$ or $E \in 2^S$), subject to the requirement that the set of all possible events $\mathscr{E} \subseteq 2^S$ be closed under complementation, union, and intersection; such a set $\mathscr{E}$ is called a *Borel field*. We remark that 2^S is a Borel field for any sample space S. An event is said to 'occur' if the process enters any of the states in the event. The sample space S is called the *certain* event. The empty set is called the *null* event. We define the *probability of an event*, $\mathbf{P}(E)$,[†] as the sum of the probabilities of the states in the event; i.e., $\mathbf{P}(E)=\Sigma_{s \in E}\mathbf{Pr}(s)$.

It should be clear that $\mathbf{P}(S)=1$, $\mathbf{P}(\emptyset)=0$, and $0 \leq \mathbf{P}(E) \leq 1$ for all events $E \subseteq S$. In other words, $\mathbf{P}: 2^S \to [0,1]$. Furthermore, from the definitions of probabilities of states and events, we have the fundamental theorem of discrete probability:

$$\mathbf{P}(E_1 \cup E_2)=\mathbf{P}(E_1)+\mathbf{P}(E_2)-\mathbf{P}(E_1 \cap E_2).$$

We say events E_1 and E_2 are ***mutually exclusive*** if the sets have no common states (i.e., if $E_1 \cap E_2=\emptyset$); since $\mathbf{P}(\emptyset)=0$, we have

$$\mathbf{P}(E_1 \cup E_2)=\mathbf{P}(E_1)+\mathbf{P}(E_2) \qquad \text{if} \quad E_1 \cap E_2=\emptyset. \tag{6.2.2(b)}$$

This is often taken to be an axiom in lieu of Eq.(6.2.2(a)).

In many cases, probabilities are defined (or 'known') only for certain events rather than for each possible state of a process. For example, in rolling an unbalanced die, we may know that even and odd outcomes are equiprobable, but that the six states (faces) are not. Here, $S=\{1,2,3,4,5,6\}$, and the set of events for which probabilities are known $\mathscr{E}=\{\emptyset, \{1,3,5\}, \{2,4,6\}, S\}$. Note then that $\mathbf{P}: \mathscr{E} \to [0,1]$. Note further that if the set of events $\mathscr{E}$ for which probabilities are defined is not a Borel field, then a probability function on $\mathscr{E}$ would not be well defined and in particular would not satisfy Eq.(6.2.2(b)). For example, if $\mathscr{E}=\{\emptyset,\{1\},\{2\},\{1,2,3,4\}\}$, so that $\{1,2\}$ is not an event, then $\mathbf{P}(\{1\} \cup \{2\})$ is not defined. (In view of this, whenever we use the term 'event', we implicitly assume the existence of a Borel field.) We remark also that, if a sample space is partitioned into mutually exclusive sets, then the partition may be regarded as a new sample space.

Remark Infinite sample spaces lead to difficulties since a probability function $\mathbf{Pr}(s)$ which is nonzero for an infinite number of states must still satisfy Eq.(6.2.2(a)). This is not possible if the states are equiprobable. Furthermore, the Borel field requirement that $\mathscr{E}$ be closed under complementation,

[†]We will also denote $\mathbf{P}(E)$ by $\mathbf{Pr}\{s \in E\}$, especially when the single letter $\mathbf{P}$ may be misinterpreted.

union, and intersection must be generalized to permit a countably infinite number of unions and intersections. The infinite analogue of Eq.(6.2.2(b)) must then also hold. (See Papoulis [6.26].) Fortunately, in computer science, we deal mainly with finite sample spaces, where the problem generally involves counting the number of states in a space S. For equiprobable states, probabilities of events are determined by the sizes of the various events (subsets of S). Combinatorial theory aids us in these calculations.

Example Let a sample space S be the set of Boolean triples

$$\{(0,0,0),(0,0,1),(0,1,0),(0,1,1),(1,0,0),(1,0,1),(1,1,0),(1,1,1)\},$$

and denote these states by $s_1, s_2, \ldots, s_8$, respectively. An event is any subset of S; e.g., the event where exactly one coordinate of a state is 1 is the set

$$E_a=\{(1,0,0),(0,1,0),(0,0,1)\},$$

and the event where the second coordinate of a state is 1 is the set

$$E_b=\{(0,1,0),(0,1,1),(1,1,0),(1,1,1)\}.$$

Let us assume equiprobable states, i.e., that $\mathbf{Pr}(s_i)=\mathbf{Pr}(s_j)$ for each pair (i,j), hence $\mathbf{Pr}(s_i)=1/8$ for each i. Then

$$\mathbf{P}(E_a)=\mathbf{P}(\{s_5,s_3,s_2\})=\mathbf{Pr}(s_5)+\mathbf{Pr}(s_3)+\mathbf{Pr}(s_2)=3/8,$$

and

$$\mathbf{P}(E_b)=\mathbf{P}(\{s_3,s_4,s_7,s_8\})=\mathbf{Pr}(s_3)+\mathbf{Pr}(s_4)+\mathbf{Pr}(s_7)+\mathbf{Pr}(s_8)=1/2.$$

Now

$$\mathbf{P}(E_a \cup E_b)=\mathbf{P}(\{s_2,s_3,s_4,s_5,s_7,s_8\})=3/4,$$

and

$$\mathbf{P}(E_a \cap E_b)=\mathbf{P}(\{s_3\})=1/8,$$

which verifies that

$$\mathbf{P}(E_a \cup E_b)=\mathbf{P}(E_a)+\mathbf{P}(E_b)-\mathbf{P}(E_a \cap E_b).$$

Events E_a and E_b are not mutually exclusive since they have state s_3 in common. $E_c=\{(1,1,1)\}$ is mutually exclusive of E_a, but not of E_b.

Conditional Probabilities Let $A \subseteq S$ be an event. If B is an event with nonzero probability, then the 'probability of A *given* B,' or the *conditional probability* of A 'assuming' B, denoted $\mathbf{P}(A \mid B)$, is defined by the formula

$$\mathbf{P}(A \mid B)=\mathbf{P}(A \cap B)/\mathbf{P}(B).$$

$\mathbf{P}(A \mid B)$, as a function of A, satisfies the properties required of probabilities: namely, $\mathbf{P}(A \mid B)\geq 0$, $\mathbf{P}(S \mid B)=1$, and Eq.(6.2.2(a)). Note that if A and B are mutually exclusive ($A \cap B = \emptyset$), then $\mathbf{P}(A \mid B) = 0$. Furthermore, if $A \subseteq B$ (or $A \cap B = A$), then $\mathbf{P}(A \mid B) = \mathbf{P}(A)/\mathbf{P}(B)$, while if $B \subseteq A$ (or $A \cap B = B$), then $\mathbf{P}(A \mid B) = 1$.

Let $\{E_1, E_2, \ldots, E_n\}$ be a partition of S: i.e., $E_i \cap E_j = \emptyset$, for $i \neq j$, and $\cup_i E_i = S$. Then, for any event $A \subseteq S$,

$$\mathbf{P}(E_i \mid A) = \mathbf{P}(A \mid E_i)\mathbf{P}(E_i)/\mathbf{P}(A)$$

where $\mathbf{P}(A) = \mathbf{P}(A \mid E_1)\mathbf{P}(E_1) + \cdots + \mathbf{P}(A \mid E_n)\mathbf{P}(E_n)$. This is known as *Bayes' theorem*. From the foregoing definition, we may rewrite the theorem in the form

$$\mathbf{P}(E_i \mid A) = \frac{\mathbf{P}(E_i \cap A)}{\mathbf{P}(E_1 \cap A) + \cdots + \mathbf{P}(E_n \cap A)}. \tag{6.2.2(c)}$$

Example Continuing the foregoing example, we have

$$\mathbf{P}(E_a \mid E_b) = (1/8)/(1/2) = 1/4;$$
$$\mathbf{P}(E_b \mid E_a) = (1/8)/(3/8) = 1/3.$$

Independent Events Conditional probabilities are defined for events which are statistically dependent. Two events A and B [in the same sample space] are said to be (pairwise) *independent* if $\mathbf{P}(A \cap B) = \mathbf{P}(A)\mathbf{P}(B)$. If A and B are independent, then $\mathbf{P}(A \mid B) = \mathbf{P}(A)$. A set of n events $\{E_1, \ldots, E_n\}$ is said to be (mutually) independent if

$$\mathbf{P}(E_{i_1} \cap E_{i_2} \cap \cdots \cap E_{i_r}) = \mathbf{P}(E_{i_1})\mathbf{P}(E_{i_2}) \cdots \mathbf{P}(E_{i_r}) \tag{6.2.2(d)}$$

where $\{i_1, i_2, \ldots, i_r\}$ is any subset of $\{1, 2, \ldots, n\}$: pairwise independence (i.e., where $r=2$) and 'n-wise' independence (where $r=n$) are not sufficient for Eq.(6.2.2(d)) to hold.

Recalling that if events A and B are mutually exclusive, then $\mathbf{P}(A \cap B) = 0$, but if A and B are independent, then $\mathbf{P}(A \cap B) = \mathbf{P}(A)\mathbf{P}(B)$, it is clear that events with nonzero probability cannot be both mutually exclusive and independent. The fact that one event excludes another is a form of dependence.

Compound Events In the foregoing, we have defined probabilities for states or events associated with a single process or experiment. In many cases, it is of interest to consider n repetitions of the same experiment (each then called a 'trial'), or a combination of n different experiments. Let S_i be the sample space of the ith experiment, E_i denote an event (subset) of S_i, and $\mathbf{P}_i$ be an associated probability function. Consider then the Cartesian product $S = S_1 \times \cdots \times S_n$. This set S is the sample space associated with the 'compound' (multi-experiment) process, and we say that the members of S are 'compound states'. A *compound event* E is a subset of S; not all subsets may be possible.[†]

The probability $\mathbf{P}$ of a compound event $E=(E_1,E_2,\ldots,E_n)$ need not depend upon the probabilities of individual events $\mathbf{P}_i(E_i)$. However, if the experiments are independent, i.e., if for any i the outcome of the ith experiment does not depend upon the other experiments, then $\mathbf{P}(E)$ is given by the relation

$$\mathbf{P}(E_1,E_2,\ldots,E_n)=\mathbf{P}_1(E_1)\mathbf{P}_2(E_2)\ldots\mathbf{P}_n(E_n).$$

(This relation may also be regarded as defining the concept of independent experiments.) Each set of events $\{E_1,E_2,\ldots,E_n\}$ is then also mutually independent. Otherwise, $\mathbf{P}$ may be defined 'by cases' for each member E in $\mathscr{E}$. Note that, for n repetitions (trials) of the same experiment, the probability of n occurrences of the same event is p^n, where p is the probability of the event's occurrence in one trial.

$\mathbf{P}(E_1,E_2,\ldots,E_n)$ is the probability that the events E_1, E_2, ..., *and* E_n all occur in their respective experiments. Suppose we consider instead the probability that one (and only one) of the events (E_1, E_2, ..., *or* E_n) occurs in a single experiment chosen at random from the n processes. Let $\mathbf{P}(S_i)$ denote the probability of an event in S_i, i.e., of choosing the ith process, and let $\mathbf{P}_i(E_i)$ be the probability of event E_i in S_i. Since $E_i \subseteq S_i$, $\mathbf{P}(E_i|S_i)=\mathbf{P}_i(E_i)$. Then $\mathbf{P}(E_i)$—the probability of event E_i in the sample space $S'=\cup_i S_i$, which occurs by choosing S_i (or i) first, followed by the performance of the ith experiment—is given by

$$\mathbf{P}(E_i)=\mathbf{P}(E_i|S_i)\mathbf{P}(S_i)=\mathbf{P}_i(E_i)\mathbf{P}(S_i). \qquad (6.2.2(e))$$

Hence, for mutually exclusive events, we have

$$\mathbf{P}(E_1\cup E_2\cup\cdots\cup E_n)=\mathbf{P}_1(E_1)\mathbf{P}(S_1)+\mathbf{P}_2(E_2)\mathbf{P}(S_2)+\cdots+\mathbf{P}_n(E_n)\mathbf{P}(S_n),$$

for all E_i. If the choice of experiments is made with equal likelihood, i.e., if $\mathbf{P}(S_i)=\mathbf{P}(S_j)=1/n$, then

$$\mathbf{P}(E_1\cup\cdots\cup E_n)=\frac{1}{n}(\mathbf{P}_1(E_1)+\cdots+\mathbf{P}_n(E_n)).$$

6.2.3 Combinatorics

In Section 2.1.2, we defined the concepts of combination, permutation, selection, and arrangement of members of a set S. We previously noted that

[†]The set $\mathscr{E}$ of compound events for which probabilities are defined must constitute a 'Borel field', as noted earlier.

the number of r-arrangements of a set S of size N is given by

$$\mathbf{A}(N,r)=N^r. \qquad (6.2.3(a))$$

Recall that this is also the number of distinct strings of length r over an alphabet of size N. The number N^r can be derived by observing that there are N ways to choose the first member, N ways to choose the second, etc., the product of r N's equaling N^r.

Now consider permutations, i.e., arrangements without repetitions. The number of permutations (i.e., N-permutations) of S is $N!$; this can be derived by observing that there are N ways to choose the first member, $N-1$ ways to choose the second (since the first cannot be repeated), $(N-r+1)$ ways to choose the rth, and only one way to choose the Nth. The number of r-permutations (for $r<N$) is therefore $N\times(N-1)\times\cdots\times(N-r+1)$. In general, the number of r-permutations of N distinct objects, denoted $\mathbf{P}(N,r)$, is given by

$$\mathbf{P}(N,r)=\frac{N!}{(N-r)!}. \qquad (6.2.3(b))$$

Note that $\mathbf{P}(N,N)=N!$ and $\mathbf{P}(N,0)=1$.

Suppose now that different orderings are not regarded as being distinct, so that, for example, the $N!$ permutations of S are considered indistinct, there being only one distinct N-combination of N objects. (Recall that a combination is an unordered multiset without repetition, i.e., a subset.) From another point of view, each r-combination of N objects has $r!$ different permutations. Therefore, we conclude that $\mathbf{P}(N,r)=r!\mathbf{C}(N,r)$, where $\mathbf{C}(N,r)$ is the number of r-combinations of N objects; in other words,

$$\mathbf{C}(N,r)=\frac{N!}{r!(N-r)!}. \qquad (6.2.3(c))$$

$\mathbf{C}(N,r)$ is commonly denoted $\binom{N}{r}$; note that $\binom{N}{r}=\binom{N}{N-r}$, and $\binom{N}{N}=\binom{N}{0}=1$. The values of $\binom{N}{r}$ are called *binomial coefficients* because they are the coefficients in the **binomial series** $\Sigma_{r=0}^{N}\binom{N}{r}x^{N-r}y^r$. It can be shown that

$$(x+y)^N=\Sigma_{r=0}^{N}\binom{N}{r}x^{N-r}y^r.$$

When $x=1$, this reduces to $(1+y)^N=\Sigma_{r=0}^{N}\binom{N}{r}y^r$, a result known as the *binomial theorem*.

Consider finally combinations of N objects with repetition, i.e., r-selections. The number of r-selections of N objects is given by

$$\mathbf{S}(N,r)=\mathbf{C}(N+r-1,r). \qquad (6.2.3(d))$$

That this is true follows by observing that there is a 1–1 correspondence between r-selections of $\{1,2,\ldots,N\}$ and r-combinations of $\{1,2,\ldots,N+r-1\}$: $\{x_1,x_2,\ldots,x_r\}$ is an r-selection of $\{1,2,\ldots,N\}$ iff $\{x_1,x_2+1,\ldots,x_r+r-1\}$ is an

r-combination of $\{1,2,\ldots,N+r-1\}$. (Since different orderings do not count as being distinct in selections and combinations, we need only consider r-selections where $x_1 \leq x_2 \leq \cdots \leq x_r$.) For further information, see Liu [6.23].

Examples Examples of applications of the formula N^r for the number of r-arrangements of a set S of size N include the following:

(a) There are b^L L-digit base b (unsigned) integers.
(b) 2^L values can be represented in an L-bit binary byte or word.
(c) A truth table for an r-variable expression has 2^r rows.
(d) A decision table having r truth-valued conditions has 2^r simple rules.
(e) There are 26^r alphabetic strings of length r.
(f) There are b^a total functions on a set A of size a into a set B of size b.
(g) The r-fold product S^r of a set S of size N has size N^r. This implies, for the case $r=2$, that the universal relation $S \times S$ has size N^2; hence, N^2 is also the number of distinct branches in a complete directed graph having N nodes.

Examples of applications of the formula $N!$ for the number of permutations of a set of N items include the following:

(a) There are $N!$ permutations of the rows and/or columns of an $(N \times N)$ matrix.
(b) There are $N!$ different orders in which N predicates can be sequentially tested.
(c) There are $N!$ different orders in which N requests (for resources, or for service) can be sequentially processed.
(d) There are $N!$ Hamiltonian paths in a complete N-node graph.
(e) There are 26! simple (1–1 substitution) ciphers of the alphabet.
(f) There are $N!$ 1–1 functions as well as $N!$ onto functions on a set A of size N into itself.

Examples of applications of the formula $\binom{N}{r}$ for the number of r-combinations of a set of size N include the following:

(a) A set S of size N has $\binom{N}{r}$ subsets of size r; this implies that the power set 2^S has size $\Sigma_{r=0}^{N}\binom{N}{r}$, which by the binomial theorem equals 2^N.
(b) There are $\binom{N}{r}$ different ways to select r requests from a pool of size N; there are $r!$ different orders in which these r requests can be sequentially processed. Since $r!\binom{N}{r}=N!/(N-r)!=\mathbf{P}(N,r)$, we conclude that there are $\mathbf{P}(N,r)$ different ways to sequentially process r requests from a pool of size N.

Examples of applications of the formula $\binom{N+r-1}{r}$ for the number of r-selections of a set of size N include the following:

(a) There are $\binom{N+2-1}{2}$ distinct branches in a complete undirected graph having N nodes (counting the N reflexive branches); note that $\binom{N+1}{2}=N(N+1)/2$.
(b) There are $\binom{N+r-1}{r}$ ways to select r jobs from N job queues. (For example, if there are three job classes, A, B, and C, then there are 10 ways to select three jobs, the possibilities being from the set

$$\{AAA,AAB,AAC,ABB,ABC,ACC,BBB,BBC,BCC,CCC\},$$

where ABB means that one class A, two class B, and no class C jobs are selected.)

6.2.4 Random Variables

Let $S=\{s_1,s_2,\ldots\}$ be a sample space (set of states). We define a *random variable* as a function $X:S\to R$, where R is the set of real numbers. Each state s_i has an associated value $X(s_i)$, and each value $r\in R$ has an associated (possibly empty) set of states. Let $\mathcal{R}_X$ denote the range of X. If $\mathcal{R}_X$ is finite or countable, we say X is a *discrete* random variable, otherwise we say X is *continuous*. For any $r\in\mathcal{R}_X$, the sets $\{s\in S\,|\,X(s)=r\}$, abbreviated $\{X=r\}$, partition S into mutually exclusive sets, which may be regarded as a new sample space. The probabilities $\mathbf{Pr}\{X=r\}$ will be denoted $\mathbf{f}(r)$. Note that $\mathbf{f}:\mathcal{R}_X\to[0,1]$, where $[0,1]$ denotes the interval of reals from 0 to 1. For any discrete random variable X, $\Sigma_{r\in\mathcal{R}_X}\mathbf{f}(r)=1$, and so $\mathbf{f}$ may be regarded as a probability function defined on the sample space $\mathcal{R}_X$. We call $\mathbf{f}$ the *probability distribution* of the random variable X.

For a continuous random variable, $\mathcal{R}_X$ (hence also S) is uncountably infinite, hence a probability function cannot be defined in the same fashion. Consider instead sets of the form $\{s\in S\,|\,X(s)\leq r\}$, abbreviated $\{X\leq r\}$. If these sets form a Borel field, we define $\mathbf{F}:\mathcal{R}_X\to[0,1]$ by $\mathbf{F}(r)=\mathbf{Pr}\{X\leq r\}$; $\mathbf{F}$ is called the *cumulative distribution function* of X. We remark that $X:S\to R$ is a (continuous) random variable only if $\{X\leq r\}$ are events for any $r\in\mathcal{R}_X$ (so that $\mathbf{F}(r)$ is always defined); furthermore, if the events $\{X=+\infty\}$ and $\{X=-\infty\}$ are possible, then they must have zero probability. If $\mathbf{f}$ is continuous and differentiable [except at most a countable number of points], then its derivative $\mathbf{f}(y)=d\mathbf{F}(y)/dy$ [wherever it exists] is defined as the *density* function of X. Note that $\mathbf{F}(r)=\int_{-\infty}^{r}\mathbf{f}(y)dy$, and $\mathbf{F}(\infty)=1$. We emphasize that the densities $\mathbf{f}(r)$ are not probabilities; in fact, for a continuous random variable, $\mathbf{Pr}\{X=r\}=0$ for any r. However, for a discrete random variable with probability distribution function $\mathbf{f}(r)$, a cumulative distribution function $\mathbf{F}(r)$ can be defined in terms of $\mathbf{f}$ as follows:

$$\mathbf{F}(r)=\Sigma_{r'\leq r}\mathbf{f}(r')$$

where the sum is over the set $\{r'\in\mathcal{R}_X\,|\,r'\leq r\}$.

We shall also have occasion to refer to conditional probability distributions of a discrete random variable X. For any event E, we define

$$\mathbf{f}(r\,|\,E)=\mathbf{Pr}\{X=r\,|\,E\}=\mathbf{Pr}\{(X=r)\cap E\}/\mathbf{P}(E),$$

where $\mathbf{P}(E)$ is the probability of E in a sample space S' (which need not be the same as S).

The *expected value* of a discrete random variable $X:S \rightarrow R$ is defined as

$$\mathbf{Exp}(X)=\Sigma_{r\in\mathcal{R}_X} r\times\mathbf{f}(r), \tag{6.2.4(a)}$$

where **f** is the probability distribution of X. The *variance* of the random variable X is defined as

$$\mathbf{Var}(X)=\mathbf{Exp}[(X-\mathbf{Exp}(X))^2]. \tag{6.2.4(b)}$$

Often, $\mathbf{Exp}(X)$ is denoted μ_X and $\mathbf{Var}(X)$ by $\sigma_X{}^2$; σ_X, the square root of $\mathbf{Var}(X)$, is called the standard deviation. We note that

$$\begin{aligned}\sigma_X{}^2&=\Sigma_{r\in\mathcal{R}_X}(r-\mu_X)^2\times\mathbf{f}(r)\\ &=\Sigma_{r\in\mathcal{R}_X}(r^2-2r\mu_X+\mu_X{}^2)\times\mathbf{f}(r)\\ &=[\Sigma_{r\in\mathcal{R}_X}r^2\times\mathbf{f}(r)]-2\mu_X\,[\Sigma_{r\in\mathcal{R}_X}r\times\mathbf{f}(r)]+\mu_X{}^2\times 1\\ &=[\Sigma_{r\in\mathcal{R}_X}r^2\times\mathbf{f}(r)]-\mu_X{}^2.\end{aligned}$$

Observe that, under the equilikelihood assumption,

$$\mathbf{f}(r)=\mathbf{f}(r')=1/N \qquad \text{for each pair } r \text{ and } r',$$

hence

$$\mu_X=\frac{1}{N}\Sigma r$$

and

$$\sigma_X{}^2=\left(\frac{1}{N}\Sigma r^2\right)-\mu_X{}^2.$$

This should be compared with the formula for the standard deviation as given in Section 6.1.2. It should also be noted that **Exp** is a linear function in the sense that

$$\mathbf{Exp}[X+Y]=\mathbf{Exp}[X]+\mathbf{Exp}[Y].$$

We may use this linearity property to show that $\mathbf{Var}(X)=\mathbf{Exp}(X^2)-\mathbf{Exp}^2(X)$, as above. $\mathbf{Exp}(X)$ and $\mathbf{Exp}(X^2)$ are called the first and second 'moments' of X, respectively. In general, the rth *moment* of X is defined as $\mathbf{Exp}(X^r)$, and the rth moment 'about the mean' is defined as $\mathbf{Exp}[(X-\mathbf{Exp}(X))^r]$. We remark that, if X and Y are independent random variables, then

$$\mathbf{Exp}_{X,Y}[\mathbf{g}(X,Y)]=\mathbf{Exp}_X[\mathbf{Exp}_Y[\mathbf{g}(X,Y)]].$$

6.2.5 Probability Distributions

Probability distributions are distinguished (defined) by the form of the function $\mathbf{f}: \mathscr{R}_X \rightarrow [0,1]$. The distribution is said to be *discrete* if $\mathscr{R}_X$ is a finite (or countable) set, and *continuous* if $\mathscr{R}_X$ is not. In the former case, $\mathbf{f}$ is a probability function; in the latter case, $\mathbf{f}$ is a density function. We may think of the graph of $\mathbf{f}$ as a generalization of the histogram concept discussed in Section 6.1.3.

Uniform Distribution In the case where $\mathscr{R}_X$ has finite size N, the *uniform distribution* is defined by $\mathbf{f}(r)=1/N$ for any $r \in \mathscr{R}_X$; this is also known as the 'equilikelihood' distribution, which was used earlier in this chapter. In the case where $\mathscr{R}_X$ is the infinite interval of reals $[a,b]$, the uniform distribution is defined by the density function $\mathbf{f}(y)=1/(b-a)$.

Binomial Distribution The discrete *binomial distribution* is defined by

$$\mathbf{f}(r)=\binom{n}{r}p^r(1-p)^{n-r} \qquad (6.2.5(a))$$

where $\binom{n}{r}$ is the binomial coefficient (cf. Eq.(6.2.3(c))), p is a constant between 0 and 1, n is any positive integer, and r is an integer such that $0 \le r \le n$. If, in a process (known as a **Bernoulli process**), the event $X=1$ ('success') has probability p while the event $X=0$ ('failure') has probability $1-p$ at each trial, then $p^r(1-p)^{n-r}$ is the probability that r successes (and $n-r$ failures) occur in a sequence of n independent trials. Since $\binom{n}{r}=n!/[r!(n-r)!]$ is the number of different ways r successes out of n trials can occur, we conclude that the probability of r successes, $\mathbf{f}(r)$, is the product given by Eq.(6.2.5(a)). Since $\mathbf{f}(r)$ also depends on p and n, we may denote it by $\mathbf{b}(r,n,p)$ instead. Comparing $\mathbf{b}(r,n,p)$ with the terms in the binomial series (see Section 6.2.3), we can immediately verify that

$$\sum_{r=0}^{n} \mathbf{f}(r)=\sum_{r=0}^{n} \mathbf{b}(r,n,p)=1.$$

The mean $(=\sum_{r=0}^{n} r\mathbf{f}(r))$ and variance of $\mathbf{f}(r)$ are np and $n(p-p^2)$, respectively. Observe that, in a single trial $(n=1)$,

$$\mathbf{Exp}(X)=1 \cdot p+0 \cdot (1-p)=p,$$

and

$$\mathbf{Var}(X)=1^2 \cdot p+0^2 \cdot (1-p)-[\mathbf{Exp}(X)]^2=p-p^2.$$

Later, we will use the facts that

$$\sum_{r=0}^{n} r\mathbf{b}(r,n,p)=np$$

and

$$\sum_{r=0}^{n} r^2\mathbf{b}(r,n,p)=n(p-p^2)+(np)^2.$$

Therefore, in general, the binomial distribution has mean np and variance $n(p-p)^2$.

Poisson distribution The discrete *Poisson distribution* is defined by

$$\mathbf{f}(r)=\frac{\lambda^r e^{-\lambda}}{r!} \tag{6.2.5(b)}$$

where λ is a positive constant, and r is any nonnegative integer. The Poisson distribution is of special importance, and will be used later in this chapter when queueing systems are discussed. If the probability of r occurrences of an event (called an 'arrival') during a randomly chosen time unit is given by $\lambda^r e^{-\lambda}/r!$, then we have a Poisson arrival process. Here, λ, called the arrival rate, is a constant equal to the mean number of arrivals per unit time, which, in this case, also happens to be the variance of the number of arrivals per unit time. The mean time between arrivals (or 'interarrival' time) $\hat{T}_a$ is $1/\lambda$. Arrival processes satisfy the Poisson distribution if arrivals occur independently of each other and of the point in time at which we start to count arrivals;, of course, the number or probability of arrivals depends upon the length of the interval of time δt during which we count arrivals. For small δt, the probability of a single arrival during an interval $(t, t+\delta t)$ is given by $\lambda\delta t+\mathbf{O}(\delta t)$, and the probability of more than one arrival is negligible. A useful property of Poisson processes is that the interarrival distribution is exponential, which is discussed below. Finally, it should be noted that the Poisson distribution, also denoted $\mathbf{p}(r,\lambda)$, serves as a useful approximation to the binomial distribution $\mathbf{b}(r,n,p)$ when n is large, p is small, and $\lambda=np$. Another useful approximation to the binomial distribution is the normal distribution, which is also discussed below.

Geometric Distribution The discrete *geometric distribution* is defined by

$$\mathbf{f}(r)=p(1-p)^{r-1} \tag{6.2.5(c)}$$

where p is a constant between 0 and 1, and r is any positive integer; its mean is $1/p$. If p is the probability of 'success', then $\mathbf{f}(r)$ is the probability of $r-1$ 'failures' before the next success. We emphasize that $\mathbf{f}(r)$ is the same regardless of the instant of time at which we start the trials.

Exponential Distribution The continuous *exponential distribution* is defined by

$$\mathbf{f}(t)=\mathbf{Pr}\{T_a<t\}=1-e^{-\lambda t}, \qquad \text{for } t\geq 0,$$

where λ is a constant (the arrival rate) and T_a is the interarrival time. The mean of the exponential distribution is $1/\lambda$ and its variance is $1/\lambda^2$.

Normal Distribution The continuous *normal* (or *Gaussian*) *distribution* is defined by the density function

$$\mathbf{f}(y)=\frac{1}{\sqrt{(2\pi)}\,\sigma}e^{-(y-\mu)^2/2\sigma^2} \qquad \text{for } -\infty<y<\infty$$

where μ and σ are constants; μ is the mean and σ^2 is the variance of the probability distribution. The density function $\mathbf{f}(y)$ has the familiar 'bell-shaped' curve, whose peak (mode) is at the mean μ, and whose dispersion is such that about 68% of the area under the curve lies between $\mu-\sigma$ and $\mu+\sigma$, and about 99.7% lies between $\mu-3\sigma$ and $\mu+3\sigma$. However, $\mathbf{f}(r)$ is never exactly zero. It can be shown that, for large n, the binomial distribution $\mathbf{b}(r,n,p)$ with mean np and variance $n(p-p^2)$ can be approximated by a normal distribution having the same mean and variance.

We remark that it is common to fit a normal distribution to a given finite set of data S. It is natural to let the mean μ and variance σ^2 of the normal distribution be the mean m and variance s^2, respectively, of the set. When the variance s^2 is not known, or the size of the set is too small for the variance to be very meaningful, but where minimum and maximum values of S are known instead, it is common to assume that μ is the mean (or sometimes the mode) of S, and that $3\sigma=(s_{\max}-s_{\min})/2$.

Others It is sometimes desirable to use a continuous probability distribution which is *exactly* zero beyond specified minimum and maximum values. Recall that a normal distribution is never exactly zero. A *beta distribution* is a continuous one characterized by a minimum value a, a maximum value b, and a most likely value (mode) m. Its density function $\mathbf{f}:\mathcal{R}_X\rightarrow[0,1]$ is such that $\mathbf{f}(x)=0$ for $x<a$ and $x>b$, $\mathbf{f}(x)>0$ for $a<x<b$, and $\int_a^b \mathbf{f}(x)dx=1$. The mean of $\mathbf{f}$ is approximately $(a+4m+b)/6$, and its variance is approximately $[(b-a)/6]^2$. A *triangular distribution* is also a continuous one characterized by a minimum value a, a maximum value b, and a mode m. Its mean equals $(a+m+b)/3$, and its variance is $[(b-a)^2+(m-a)(m-b)]/18$. See Hillier [6.14].

6.2.6 Estimation

As noted earlier, one of the major uses of statistics is to *infer* something about the source of a set of data. For example, given a sample set of numbers $S=\{1, 10, 17, 3, 14, 8, 10\}$ which come from, say, a random number generator, we may wish to characterize the generator—i.e., to determine the form of its probability distribution $\mathbf{f}$, or to determine its moments, given the form. The simplest case is that of inferring or 'estimating' the (true) mean of the distribution, denoted $\mu_{\mathbf{f}}$. One obvious way to estimate the mean is simply to calculate the mean of a sample $S=\{s_1,s_2,\ldots,s_n\}$, where each sample or 'observed' value s_i is in the set of reals R. We write $\mathbf{M}(s_1,s_2,\ldots,s_n)=\frac{1}{n}\sum_{i=1}^{n}s_i$.

Note that the sample mean **M** is a random variable, **M**: $R^n \to R$. Its expected value is given by

$$\mathbf{Exp}[\mathbf{M}] = \mathbf{Exp}[\frac{1}{n}\sum_{i=1}^{n} s_i] = \frac{1}{n}\sum_{i=1}^{n} \mathbf{Exp}[s_i]$$

$$= \frac{1}{n}\sum_{i=1}^{n} \mu_{\mathbf{f}} = \mu_{\mathbf{f}}.$$

In other words, the expected value of the sample mean equals the true mean. While the expected value of **M** equals $\mu_{\mathbf{f}}$ for an arbitrarily given S, the sample mean may not be close to $\mu_{\mathbf{f}}$, in which case the sample is said to be 'not representative'.

The sample mean is only one of several possible *estimators* of the true mean of a distribution given a sample set S. (Other estimators include, for example, the median and mode of S, or for that matter just one arbitrarily chosen value s_i.) It is, however, a good estimator because it is 'unbiased' and 'consistent'. An unbiased estimator is one whose expected value equals the true value. Formally, we say that **F**: $R^n \to R$ is an *unbiased* estimator of a parameter $\theta_{\mathbf{f}}$ (of a distribution **f** from which n samples are taken) if $\mathbf{Exp}[\mathbf{F}] = \theta_{\mathbf{f}}$. Most estimators for a mean are unbiased. A consistent estimator is one which improves as the sample size n increases. Formally, we say that **F**: $R^n \to R$ is a *consistent* estimator of a parameter $\theta_{\mathbf{f}}$ if

$$\mathbf{lim}_{n\to\infty}\mathbf{Pr}\{|\mathbf{F}(s_1, s_2, \ldots, s_n) - \theta_{\mathbf{f}}| < \epsilon\} = 1 \qquad \text{for any } \epsilon < 0.$$

That this holds for the sample mean follows from the law of large numbers (cf. Section 6.2). It clearly does not hold for some other estimators, such as $\mathbf{F}(s_1, s_2, \ldots, s_n) = s_i$ for an arbitrarily chosen i.

Consider now the sample variance

$$\mathbf{V}(s_1, s_2, \ldots, s_n) = \frac{1}{n}\sum_{i=1}^{n} (s_i - \mathbf{M})^2$$

computed using the sample mean $\mathbf{M} = \frac{1}{n}\Sigma_{i=1}^{n} s_i$. It can be shown that **V** is not an unbiased estimator for the true variance $\sigma_{\mathbf{f}}^2$. However,

$$\frac{1}{n-1}\sum_{i=1}^{n} (s_i - \mathbf{M})^2$$

is an unbiased estimator for $\sigma_{\mathbf{f}}^2$ (to which we alluded in Section 6.1.2). On the other hand, if the true mean $\mu_{\mathbf{f}}$ is known, then

$$\frac{1}{n}\sum_{i=1}^{n} (s_i - \mu_{\mathbf{f}})^2$$

is also an unbiased estimator of $\sigma_{\mathbf{f}}^2$. For details, see Winkler [5.34].

6.2.7 Application: Reliability

Computer systems, like other large systems, consist of many interconnected components, each of which may fail to function properly for any of a variety of hardware or software reasons. These components may be connected in 'series' or in 'parallel', as illustrated in Figures 6.2.7(a) and (b), respectively. Let r_i denote the probability that the ith component works during some specified time interval; then $1-r_i$ is the probability that the ith component fails. These probabilities may also be regarded as the percentages of identical components that work or fail.

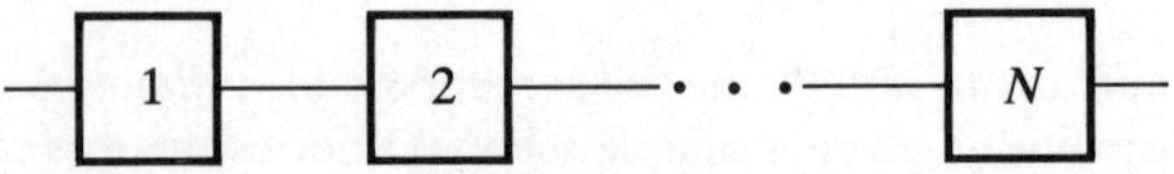

Figure 6.2.7(a)

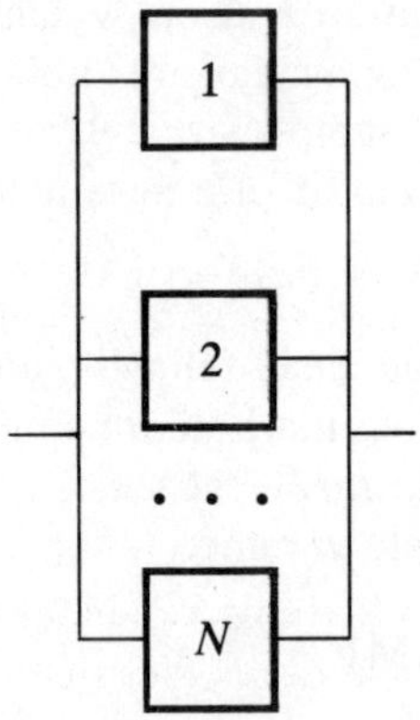

Figure 6.2.7(b)

The *reliability* of the series system is defined as the probability that all components work (or that none fail), while the reliability of the parallel system is defined as the probability that at least one of the components works (or that not all fail). Thus

$$R_{\text{series}}=\prod_{i=1}^{N} r_i$$

$$R_{\text{parallel}}=1-\prod_{i=1}^{N} (1-r_i).$$

Note that R_{series} decreases while R_{parallel} increases as the number of components increases; this is consistent with our intuitive notion that redundancy (which provides alternative ways of working) should enhance reliability. The reliability of larger more complex systems may be determined by regarding them as combinations of series and parallel subsystems. For example, consider the system shown in Figure 6.2.7(c). Its reliability is

$$1-(1-r_1[1-(1-r_2)(1-r_3)])(1-r_4).$$

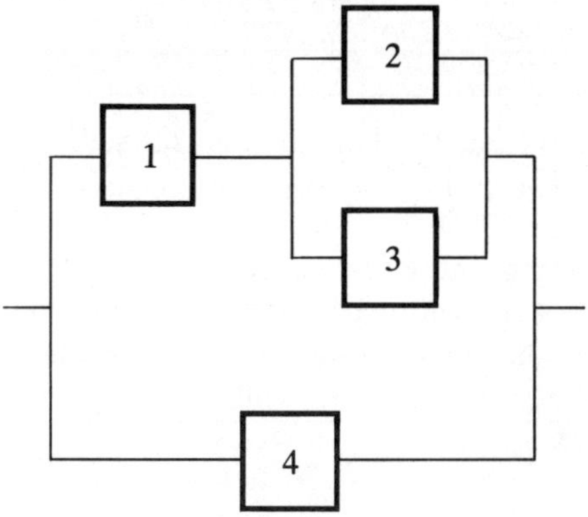

Figure 6.2.7(c)

Suppose now that all components have the same probability of working, r. Then $R_{\text{series}}=r^N$ and $R_{\text{parallel}}=1-(1-r)^N$, from which it is clear that the reliability of the series system is less than that of the parallel system. It is also clear that reliability of the series system can be increased by duplicating it in parallel as shown in Figure 6.2.7(d); the reliability of this redundant system is $1-(1-r^N)^2$. Even greater reliability can be achieved if N additional components are used to make each of a series of N components redundant, as

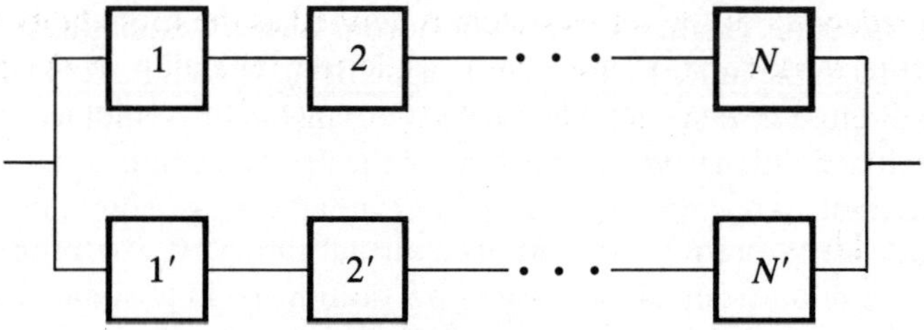

Figure 6.2.7(d)

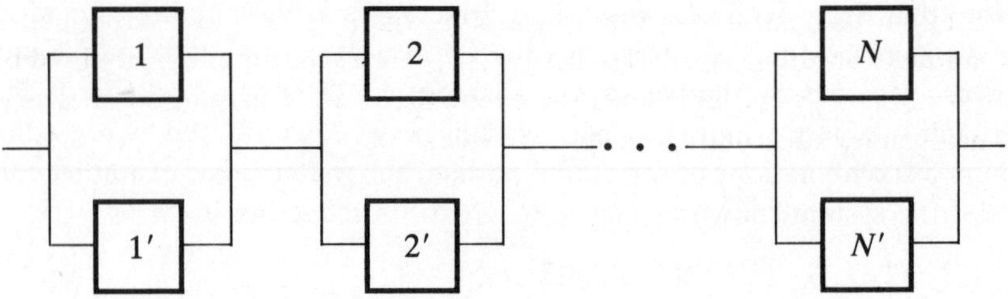

Figure 6.2.7(e)

illustrated in Figure 6.2.7(e); the reliability of this componentwise redundant system is $(1-(1-r)^2)^N$. This greater reliability is achieved at the expense of added connection costs (since, for example, components 1 and 2′ must be connected).

Finally, consider the more realistic situation where r is variable, or more specifically where the reliability, $\mathbf{r}(t)$, of a system component decreases with time t. The question then arises: when can it be expected to fail? A common assumption is that reliability, the probability of working at time t, decreases exponentially—i.e.,

$$\mathbf{r}(t)=e^{-ct},$$

where c is a constant. For this assumption, it can be shown that the *mean time to failure* (MTF) equals $1/c$. Furthermore, for a series of N components, each with reliability e^{-ct}, $\text{MTF}_{\text{series}}=1/(cN)$, and for N like parallel components, $\text{MTF}_{\text{parallel}}=(1/c)\cdot\Sigma_{j=1}^{N}(1/j)$. Derivations of these results appear in Hellerman [6.12].

6.3 MARKOV CHAINS

Let $S=\{s_1,s_2,\ldots,s_N\}$ be a finite set of states (or sample space). An important kind of probabilistic process defined on S is a 'dynamic' one in the sense that the process moves (makes a 'transition') from state to state as a function of time (although the unit of time is left unspecified). In fact, it is convenient to define a discrete time scale in terms of state transitions: 0 is some arbitrarily chosen initial point in time, 1 is the time of the first transition, 2 is the time of the next transition, and so forth. (These numbers are sometimes called *virtual* times.) We require that at each virtual time $t>0$, the process *must* make a transition from its 'old' state $s_i \in S$ (at time $t-1$) to some 'new' state $s_j \in S$ (at time t), where j may be i.

Let S_0 denote the initial state of the process, at (virtual) time 0, and let S_t

denote the state entered at time t. We may regard $S_0,S_1,S_2,\ldots$ as a sequence of random variables, where S_t: $\{s_1,s_2,\ldots,s_N\}\rightarrow[1,N]$ is, for each t, defined by $S_t(s_i)=i$. $\{S_t=i\}$ thus represents the event that state s_i is entered at time t. Consider now the conditional probabilities

$$\mathbf{Pr}\{S_{t+1}=i_{t+1}\,|\,S_0=i_0,\ S_1=i_1,\ldots,\ S_t=i_t\}.$$

If, for any $t\geq0$, this is equal to

$$\mathbf{Pr}\{S_{t+1}=i_{t+1}\,|\,S_t=i_t\}$$

—i.e., if the probability of entering a given 'next' (new) state at time $t+1$ depends upon the 'current' (old) state at time t, but not upon preceding states (which we call the 'Markov property')—then the sequence $S_0,S_1,S_2,\ldots$ is called a (finite) *Markov chain*. We say the Markov chain is finite when the set of states S is finite; the sequence, however, is infinite. (Markov chains having an infinite number of states are used in Section 6.4.)

Let $\mathbf{f}_t$ denote the probability distribution of S_t,

$$\mathbf{f}_t(i)=\mathbf{Pr}\{S_t=i\},$$

and let

$$\mathbf{P}_t^{(k)}(i,j)=\mathbf{Pr}\{S_{t+k}=j\,|\,S_t=i\}$$

for $k\geq1$, $t\geq0$. (We remark that our ordering of the subscripts i and j is not universally adopted.) We call these conditional probabilities, $\mathbf{P}_t^{(k)}$, the *k-step transition probabilities*: $\mathbf{P}_t^{(k)}(i,j)$ is the probability of moving from state s_i at time t to state s_j at time $t+k$, via a sequence of k transitions. It can be shown that

$$\mathbf{f}_{t+k}(j)=\sum_{i=1}^{N}\mathbf{f}_t(i)\mathbf{P}_t^{(k)}(i,j)$$

and that

$$\mathbf{P}_t^{(k_1+k_2)}(i,j)=\sum_{\ell=1}^{N}\mathbf{P}_t^{(k_1)}(i,\ell)\mathbf{P}_{t+k_1}^{(k_2)}(\ell,j)$$

where $k_1,k_2\geq1$ and $t\geq0$. (The latter are known as the Chapman–Kolmogoroff equations; see Kemeny [6.18].)

If $\mathbf{f}_t$ is regarded as a row vector, and $\mathbf{P}_t^{(k)}$ as a matrix, then the foregoing can be written as

$$\mathbf{f}_{t+k}=\mathbf{f}_t\circledast\mathbf{P}_t^{(k)}$$

and

$$\mathbf{P}_t^{(k_1+k_2)}=\mathbf{P}_t^{(k_1)}\circledast\mathbf{P}_{t+k_1}^{(k_2)},$$

respectively. Note that $\mathbf{f}_t$ and each row of $\mathbf{P}_t^{(k)}$ must be *probability vectors*, which are defined as N-tuples $(x_1,x_2,\ldots,x_N)$ satisfying $x_i\geq0$ and

$x_1+x_2+\cdots+x_N=1$; each $\mathbf{P}_t^{(k)}$ is an $(N\times N)$ stochastic matrix (see Section 3.2.2).

If the conditional probabilities $\mathbf{P}_t^{(k)}$ are all independent of t, we say that the Markov chain is *homogeneous,* and denote $\mathbf{P}_t^{(k)}$ simply by $\mathbf{P}^{(k)}$. Regarded as matrices, it is clear that $\mathbf{P}^{(k)}$ are the kth powers of $\mathbf{P}^{(1)}$; if the *one-step transition probability matrix* is denoted by Π, then $\mathbf{P}^{(k)}=\Pi^k$, $k\geq 1$, and we have

$$\mathbf{f}_{t+k}=\mathbf{f}_t \circledast \Pi^k.$$

This equation permits us to compute the probability distribution of S_t given the *initial distribution* $\mathbf{f}_0$ and Π:

$$\mathbf{f}_t=\mathbf{f}_0 \circledast \Pi^t.$$

Note that Π and $\mathbf{f}_0$ then characterize the Markov chain, i.e., possible sequences of states S_0, S_1, S_2,... . Often, we will just use Π to 'define' a homogeneous Markov chain.

6.3.1 Stationary Distributions

A Markov chain is said to be *stationary* if $\mathbf{f}_t=\mathbf{f}_0$ for each $t\geq 0$. This is clearly the case when $\mathbf{f}_0$ satisfies the matrix equation

$$\mathbf{f}=\mathbf{f} \circledast \Pi.$$

In fact, given Π, we define a *stationary distribution* for a Markov chain as such an $\mathbf{f}$. We remark that every homogeneous Markov chain has at least one stationary distribution $\mathbf{f}^*$, but that if $\mathbf{f}_0$ (hence the chain) is not stationary, then $\mathbf{f}_t$ may or may not tend to $\mathbf{f}^*$ (i.e., 'converge' for large t) depending upon the properties of Π (or Π^t). We remark also that an arbitrarily chosen initial distribution $\mathbf{f}_0$ will generally not be stationary, especially if we make the common assumption that the process starts in a specific state, (say) s_1, in which case $\mathbf{f}_0=(1,0,...,0)$.

To find $\mathbf{f}^*=(x_1^*,...,x_N^*)$ given the matrix Π, we solve the matrix equation $\mathbf{f}=\mathbf{f}\circledast\Pi$, which is a linear system of N equations in N unknowns, $\{x_1,...,x_N\}$, together with an additional 'normalizing' equation $x_1+x_2+\cdots+x_N=1$. We have $(N+1)$ equations since the original system of equations is not independent. (Furthermore, without the normalizing equation, $\mathbf{f}^*=(0,...,0)$ would be a solution.) That a solution to the foregoing exists is a consequence of our (implicit) requirement that Π be a 'legal' transition probability matrix; in particular, we require that each row of Π be a probability vector. For a unique solution to exist requires the additional assumption of ergodicity (see below).

For a stationary Markov chain, the stationary distribution

$\mathbf{f}^* = (x_1^*, \ldots, x_N^*)$ is of importance because it permits us to associate with each state $s_i \in S$ a numerical constant x_i^*, equal to the probability that the process is in state s_i at an arbitrary point in time. Note that, since the process is regarded as infinite, each state may be entered infinitely often (so that frequencies are infinite). However, if we know (or assume) that the number of times state s_i is entered equals n_i, then the expected number of times any other state s_j is entered is given by $n_j = (n_i / x_i^*) \cdot x_j^*$. Note also that

$$n_j = \sum_{i=1}^{N} n_i \cdot \Pi(i,j),$$

i.e., $(n_1, n_2, \ldots, n_N)$ is a solution of $\mathbf{f} = \mathbf{f} \circledast \Pi$.

Finally, since $\Pi(i,i)$ is the probability of not making a transition from state s_i to some other state s_j, $j \neq i$, the probability of remaining in state s_i for a duration $\tau_i = k$ units of time before making a transition to some other state is given by

$$\mathbf{Pr}\{\tau_i = k\} = \Pi^{k-1}(i,i) \cdot (1 - \Pi(i,i)).$$

Note that this is the form of the geometric probability distribution, which has mean $1/(1-\Pi(i,i))$. We emphasize that, at each instant, the expected time until a transition to a state other than s_i is independent of the time that has elapsed since s_i was last entered.

6.3.2 Digraph Representation

It is common to represent a homogeneous Markov chain by a labeled directed graph, where the sample space $S = \{s_1, s_2, \ldots, s_N\}$ corresponds to a set of nodes and the branch (s_i, s_j) has the one-step transition probability $\Pi(i,j)$ as its label. Π may be thought of as a weighted adjacency matrix, where the branch-label weights are probabilities. It is often convenient to define the adjacency matrix $\mathbf{C}$ by

$$\mathbf{C}(i,j) = \begin{cases} 1 & \text{if } \Pi(i,j) > 0 \\ 0 & \text{if } \Pi(i,j) = 0 \end{cases}$$

which represents the digraph (S, C) containing a branch (s_i, s_j) iff there is a nonzero probability of moving from state s_i directly to state s_j. From the path matrix (defined as $\mathbf{C} ⓥ \mathbf{C}^2 ⓥ \cdots ⓥ \mathbf{C}^N$), it is then possible to tell whether one state is reachable from another. See Pearl [6.27] for details.

If the digraph (S, C) representing a Markov chain is not connected, then each of its pieces may be treated separately. Henceforth, we assume the digraph is connected, possibly only weakly. If the digraph is strongly connected, then the Markov chain is said to be *ergodic*. Otherwise, the set of states may be partitioned into at least two maximal strongly connected regions, and the graph may be condensed. The sets of states corresponding to the sinks of the condensed graph are called *absorbing* sets, and if such a set consists of a single state, that state is said to be absorbing. (State s_i is absorbing iff $\Pi(i,i)=1$.) A state that is not in an absorbing set is called *transient*.

6.3.3 Ergodic Chains

A homogeneous Markov chain is *ergodic* (or *irreducible*) if it is possible to reach any state from any other state via a sequence of transitions. Let $\mathbf{f}=(x_1,\ldots,x_N)$. For ergodic chains, the system of equations

$$\mathbf{f}=\mathbf{f}\circledast\Pi$$

and

$$x_1+x_2+\cdots+x_N=1$$

has a unique solution $\mathbf{f}^* = (x_1^*, x_2^*, \ldots, x_N^*)$, a stationary distribution, where x_i^* is the probability that the process is in state s_i. We emphasize that it is the ergodicity which ensures a unique solution.

There are two subclasses of ergodic chains: regular ones, and periodic ones. For a *regular* chain, powers of Π converge to a matrix whose rows are identical, each being equal to the stationary distribution $\mathbf{f}^*$. We call a transition probability matrix whose powers Π^k converge in this way a 'regular' matrix. A *periodic* chain is one whose transition probability matrix has powers which do not converge; instead, these powers are ultimately periodic in that, for large k, $\Pi^{k+r}=\Pi^k$, where r is the period. For example, suppose Π is the $(N\times N)$ matrix:

$$\begin{bmatrix} 0 & 1 & 0 & 0 & \ldots & 0 \\ 0 & 0 & 1 & 0 & \ldots & 0 \\ 0 & 0 & 0 & 1 & \ldots & 0 \\ \vdots & \vdots & \vdots & \vdots & \ldots & \vdots \\ 0 & 0 & 0 & 0 & \ldots & 1 \\ 1 & 0 & 0 & 0 & \ldots & 0 \end{bmatrix}.$$

In this special case, the stationary distribution clearly consists of equal values ($x_i^* = 1/N$ for $i=1, 2, \ldots, N$). Moreover, for an initial distribution $\mathbf{f}_0=(1, 0, 0, \ldots, 0)$, subsequent distributions behave as follows:

$$\mathbf{f}_1=(0, 1, 0, ..., 0)$$
$$\mathbf{f}_2=(0, 0, 1, ..., 0)$$
$$\cdots$$
$$\mathbf{f}_{N-1}=(0, 0, 0, ..., 1)$$
$$\mathbf{f}_N=(1, 0, 0, ..., 0)$$
$$\mathbf{f}_{N+1}=(0, 1, 0, ..., 0).$$

Note that each state can then be entered only periodically, with period N. In fact, state i is entered at precise times, $t=i-1$, $i-1+N$, $i-1+2N$, etc.

For regular chains, the times at which states can be entered is no longer predictable. What is commonly of interest in this case is the probability that state j is entered for the first time at time t, given that the system starts in state i at time 0; we denote this probability $a_{ij}^{(t)}$. For the case $i=j$, $a_{ii}^{(t)}$ is the probability that state i is reentered for the *first* time at time t, given that the system starts in that state; $a_i=\Sigma_{t=1}^{\infty}a_{ii}^{(t)}$, the probability that state i will ever be reentered, is also called the ***recurrence*** probability of state i. For an ergodic chain, $a_i=1$ for each state; $a_i<1$ implies that state i may not *ever* be reentered, a possibility only for transient states. The expected time of first reentry for state i, also called the ***mean recurrence time*** of state i, is given by

$$\tau_i=\sum_{t=1}^{\infty} t\cdot a_{ii}^{(t)};$$

τ_i is necessarily finite (unless there are an infinite number of states).

Remark Suppose that a chain is not ergodic, and that state s_i cannot be reached from state s_j. In addition, suppose that the process starts out in state s_i, reenters it occasionally, and eventually enters state s_j. Thereafter, the states reachable from s_j will be reentered exclusively; hence, as the number of transitions increases indefinitely, the probability that the process will ever be in state s_i will approach zero.

6.3.4 Absorbing Chains

If a homogeneous Markov chain, defined by a one-step transition probability matrix Π, has one or more absorbing sets of states, we say that the chain is ***absorbing*** (or ***reducible***). For some ordering of the states, the matrix Π then has the block-triangular form

$$\begin{bmatrix} Q & R \\ 0 & A \end{bmatrix}$$

where Q contains the (one-step) transition probabilities between transient states, R contains the transition probabilities from transient states to absorbing sets, A contains transition probabilities from states in absorbing sets to themselves (and so is the identity matrix when all absorbing sets have size one), and 0 contains transition probabilities from absorbing sets to transient states (and so must be a matrix of zeroes). Clearly, if the process starts out in a transient state, a sequence of transitions from transient states to transient states will take place, until a transition to an absorbing set occurs, after which time (the 'absorption' time) the process remains in that set. It can be shown that $(I-Q)$ is invertible, where I is the identity matrix, and that the element $T(i,j)$ of the matrix

$$T=(I-Q)^{-1}$$

is the expected number of times the process enters state s_j before absorption, given that the process initially enters state s_i.

Recall the remark made in the preceding section on ergodic chains. Since absorbing chains are not ergodic, and transient states cannot be reached from absorbing sets, the probability that the process is in a transient state will eventually become zero, while the probability that the process is in an absorbing set will become 1. For this reason, frequencies with which transient states are entered *before absorption* become relevant instead.

If an absorbing chain having N states has only one absorbing state (the other states being transient), then there is a unique stationary distribution $\mathbf{f}^* =(0, 0, ..., 0, 1)$, assuming that the absorbing state is s_N. If the process is in state s_N, it reenters it on each transition. If there are two separate absorbing states, say s_{N-1} and s_N, then (0, 0, ..., 1, 0) and (0, 0, ..., 0, 1) are both stationary distributions, so uniqueness is lost. It should be noted that the submatrix A in the above is of the block diagonal form

$$A=\begin{bmatrix} A_1 & 0 & \cdots & 0 \\ 0 & A_2 & \cdots & 0 \\ \cdots & \cdots & \cdots & \cdots \\ 0 & 0 & \cdots & A_m \end{bmatrix}$$

where each A_i is an identity matrix (corresponding to separate absorbing states) or is a regular or periodic matrix (corresponding to an absorbing set).

6.3.5 Examples

(a) For

$$\Pi=\begin{bmatrix} 0 & 1 & 0 \\ 0 & 0 & 1 \\ 1 & 0 & 0 \end{bmatrix},$$

the chain is ergodic. The unique solution of $\mathbf{f}=\mathbf{f}\circledast\Pi$, $x_1+x_2+x_3=1$, is $\mathbf{f}^* =(1/3, 1/3, 1/3)$. If $\mathbf{f}_0=(1, 0, 0)$, then $\mathbf{f}_1=(0, 1, 0)$, $\mathbf{f}_2=(0, 0, 1)$, $\mathbf{f}_3=(1, 0, 0)$, etc., which illustrates why such a chain is called periodic.

(b) For

$$\Pi = \begin{bmatrix} 0 & 1 & 0 \\ 0 & 0 & 1 \\ 0 & 0 & 1 \end{bmatrix},$$

the chain is absorbing (state s_3 is absorbing), and

$$Q = \begin{bmatrix} 0 & 1 \\ 0 & 0 \end{bmatrix}, \qquad T=(I-Q)^{-1}= \begin{bmatrix} 1 & 1 \\ 0 & 1 \end{bmatrix}.$$

If $\mathbf{f}_0=(1, 0, 0)$, then state s_2 will be entered exactly once before absorption (into state s_3). If $\mathbf{f}_1=(0, 1, 0)$, then state s_1 will not be entered and state s_2 will not be reentered before absorption. $\mathbf{f}^* =(0, 0, 1)$ is the unique stationary distribution.

(c) For

$$\Pi = \begin{bmatrix} 0 & 1/2 & 1/2 \\ 0 & 1 & 0 \\ 0 & 0 & 1 \end{bmatrix},$$

the chain is absorbing (states s_2 and s_3 are absorbing), and

$$Q=[0], \qquad T=(I-Q)^{-1}=[1].$$

State s_1, the only transient one, can be entered only once (initially). There are an infinite number of stationary distributions, including (0, 1, 0), (0, 0, 1), and (0, 1/2, 1/2).

(d) For

$$\Pi = \begin{bmatrix} 0 & 1 & 0 \\ 0 & 0 & 1 \\ 0 & 1 & 0 \end{bmatrix},$$

the chain has an absorbing set of states, $\{s_2, s_3\}$. This set ultimately has a probability of one, split equally, hence $\mathbf{f}^* =(0, 1/2, 1/2)$.

(e) For

$$\Pi = \begin{bmatrix} 0 & 1/2 & 1/2 \\ 0 & 0 & 1 \\ 1 & 0 & 0 \end{bmatrix},$$

the chain is ergodic. The unique solution of $\mathbf{f}=\mathbf{f}\circledast\Pi$, $x_1+x_2+x_3=1$, is $\mathbf{f}^* =(2/5, 1/5, 2/5)$. This chain is also regular, so $\mathbf{f}^*$ can alternatively be determined by computing the powers Π^k:

$$\Pi^2 = \begin{bmatrix} .5 & 0 & .5 \\ 1 & 0 & 0 \\ 0 & .5 & .5 \end{bmatrix}, \qquad \Pi^3 = \begin{bmatrix} .5 & .25 & .25 \\ 0 & .5 & .5 \\ .5 & 0 & .5 \end{bmatrix},$$

$$\Pi^4 = \begin{bmatrix} .25 & .25 & .5 \\ .5 & 0 & .5 \\ .5 & .25 & .25 \end{bmatrix}, \qquad \Pi^5 = \begin{bmatrix} .5 & .125 & .375 \\ .5 & .25 & .25 \\ .25 & .25 & .5 \end{bmatrix},$$

$$\Pi^6 = \begin{bmatrix} .375 & .25 & .375 \\ .25 & .25 & .5 \\ .5 & .125 & .375 \end{bmatrix}, \quad \Pi^7 = \begin{bmatrix} .375 & .1875 & .4375 \\ .5 & .125 & .375 \\ .375 & .25 & .375 \end{bmatrix}, \ldots,$$

$$\Pi^{12} = \begin{bmatrix} .39 & .20 & .41 \\ .41 & .19 & .41 \\ .41 & .20 & .39 \end{bmatrix}, \ldots$$

which converges to the matrix $\begin{bmatrix} \mathbf{f}^* \\ \mathbf{f}^* \\ \mathbf{f}^* \end{bmatrix}$, where $\mathbf{f}^* = (.4, .2, .4)$.

6.3.6 Application: Analysis of Algorithms

Consider Figure 6.3.6(a), which represents a program flowchart having four blocks (or nodes), with possible branches between blocks as shown. Having no better information, we assume equiprobable branches: we assume the three branches from node a each has probability $1/3$, the two branches from node b each has probability $1/2$, and the single branch from node c has probability 1.

In order to use an absorbing chain model, the program exit node must be an absorbing state. As shown in Figure 6.3.6(b), this is accomplished by adding a branch from node d to itself, having probability 1. Then the transition probability matrix

$$\Pi = \begin{bmatrix} 0 & 1/3 & 1/3 & 1/3 \\ 0 & 0 & 1/2 & 1/2 \\ 0 & 0 & 0 & 1 \\ 0 & 0 & 0 & 1 \end{bmatrix}$$

has the required block triangular form, with

$$Q = \begin{bmatrix} 0 & 1/3 & 1/3 \\ 0 & 0 & 1/2 \\ 0 & 0 & 0 \end{bmatrix},$$

whence

$$T = (I - Q)^{-1} = \begin{bmatrix} 1 & 1/3 & 1/2 \\ 0 & 1 & 1/2 \\ 0 & 0 & 1 \end{bmatrix}.$$

From the top row of T, we conclude, given that the program starts at node a, that:

expected no. of executions of node a before absorption $= 1$
expected no. of executions of node b before absorption $= 1/3$
expected no. of executions of node c before absorption $= 1/2$.

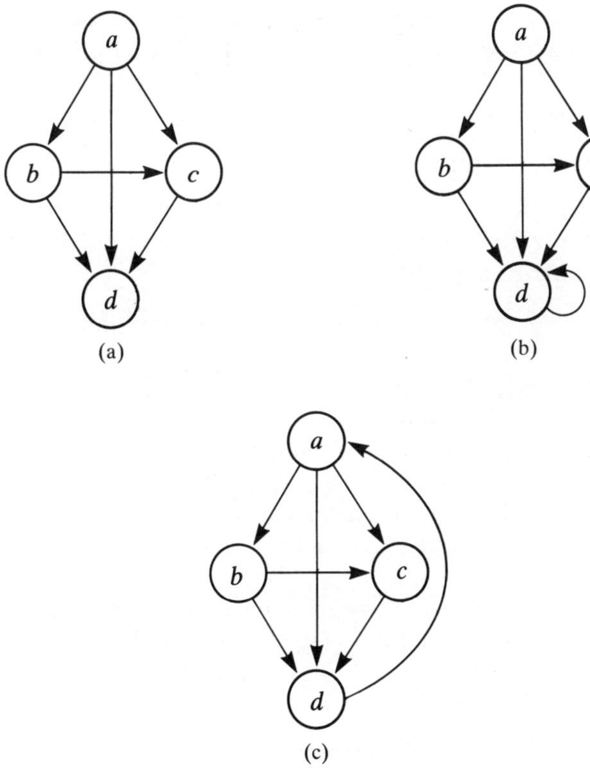

Figure 6.3.6

In order to use an ergodic chain model, the program must not be permitted to exit. This is prevented by adding a branch, having probability 1, from node d back to the start node a, as shown in Figure 6.3.6(c). Then the transition probability matrix is

$$\Pi = \begin{bmatrix} 0 & 1/3 & 1/3 & 1/3 \\ 0 & 0 & 1/2 & 1/2 \\ 0 & 0 & 0 & 1 \\ 1 & 0 & 0 & 0 \end{bmatrix}.$$

Now, letting $\mathbf{f}=(x_1,x_2,x_3,x_4)$, and solving $\mathbf{f}=\mathbf{f}\circledast\Pi$ with the normalizing equation $x_1+x_2+x_3+x_4=1$, we have $x_1=6/17$, $x_2=2/17$, $x_3=3/17$, $x_4=6/17$. The x_i are the relative execution frequencies of the four program nodes. Hence if node a is executed once, we conclude that:

expected no. of executions of node $b=1/3$
expected no. of executions of node $c=1/2$
expected no. of executions of node $d=1$.

Consider now the more explicit practical problem of estimating the execution time of the program of Figure 6.3.6(a). If T_a, T_b, T_c and T_d denote the execution times of the program blocks, from either of the absorbing or ergodic chain analyses, we may conclude that, for the program,

$$\text{expected execution time} = T_a + T_b/3 + T_c/2 + T_d.$$

For further information on the use of Markov chains to analyze computer programs, see Ramamoorthy [6.28]. For some related concepts, see Beizer [6.3] and Spirn [6.30].

6.4 QUEUEING SYSTEMS

A *queueing system* consists of one or more *servers* (or processing facilities) and a source of *customers* (or users), not all of which may be ready for service at any given time, and only one of which may be processed by a given server at a time. As a consequence, a customer when ready for service may have to wait in a line or *queue* for its turn. An example of a single-server queueing system is depicted in Figure 6.4(a).

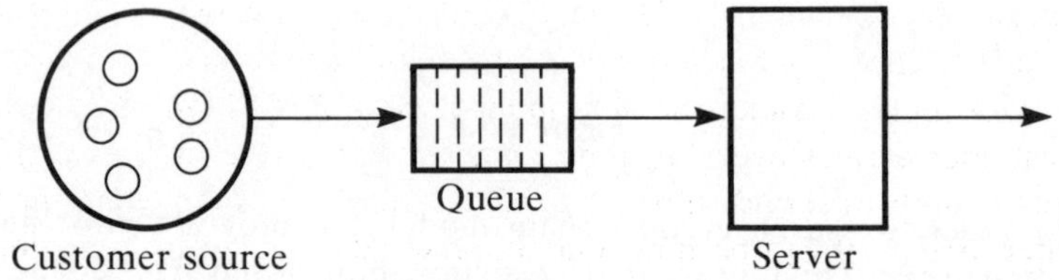

Figure 6.4(a)

We say that a customer *arrives* when it is ready for service; at this time it joins the queue, but at the same instant it may enter the processing facility if the server is free and there are no others with greater priority waiting in the queue. The customer in the queue which is selected for service then leaves the queue and enters the processing facility for a duration that depends upon the customer's *service time* requirement T_s. The server is said to be *busy* for this duration. After its service time requirement is satisfied, the customer leaves the server (and the queueing system) at what is called its *departure time*. The server is then immediately free to provide service to any other member of the queue (including a newly arriving customer).

There are several parameters which affect the operation of such a queueing system. Among these are (**a**) the times or rate λ at which customers arrive, (**b**) the service time requirements of the customers, (**c**) the number of individual servers in the system, (**d**) the number of customers permitted to wait in the queue at any one time, and (**e**) the priority scheme or 'scheduling' rule by which the server selects members of the queue for service. Parameters **a** and **b** are ordinarily specified by assuming certain probability distributions for the interarrival times and service times. A given queueing system will be denoted by a 5-tuple (**a**/**b**/**c**/**d**/**e**) specifying the assumptions made for each parameter. For example, if we let **G** denote any probability distribution, and FCFS† denote the 'first-come–first-served' scheduling rule, then a general queueing system with N servers and a limit on the size of the queue of S will be denoted (**G**/**G**/N/S/FCFS). **M** will be used instead of **G** to denote the exponential probability distribution, and **C** will be used to denote the constant case where interarrival or service times are fixed. In this chapter, we restrict our attention to FCFS; other scheduling rules are discussed in Chapter 9. General introductions to queues appear in Cox [6.5] and Takacs [6.31].

It should be clear that the five parameters affect performance measures such as the average or maximum 'queue size' L_w (i.e., the number of customers waiting in the queue at any one time), 'waiting time' T_w (i.e., the amount of time customers spend in the queue), 'turnaround time' (i.e., the time between a customer's departure and arrival times), or 'utilization factor' ρ (i.e., the percentage of time or probability that the server is busy).

For example, consider the deterministic case (**C**/**C**/1/∞/FCFS) where one customer arrives every T_a time units, i.e., at a rate of $\lambda=1/T_a$, and requires T_s units of service time. If $T_a=T_s=1$ time unit, then when the first customer arrives, it can be immediately serviced. During the next unit of time, the server is busy and the queue is empty. When this service duration is over, the first customer leaves the queueing system as the second customer arrives and is given immediate service. For this case, the server is continually busy, and the queue is continually empty, except for the times of zero duration when a customer arrives and instantaneously passes through the queue to the server.

In the case where the interarrival time T_a exceeds the service time T_s, the server will be idle when the first customer completes service, for a duration T_a-T_s. The server would then be busy only $T_s/T_a\times 100$ percent of the time; hence $\rho=\lambda T_s$. In the case where the interarrival time is less than

†FCFS corresponds to implementing a queue (in a waiting-line sense) as a 'queue' in a FIFO data-structure sense; in contrast, 'last-come–first-served' corresponds to implementing a waiting-line queue as a 'stack'.

the service time, a second customer will arrive before the first customer completes service, so while the server is busy the second customer must wait in the queue for a duration $T_s - T_a$. Similarly, the third customer must wait in the queue twice as long, and so forth. If this continues, the server will not be able to keep up with the arrivals, and will get further and further behind as the queue size increases indefinitely. This situation is said to be *unstable*. If we impose a limit on the queue size, then instability will eventually lead to an *overflow* of the queue.

The foregoing conclusions can be verified by drawing timing charts as shown in Figure 6.4(b); the horizontal axis represents time, increasing from left to right. In the following section, we show how a queueing system can be modeled as a Markov chain having an infinite number of states.

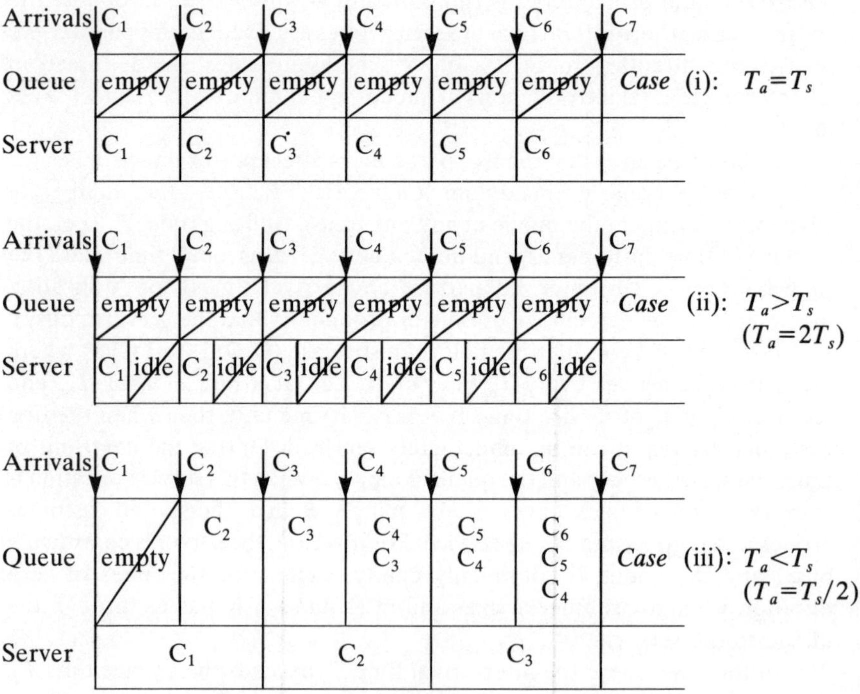

Figure 6.4(b)

6.4.1 Probabilistic Models

In the preceding section, we assumed that arrivals occur regularly each T_a time units, and that service times T_s are the same for each customer. Suppose now that arrival and/or service times are determined nondeterministically, with the times between arrivals and for service varying according to given probability distributions. Whatever the nature of the probability distributions, it is apparent that if the average interarrival time is less than the average service time, then in the long run the queue size will increase indefinitely. On the other hand, if the average interarrival time exceeds the average service time, then we would expect the queuing system to remain stable with temporary increases in queue size caused by 'runs' of arrivals being offset by periods when no new customers arrive, giving the server a chance to empty the queue and to occasionally become idle. The stable case is the one of interest, and our objective will be to analyze the performance of the queuing system with respect to measures such as average waiting time, average queue size, probability that the queue size exceeds a specified limit, and utilization factor. For analytical simplicity, we must make some assumptions about the nature of the interarrival and service time distributions.

As an initial example, let us consider the (**G**/**C**/1/∞/FCFS) case, where we assume that service times are constant ($=T_s$) and that the server can initiate service to a customer only at fixed instants of time: $0, T_s, 2T_s, 3T_s, \ldots$. If the server is idle, say, between times $2T_s$ and $3T_s$, then any new arrivals during that third period will have to wait at least until time $3T_s$ (rather than being given immediate service). With the durations and starting times for service so fixed, the only randomness left is in the arrival times. We define the random variable X_i to be the number of customers that arrive during the ith period (i.e., from $(i-1)T_s$ to iT_s, including arrivals at exactly time iT_s), and let $a_k=\mathbf{Pr}\{X_i=k\}$,† which we assume is independent both of i and of X_j for $j\neq i$. Finally, we define the *state* L_i of the queuing system at time iT_s to be the number of customers waiting in the queue at that time. The number of customers waiting, at the next instant of time $(i+1)T_s$, should clearly increase by the number of new arrivals X_{i+1} and should decrease by one if a customer was served in the interim: i.e.,

$$L_{i+1}=L_i+X_{i+1}-1.$$

This formula applies except when a customer is not served during the interval from iT_s to $(i+1)T_s$ ($L_i=0$ and $X_i=0$) and no new customers arrive ($X_{i+1}=0$); in this case, $L_{i+1}=0$.‡ We may, as an alternative which applies with

†Recall that $0\leq a_k\leq 1$ for each k, and $\Sigma_{k=0}^{\infty}\, a_k=1$, since the a_k are probabilities.

‡If $L_i=0$ and $X_i=1$, then the newly arriving customer is immediately served; if $L_i=1$ and $X_i=0$, then the waiting customer is provided service. In both cases, the queue is empty from iT_s to $(i+1)T_s$, so that $L_{i+1}=X_{i+1}-1$ if $X_{i+1}>0$, but $L_{i+1}=0$ if $X_{i+1}=0$.

no exceptions, write

$$L_{i+1}=\mathbf{max}\{L_i+X_{i+1}-1,\ 0\}$$

where $L_0=\mathbf{max}\{X_0-1,\ 0\}$, one less than the number of arrivals at time 0, if any. The behavior of this queueing system is then modeled by an infinite Markov chain whose state transition probability matrix, $\Pi=(p_{ij})$, is given by

$$p_{ij}=\begin{cases} a_{j-i+1} & \text{if } i-1\leq j \\ 0 & \text{if } i-1>j \end{cases}$$

for $i\geq 0$, $j\geq 0$, except that $p_{00}=a_0+a_1$. The expected number of arrivals during the ith period is given by

$$\mathbf{Exp}(X_i)=\sum_{k=0}^{\infty} k\times\mathbf{Pr}\{X_i=k\}=\sum_{k=0}^{\infty} ka^k,$$

which clearly must be less than 1 for the system to be stable since only one customer can be served during that period. Finally, since X_i is independent of i, we define λ to be the expected number of arrivals [$=\mathbf{Exp}(X_i)$] per unit time [$=T_s$]. Now suppose that $\lambda=1/2$, i.e., that one customer arrives every other time unit *on the average*. Case (ii) of Figure 6.4(b) depicts a deterministic version of this case, where $\lambda=1/2$, $T_s=1$, and $T_a=2$; in the present probabilistic example, the interarrival time T_a has a mean of $1/\lambda=2$, but a nonzero variance $\sigma_a{}^2$. For small variations in T_a, each customer will usually not have to wait for service, and so will be in the queueing system for exactly one unit of time. Under these circumstances, the system will alternate between having zero and one customer present (in the server) for an average of 1/2. Hence,

$\hat{L}_w$=expected no. waiting in line=0+
$\hat{L}$ =expected no. in queueing system=1/2+
$\hat{T}_w$=expected waiting time=0+
$\hat{T}$ =expected time in system=1+

where the + indicates that these values should increase by amounts dependent upon the variance $\sigma_a{}^2$. For, if T_a should for a while be small, several customers may arrive during the same service interval, and have to wait in the queue for some nonnegligible amount of time; $\hat{L}_w$ and $\hat{T}_w$ would then be nonzero. These periods would be offset by those for which T_a is large, during which the server can catch up.

In summary, for constant service times, $\hat{L}_w$, $\hat{L}$, $\hat{T}_w$ and $\hat{T}$ depend upon variations in arrival time. On the other hand, for a deterministic arrival process, $\hat{L}_w$, $\hat{L}$, $\hat{T}_w$, and $\hat{T}$ depend upon variations in the service time T_s. When both arrivals and service times are determined probabilistically, variations in both T_a and T_s must be considered.

6.4.2 Steady-state Queueing Results

In this section, we state (without proof) the basic theorems which allow the performance of queueing systems to be evaluated. Of primary importance is *Little's theorem*:

$$\hat{L}_w=\lambda\hat{T}_w \qquad \text{and} \qquad \hat{L}=\lambda\hat{T}$$

which holds for a queue in *steady state,* i.e., where the arrival rate λ (the number of arrivals per unit time) equals the departure rate γ (the number of customers served per unit time, i.e., the 'throughput'). (Note that in the example at the end of the preceding section, $\lambda=1/2=\gamma$.) Little's theorem applies to general queueing systems, without restrictions on the nature of the arrival process or service time distributions, or on the scheduling rules. The steady-state requirement is based upon an intuitive 'conservation' principle: departures cannot exceed arrivals, on the average; if arrivals exceed departures, on the average, then the queueing system is unstable (and $\hat{L}_w$ and $\hat{T}_w$ will tend to infinity). Some additional general formulas for stable steady-state queues include:

$$\rho=\lambda\hat{T}_s \qquad (6.4.2(a))$$
$$\hat{L}=\hat{L}_w+\rho \qquad (6.4.2(b))$$
$$\hat{T}=\hat{T}_w+\hat{T}_s \qquad (6.4.2(c))$$

where ρ is the utilization factor and $\hat{T}_s$ is the average service time. These formulas have intuitive interpretations. For example, the utilization factor can be interpreted as the ratio between the average service time and the average interarrival time, hence $\rho=\hat{T}_s/\hat{T}_a=\lambda T_s$. If we let μ denote the service rate (i.e., the mean of $1/T_s$), then

$$\mu=1/\hat{T}_s$$

and

$$\rho=\lambda/\mu.$$

Note that $\mu=\gamma$ only when $\rho=1$. Note also that a queueing system is unstable when $\rho>1$; in effect, **ceiling**(ρ) processing facilities must serve the same queue for the system to be stable. Finally, we may regard ρ as the mean number of customers in the server.

Little's theorem enables the calculation of three of the four quantities $\hat{L}_w$, $\hat{T}_w$, $\hat{L}$, $\hat{T}$ given only one of them (and a knowledge of λ and $\hat{T}_s$). To determine any one of them, however, requires additional information on the nature of the arrival process and the service time distribution. A common assumption made about the arrival process is that customers arrive according to a Poisson distribution with mean λ: i.e., that the probability that r customers arrive in any unit time period equals $\lambda^r e^{-\lambda}/r!$, or equivalently that interarrival times are exponentially distributed (see Section 6.2.5). For this (**M**/**G**/1/∞/FCFS) case,

$$\hat{L}_w = \frac{\lambda^2(\sigma_s^2+\hat{T}_s^2)}{2(1-\rho)}$$

where σ_s^2 is the variance in service time. Hence

$$\hat{T}_w = \hat{L}_w/\lambda = \frac{\lambda(\sigma_s^2+\hat{T}_s^2)}{2(1-\rho)}$$

$$\hat{L} = \hat{L}_w + \rho$$

$$\hat{T} = \hat{T}_w + \hat{T}_s = \hat{L}/\lambda .$$

For the special case (**M**/**C**/1/∞/FCFS), where service times are constant, $\sigma_s^2=0$; hence

$$\hat{L}_w = \frac{\lambda^2 T_s^2}{2(1-\rho)} = \frac{\rho^2}{2(1-\rho)}$$

$$\hat{T}_w = \frac{\rho}{2(1-\rho)} \cdot T_s \; .$$

For the special case (**M**/**M**/1/∞/FCFS), where service times are also exponentially distributed, $\sigma_s^2=\hat{T}_s^2$; hence

$$\hat{L}_w = \frac{\lambda^2 \hat{T}_s^2}{(1-\rho)} = \frac{\rho^2}{(1-\rho)} \qquad \hat{T}_w = \frac{\rho}{(1-\rho)} \cdot \hat{T}_s$$

$$\hat{L} = \frac{\rho}{(1-\rho)} \qquad \hat{T} = \frac{\hat{T}_s}{(1-\rho)} \; .$$

For this case, **Pr**{i in system}=**Pr**{$i-1$ waiting in queue}$=(1-\rho)\rho^i$ for $i>0$, which is the geometric distribution.

6.4.3 Feedback Queues

A *feedback queue* is one where, following a service period of $1/\mu$ time units, a customer may with probability p return (be 'fed back') to the waiting line for another service period. We assume this probability is the same for each customer, whether or not the customer has been previously served. Without feedback, the number of customers that leave the server per unit time, μ, equals the number of customers that enter the system per unit time, λ. With feedback, there are in effect $p\mu$ additional customers per unit time, hence $\mu=\lambda+p\mu$ customers leave the server per unit time, which yields

$$\mu = \frac{\lambda}{1-p} .$$

Of course, the number of customers that leave the system per unit time, $\gamma=(1-p)\mu$, must still be λ. However, the utilization factor

$$\rho=(\lambda+p\mu)\hat{T}_s=(\frac{\lambda}{1-p})\hat{T}_s$$

will generally increase, since customers may make several passes through the server. The expected number of passes is given by

$$\hat{N}=1\cdot(1-p)+2\cdot p(1-p)+3\cdot p^2(1-p)+\cdots \quad .$$

(We leave, as an exercise, the proof that the sum of this series is $1/(1-p)$.) Therefore, there is, in effect, a net arrival rate of $\hat{N}\lambda$; $\hat{T}$ can be calculated accordingly. For example, for exponential service,

$$\hat{T}=\frac{1}{1-p}\,(\frac{\hat{T}_s}{1-\rho}).$$

6.4.4 Finite Queues

An important class of queueing system is that in which the size of the waiting line is limited, i.e., where some maximum number of customers can wait at a time. If additional customers arrive when the waiting line is full, they are assumed to be 'lost' (i.e., they immediately leave the system without returning), or equivalently it is assumed that no arrivals can occur when the queue is full. Let P_i denote the probability that there are i customers in the system (counting the one being served), and let J be the limit on i. Assuming exponentially distributed arrivals and service times with parameters λ and μ, respectively, (except that $\lambda=0$ when the queue is full), we have

$$P_i=\frac{\rho^i(1-\rho)}{1-\rho^{J+1}} \qquad \text{for } i=0,\ 1,\ \ldots,\ J$$

where $\rho=\lambda/\mu$. Note that P_J is the probability that the system is full. For this (**M**/**M**/1/J/FCFS) case, the mean number of customers in the system, as a function of J, is then given by

$$\hat{L}(J)=\sum_{i=0}^{J} iP_i=\sum_{i=0}^{J}\frac{i\rho^i\,(1-\rho)}{1-\rho^{J+1}}=\frac{\rho}{1-\rho}-\frac{(J+1)\rho^{J+1}}{1-\rho^{J+1}} \quad .$$

Furthermore,

$$\hat{L}_w(J)=\sum_{i=1}^{J}(i-1)P_i=\hat{L}(J)-(1-P_0)$$

$$=\frac{\rho}{1-\rho}-1-\frac{1-\rho+(J+1)\rho^{J+1}}{1-\rho^{J+1}} \quad .$$

An application is given in Section 7.3.4.

6.4.5 Applications

The formulas given above assume Poisson arrivals. It is fortuitous that the assumption of random independent arrivals upon which the Poisson process is based is a realistic one. In addition, the assumption permits relatively easy analysis. When arrivals are not Poisson, then, except for relatively restrictive assumptions on service times, it is not easy to derive $\hat{L}_w$ or the other three parameters. Nevertheless, more general queueing systems are of importance, and their analyses and applications may be found in various texts (e.g., Kleinrock [6.19]). We remark here that among the generalizations that have been treated are:

(a) non-Poisson arrival processes (e.g., a 'renewal' process)
(b) general service time distributions
(c) multiple servers (i.e., additional processing facilities, in parallel)
(d) a limit on the size of the waiting line
(e) a limit on the number of arrivals
(f) scheduling policies (where customers are not considered identical, so that some have greater priority for service than others)
(g) preemption (where a customer while in service can be interrupted and forced to wait again for completion of service).

The latter, of course, corresponds to 'time-sharing' systems, which we discuss further in Chapter 9. For related applications, see Coffman and Denning [6.4], Kobayashi [6.21], and Everling [6.6].

EXERCISES

1. For $X=\{3,1,4,1,5,9,2,6,5,3,5,9\}$ and $Y=\{2,7,1,8,2,8,1,8,2,8,4,6\}$, compute

(a) m_X, m_Y
(b) s_X, s_Y
(c) r_{XY}
(d) the regression line $y_0+b \cdot x$.

2. Suppose a record of data has from 1 to 6 characters. In a file of 100 records, a count of the lengths gave the following:

no. of records with length 1=25
no. of records with length 2=25
no. of records with length 3=20
no. of records with length 4=15
no. of records with length 5=10
no. of records with length 6= 5

(a) What are the mean, mode, and median of the record lengths?
(b) What is the standard deviation of the record lengths?
(c) What are the range and modality?

3. Compare the space requirements for the file of Exercise 2 assuming:

(a) a fixed length of 6 characters is allocated for each record
(b) a length of $L+1$ characters is allocated for each record, L for the data and one for the counter within each record.

4. Let $\{a,s,d,f\}$ be a set of single-letter data items which we wish to search.

(a) For an equilikelihood assumption, and assuming that the item we wish to find is in the set, what is the expected number of comparisons needed to find a randomly chosen item?
(b) For an equally likely letter assumption, and assuming that the item we wish to find may not be in the set (with probability 22/26), what is the expected number of comparisons needed to find a randomly chosen item?

5. Assuming $\mathbf{Pr}\{K=a\}=.3$, $\mathbf{Pr}\{K=s\}=.4$, $\mathbf{Pr}\{K=d\}=.1$, $\mathbf{Pr}\{K=f\}=.2$, what is the expected number of comparisons needed to find a randomly chosen item:

(a) if the items are ordered by descending probability (i.e., s first)?
(b) if the items are ordered by ascending probability?

6. Refer to Exercise 2 above. Interpret the numbers of records with lengths 1 through 6 given there as probabilities, e.g., $\mathbf{Pr}\{length=1\}=0.25$, etc. Suppose a fixed length of 4 characters is allocated for each record, 3 for data and 1 for an 'overflow' flag that indicates whether or not the next record contains a continuation of the data.

(a) What is the probability of an overflow?
(b) What percentage of the total space occupied by the file will contain actual data?

7. Suppose there are two files which are to be accessed in alternation. Compare average seek distances, making equilikelihood assumptions, for the cases where:

(a) the two files are stored adjacent to each other;
(b) the smaller file is stored in the middle of the larger file;
(c) the larger file is stored in the middle of the smaller file.

8. (Section 6.2.1) Show that $\mathbf{Exp}(t)=(W+1)/3$ if $t_i=i$ for each i.

9. How many legal FORTRAN variable names are there?

10. Show that $\mathbf{Exp}\ \mathbf{f}(r)=np$ for the binomial distribution.

11. For example (e) of Section 6.3.5, calculate Π^8.

12. Flowchart a sorting algorithm. What are the minimum, maximum, and expected number of times through each loop, when sorting an array of N elements? (Use Markov chain analysis to compute the latter.)

13. Find the stationary distribution for

$$\Pi = \begin{bmatrix} 1 & 1 & 0 \\ 1/2 & 0 & 1/2 \\ 0 & 1 & 0 \end{bmatrix}$$

14. Given $\mathbf{f}_{t+k} = \mathbf{f}_t \circledast \Pi^k$, show that

$$\mathbf{f}_t = \mathbf{f}_0 \circledast \Pi^t.$$

15. Consider the (**C**/**G**/1/∞/FCFS) case, a variation of the example of Section 6.4.1. Assume that the state of the process is the length of the waiting line and that, in each unit of time, one customer arrives and K customers are served with probability $a_K > 0$ ($K = 0, 1, 2, \ldots$) if there are at least K in the waiting line. Give the state transition probability matrix.

16. For the feedback queue of Section 6.4.3, show that the expected number of passes through the server equals $1/(1-p)$.

***17.** Let X and Y be independent random variables. Show that

$$\mathbf{Exp}_{X,Y}[\mathbf{g}(X,Y)] = \mathbf{Exp}_X[\mathbf{Exp}_Y[\mathbf{g}(X,Y)]].$$

18. (Section 6.1.5) An $\mathbf{O}(N)$ algorithm was executed twice, for $N = 10$ and $N = 20$, with execution times of 81 ms and 151 ms, respectively. What are y_0 and b?

PROGRAMMING ASSIGNMENTS

1. Write a program which computes the mean, mode, median, and standard deviation of an array of numbers.

2. Write a program which computes the correlation r_{XY} of two sets of numbers X and Y.

3. If a statistical package (such as SPSS, SAS, BMD) is available to you, use it to

(a) compute the mean and standard deviation of an array of numbers;
(b) compute the correlation of two sets of numbers;
(c) compute the regression line given two sets of numbers.

4. Write a program which inputs the transition probability matrix of a Markov chain and determines whether it is ergodic or absorbing.

5. Given a uniform random number generator (e.g., RANDU of Chapter 1), write subroutines which will generate random numbers which are

(a) normally distributed
(b) exponentially distributed.

6. Write a program which simulates a (**M**/**M**/1/∞/FCFS) queueing system. Use a random-number generator to determine at each instant of time whether an arrival occurs. For each arrival, determine the service time required by generating another random number.

7. Write a program which simulates a Markov chain, printing out the identity of each state as it is entered. Now partition the set of states in several blocks, and repeat the simulation printing out the identity of each block whenever a transition is made from a state in one block to another block. Compare the lengths of the printed sequences of blocks for various partitioning strategies. (Note: this is related to paging sequences, which we discuss in Chapter 9.)

7 APPLICATIONS—INFORMATION STRUCTURES

7.1 INFORMATION PROCESSING

An important body of knowledge, known as *information theory,* is associated with the general problem of communicating information, and with the particular problems of selecting a *code* and of detecting and correcting errors. First, we introduce the problems of data compression. An important example is that of storing data whose representational lengths may vary. It is clearly inefficient to reserve space for the maximum possible length if the average length is significantly less. On the other hand, reserving space for just the average length, or of the average plus one standard deviation length, can lead to greater inefficiencies if excessive handling of overflows seriously degrades performance. Next, because errors can be introduced into data in various ways (especially when data is read or written or otherwise transmitted), we discuss the use of parity bits for error-checking purposes, a good precaution to take for both hardware and software reasons since there is, in general, a nonzero probability that any given bit of a data item will be incorrect. Some interesting applications, which we shall not discuss here, may be found in Meadow [7.26], Ganapathy [7.11], and Taylor [7.32].

7.1.1 Codes

Suppose we wish to transmit a message from a 'source' to a 'destination' by means of a 'channel' (which may introduce errors). The message, which consists of a sequence of source characters (from a source alphabet A), must be encoded (translated) into a sequence of code characters (from a code alphabet B). Generally, each source character may be encoded into more than one code character. We shall say that a *coding* **C** is a function relating each character of the source alphabet with a fixed sequence of characters of the code alphabet; i.e., $\mathbf{C}: A \to B^*$. The range of **C**, denoted $\mathcal{R}_{\mathbf{C}}$, will be called the *code*, each member of which is called a *code word*. Source strings, $a_1a_2...a_\ell \in A^*$, may be coded by the homomorphic function $\mathbf{C}': A^* \to B^*$ defined by

$$\mathbf{C}'(a_1a_2...a_\ell)=\mathbf{C}(a_1)\cdot \mathbf{C}(a_2)\cdot ... \cdot\mathbf{C}(a_\ell),$$

the concatenation of corresponding code words. $\mathbf{C}'$ is homomorphic because $\mathbf{C}'(x\cdot y)=\mathbf{C}'(x)\cdot\mathbf{C}'(y)$ for all $x,y\in A^*$.

A code is *binary* if the code alphabet contains exactly two characters (denoted 0 and 1). The code is *nonsingular* if **C** is 1–1. A nonsingular code is *uniquely decodeable* if $\mathbf{C}'$ is also 1–1. A uniquely decodeable code is *instantaneous* if it is possible, in scanning a sequence of code characters from left to right, to decode each code word without scanning beyond its rightmost character; this requires that no code word be a prefix of another code word. Codes for computer character sets are discussed in detail in Mackenzie [7.23].

Examples Consider the examples of code words for the decimal source digits given in Table 7.1.1. We note that Code 1 (EBCDIC) is instantaneous, but 'inefficient' in that the common prefix '1111' does not serve to distinguish the source characters. (They are used to distinguish digits from letters, etc., in an extended alphabet.) Omission of this prefix yields the packed BCD 4-bit code, Code 2, which is also instantaneous. Suppose we proceed even further and omit leading zeroes, yielding Code 3. This code is no longer instantaneous, since it is not uniquely decodeable: for example, the two source strings '56' and '232' would both be encoded into '101110'. Care must evidently be exercised in the selection of 'efficient' (short) codes.

Source character	*Code* 1	*Code* 2	*Code* 3	*Code* 4	*Code* 5	*Code* 6
0	11110000	0000	0	10	000	00000
1	11110001	0001	1	110	001	00011
2	11110010	0010	10	1110	010	00101
3	11110011	0011	11	11110	011	00110
4	11110100	0100	100	111110	100	01001
5	11110101	0101	101	1111110	101	01010
6	11110110	0110	110	11111110	1100	01100
7	11110111	0111	111	111111110	1101	01111
8	11111000	1000	1000	1111111110	1110	10001
9	11111001	1001	1001	11111111110	1111	10010

Table 7.1.1

Consider now Code 4, which is essentially a monadic code, but with the addition of a code character (called a 'comma', but denoted by 0) to mark the end of a code word, which thus makes it instantaneous. We observe that average word length for Code 2 is 4 bits, whereas for Code 4 it is 6.5 bits. (For Code 3, it is 2.6, but we reject it because it is not instantaneous.) In this sense, we can say that Code 2 is more efficient than Code 4. However, suppose some source characters are much more likely to be used than others. (For example, the letter **E** is more likely than **Z** in English text; the letter **X** is more likely than **Q** in computer programs.) More specifically, suppose that the decimal digits 0 and 1 are much more likely than the others. Then the weighted-average length of a message using Code 4 may be shorter than one using Code 2.

7.1.2 Information Content (Entropy)

Some property of information communicated in a message certainly depends upon the probability of the message. If the message is not very probable (e.g., 'it snowed in Waikiki'), then the message conveys much more information than otherwise. In other words, if M is a message (or set of messages), the *information content* of M, denoted $\mathbf{I}(M)$, is inversely proportional to the probability of the event that M occurs, denoted $\mathbf{P}(M)$. Mathematically, this is written $\mathbf{I}(M) \propto 1/\mathbf{P}(M)$. So that $\mathbf{I}(M)$ will be additive, i.e., so that $\mathbf{I}(M_1, M_2) = \mathbf{I}(M_1) + \mathbf{I}(M_2)$ for an arbitrary pair of message events M_1 and M_2, we formally define

$$\mathbf{I}(M) = \mathbf{log}(1/\mathbf{P}(M)),$$

where **log** is the logarithm function (to some unspecified base r). The units by which we measure information depends upon the choice of r. In order for one binary character to carry one unit of information, we shall choose r to be 2, in which case we say $\mathbf{I}(M)$ is in units of *bits*; the information content of a message with probability equal to 1/2 is then 1 bit. The *entropy* H of a source of information is defined as the expected value of the information contents of the source characters, and we should expect to require at least H bits, on the (weighted) average, for our code words. For an elaboration, see Abramson [7.1].

Suppose that we have 10 source characters $\{0,1,\ldots,9\}$ and that their probabilities of occurring in a message are $\{p_0, p_1, \ldots, p_9\}$, respectively. Their information contents are then $\mathbf{log}(1/p_0), \ldots, \mathbf{log}(1/p_9)$, respectively. The entropy is

$$H = \mathbf{log}(1/p_0) \times p_0 + \mathbf{log}(1/p_1) \times p_1 + \cdots + \mathbf{log}(1/p_9) \times p_9,$$

which, for an equilikelihood assumption ($p_i = 1/10$), turns out to be $\mathbf{log}_2(10) = 3.322$ bits per source character. We conclude from this that the best instantaneous binary code for 10 equally likely characters must have a mean word length of *at least* 3.322 bits. This lower bound may not be achievable. For example, for the given assumptions, we can do no better than Code 5 in Table 7.1.1, which has a mean word length of 3.4 bits. (Code 5 is called a 'Huffman code', and is derived in Section 9.1.1.)

7.1.3 Application: Data Compression

We observe, from the foregoing example, that there are more space-efficient BCD representations for numerical (decimal) data than packed strings (Code 2); in fact, Code 5 is 15% shorter, on the average. For large amounts of data, this may be very significant. We do not, however, recommend Code 5 for general use. One reason is that computers are not designed to handle sequences of variable-length bit strings very efficiently, and use of Code 5 may prove impractical as a consequence. However, this and other sophisticated encodings may become attractive as the cost of computer processing time continues to decrease.

'Data compression' refers to the storage of data in as little space as possible. With few exceptions, both numerical and nonnumerical data are presented to and obtained from a computer in character string form, internally represented in the computer as an extended BCD string. Alphanumerical strings require at least a 6-bit code, and a 7-bit code is required to handle both upper- and lower-case letters as well as the usual special characters. However, numerical strings need only a 4-bit code, and (upper-case) alphabetic strings need only a 5-bit code, so use of a 7- or 8-bit code for them would waste 40–50% of their allotted space. Unfortunately, computer hardware is ordinarily designed with just general alphanumeric strings in mind; 'byte-oriented' instructions make their processing time reasonably efficient. Special instructions to handle 5-bit alphabetic strings, or in some cases 4-bit numerical strings, are not ordinarily provided. Some computers, at additional cost, have 'decimal arithmetic' instructions which make processing of 4-bit 'packed decimal' numerical strings feasible; otherwise, or simply for greater efficiency, a numerical [BCD] *string* should be converted to a binary *number* [in, say, 2's complement] for computation. In any event, when hardware facilities are not available, it is possible to use software facilities instead.

Character codes need not employ a fixed number of bits. If some characters are much more likely to occur than others, and the various occurrence probabilities are known, then the more likely characters may be given shorter codes while the less likely characters may be given longer codes. For example, Code 5 of Table 7.1.1 would not be efficient if the higher digits were more likely than the lower digits. In Chapter 9, we show how the optimal code for any given set of probabilities can be found.

When data items are known to have a small predeterminable set of possible values, space can be saved by using abbreviations. For example, storage of M and F or HI and ME, for MALE and FEMALE or HAWAII and MAINE, respectively, can save much space. While abbreviations can result in a significant reduction in space, they can lead to mistakes too if the user must supply or interpret them; e.g., the user may erroneously supply MI for MISSOURI on input, or misinterpret MO as MONTANA on output. To prevent such mistakes, the full words may be used for input and output, while their abbreviations may be used internally as part of the stored data; any referencing program would have to perform the required translations. If the internal form for data need not coincide with the external form, then other translations, e.g., into numerical codes, may just as well be used. For example, M and F may be represented by a single bit, and any of the 50 American states may be represented by a 6-bit integer.

The problems of data compression are much larger, both qualitatively and quantitatively, than can be seen from the foregoing examples. (See Held [7.14].) We shall leave the subject, however, but with one final reminder: namely, that usually (but not always) storage space can be saved only at the expense of processing time.

7.1.4 Parity Checking

Information should not necessarily be stored in computers in the most efficient fashion. Of great importance is that errors in data be detectable, and ideally correctable, if at all possible. This objective generally requires that some additional information be coded along with the data to provide some 'redundancy'. This redundant information can range from a complete second copy of each item of data, to just a single 'parity' digit. Some version of the latter is generally advisable to protect against 'typographical' errors, at the least.

The 'parity' concept has long been used by computer designers to guard against hardware failures. Whenever a binary source word (or a binary encoding of a source character) is processed, an additional bit, called a *parity bit,* is appended with value dependent upon the number of nonzero bits in the source word; for *even* (*odd*) parity systems, the parity bit is set so that the number of 'one' bits in the extended word is even (odd). (Code 6 of Table 7.1.1 is an error-detecting code, where the rightmost bit is chosen for even parity.) The setting of parity bits and the checking of parity of extended words are built into all computer systems; if the parity of an extended word is ever found to be wrong (i.e., opposite to that expected), a hardware *fault* occurs, and some system-dependent action takes place. We remark that parity errors most commonly occur during I/O operations (especially during the reading of characters from dirty or worn magnetic tapes, or during telecommunications).

The use of a parity bit for error detection involves coding m-bit binary strings (source characters) by $(m+1)$-bit binary strings (code words). We formalize the procedure as follows. Let $B=\{0,1\}$, and let B^n be the set of n-bit binary strings (i.e., Boolean n-vectors). An even parity coding of Boolean m-vectors $a=(a_1,\ldots,a_m)$ is a function $\mathbf{C}$: $B^m \to B^{m+1}$ such that $\mathbf{C}(a_1,\ldots,a_m)=(a_1,\ldots,a_m,a_{m+1})$, where $a_{m+1}=\mathbf{wgt}(a)\ \mathbf{mod}\ 2$.[†] $\mathbf{C}$ is 'partially invertible' in the sense that the *decoding* function $\mathbf{C}^{-1}$: $\mathcal{R}_\mathbf{C} \to B^m$ is well defined [$\mathbf{C}^{-1}(b)=a$ is such that $\mathbf{C}(a)=b$], but $\mathbf{C}^{-1}(b)$ is undefined for b not in $\mathcal{R}_\mathbf{C}$, i.e., for $b \in B^{m+1}$ having odd weight.

In summary, the set of code words, $\mathcal{R}_\mathbf{C}$, consists *only* of those members of B^{m+1} having even weight. Hence the code is error-detecting in that no member of B^{m+1} having odd weight can be the encoding of an m-bit source string. Given $b \in B^{m+1}$ not in the code, we conclude that at least one single-digit substitution (0 for 1, or 1 for 0) error must have occurred to transform a code word $b' \in \mathcal{R}_\mathbf{C}$ to b. However, we cannot tell which digit is erroneous. Furthermore, given $b \in \mathcal{R}_\mathbf{C}$, we cannot conclude that there was no error; two or any even number of errors may have occurred. The even parity coding scheme is therefore called a single-error error-detecting code. So is the analogous odd parity code.

Check Characters In addition to associating parity or 'check' bits with each source word or character (also called 'vertical' checking), we may associate a *check character* with a set or sequence of source characters (called 'longitudinal' checking). To be specific, suppose our data consists of the sequence of digits '235' encoded in BCD as the sequence 0010, 0011, 0101. Appending an even-parity bit for vertical checking, we have the sequence 00101, 00110, 01010. To this sequence, we append the longitudinal check character 01001, which is chosen so that the sets of corresponding (first to fifth) bits—i.e., the columns of the array

$$\begin{bmatrix} 00101 \\ 00110 \\ 01010 \\ 01001 \end{bmatrix}$$

—have even parity. '235' is therefore coded as the extended sequence of digits '2354'.

Suppose our data consists of sequences of letters, e.g., the string

[†]Recall that the *weight* of a in B^n is the number of coordinates in a which are 1, and the Hamming distance $\mathbf{d}(a,b)$ between a and b in B^n is the number of coordinates where a and b differ. Thus, $\mathbf{wgt}(a)=\mathbf{d}(a,0)$.

'ALPHABET'. If this string is binary coded (say, using EBCDIC), a longitudinal check character may be defined in the same fashion as above. However, let us instead associate with each letter a number equal to its ordinate position in the alphabet, A=1, B=2,..., Z=26, and associate with each string the sum of the numbers associated with each of its letters: i.e., with 'ALPHABET' we associate the number

$$1+12+16+8+1+2+5+20=65.$$

Now let us take this number, modulo 26, and add one; the check character associated with the string is then defined as the letter associated with the resulting number; since

$$(65 \ \mathbf{mod}\ 26)+1=13+1=14,$$

the check character is 'N'. Finally, we append this character to the string, obtaining 'ALPHABETN'. It can easily be shown that if any single character of an extended string (including the check character) is misspelled, i.e., if another letter is substituted for it, then this misspelling can be detected by verifying the check character. More sophisticated 'redundancy' schemes are necessary to detect (or automatically correct) other common kinds of errors, such as multiple misspellings, omission of letters, insertion of stray letters, and transposition of adjacent letters. For example, instead of the simple sum above, we may compute a weighted sum or perhaps a Gödel number equivalent, or use an algebraic (group-theoretic) approach, in the way we discuss below. We emphasize that the use of check characters is suitable not only for hardware (modern I/O devices do in fact utilize the concept) but also for software (to check sets of numerical or nonnumerical data for, say, typographical errors in their entry).

7.1.5 Application: Error Detection and Correction

We now address two questions. Firstly, can we extend the foregoing to detect more than one substitution error? Secondly, can we correct an error in the sense of determining the member $b' \in \mathcal{R}_\mathbf{C}$ which differs from $b \notin \mathcal{R}_\mathbf{C}$ by such an error? Both questions can be answered in the affirmative, provided enough $(r>1)$ 'parity' or 'check' bits are used.

Let $\mathbf{C}: B^m \rightarrow B^{m+r}$, $\mathcal{R}_\mathbf{C} \subseteq B^{m+r}$ and $\mathbf{C}^{-1}: \mathcal{R}_\mathbf{C} \rightarrow B^m$. The minimal distance of such a code is defined as the minimum Hamming distance between the distinct code words it contains. If the Hamming distance between a and b in B^{m+r} equals k, then it takes at least k substitutions to transform a to b. Hence, if the minimum Hamming distance between distinct code words equals $k+1$, it would take at least $k+1$ substitutions to transform one to another. We conclude that k or fewer substitutions must necessarily transform any code word $a \in \mathcal{R}_\mathbf{C}$ to an erroneous word $b \notin \mathcal{R}_\mathbf{C}$, and that an equal number of substitutions are required to transform b back to a. However, only one substitution may be required to transform b to another $a' \in \mathscr{R}_\mathbf{C}$ unless $\mathbf{d}(a,a') > 2k$. In other words:

> A code having minimal distance $k_\mathbf{d}+1$ can *detect* $k_\mathbf{d}$ (or fewer) substitution errors. Furthermore, a code having minimal distance $2k_\mathbf{c}+1$ can *correct* $k_\mathbf{c}$ (or fewer) errors.

We note that the single parity bit code ($r=1$) has a minimal distance of 2; hence it can detect single errors and correct none. To be a double-error detecting [$k_\mathbf{d}=2$] and single-error correcting [$k_\mathbf{c}=1$] code, $\mathscr{R}_\mathbf{C} \subseteq B^{m+r}$ must have minimal distance of [$k_\mathbf{d}+1=2k_\mathbf{c}+1=$] 3, hence the number of check digits r must be greater than one. How much greater depends upon m. The problem then is to choose these check digits so that $\mathscr{R}_\mathbf{C}$ has minimal distance of three, and so that the detection and correction processes are not too complex. For the latter purpose, $\mathscr{R}_\mathbf{C}$ will be constructed to satisfy subgroup properties. We shall define $\mathscr{R}_\mathbf{C}$ in terms of a 'parity check matrix'.

Parity Check Matrices Given m, we wish to construct a Boolean $((m+r)\times r)$ matrix [each of whose elements are in $B=\{0,1\}$] having no row of zeroes and no pair of identical rows, and such that its bottom r rows is the $(r\times r)$ identity matrix. A matrix having these properties can have at most 2^r-1 distinct rows (i.e., $m+r \leq 2^r-1$); this requirement imposes a lower bound on r for a given m. For this smallest r, let $n=m+r$. The rows of a corresponding smallest matrix H, interpreted as binary numbers, must be between 1 and n. We call a matrix H satisfying the foregoing properties a *parity check* matrix. For example, $m=3$ implies $r\geq 3$, so H is a (6×3) matrix, such as

$$\hat{H}=\begin{bmatrix} 0 & 1 & 1 \\ 1 & 0 & 1 \\ 1 & 1 & 0 \\ 1 & 0 & 0 \\ 0 & 1 & 0 \\ 0 & 0 & 1 \end{bmatrix}.$$

For any $((m+r)\times r)$ parity check matrix H, let $C_H=\{b\in B^n \mid b \circledast H=\theta\}$, where θ is an r-vector of zeroes. Here, an n-vector is regarded as a $(1\times n)$ matrix, and $\circledast$ is the Boolean matrix product, hence $b\circledast H\in B^r$. For $H=\hat{H}$,

$$C_H=\hat{C}=\begin{bmatrix} (0,0,0,0,0,0) \\ (0,0,1,1,1,0) \\ (0,1,0,1,0,1) \\ (0,1,1,0,1,1) \\ (1,0,0,0,1,1) \\ (1,0,1,1,0,1) \\ (1,1,0,1,1,0) \\ (1,1,1,0,0,0) \end{bmatrix}.$$

An important property of $\hat{C}$ is that it has minimal distance of at least 3, as was shown in Section 2.5.2. This property also holds for the C_H associated with any parity check matrix H (as defined above). Therefore, any such C_H can be used for double-error detection and single-error correction. Furthermore, $(C_H, \oplus, \theta)$ is a subgroup of $(B^n, \oplus, \theta)$.

The subgroup property is the key to simplifying error detection and correction. By definition, $bC_H = \{b \oplus a \mid a \in C_H\}$ is a left coset of C_H in B^n. Furthermore, $\#(C_H) = 2^m$, hence (by the group decomposition theorem and Lagrange's theorem) the left cosets partition B^n into 2^r equivalence classes; since there are 2^r different Boolean r-tuples, we may uniquely associate each equivalence class with an r-tuple $\rho \in B^r$. Finally, since b and b' are in the same left coset (of C_H in B^n) iff $b \circledast H = b' \circledast H$,[†] the set $\{b \in B^n \mid b \circledast H = \rho\}$ defines the equivalence class associated with ρ; we denote this set **coset** $[\rho]$ for $\rho \in B^r$. Observe that **coset**$[\theta] = C_H$.

In Section 2.5.2, we also observed that there are 8 distinct cosets $b\hat{C}$. For b equal to each of the 8 leaders given in that section, $b \circledast \hat{H}$ equals 000, 011, 101, 110, 100, 010, 001, and 111, respectively. We may use these 3-bit numbers to order the leaders, defining an array L of leaders as follows:

$L(0) = 000000$
$L(1) = 000001$
$L(2) = 000010$
$L(3) = 100000$
$L(4) = 000100$
$L(5) = 010000$
$L(6) = 001000$
$L(7) = 100100$

Then, given b, we may calculate $b \circledast \hat{H} = \rho$; interpreting ρ as a 3-bit binary number, $L(\rho)$ is the leader of the coset containing b. Note that no coset leader has weight greater than 2. This property, which by theorem also holds for the C_H associated with any parity check matrix H, makes it easy to find the L array for any such C_H. Thus, we can easily determine the leader of $b\hat{C}$ for any $b \in B^6$ using Boolean matrix operations.

Coding and Decoding Let H be an $((m+r) \times r)$ parity check matrix. Recall that r is determined by m, and $n = m+r$. We define $\mathbf{C}$: $B^m \rightarrow B^{m+r}$ such that

$$\mathbf{C}(a_1, \ldots, a_m) = (a_1, \ldots, a_m, a_{m+1}, \ldots, a_{m+r})$$

where the r check digits are given by

$$a_{m+j} = a_1 \times h_{1j} \oplus a_2 \times h_{2j} \oplus \cdots \oplus a_m \times h_{mj}, \qquad j = 1, \ldots, r$$

(where $\oplus$ and $\times$ denote the Boolean sum and product). Finally, we note that

[†]Proof of this is left as an exercise.

$\mathcal{R}_\mathbf{C}$ equals

$$C_H=\{b\in B^{m+r}\,|\,b\circledast H=\theta\};$$

this code was seen earlier to have minimal distance of at least 3, hence it can detect 2 (or fewer) errors and can correct 1 error. Detection is as before: if $b\in B^{m+r}$ is not in C_H, then b differs from a member of C_H by a single or a double error. We may proceed by computing $b\circledast H$, an r-tuple. If $b\circledast H=\theta$, no error is detected and b can be decoded by dropping the r check digits. If $b\circledast H=\rho\neq\theta$, then we determine the equivalence class, **coset**$[\rho]$, associated with ρ. For any member ℓ of **coset**$[\rho]$, $a=b\oplus\ell$ may be used as a 'correction' to b provided it is in C_H. Recalling that $b\circledast H=\rho=\ell\circledast H$ implies b and ℓ are in the same coset

$$bC_H=\{a\oplus b\,|\,a\in C_H\}=\ell C_H=\{a\oplus\ell\,|\,a\in C_H\}$$

[so $b=\theta\oplus b\in\ell C_H$], we conclude that there must exist an $a\in C_H$ such that $a\oplus\ell=b$, or $b\oplus\ell=a$; this $a\in C_H$ is the correction and may be decoded by dropping the r check digits. The remaining problem is to decide which member of **coset**$[\rho]$ to select for ℓ. Since $\mathbf{d}(a,b)=\mathbf{wgt}(a\oplus b)=\mathbf{wgt}(\ell)$, we observe that a and b differ in each non-zero coordinate position of ℓ. Therefore, in order for the correction $a\in C_H$ to be closest to $b\notin C_H$, the member ℓ of **coset**$[\rho]$ which should be selected should have minimal weight. We conclude that ℓ should be the leader of **coset**$[b\circledast H]$. Note that if $b=\ell$, then $a=\theta$. For additional information, see Peterson and Weldon [7.29].

7.2 SORTING AND SEARCHING

Sorting is the process of placing or rearranging data within a structure in a specified manner, usually in lexicographical order of some portion of the data (called its 'key'). *Searching* is the process of finding data items which have been placed in a structure, given their keys. The main reason for sorting is to facilitate searching. In other words, if data is placed in specified locations within a structure, then it is easier to find the data.

There are numerous algorithms for sorting and for searching. Entire books have been devoted to these subjects, including those by Knuth [7.20], Flores [7.10], and Lorin [7.22]. In this section, we discuss only some of the most well-known algorithms. We distinguish between *internal* sorting algorithms, which arrange data in main memory, and *external* sorting algorithms, which arrange data in external storage. For a survey, see Martin [7.24].

7.2.1 Internal Sorting (of Arrays)

The most common internal sorting algorithms rearrange (permute!) the elements of an array A of data values in ascending (or descending) order, if the values are numerical (or in lexicographical order, if the values are character strings). Such algorithms may be classified according to whether or not a second array B of equal size N is used in the sorting process. If no extra storage locations are used (except perhaps a few for 'bookkeeping'†), we say the sorting is done 'in place'. An example of an in-place sorting algorithm is the 'bubble sort' (described below). One 'out-of-place' sorting algorithm proceeds by successively finding the smallest value in a given array A, deleting it from A, and placing it sequentially in another array B; $B(i)$ then contains the ith smallest value in A, for $i=1, ..., N$. (To delete a value from A, it must be 'marked' if its position within the array cannot be physically removed; a special value, say 'infinity', is commonly used as the mark.) These two algorithms have $\mathbf{O}(N^2)$ complexity (see Section 3.1.6) because, in effect, they require N passes through the array A, and make $\mathbf{O}(N)$ comparisons per pass. More efficient algorithms are given in Knuth [7.20].

Example*: *Bubble Sort Bubble-sorting is based on systematically exchanging adjacent elements if they are not in the right order. In other words, we would perform

if $A(j)>A(j+1)$ **then** exchange $A(j)$ with $A(j+1)$

for $j=1,...,N-1$, repeating the process until there are no further exchanges. After the jth pass, the largest element in $\{A(1),...,A(N-j+1)\}$ is 'bubbled' (moved) to the right, i.e., the subarray is permuted so that $A(N-j+1)$ is its largest element.

Example*: *Quicksort 'Quicksorting' is based on recursively partitioning an array into two subarrays, the left subarray containing only elements less than, and the right subarray containing only elements greater than, one of the array's elements. In a sense, elements are 'bubbled' both to the left and to the right depending upon their values; thus, quicksorting is faster than bubble-sorting, in which elements are bubbled in only one direction. For details, see Hoare [7.16].

7.2.2 External Sorting (of Files)

In principle, any internal sorting algorithm can be used to sort files (regarded as arrays) which are stored in external memory. In practice, internal sorting algorithms are not used to sort external files because it is too inefficient to process the file other than sequentially, especially if stored data values must be exchanged. Furthermore, files are ordinarily too large to fit within main memory, so we exclude the possibility of reading an entire file into an array,

†Specifically, extra locations used for 'bookkeeping' may include those for loop controlling variables and a temporary variable used when exchanging two values.

sorting it internally, and then writing out the sorted array. Hence, our objective is to sort an external file by reading sequentially only one portion at a time, in order to create longer and longer *runs* (sorted consecutive subsequences). We say a run is *natural* if it is maximal, i.e., it is not a subrun of a longer run; a run is *artificial* if it has a prescribed length. Of course, in sorting a file, we wish to require as few passes as possible. Most external sorting algorithms are based on the principle of *merging*. See Knuth [7.20] for a definitive treatment.

Example*: *2-way Merge The simplest version of merging requires reading the file once, in order to split or *distribute* its natural runs into two files, A and B, placing the first half in A and the second half in B. Files A and B are then combined by reading files A and B sequentially, one data item at a time, while writing onto a third file C the smaller of the leading data items in files A and B, unless only the larger one will continue a run on file C. When a data item from a file is written, the next data item from that file is read. When the ends of both files A and B are reached, another pass is made after redistributing C into A and B. We remark that this process merges the runs which appear naturally in files A and B to produce longer and longer runs, until one run containing all of the data items is produced. The number of passes required depends upon the initial 'randomness' of the file.

An alternative, called *balanced* (vs. natural) 2-way merging, takes a predictable number of passes (see Exercise 9), each of which merges two runs of length i to produce a run of length $2i$ (where $i=1$ initially). In balanced merging, the artificial run boundaries (determined by counters) control the merge process. In contrast to natural merging, files A and B are combined by merging one run of prescribed length from each, reading sequentially from each file one data item at a time while writing onto file C the smaller of the leading data items in the respective runs, until both runs are exhausted (even if a longer run can be produced by continuing past one of the artificial run boundaries).

The distribution phase required of the balanced 2-way merge is wasted effort in that the file does not become increasingly sorted in the process. (It also requires lengthy rewind or seek operations if the sorting uses magnetic tape or disk files.) Distribution is necessary after each pass because all the runs end up in file C. One means of eliminating the distribution phase is to initially split the file unevenly so that file A contains, say, x runs and file B contains y runs, with $x<y$. Then merging the x runs of A with the first x runs of B would produce x runs in C, leaving 0 runs in A and $y-x$ runs in B. Since A has been completely read, file C and the remainder of B can be merged to produce runs back in A. This process can be repeated until all the runs end up in a single file. If there is only one run in this file, then it must be completely sorted, as desired; but, if not, a distribution is required, as would be the case if $y=2x$ in the above (when merging $y-x=x$ runs in B with x runs in C would produce x runs in A and no runs in B and C). To avoid this situation, the initial distribution of runs cannot be chosen randomly; a good choice is for x and y to be the first two successive numbers in the 'Fibonacci' sequence $\langle 1,1,2,3,5,8,\ldots\rangle$ whose sum equals (or exceeds) the number of runs initially in the file (where it may be necessary to insert some 'dummy' runs during the initial distribution). This is an example of *polyphase* merging.

7.2.3 Linear Search

Given an ordered set $\{a_1, a_2, \ldots, a_n\}$, a *linear search* for a member of the set equal to a given key K is performed by first testing whether $K=a_1$, and if not then testing whether $K=a_2$, and if not then testing whether $K=a_3$, etc.; the order of the search is controlled by varying the value of a subscript I in increasing order from 1 to N. If the value K is in the set, the search terminates with I such that $K=a_I$, and $K\neq a_j$ for $j<I$; the latter condition ensures that if K appears more than once in the set, then a_I is its first occurrence. If K is not in the set, then I is undefined following the search; observe that it takes only one comparison to determine whether $K=a_1$, but N comparisons to determine that $K=a_N$ or that K is not in the set. We would expect the 'average' number of comparisons needed to find a randomly chosen K in the set to be about $(1+N)/2$.

To be precise, let p_i denote the probability that the desired value K is the ith member of the set—i.e., $p_i=\mathbf{Pr}\{K=a_i\}$. Then the expected number of comparisons needed to find K (or to find that K is not in the set) is given by the sum

$$\mathbf{Exp}(I)=\sum_{i=1}^{N} p_i\times i+\left(1-\sum_{i=1}^{N} p_i\right)\times N.$$

The latter term accounts for the probability $(=1-\sum_{i=1}^{N}p_i)$ that K is not in the set. In the special and commonly assumed case that K is in the set, and that K is equally likely to be anywhere within the set, then $p_i=1/N$ for each i, and

$$\mathbf{Exp}(I)=\sum_{i=1}^{N} (1/N)i=(N+1)/2.$$

This is consistent with our intuitive notion that if K is as likely to be in the first half of the set as in the second half, then we need to search about half-way through the set on the average. However, if equilikelihood does not hold, then it should be clear that $\mathbf{Exp}(I)$ is minimal when the values in the set are ordered by descending probability. (This rather obvious result will be formally proved in Chapter 8.)

We remark that a linear search program would take $a(N+1)/2+b$ units of execution time, where a is the time required per comparison, and b is the 'overhead' time required for initiation and termination of the program. The values of a and b are generally small, so linear search is efficient except for large N.

7.2.4 Binary Search

In this section, we analyze the use of binary searching for keys in a sorted

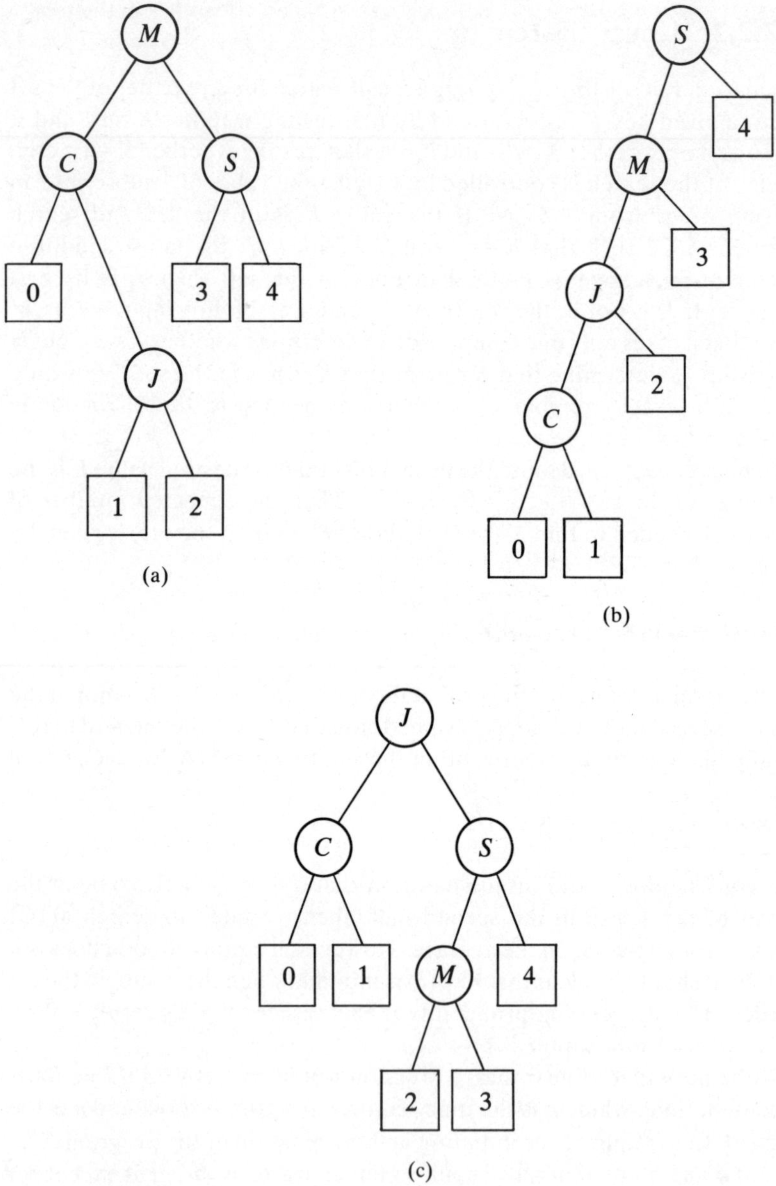

Figure 7.2.4

binary tree, e.g., in the one shown in Figure 7.2.4(a). The square nodes represent null leaves, as discussed in Section 4.2. In Section 5.3.3, we showed that it takes **depth**$(x)+1$ comparisons to find data node x in the tree, and **depth**(y) comparisons to find that the desired key is not in the tree, but is instead in the range represented by null leaf y. Suppose now that we associate probabilities with each node. Let $p(x)$ denote the probability that a data node x in the tree is the desired one, and let $q(y)$ denote the probability that the desired key is in the range represented by a square node y. Then the *expected number of comparisons to search the tree for an arbitrarily given key* (we call this the *cost* of the tree, for brevity) is

$$\hat{C}=\Sigma_x[\mathbf{depth}(x)+1]\cdot p(x)+\Sigma_y \mathbf{depth}(y)\cdot q(y)$$

where the first summation is over all (round) data nodes, and the second summation is over all (square) null leaves. Note that $\Sigma_x p(x)+\Sigma_y q(y)=1$. Furthermore, if 'missing keys' are not possible (i.e., $q(y)=0$ for each y), then the second term vanishes.

Example 1 Consider the sorted binary tree shown in Figure 7.2.4(a) (cf. Figure 4.2(c)), where the square nodes have been numbered. The cost of this tree is

$$\begin{aligned}\hat{C}&=\Sigma_{x\in\{C,J,M,S\}}[\mathbf{depth}(x)+1]\cdot p(x)+\Sigma_{y\in\{0,1,2,3,4\}}\mathbf{depth}(y)\cdot q(y)\\&=2p(C)+3p(J)+p(M)+2p(S)+2q(0)+3q(1)+3q(2)+2q(3)+2q(4).\end{aligned}$$

If the desired key must be one of 26 equally likely letters, then

$p(C)=p(J)=p(M)=p(S)=1/26$,
$q(0)$ $[=p(A)+p(B)]=2/26$,
$q(1)$ $[=p(D)+p(E)+\ \dots\ +p(I)]=6/26$,
$q(2)=2/26$, $q(3)=5/26$, $q(4)=7/26$,
and $\hat{C}=60/26=2.31$.

For the simpler equilikelihood assumption where the data nodes are equiprobable and missing keys are not possible,

$$p(C)=p(J)=p(M)=p(S)=1/4,\ \ q(0)=q(1)=q(2)=q(3)=q(4)=0,$$

and

$$\hat{C}=8/4=2.$$

Example 2 Consider the sorted binary tree shown in Figure 7.2.4(b). For the two equilikelihood assumptions made above, $\hat{C}$ for this second tree equals $65/26=2.5$ and $10/4=2.5$, respectively; in both cases, $\hat{C}$ is larger for the second tree. However, for the tree shown in Figure 7.2.4(c), the 'equally likely letters' assumption yields $\hat{C}=59/26=2.27$, which is smaller than that for the first tree. The reason for this is that, in the third tree, a more probable square node ($y=1$, as opposed to $y=3$) appears higher where fewer comparisons are needed.

In other words, a good heuristic is to place high probability nodes high in the tree. We are constrained in doing so, of course, by the requirement that the tree be sorted. Furthermore, in doing so, the tree may become very off-balanced, which increases the height of the tree; there will then be more nodes at greater depths, which if not too improbable will add significantly to the cost of the tree. This is the case for the tree in Figure 7.2.4(b), where the most probable square node is highest in the tree but the next most probable square node is lowest. We show how to find the least-cost tree for arbitrarily given probabilities in Section 9.1.2. When nodes are equiprobable, there is no advantage in having an off-balanced tree. In fact, for the equiprobable node assumption, the least-cost tree is depth-balanced.

Recall from Section 4.2.6 that the height H of a depth-balanced binary tree having N nodes is **int** ($\mathbf{log}_2(N)$). In such a tree, there is one node with depth 0, two nodes with depth 1, four nodes with depth 2,..., 2^d nodes with depth d for $d<H$, and $N-(2^H-1)$ nodes with depth H. Note that (2^H-1) is the number of nodes in a complete tree of height $H-1$, which equals the number of nodes at depth less than H; i.e., $\Sigma_{d=0}^{H-1} 2^d=2^H-1$. Hence the cost of a depth-balanced binary tree having N equiprobable data nodes (and impossible missing keys) is given by

$$\hat{C}=\sum_{d=0}^{H-1}(d+1)\cdot\frac{2^d}{N}+(H+1)\cdot\frac{N-2^H+1}{N},$$

where $H=\mathbf{int}(\mathbf{log}_2 N)$. Using the identity (from Section 1.3.3)

$$\Sigma_{i=1}^{k} i\cdot 2^i=2+2^k(2k-2),$$

we obtain

$$\hat{C}=\left(\frac{N+1}{N}\right)(H+1)-\left(\frac{2^{H+1}-1}{N}\right).$$

Note that when the tree is complete, $\mathbf{log}_2(N+1)=H+1$, so that

$$\hat{C}=\left(\frac{N+1}{N}\right)\mathbf{log}_2(N+1)-1.$$

We conclude that for large N the expected number of comparisons required to binary search a balanced tree of N data items is approximately $\mathbf{log}_2 N$. (This should be contrasted with the expected number of comparisons for linear searching a set of N data items, $(N+1)/2$, as derived in the preceding section.) For additional information, see Nievergelt [7.27] and Comer [7.8].

7.2.5 Hashing (with Chaining)

When N data items are stored in a hash table (see Section 3.2.7), the expected number of comparisons required to access a data item depends not just upon N, but also upon the amount of space S allocated for the table. The dependence is indirect in that what really matters is the likelihood of ‘collisions' (relatively small if S is much greater than N) and the placement of 'synonyms'. There are S^N possible hash functions which map $[1,N]$ to $[1,S]$, only $S!/(S-N)!$ of which avoid collisions. Placement strategies include: 'linear' (also called 'open addressing') where a colliding data item is placed in the next open location in the table; 'random' where a colliding data item is placed randomly in the table, and a pointer is set to help locate it; and 'chaining' where colliding data items are placed in a separate *overflow* area (usually with synonyms chained together). See Maurer [7.25]. Note that the latter two strategies require extra space, which if available could instead be used to increase S.

The easiest of these strategies to analyze is hashing with chaining. Consider any location in the table. The probability that an arbitrary data item will map to this location is $1/S$ assuming equilikelihood. The probability that k out of the N data items will map to this location—i.e., the probability that an arbitrary chain will have length k—is given by the binomial distribution (see Section 6.2.5):

$$\mathbf{b}(k,N,p)=\binom{N}{k}p^k(1-p)^{N-k}$$

where $p=1/S$. The expected length of a chain is then

$$\hat{L}=\Sigma_{k=0}^{N}k\mathbf{b}(k,N,1/S)$$

which (by the property of the binomial distribution) equals N/S. (For example, if $N=S$, then $\hat{L}=1$, since chains of length greater than one are offset by chains of length zero for the table as a whole.) Note that the expected number of comparisons required to access an arbitrary data item in the table, denoted $\hat{C}$, is not just that required to (linear) search an average-length chain; i.e., $\hat{C}\neq(\hat{L}+1)/2$. (This should be clear from the fact that $\hat{L}<1$ for $N<S$, but $\hat{C}>1$ if any data item overflows.)

To calculate $\hat{C}$, we must consider chains of each possible length $k=0,1,2,\ldots,N$. The *total* number of comparisons required to find each data item on a chain of length k equals $0+1+2+\cdots+k=k(k+1)/2$; the probability of a length k chain is $\mathbf{b}(k,N,1/S)$. Hence, the *expected total* number of comparisons per chain, averaged over each possible chain length, is

$$\Sigma_{k=0}^{N}(k(k+1)/2)\mathbf{b}(k,N,1/S).$$

Again, this is the expected total number of comparisons, summed over all the data items in an arbitrary chain; the expected number of comparisons *per* data item is this expected total divided by the expected length of the chain. In other words,

$$\hat{C}=\frac{\Sigma_{k=0}^{N}\frac{1}{2}k(k+1)\mathbf{b}(k,N,1/S)}{\Sigma_{k=0}^{N}(k)\mathbf{b}(k,N,1/S)}=\frac{1}{2}\left(\frac{\Sigma_{k=0}^{N}k^2\mathbf{b}(k,N,1/S)}{\Sigma_{k=0}^{N}k\mathbf{b}(k,N,1/S)}+1\right)$$

$$=\tfrac{1}{2}(1-1/S+N/S+1)=1-\tfrac{1}{2}(1/S)+\tfrac{1}{2}\,(N/S)$$

which for large S is approximately equal to $1+(N/S)/2$. For a *loading factor*, defined as the ratio N/S, in the range from 50% to 90%, $\hat{C}$ ranges from 1.25 to 1.45. Thus, except for a heavily loaded hash table, usually only one or two comparisons are required to access an arbitrary data item when chaining is used.

Analyses for other placement strategies, such as linear and random, are much more difficult since placement of a colliding data item from one location may cause further collisions elsewhere. It has been shown that

$$\hat{C}\ (\text{linear}) \approx 1+\frac{1}{2}\left(\frac{N/S}{1-N/S}\right)$$

and that

$$\hat{C}\ (\text{random}) \approx -\frac{1}{N/S}\mathbf{log}_e(1-N/S);$$

see Knuth [7.20]. For N/S in the same 50–90% range, hashing performance is not quite as good as before, but is still much better than non-hashing methods. Recall our assumption that hash functions map keys to locations in an equally-likely fashion; random number generating algorithms can be used to obtain such uniform distributions (see Knott [7.18]).

7.3 DATA MODELS

7.3.1 Data Flow

The flowgraphs defined in Section 5.1.1 have nodes corresponding to instructions and branches corresponding to transfers of control, but do not incorporate information about data references. If an instruction or processing node I uses the value D_1 to compute $D_2=\mathbf{f}_I(D_1)$, we say D_1 is an input datum and D_2 is an output datum of I. Let each separate datum be represented by a node (drawn square to distinguish data nodes from process nodes), and let there be a branch from/to a process node I to/from the datum node D if D is an output/input datum of I. We say D is in the *domain* (is an input) of I [also denoted $D \in \mathbf{domain}(I)$] if I reads (fetches) the value of D, and D is in the *range* (is an output) of I [also denoted $D \in \mathbf{range}(I)$] if I writes (stores) the value of D. For example, in Figure 7.3.1, process node I_1 reads the value of D_1 and writes the value of D_2, and the successor I_2 of I_1 reads the values of D_1 and D_2 and writes the values of D_2 and D_3. We call such an augmented graph the *I–D* flowgraph of a program.

A few important conclusions can be drawn from a simple analysis of the *I–D* flowgraph. First, data nodes with zero in-degree can never be written (so

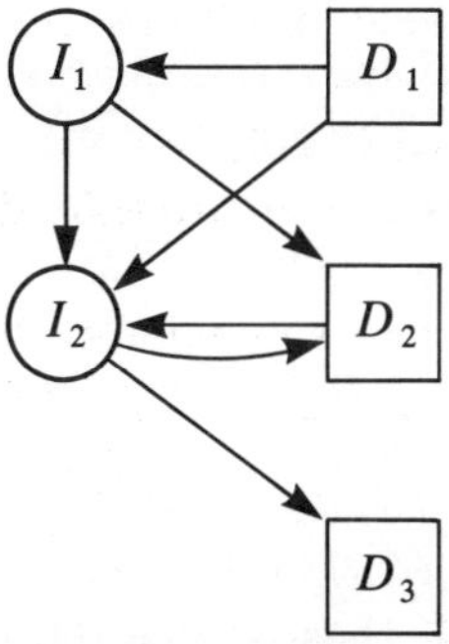

Figure 7.3.1

they must be inputs of the program as a whole) and data nodes with zero out-degree can never be read (so they must be outputs). Furthermore, two data nodes are 'dependent' if there is a path from one to the other in the *I–D* flowgraph. For example, in Figure 7.3.1, D_2 depends upon D_1 and itself, while D_3 depends upon D_1 and D_2; D_1 must be an input node, D_3 must be an output node, and D_2 is an intermediate node (which serves as both input and output). This kind of information can be beneficial in determining how data should be structured and stored. When employing a sequential file structure, for example, data should be stored as much as possible according to their dependence relations. Common outputs of a process node should also be close together. Of course, real processes are usually so complex that many compromises become necessary, but the foregoing still provides the basis for useful heuristic policies.

Of related interest is the case where the process nodes are associated with 'independent' programs (i.e., programs whose order of execution is arbitrary). If these programs do not share the same data, then they can be executed in any order; but if D is an output of both, then the final value of D depends upon the order of execution. See Bernstein [7.3]. We discuss this further in Section 9.2.1.

7.3.2 Information Retrieval

The general problem of 'information retrieval' is that of finding a piece of information given some description of it. Linear and binary search are examples of simple retrieval techniques. We discuss here the problem of retrieving information from a graph-structured set of data or 'data base'.

To retrieve a data item given a key, any technique which completely traverses the graph can be used (e.g., depth-first or breadth-first search). Of course, some techniques may be more efficient than others, at least for some classes of graphs, but this will not be our concern here. Instead, we describe just two of the many graph-related retrieval problems as a brief introduction to the myriad of applications. The problems discussed here were chosen for their simplicity and as illustrations of some of the theory of graphs. For a variety of other concepts, see Wiederhold [7.35] and Heaps [7.13].

Dominating Sets Suppose each data item (or 'document') in a data base can have several keys (or 'keywords'). We wish to find a minimal set of keys such that all the documents can still be retrieved. For example, if document A has the keys {LANGUAGE,SOFTWARE}, and document B has the keys {SOFTWARE,STATISTICS}, then we can retrieve both documents using the single key SOFTWARE. We model our data base by a (undirected) graph whose nodes are the documents, and whose branches indicate one or more common keys: i.e., (D_i,D_j) is a branch of the graph, with label w, if documents D_i and D_j share a key w. We observe that the minimal dominating sets of nodes in the graph are of significance. Let L_S be the set of labels of branches incident at a minimal dominating set S together with the labels of any isolated nodes in S. The labels in L_S are sufficient to retrieve all the documents. We wish to find the smallest L_S among all minimal dominating sets S.

Example In Figure 7.3.2, the minimal dominating sets are

$$S_1=\{D_1,D_2,D_3\},\ S_2=\{D_1,D_4\},\ S_3=\{D_2,D_5\},\ S_4=\{D_3,D_5\},\ S_5=\{D_4,D_5\};$$

the sets of labels are

$$L_{S_1}=\{w_1,w_2,w_3,w_4,w_5\},\qquad L_{S_4}=\{w_1,w_3,w_4,w_5\},$$
$$L_{S_2}=\{w_1,w_2,w_3\},\qquad L_{S_5}=\{w_1,w_2,w_3,w_4,w_5\}.$$
$$L_{S_3}=\{w_1,w_2,w_3,w_4,w_5\},$$

Therefore, the keys $\{w_1,w_2,w_3\}$ are sufficient to retrieve all five documents, although only $\{w_1,w_3\}$ are necessary. (We leave as an exercise the problem of finding a necessary subset.)

Graph Matching Given a data base which is structured as a graph, there are many retrieval problems which require finding within the data base a 'matching' node or set of nodes. We say a digraph $G_1=(S_1, R_1)$ matches a subgraph $G_2=(S_2,R_2)$ of another digraph $G=(S,R)$ if G_1 and G_2 are isomorphic. In some applications, we require that the labels of S_1 and S_2 and/or those of R_1 and R_2 also match. In principle, this matching problem can be solved by comparing G_1 (or its matrix representation) with all possible subgraphs (of G) and their permutations which have equal size. In practice, some means of reducing the number of comparisons is necessary, as suggested in Section 5.2.4. See Salton [7.31] for a discussion of Sussenguth's algorithm for so doing, as well as other concepts.

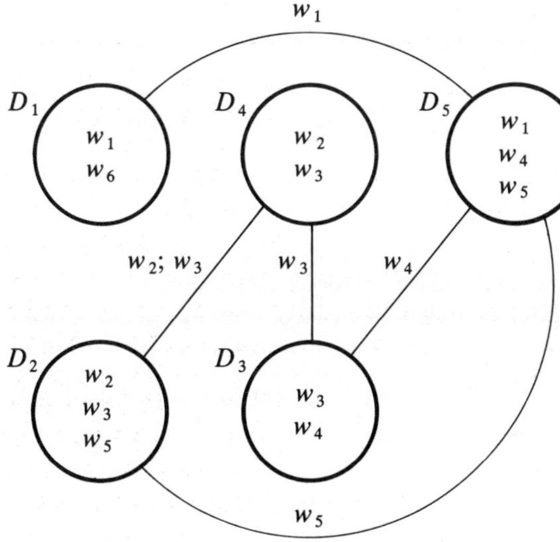

Figure 7.3.2

7.3.3 Data-base Systems

We view a *data base,* informally, as a set of data items organized in a fashion which facilitates computer processing and management (storage, retrieval, modification, security). We say that the data items are *entities* having certain properties, called *attributes,* whose values serve to identify the entities. If an entity is uniquely identified by an attribute or a set of attributes, the attribute set is called the entity's *key*. An *entity set* is a collection of entities which are grouped together for some generally application-dependent reason. Let E_1, E_2, ..., E_n be sets of entities. A *relationship* among these entity sets is a relation $R \subseteq E_1 \times E_2 \times \cdots \times E_n$; the constituent entity sets need not be unique (i.e., E_i may be identical to E_j). One very useful model for data-base systems is the algebraic system consisting of *sets* of relationships, together with various operations on these sets.

Relational Algebra A *relational algebra* is an algebraic system $(\mathscr{R}, \cup, -, \times, \Pi, \sigma)$, where $\mathscr{R}$ is a set of n-ary relations (not necessary all of the same n), $\cup$ is the union operation, $-$ is the set difference operation, $\times$ is the Cartesian product, Π is a *projection* operation, and σ is a *selection* operation. We say two n-ary relations are *like* if they have the same n. The union and set difference operations are defined only between like relations. The Cartesian product operation takes an n_1-ary and n_2-ary relation and produces an (n_1+n_2)-ary relation. The projection operation Π_β takes an n-ary relation

$R = s_1 \times s_2 \times \cdots \times s_n$ and produces an m-ary $(m<n)$ relation $R' = s_{i_1} \times \cdots \times s_{i_m}$, where $\beta = (i_1, \ldots, i_m)$ is a subset of $\{1, 2, \ldots, n\}$: if $(a_1, a_2, \ldots, a_n) \in R$, then $(a_{i_1}, \ldots, a_{i_m}) \in R'$. The selection operation $\sigma_{\mathbf{Q}}$, for a *query* **Q** of the form '$a_i = K$', produces that subset of R whose members have components satisfying **Q**. For further information, see Codd [7.6].

Entity-relationship Diagram An '*E–R*' diagram is a graph having three kinds of nodes:

1. entity-set nodes (drawn as rectangles);
2. attribute nodes (drawn as circles);
3. relationship nodes (drawn as diamonds).

An undirected branch links each entity set to its attributes, and each relationship to its constituent entity sets. *E–R* diagrams have been shown to be a very useful and general data-base model (see Chen [7.5]). In a *relational* data-base system, the entity sets and relationships are represented by relations. In a *network* data-base system, relationships are restricted to be binary, hence digraph algorithms are applicable. In a *hierarchical* data-base system, networks are restricted in addition to be forests. Practical data-base systems have been developed using each of these three models. See Ullman [7.34] and Date [7.9] for further discussion.

7.3.4 Buffering

A *buffer* is a data area which is alternately 'filled' and 'emptied' by *Producer* and *Consumer* processes, respectively. In input buffering, a Producer reads data into a buffer which a Consumer then processes. In output buffering, a Producer generates (computes) data, and a Consumer then writes the computed data. In 'message' buffering, a Producer sends while a Consumer receives. In each case, a Producer cannot fill a buffer which has not been emptied, and a Consumer cannot empty a buffer which has not been filled. To ensure this, the status of each buffer area is given by its buffer flag, which alternates between being EMPTY and FILLED, and each Producer and Consumer must test and set these flags appropriately.

Let us define the state of a multiple-buffering system as the K-tuple of buffer-flag values $\langle BF_1, BF_2, \ldots, BF_K \rangle$, together with pointers to the buffer areas to be referenced next by a Producer or a Consumer. The Producer and Consumer are to use the buffer cyclically (i.e., as a circular list). We will use an underline under BF_i to indicate that the Consumer is to next use (and make empty) the ith buffer, and we will use an overbar over BF_j to indicate that the Producer is to next fill the jth buffer. (The values of i and j serve as 'next-pointers' in the foregoing; for a two-buffer system, i and j will alternate between the two.) For example, $\langle \mathbf{F},\bar{\mathbf{E}},\mathbf{E},\mathbf{E},\underline{\mathbf{F}} \rangle$ represents the state of a 5-buffer system where the first and fifth buffers are **F**(illed) and the other three are **E**(mpty); the second buffer is the next to be filled by the Producer, and the fifth is the next to be emptied by the Consumer. A complete state transition diagram for a two-buffer system is shown in Figure 7.3.4; from any state, the Producer (when filling a buffer) causes a transition to the right, and the Consumer (when emptying a buffer) causes a transition downwards. Other state transition models appear in Hellerman [7.15], Knuth [7.19], and Hsiao [7.17].

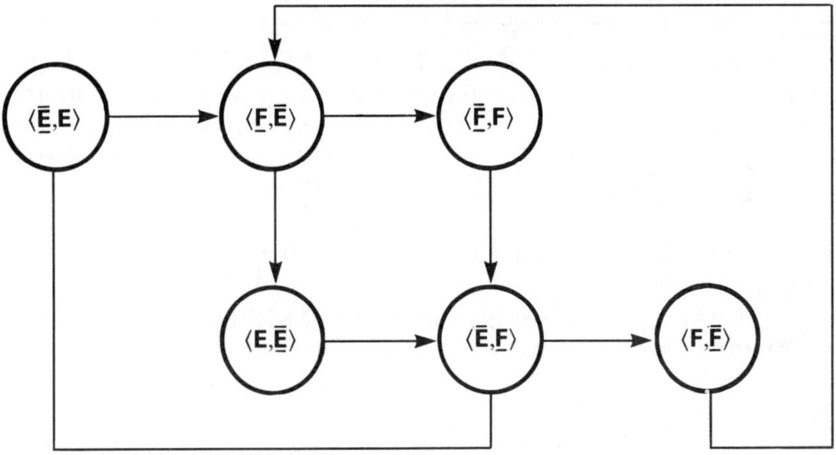

Figure 7.3.4

Buffer Queues One way of looking at buffering is in terms of waiting-line queues (cf. Section 6.4). If a server (e.g., the System or a user program) is not immediately ready to process or consume a block of data (produced by the user for the System, or by the System for the user), then this block of data is placed in a buffer, so that the System or user, respectively, may continue to produce more. These buffers are customers that wait in a queue for service (or consumption). Two important parameters of this model are the sizes of the buffer areas and the number of buffer areas. If the amount of data produced at one time is of fixed size, then this size should be selected for the buffer size. We assume here that buffer sizes are fixed at one unit, but that a variable number of buffer areas may be needed at any one time.

To be more specific, suppose that the Producer generates unit data blocks at intervals that are independent and exponentially distributed with parameter λ, and that when all J buffer areas are filled (i.e., when the buffer queue is full), the Producer waits until one becomes empty. Also suppose that the time required to consume a block of data is exponentially distributed with mean $1/\mu$. Then the mean number of data blocks waiting in the buffer queue is given, from Section 6.4.4, by

$$\hat{L}_w = \frac{\rho}{1-\rho} - 1 - \frac{(1-\rho)+(J+1)\rho^{J+1}}{1-\rho^{J+1}},$$

where $\rho = \lambda/\mu$. The ratio $\hat{L}_w/J$ is the percentage of allocated buffer space that is actually filled on the average, i.e., the 'storage utilization'. Note that, for a stable system ($\rho<1$), storage utilization increases as J decreases; but

$$P_J = \frac{\rho^J(1-\rho)}{1-\rho^{J+1}},$$

which is the probability that the Producer will have to wait (for an empty buffer), also increases, thereby decreasing 'processor utilization' (or 'throughput') defined as $\lambda(1-P_J)$. Ideally, J should be chosen to balance the objectives of storage and processor utilization. For other queueing models of buffering systems, see Coffman and Denning [7.7].

Buffer Size In I/O buffering, buffer size ideally should equal the length of the physical record or block that is read or written. When blocks are of variable length, the buffer size *must* be chosen equal to (or greater than) the maximum block size, otherwise 'excess' information in the block is lost. Buffers should be large in order to accommodate large blocks or blocking factors. However, buffers should be small because they occupy valuable main memory space. Furthermore, for a fixed amount of space, small buffers permit a greater number of individual buffer areas which can be pooled and shared, although at an increased overhead cost.

Timing is also a factor which influences the choice of buffer size. Ideally, when I/O activity and CPU processing overlap, the time required to read or write a buffer equals the time required to process the data. Since increasing blocking factors generally decreases per record I/O times but has no significant effect on per record processing times, increasing block size for I/O-bound programs would lead to greater overlap. (Decreasing block size for CPU-bound programs would not just increase overlap, but also increase total I/O activity, hence large blocks are also advantageous in this case.) One situation where a smaller block may be preferable occurs when an increase in CPU processing time would delay the start of the next I/O operation past a certain point, so that a tape may have to be restarted or a disk may have to complete another revolution. For a further discussion, see Hellerman [7.15] and Peterson [7.28].

It should be clear that the problem of determining the 'optimal' buffer size is very complex, and no general solution is available. However, some simpler subproblems of importance can be solved, using techniques to be discussed in Chapter 8.

EXERCISES

1. Code 5 is not the only nonsingular binary code with a mean word length of 3.4 bits. Find another such code (other than a permutation of Code 5).

2. Determine a set of nonzero probabilities for the source characters such that Code 4 is better than Code 5.

3. Give a nonsingular binary code that has mean word length less than Code 3. (Note that it cannot be uniquely decodeable.)

4. Determine the 'modulo 26+1' check character for the string '**COMPUTER**'.

5. For $\hat{H}$ and $\hat{C}$ as defined in Section 7.1.4,

(a) give an example of a double substitution error, and show that it is detectable;

(b) give an example of a triple substitution error, and show that it is not detectable;

(c) give an example of a single substitution error, and show that it is correctable (give the correction!);

(d) give an example of a double substitution error, and show that it is not correctable.

6. In Section 7.1.4, we added three parity bits to each three bit number to obtain 6-bit code words. Why aren't these three parity bits chosen so as to just *duplicate* the data (i.e., so that the code words are {000000, 001001 ,..., 111111})?

7. Show that b and b' are in the same left coset (of C_H in B^n) iff $b \circledast H = b' \circledast H$.

***8.** Show that quicksorting has complexity less than $\mathbf{O}(N^2)$.

9. Determine the number of passes required to sort a file using a balanced 2-way merge.

10. The time per comparison and other overhead for binary search is greater than that for linear search, hence for very small arrays linear search is better. At what array size does binary search become better than linear search? (Cf. Programming Assignment 8.)

11. For an equally likely letter assumption, determine the optimal sorted binary tree containing the keys $\{K,E,Y,S\}$.

12. Consider hashing with chaining for the case $N=2$, $S=4$. What is the probability that the two data items will not collide? If they don't collide, it takes one comparison to find each; if they do collide, it takes an average of 1.5 comparisons per data item. Use this information to compute $\hat{C}$.

13. Observe that the expected chain length $\hat{L}=N/S$ is generally less than one, since chains of length zero are included. Suppose we discount such chains, and compute

$$\tilde{L}=\Sigma_{k=1}^{N} k \times \mathbf{b}'(k,N,1/S)$$

where $\mathbf{b}'$ is the probability that a chain has length k given that $k>0$. (Note that $\tilde{L}>1$ for $N>1$.) Show that $\tilde{C}$, defined by

$$\tilde{C}=\frac{\Sigma_{k=1}^{N}(k)(k+1)/2 \times \mathbf{b}'(k,N,1/S)}{\Sigma_{k=1} k \times \mathbf{b}'(k,N,1/S)}$$

is equal to $\hat{C}$.

14. Given an *I–D* flowgraph, formalize the conditions under which instructions/program nodes are independent in the sense that their order of execution is arbitrary.

15. Show that the smallest set of labels (L_S) among all dominating sets (S) need not coincide with the smallest dominating set.

***16.** Show how a 'telephone directory' file can be represented in a (a) relational, (b) network, and (c) hierarchical data-base system.

17. Draw a complete state transition diagram for a 3-buffer system (cf. Figure 7.3.4).

PROGRAMMING ASSIGNMENTS

1. Write a program which reads a set of alphabetic strings, and prints each of them augmented by a '**modulo** 26+1' check character.

2. Write a program which implements the double-error-detecting (single-error-correcting) decoding scheme.

3. Determine, using a computer system available to you, the maximum amount of data that can fit on one track of a disk for different block sizes. (What error message results when this limit is exceeded?)

4. Implement quicksort.

5. Implement a polyphase sort.

6. Implement linear search

(a) of an array; (b) of a linear linked list.

7. Implement binary search

(a) of an array; (b) of a binary tree.

8. Use the programs of the foregoing two assignments to determine experimentally the point at which binary search becomes better than linear search. (Cf. Exercise 10.)

9. Implement hashing with chaining.

10. Implement hashing with open addressing.

11. Write a program which simulates the buffer system of Fig. 7.3.4. (At each point in time, a decision to move to the right or downwards should be made randomly.)

12. Implement a 'telephone directory' information retrieval system with facilities for insertions, deletions, updating, sequential printouts, and lookup by name, address, or telephone number.

***13.** Use a commercial data base system to implement a telephone directory system.

8 OPTIMIZATION AND DYNAMIC PROGRAMMING

8.1 MAXIMA AND MINIMA

The *optimum* of a set $S=\{s_1,s_2,\ldots\}$ of objects, which we denote s^* and also call the *optimal* object, is the 'best' one with respect to some *criterion* (or set of criteria, in which case we say that the criterion is multiple- or vector-valued). Note that it is possible, in general, for no single object to be optimal for a given criterion. In this event, we will usually be content to find any optimum. For convenience, we often use the term '*the* optimum' (rather than '*an* optimum') even when it is not unique.

Mathematical optimization techniques, i.e., methods of finding the optimum of S, require that the criterion be a quantitative measure of some property of the objects, or more formally that the criterion be a function $\mathbf{C}:S\to R$, where $(R,\le)$ is a partially ordered set. Then the optimization problem involves finding the least (unique minimal) member of the range of $\mathbf{C}$, $\mathcal{R}_\mathbf{C}=\{\mathbf{C}(s)\,|\,s\in S\}$, and the desired optimum of S is a member of the set

$$S^* = \{s\in S\,|\,\mathbf{C}(s)\le\mathbf{C}(s') \text{ for all } s'\in S\}.$$

We also write $\mathbf{min}_{s\in S}\{\mathbf{C}(s)\}$ to denote the least member c^* in the set $\mathcal{R}_\mathbf{C}=\{\mathbf{C}(s)\,|\,s\in S\}$; $\mathbf{C}(s)$ is called the *minimand*, c^* the *minimum*, and s^* (such that $\mathbf{C}(s^*)=c^*$) a *minimizing* value of S. (When S is a Cartesian product $S_1\times\ldots\times S_n$, we write $\mathbf{min}_{s_1,\ldots,s_n}\{\mathbf{C}(s_1,\ldots,s_n)\}$ where $(s_1,\ldots,s_n)\in S$.) Usually, R is taken to be a set of numbers, so that the least member of $\mathcal{R}_\mathbf{C}$ is the minimum number in the set. However, for some applications, it may be convenient to regard 'best' as 'most', in which case it is more natural to seek the maximum. Note, however, that the maximum number in the set $\{\mathbf{C}(s)\,|\,s\in S\}$ is the negative of the minimum number in the set $\{-\mathbf{C}(s)\,|\,s\in S\}$.

Hence our descriptions of minimization techniques below apply also to maximization analogues.

8.1.1 Set Minima

We shall consider first the problem of determining the minimum number c^* in a finite set $\{c_1,c_2,\ldots,c_N\}$. A solution to this problem requires the use of the **min** operation on a pair of numbers, defined by

$$\mathbf{min}(a,b)=\begin{cases} a & \text{if } a\leq b \\ b & \text{if } a>b. \end{cases}$$

The minimum number in the set $\{c_1,c_2,\ldots,c_N\}$ is then given by the sequence of minimizations

$$c^* = \mathbf{min}(\mathbf{min}(\mathbf{min}(\mathbf{min}(c_1,c_2),c_3),\ldots),c_N). \tag{8.1.1}$$

The pairwise minimizations may be taken in any order.

We remark that a set minimum may be defined for any partially ordered set $(S,\leq)$ as a 'minimal' member of S or (if unique) as the 'least' member of S; these are also called 'local' (or 'relative') minima, or the 'global' (or 'absolute') minimum, respectively. Because, in general, the pairwise minimizations may not all be defined, sequential minimizations as in Eq. (8.1.1) may not be possible if $(S,\leq)$ is not a linear order; however, this generality is rarely needed in practice, and so we assume $(S,\leq)$ is a linear order in what follows, in which case the set minimum is global.

8.1.2 The Minimization Operator

In this section, we exhibit several properties of the **min** operation defined on sets of numbers. First, we have

$$\mathbf{min}_x\{c\cdot \mathbf{f}(x)\}=c\cdot \mathbf{min}_x\{\mathbf{f}(x)\} \tag{P1}$$

for any nonnegative real number c and numerical-valued function **f**. Furthermore,

$$\mathbf{min}_x\ \{\mathbf{f}(x)+\mathbf{g}(x)\}\geq \mathbf{min}_x\ \{\mathbf{f}(x)\}+\mathbf{min}_x\ \{\mathbf{g}(x)\} \tag{P2}$$

for any numerical-valued functions **f** and **g**; equality does not generally hold (i.e., **min** is not additive) since the minimizing values for **f** and **g** need not be identical. On the other hand, if one of the functions (say, **f**) is independent of x, i.e., is a constant c with respect to x, then the equality

$$\mathbf{min}_x \{c+\mathbf{g}(x)\}=c+\mathbf{min}_x \{\mathbf{g}(x)\} \tag{P3}$$

holds. Consider now the simultaneous minimization of a function $\mathbf{F}$ of two variables x and y; note that $\mathbf{min}_y \{\mathbf{F}(x,y)\}$ is a function of x alone. Hence, we have

$$\mathbf{min}_{x,y}\{\mathbf{F}(x,y)\}=\mathbf{min}_x[\mathbf{min}_y\{\mathbf{F}(x,y)\}]=\mathbf{min}_y[\mathbf{min}_x\{\mathbf{F}(x,y)\}]. \tag{P4}$$

If $\mathbf{F}(x,y)=\mathbf{f}(x,y)+\mathbf{g}(x,y)$, then (P4) yields

$$\mathbf{min}_{x,y}\{\mathbf{f}(x,y)+\mathbf{g}(x,y)\}=\mathbf{min}_x[\mathbf{min}_y\{\mathbf{f}(x,y)+\mathbf{g}(x,y)\}].$$

If, in addition, $\mathbf{f}$ is independent of y, then using (P3) we have

$$\mathbf{min}_{x,y}\{\mathbf{f}(x)+\mathbf{g}(x,y)\}=\mathbf{min}_x[\mathbf{f}(x)+\mathbf{min}_y\{\mathbf{g}(x,y)\}]. \tag{P5}$$

(P5) is called the *separability* property of the minimization operator; note that (P4) also can be considered a special case of (P5). With the further assumption that $\mathbf{g}$ is independent of x, (P5) reduces to

$$\mathbf{min}_{x,y}\{\mathbf{f}(x)+\mathbf{g}(y)\}=\mathbf{min}_x\{\mathbf{f}(x)\}+\mathbf{min}_y\{\mathbf{g}(y)\};$$

this should be contrasted with the inequality (P2).

8.1.3 The Inverse Problem

While, for some applications, we may only be interested in knowing the minimum value c^*, it is more common to seek a minimizing value, i.e., an $s \in S$, denoted by s^*, satisfying $\mathbf{C}(s)=c^*$. The problem of determining s^* is equivalent to that of evaluating an 'inverse function' $\mathbf{C}^{-1}:\mathcal{R}_\mathbf{C}\rightarrow S$; i.e., $s^*=\mathbf{C}^{-1}(c^*)$. Unfortunately, if $\mathbf{C}$ is not 1–1, then $\mathbf{C}^{-1}$ is not a function (i.e., not single-valued); this is the case, for example, when $\mathcal{R}_\mathbf{C}$ is finite but S is not.

In the event that $\mathbf{C}$ is not necessarily 1–1, but S is finite, we may proceed as follows. Consider S as an ordered set, $\{s_1,s_2,\ldots,s_N\}$, and $\mathcal{R}_\mathbf{C}$ as the ordered multiset $\{c_1,c_2,\ldots,c_N\}$, where $c_i=\mathbf{C}(s_i)$. Now extend **min** to be an operation on pairs ($\in\mathcal{R}_\mathbf{C}\times[1,N]$) as follows:

$$\mathbf{min}((a,i),\ (b,j))=\begin{cases}(a,i) & \text{if } a\le b\\ (b,j) & \text{if } a>b.\end{cases}$$

Then $s^*=s_{i^*}$, where

$$(c^*,i^*)=\mathbf{min}(\mathbf{min}(\mathbf{min}((c_1,1),(c_2,2)),\ldots),(c_N,N)).$$

Note that according to this definition, if there is no unique minimizing value, then s_i^* is the 'first' minimizing value with respect to the listing ordering of S.

8.1.4 Cost Functions

A convenient restriction to impose on a criterion function $\mathbf{C}:S \rightarrow R$ is that its range $\mathscr{R}_{\mathbf{C}}$ be a set of nonnegative numbers. We call such a criterion a *cost* function, and say that a minimizing value s^* has the minimum cost c^*. For cost functions, the closer $\mathbf{C}(s)$ is to zero, the closer s is to being optimal. In fact, if $\mathbf{C}(s)=0$, then s must be a minimizing value; however, in general, the optimum s^* may have a nonzero cost c^*.

Cost functions are commonly defined on extended domains S' ($S \subseteq S'$) containing a number $s_0 \notin S$ such that $\mathbf{C}(s_0)=0$, but where $\mathbf{C}(s)>0$ for $s \in S$. For example, we may let $\mathbf{C}(s)=\mathbf{d}(s,s_0)$, where $\mathbf{d}$ is a distance function on S'. In this case, we also say that we wish to determine the optimum of the set S', with respect to the criterion $\mathbf{C}:S' \rightarrow R$, 'subject to the constraint' that $s \in S$. (We discuss constraints further below.)

In many applications, there may be several criteria which are to be minimized, and it is common for these criteria to be competing (e.g., time vs. space). One approach to 'balancing' multiple criteria is to make some commensurability assumption, such as measuring time and space in monetary units, thereby defining a composite cost function of several variables. See Thesen [8.50], for example.

8.1.5 Continuous Functions

Consider now the case where S (and R) is the set of all real numbers, and where $\mathbf{C}:S \rightarrow R$ satisfies certain continuity (and differentiability) properties. Then the optimal values of S, i.e., those which minimize $\mathbf{C}(s)$, may be found by using differential calculus. Let

$$\hat{R}=\{\mathbf{C}(s) \mid s \in S,\ \mathbf{C}'(s)=0,\ \mathbf{C}''(s)>0\},$$

where $\mathbf{C}'$ and $\mathbf{C}''$ denote the first and second derivatives, respectively, of $\mathbf{C}(s)$ with respect to s. (These derivatives must, of course, exist and in addition must be continuous; to find maximizing values of S, $\mathbf{C}''(s)$ must be negative instead. See Section 8.3.1 for additional details.) $\hat{R}$ is the set of local minima of $\mathbf{C}(s)$. The global minimum is the minimum number c^* in the set $\hat{R}$, and the optimal values of S that we seek are members of the set

$$S^* = \{s \in S \mid \mathbf{C}(s)=c^*\}.$$

Much of the field of mathematical analysis is devoted to extensions of the foregoing. We cite, for example, (a) the derivation of conditions under which relative minima are also absolute minima (say, when **C** is a convex function), (b) the treatment of constraints of various sorts (such as a restricted domain, say, $S=[a,b]\subseteq R$), (c) the treatment of 'vector' domains ($S=R^n$), and (d) the treatment of functional domains (i.e., where S is a set of functions). Consideration of these topics in any detail is well beyond the scope of this book. For additional information, see Saaty [8.47] and Sage [8.48].

8.1.6 Application: Approximation of Functions

Optimization techniques have so many applications because it is natural to seek the best (or at least a good) means of handling any given task. One broad subclass of applications involves finding a good approximation to the solution of some problem, where an approximation is necessary because an exact solution is impossible for theoretical reasons or is simply impractical for computational reasons. For example, the subroutines in programming system libraries for the evaluation of mathematical functions (such as the trigonometric functions) commonly utilize polynomial approximations to the functions.

Recalling Section 3.1.4, we say that a polynomial of degree n,

$$\mathbf{S}(x)=a_0+a_1x+a_2x^2+\cdots+a_nx^n,$$

is a good approximation to a function $\mathbf{F}(x)$ if $\delta(x)=\mathbf{abs}(\mathbf{F}(x)-\mathbf{S}(x))$ is 'small' for any x. In practice this requirement is interpreted to mean that $\delta(x)$ must be small for *each* value of x, or else that the *sum* of these differences (possibly squared) must be small. When x can assume a continuum of values between given bounds, the summation is replaced by integration. For a given n, we say that the polynomial whose set of coefficients is the solution to

$$\mathbf{min}_{\{a_0,\ldots,a_n\}}\{\mathbf{max}_x\ [\delta(x)]\}$$

is the optimal *uniform* approximation to **F**, and the solution to

$$\mathbf{min}_{\{a_0,\ldots,a_n\}}\{\textstyle\int[\delta(x)]^2\,dx\}$$

is the optimal *least-squares* approximation to **F**. A less common approach is to use the solution $\mathbf{S}(x)$ to a differential equation whose set of coefficients b_i and initial conditions c_i is the solution to

$$\mathbf{min}_{\{b_1,\ldots,b_n;\,c_1,\ldots,c_n\}}\{\textstyle\int[\mathbf{F}(x)-\mathbf{S}(x)]^2\,dx\}.$$

This approach is known as 'differential approximation' (see Lew [8.38]), which is advantageous in system simulation problems when an approximation to $\mathbf{F}(x)$ is needed for an ascending sequence of values of x, not just for scattered values of x.

Solutions of the minimization problems mentioned above are difficult to obtain, except when **F** has especially convenient properties and/or its domain and range are suitably restricted. For example, when the domain is finite, the problem reduces to the more familiar one of 'curve fitting.' In the further special case when $n=1$, we seek a straight-line approximation (a_0+a_1x) to a set of points: this is also known as the linear *regression* problem, which we discussed in Section 6.1.5. For details, see Blum [8.15] and Draper [8.22].

8.2 DYNAMIC PROGRAMMING

An approach to the solution of mathematical optimization problems which has very broad applicability is that of *dynamic programming* (DP, for short). This approach is based on Bellman's *Principle of Optimality,* which he phrases as follows (Bellman [8.6]):

> An optimal policy has the property that whatever the initial state and initial decision are, the remaining decisions must constitute an optimal policy with regard to the state resulting from the first decision.

More briefly, this principle asserts that 'optimal policies have optimal subpolicies'. The class of problems to which this principle can be applied may be characterized as 'sequential decision processes', in which we must make a sequence of decisions, where each decision transforms the process from one state to another. The term 'policy' in the above means a *sequence of decisions*. That the principle is valid follows from the observation that, if a policy has a subpolicy that is not optimal, then replacement of the subpolicy by an optimal subpolicy would improve the original policy. (This essentially proves the principle by contradiction.) We have presupposed here that the subpolicy is independent of the initial state and initial decision (i.e., it depends on the identity of the 'resulting' state, not on how the state resulted), so that the 'cost' of a subpolicy can be separated from that of the initial decision.

Problems for which DP is well suited involve the determination of an optimal cost function $\mathbf{f}:S\rightarrow R$, where S is a set of states and R is the set of nonnegative numbers. When S is finite or countable, we have a *discrete* optimization problem. We say $\mathbf{f}(s)$ is the cost associated with being in state s, and transitions from state to state are made at an added 'price'. More specifically, this means that there is a second cost function $\mathbf{d}:S\times S\rightarrow R$, where $\mathbf{d}(p,q)$ is the cost of 'deciding' to make a transition from state p to state q. (This transition may be denoted by the ordered pairs (p,q).) We call **f** the *minimal cost* (or *criterion*) function, and **d** the *decision* (or *transition*) *cost* function; **f** and **d** must be independent of each other. A basic assumption is that the costs for two successive decisions can be *added* together; we will call this the 'additive cost' assumption. Then, applying the principle of optimality, we may conclude that the minimal cost $\mathbf{f}(p)$ associated with making a decision while in current state p depends upon the sum of a decision cost $\mathbf{d}(p,q)$ and the *minimal* cost associated with making a decision while in the resultant next state q; in fact, $\mathbf{f}(p)$ is the minimum of these sums, for each q. Formally,

$$\mathbf{f}(p)=\mathbf{min}_{q\in S}\ \{\mathbf{d}(p,q)+\mathbf{f}(q)\}. \tag{8.2}$$

Equations of this form are called *dynamic programming functional equations*. If a transition from p to q is not possible, we may let $\mathbf{d}(p,q)=\infty$, or alternatively constrain the set over which a minimum is sought. The latter approach is more general, permitting arbitrary constraints of the form $q\in\mathbf{T}(p)$, where $\mathbf{T}(p)$ is a (proper) subset of S which varies with p ($\mathbf{T}:S\rightarrow 2^S$).

Dynamic programming is the subject of numerous books. We recommend Jacobs [8.29], Williams [8.56], and Nemhauser [8.41] for an introduction, and Bellman [8.12], Dreyfuss [8.24], and Larson [8.36] for further discussion. Books emphasizing discrete optimization include Bellman [8.11] and Kaufmann [8.32]. See also Karp [8.31] and Bonzon [8.16].

8.2.1 Multi-stage Decision Processes

In the following, we restrict ourselves to discrete optimization problems in which the set of states S is finite. A 'multi-stage process' is defined here as a sequence of states, $p_0,p_1,p_2,\ldots$, in S (or of state transitions, $(p_0,p_1),(p_1,p_2),\ldots$), where $p_n\in\mathbf{T}(p_{n-1})$ for $n>0$; n will be called the *stage* number, which corresponds to 'times' of transitions. We emphasize that, while the set S is finite, the process may be an infinite sequence. The process is said to be *deterministic* if $\mathbf{T}:S\rightarrow S$ is a (single-valued) function. Then we may write

$$p_n=\mathbf{T}(p_{n-1})=\mathbf{T}(\mathbf{T}(p_{n-2}))=\cdots=\mathbf{T}(\mathbf{T}(\ldots\mathbf{T}(p_0))),$$

which we denote $\mathbf{T}^n(p_0)$, for $n>0$; by definition, $\mathbf{T}^0(p_0)=p_0$. Observe that $\mathbf{T}^{i+j}(p)=\mathbf{T}^j(\mathbf{T}^i(p))$ for all $p\in S$, $i,j\geq 0$ (sometimes called the 'causality' property). Furthermore, p_i, for any $i>0$, depends only upon p_{i-1} and not upon p_k, for $k<i-1$; this is called the 'Markov property', especially when transitions are made probabilistically (cf. Section 6.3). (In this chapter, we consider only deterministic processes; some probabilistic applications are discussed in Chapter 9.)

Consider now a process, sometimes called a 'control system', where state transitions are governed by a function $\mathbf{T}:S\times V\rightarrow S$, where V is a set of *decision* (or *control*) variables. A sequence of states, $p_0,p_1,p_2,\ldots$, in S, is then determined not just by the initial state p_0, but also by a 'control sequence' of decisions, $v_0, v_1, \ldots$, in V, where

$$p_n = \mathbf{T}(p_{n-1}, v_{n-1}) \text{ for } n>0.$$

Processes where $\mathbf{T}: S \to S$, i.e., where there is no 'control', are said to be *autonomous*. For a given control sequence, the multi-stage decision process is still said to be deterministic, although it is no longer autonomous. The control problem we shall address is that of choosing the sequence $v_0, v_1, \ldots$, in an 'optimal' fashion, i.e., in the sense of minimizing a cost function $\Phi(p_0, p_1, p_2, \ldots; v_0, v_1, \ldots)$. The minimum is denoted by $\Phi^*(p_0)$ and the corresponding state and control sequences are denoted by $p_0^*, p_1^*, p_2^*, \ldots$, and $v_0^*, v_1^*, \ldots$, respectively. If the cost function Φ is additive, i.e., can be written in the form

$$\Phi(p_0, v_0) + \Phi(p_1, v_1) + \cdots,$$

then

$$\begin{aligned} \Phi^*(p_0) &= \mathbf{min}_{v_0, v_1, \ldots} \{\Phi(p_0, p_1, p_2, \ldots; v_0, v_1, \ldots)\} \\ &= \mathbf{min}_{v_0, v_1, \ldots} \{\Phi(p_0, v_0) + \Phi(p_1, v_1) + \ldots\} \\ &= \mathbf{min}_{v_0} \{\Phi(p_0, v_0) + \mathbf{min}_{v_1, \ldots} \{\Phi(p_1, v_1) + \ldots\}\} \\ &= \mathbf{min}_{v_0 \in V} \{\Phi(p_0, v_0) + \Phi^*(p_1)\}, \end{aligned}$$

the third equality following from the separability property of the **min** operation. If $\Phi(p_0, v_0)$ is rewritten as $\mathbf{d}(p_0, p_1)$, where $p_1 = \mathbf{T}(p_0, v_0)$ and $\mathbf{T}$ is a 1–1 function, then

$$\Phi^*(p_0) = \mathbf{min}_{p_1 \in S} \{\mathbf{d}(p_0, p_1) + \Phi^*(p_1)\}. \tag{8.2.1}$$

We have therefore proved a case of the principle of optimality.

Equation (8.2.1) is a 'functional' equation, where the unknown is the function $\Phi^*: S \to R$. The indicated minimization, and the minimizing value p_1^*, cannot be determined explicitly without knowing Φ^*. In order to solve this equation for Φ^* (or p_1^*), some 'boundary condition' must be known: usually, either a 'target' state $s_0 \in S$ which is to be reached after an unspecified number of transitions such that $\Phi^*(s_0) = C_0$, a given constant (commonly, zero), or else a terminal 'time' N which fixes the number of stages in the process.

8.2.2 Digraph Representation

It is often convenient to model a multi-stage decision process by means of a labeled directed graph. We do so by letting the nodes of the digraph represent the states S, and the branches of the digraph represent possible (finite cost) transitions from one state to another. Each branch (s_i, s_j) is labeled by the pair $\langle \mathbf{d}(s_i,s_j),v_{ij}\rangle$, where $v_{ij}\in V$ is such that $s_j=\mathbf{T}(s_i,v_{ij})$; sometimes we just write the simpler label $\mathbf{d}(s_i,s_j)$, leaving v_{ij} defined only implicitly. Each node s_i is labeled by the cost $\Phi^*(s_i)$ (in addition to its symbolic label s_i). The concepts of paths or reachability from one state to another have their obvious meanings. Of special importance is the notion that in an acyclic process it is not possible to reach a state from itself (via a path of nonzero length).

8.2.3 Solution Methodologies

Target States Let $S_0=\{s\in S \mid \Phi^*(s)=0\}$ be a set of 'target' states. Then let $\Phi^*(s)$ be the minimal cost of starting in state $s\in S$ and reaching any state in S_0 via a sequence of transitions. We assume S_0 is reachable from s. Note that if $s\in S_0$, then no transitions are necessary. Since we are interested in reaching any state in S_0, we may also assume that no transitions are possible from a state in S_0 to itself or to any other state. Let S_1 be the set of states not in S_0 from which a state in S_0 is reachable in *at most* a single transition; then for $s_1\in S_1$

$$\begin{aligned}\Phi^*(s_1)&=\mathbf{min}_{s_0\in S_0}\{\mathbf{d}(s_1,s_0)+\Phi^*(s_0)\}\\&=\mathbf{min}_{s_0\in S_0}\{\mathbf{d}(s_1,s_0)\}.\end{aligned}$$

If S_0 contains only one member s_0, then $\Phi^*(s_1)=\mathbf{d}(s_1,s_0)$. Let S_2 be the set of states not in $S_0\cup S_1$ from which a state in S_0 is reachable in *at most* two transitions; then for $s_2\in S_2$

$$\Phi^*(s_2)=\mathbf{min}_{s_1\in S_0\cup S_1}\{\mathbf{d}(s_2,s_1)+\Phi^*(s_1)\}$$

where now $\Phi^*(s_1)$ has been previously defined. This procedure can be continued until $\Phi^*(s)$ has been computed for $s=p_0$ (the initial state) or for each $s\in S$, which is certain to occur if S is finite. (S_0 is not reachable at all otherwise.) In general, we have, for $s\in S_i$, the set of states not in $\cup_{j<i}S_j$ from which a state in S_0 is reachable in *at most i* transitions,

$$\Phi^*(s)=\mathbf{min}_{s'\in \cup_{j<i}S_j}\{\mathbf{d}(s,s')+\Phi^*(s')\}. \tag{8.2.3(a)}$$

We emphasize that Eq. (8.2.3(a)) is solved recurrently, starting at the target set S_0, and then working 'backwards' towards the other states until $p_0\in S_i$.

We have implicitly assumed in the foregoing that it is not possible to go from a state to itself via a nonempty sequence of transitions. For, if state s is in a cycle, then a state in S_0 is not reachable from s in *at most* any finite number of transitions (so S_i may be empty). In other words, the target-state approach is only applicable to acyclic processes. Next, we show how cyclic processes can be treated.

Fixed-Time (N-Stage) Processes When the number of stages or transitions are fixed at N, so that the sequences $p_0, p_1, \ldots, p_N$ and $v_0, v_1, \ldots, v_{N-1}$ are required, we may proceed by redefining Φ^* to be a function of the stage number also. Let $\Phi^*(s,n)$ denote the minimal cost of being in state $s \in S$ and making exactly n transitions. We define $\Phi^*(s,0)=\mathbf{C}_0(s)$ for each $s \in S$, the cost of ending up in state s, and our objective is to determine $\Phi^*(p_0,N)$. Observe that, for each $p \in S$,

$$\Phi^*(p,1)=\mathbf{min}_{s \in S}\{\mathbf{d}(p,s)+\Phi^*(s,0)\};$$

the minimal cost of being in state p and making exactly one transition (to, say, state s) is the minimum of the sums of the transition cost plus the cost of being in state s and making no further transitions. Similarly,

$$\Phi^*(p,2)=\mathbf{min}_{s \in S}\{\mathbf{d}(p,s)+\Phi^*(s,1)\},$$

where $\Phi^*(s,1)$ has been previously defined. This procedure can be continued until $\Phi^*(p,N)$ is computed, in terms of $\{\Phi^*(p,N-1)\}$. In general,

$$\Phi^*(p,i)=\mathbf{min}_{s \in S}\{\mathbf{d}(p,s)+\Phi^*(s,i-1)\}. \qquad (8.2.3(b))$$

We emphasize that Eq.(8.2.3(b)) is solved recurrently, starting with a zero-stage process, and then working 'backwards' towards the desired N-stage process.

By using the above procedure, we can now solve the target state problem for cyclic processes by simply computing

$$\Phi^*(s)=\mathbf{min}_N \Phi^*(s,N).$$

In this context,

$$\Phi^*(s,0)=\infty \text{ for } s \notin S_0,$$
$$\Phi^*(t,0)=0 \text{ for } t \in S_0,$$
$$\Phi^*(t,i)=\infty \text{ for } t \in S_0,\ i>0,$$

where S_0 is the set of target states.

Reconstruction We have shown above how the minimal cost function Φ^* may be computed. In most cases, however, we wish also to determine an optimizing sequence of state transitions and the decisions which yield the minimal cost. If Φ^* is known, then in the course of evaluating $\Phi^*(p)$ using

$$\mathbf{min}_{s \in S}\{\mathbf{d}(p,s)+\Phi^*(s)\}, \qquad (8.2.3(c))$$

a minimizing next state $s^*(p)$, a function of the current state p, is determined automatically; the minimizing decision $v^*(p)$ is defined such that $s^*(p)=\mathbf{T}(p,v^*(p))$. Alternatively, $v^*(p)$ may be defined as the value of v which yields the minimum in

$$\mathbf{min}_{v \in V}\{\mathbf{d}(p,\mathbf{T}(p,v))+\Phi^*(\mathbf{T}(p,v))\}.$$

The optimizing control sequence $v_0^*, v_1^*, v_2^*, \ldots,$ is then given by

$$v^*(p_0),\ v^*(\mathbf{T}(p_0, v_0^*)),\ v^*(\mathbf{T}(\mathbf{T}(p_0, v_0^*), v_1^*)), \ldots,$$

respectively. The corresponding state sequence $p_0^*, p_1^*, p_2^*, \ldots,$ is given by

$$p_0,\quad s^*(p_0) = \mathbf{T}(p_0, v_0^*),\quad s^*(p_1^*) = \mathbf{T}(\mathbf{T}(p_0, v_0^*), v_1^*),\quad \ldots,$$

respectively. We emphasize that Φ^* must first be computed 'backwards' from a target state or a 0-stage process, whereas p_i^* and v_i^* are then computed 'forwards' from p_0 utilizing Φ^* in what is called a *reconstruction* process. To facilitate this reconstruction, each time $\Phi^*(p)$ is calculated using Eq.(8.2.3(c)), the minimizing value $s^*(p)$ or the minimizing decision $v^*(p)$ should be saved.

8.2.4 Shortest Path in a Graph

As an archetypal example illustrating the use of dynamic programming, consider the problem of determining a shortest path in a numerically weighted directed graph $G=(S,R)$. Denote the label of the (unique) branch from node a to node b, i.e., the weighted distance between the nodes, by $\mathbf{d}(a,b)$; if there is no branch from a to b, let $\mathbf{d}(a,b)=\infty$. The *weighted* length of a path is the sum of the labels of the branches comprising the path (which should not be confused with its 'branch length', the number of branches in the path). Our objective is to determine, among all paths from any given node s in S to any other given node t in S, a path of minimum (weighted) length. This is also known as the (optimal) *routing* problem; for further discussion, see Bellman[8.7] as well as Section 9.1.3. Regarding the nodes as states, a shortest path from s to t can be found in several ways. If the graph is acyclic, we may proceed as follows.

Acyclic Digraphs First, we regard t as our target state, and define $\mathbf{f}(a)$, $a \in S$, the cost of being in any other state a, as the minimum path length from a to t; by definition, $\mathbf{f}(t)=0$. The problem is to determine $\mathbf{f}(s)$. By the principle of optimality,

$$\mathbf{f}(a) = \mathbf{min}_{b \in S}\ \{\mathbf{d}(a,b) + \mathbf{f}(b)\}.$$

In other words, the shortest path from a to t is the shortest of the paths to each node b reachable from a and their shortest continuation to t. Hence, if $\{b_1, b_2, \ldots, b_k\}$ is the set of nodes reachable from s, then

$$\mathbf{f}(s) = \mathbf{min} \left\{ \begin{array}{c} \mathbf{d}(s,b_1) + \mathbf{f}(b_1) \\ \mathbf{d}(s,b_2) + \mathbf{f}(b_2) \\ \cdots \\ \mathbf{d}(s,b_k) + \mathbf{f}(b_k) \end{array} \right\},$$

the simple minimum of k numbers. However, we do not yet know the values of $\mathbf{f}(b_1)$, $\mathbf{f}(b_2)$,...,$\mathbf{f}(b_k)$; to determine these, we must compute

$$\mathbf{f}(b_1)=\mathbf{min}\begin{Bmatrix} \mathbf{d}(b_1,b_1')+\mathbf{f}(b_1') \\ \mathbf{d}(b_1,b_2')+\mathbf{f}(b_2') \\ \cdots \\ \mathbf{d}(b_1,b_{k'}')+\mathbf{f}(b_{k'}') \end{Bmatrix}$$

and similarly for $\mathbf{f}(b_2)$, etc. Again, we do not know the values $\mathbf{f}(b_1')$, $\mathbf{f}(b_2')$,..., $\mathbf{f}(b_{k'}')$. So we work backwards instead.

We do know that the value of $\mathbf{f}(t)$ is 0, and that if node a is connected to node t by *only* a path of at most one branch, then

$$\mathbf{f}(a)=\mathbf{d}(a,t)+\mathbf{f}(t)=\mathbf{d}(a,t).$$

(A minimization is not necessary in this case since, by our earlier assumption, there is just a single branch from any node to another.) On the other hand, if there are other paths from a to t, e.g., one from a to b to t, then the length of this path $\mathbf{d}(a,b)+\mathbf{d}(b,t)$ may be less than $\mathbf{d}(a,t)$. This means that, for any node a adjacent to node t, $\mathbf{f}(a)$ cannot be evaluated without examining all paths from a to t. Hence, initially, we can evaluate $\mathbf{f}(a)$ (by the above formula) only for nodes connected to the 'target' $S_0=\{t\}$ by just single branches; let S_1 denote the set of such nodes. Now consider $S_0 \cup S_1$ as a new target. Let S_2 be the set of nodes not in $S_0 \cup S_1$ that are connected to $S_0 \cup S_1$ (regarded as a single merged node) by only single branches; nodes in S_2 are connected to t by paths of at most two branches. For these nodes, we may then evaluate

$$\mathbf{f}(x)=\mathbf{min}_{a\in S_0 \cup S_1}\ \{\mathbf{d}(x,a)+\mathbf{f}(a)\},\ x\in S_2,$$

since each $\mathbf{f}(a)$ has been previously evaluated, and since, by definition of S_2, there can be no path from $x\in S_2$ to t via a node not in $S_0 \cup S_1$. Generalizing, let S_i be the set of nodes not in $\cup_{j<i}S_j$ which are connected to $\cup_{j<i}S_j$ by only single branches; nodes in S_i are connected to t by paths of at most i branches. We have

$$\mathbf{f}(x)=\mathbf{min}_{a\in \cup_{j<i}S_j}\ \{\mathbf{d}(x,a)+\mathbf{f}(a)\},\ x\in S_i.$$

This procedure terminates when the given node s belongs to S_i, which is guaranteed if a path from s to t exists.

Implementation of the foregoing procedure does not require that the sets S_0, S_1, S_2,... be explicitly determined. Instead, we may topologically sort the graph (cf. Section 5.2.5), obtaining an ordered set of nodes $\{x_1, x_2,..., x_N\}$ where x_i may precede x_j in the ordering only if there is no path from x_i to x_j; our acyclic assumption guarantees that such an ordering is possible. If $t=x_i$, then prior nodes $x_1,..., x_{i-1}$ may be ignored because there are no paths from these nodes to t; formally, for these nodes, $\mathbf{f}(x_j)=\infty$. For subsequent nodes x_j, $j=i+1,\ldots, N$, we write

$$\mathbf{f}(x_j)=\mathbf{min}_{x_k}\ \{\mathbf{d}(x_j,x_k)+\mathbf{f}(x_k)\};$$

$\mathbf{d}(x_j,x_k)$ is finite only for x_k that precedes x_j in the ordering. Hence we may evaluate $\mathbf{f}(x_j)$, $\mathbf{f}(x_{j+1}),\ldots,$ in sequence, without ever requiring a value not previously evaluated in the sequence. We stop when $x_j=s$. As an aside, we note that, with respect to the set S_i defined above, nodes in S_j precede nodes in S_{j+1} in the topological ordering.

Cyclic Graphs (with nonnegative distances) If the given graph $G=(S,R)$ is cyclic, we may no longer use the target-state approach. (A topological sort would no longer be possible.) Instead, to find the shortest path from s to t, we may use the fixed-time approach.

We define $\mathbf{f}(a,i)$, the cost of being in state a and having to make i transitions, as the minimum length path which has i branches and goes from a to t. By definition, $\mathbf{f}(t,0)=0$, but $\mathbf{f}(t,i)=\infty$ for $i>0$ and $\mathbf{f}(a,0)=\infty$ for $a\in S-\{t\}$. Consider now single-branch paths:

$$\mathbf{f}(x,1)=\mathbf{min}_{a\in S}\ \{\mathbf{d}(x,a)+\mathbf{f}(a,0)\},\ x\in S,$$

which can be evaluated in terms of the known values $\{\mathbf{f}(a,0)\}$. Generalizing, we may evaluate

$$\mathbf{f}(x,i)=\mathbf{min}_{a\in S}\ \{\mathbf{d}(x,a)+\mathbf{f}(a,i-1)\},\ x\in S,\ i\geq 1,$$

in terms of the previously evaluated $\{\mathbf{f}(a,i-1)\}$. If the distances $\mathbf{d}(x,a)$ are all nonnegative, we need not evaluate $\mathbf{f}(x,i)$ for $i\geq\#(S)$, since these correspond to cyclic paths which contain acyclic subpaths of shorter length. (If we allow negative distances, a minimum length path may be undefined since a path containing a cycle of negative length can be shortened indefinitely by traversing the cycle over and over; of course, negative distances are no problem for acyclic graphs.)

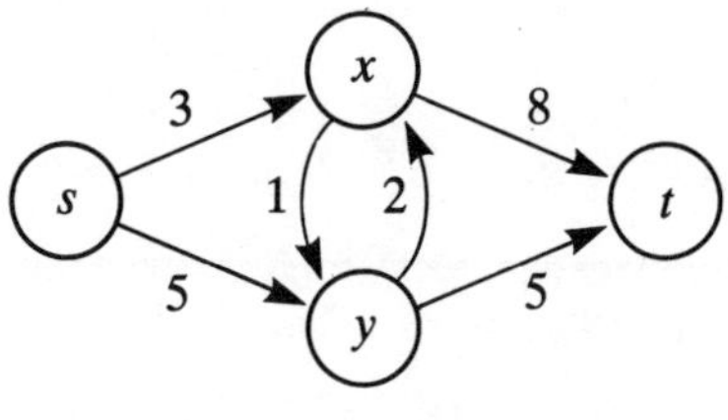

Figure 8.2.4

Example Consider the graph shown in Figure 8.2.4. We wish to find the shortest path from node s to node t. Considering t to be the target state, so that $\mathbf{f}(t)=0$, we write initially:

$$\mathbf{f}(s)=\mathbf{min}\left\{\begin{array}{c}\mathbf{d}(s,x)+\mathbf{f}(x)\\ \mathbf{d}(s,y)+\mathbf{f}(y)\end{array}\right\}=\mathbf{min}\left\{\begin{array}{l}3+\mathbf{f}(x)\\ 5+\mathbf{f}(y)\end{array}\right\}$$

$$\mathbf{f}(x)=\mathbf{min}\left\{\begin{array}{c}\mathbf{d}(x,y)+\mathbf{f}(y)\\ \mathbf{d}(x,t)+\mathbf{f}(t)\end{array}\right\}=\mathbf{min}\left\{\begin{array}{l}1+\mathbf{f}(y)\\ 8+0\end{array}\right\}$$

$$\mathbf{f}(y)=\mathbf{min}\left\{\begin{array}{c}\mathbf{d}(y,x)+\mathbf{f}(x)\\ \mathbf{d}(y,t)+\mathbf{f}(t)\end{array}\right\}=\mathbf{min}\left\{\begin{array}{l}2+\mathbf{f}(x)\\ 5+0\end{array}\right\}.$$

Note that $\mathbf{f}(x)$ is defined in terms of $\mathbf{f}(y)$ and vice versa. If we attempt to work backwards from t, we find that there are no nodes which are at most one branch from t. Consequently, the target-state approach is not applicable. (This is, of course, because the graph is cyclic. Note that letting $\mathbf{f}(x)=\mathbf{d}(x,t)=8$ would be incorrect, for in fact a shorter path of length 6 from x to t via y exists.)

Using the fixed-time process approach, we let $\mathbf{f}(s,i)$ denote the length of the shortest path from s to t that has i branches. There is no branch from s and t, hence $\mathbf{f}(s,1)=\infty$. For $i>1$,

$$\mathbf{f}(s,i)=\mathbf{min}\left\{\begin{array}{c}\mathbf{d}(s,x)+\mathbf{f}(x,i-1)\\ \mathbf{d}(s,y)+\mathbf{f}(y,i-1)\end{array}\right\}.$$

Now

$$\begin{aligned}
&\mathbf{f}(t,0)=0\leftarrow\\
&\mathbf{f}(x,1)=\mathbf{d}(x,t)+\mathbf{f}(t,0)=8\\
&\mathbf{f}(y,1)=\mathbf{d}(y,t)+\mathbf{f}(t,0)=5\leftarrow\\
&\mathbf{f}(x,2)=\mathbf{d}(x,y)+\mathbf{f}(y,1)=1+5=6\leftarrow\\
&\mathbf{f}(y,2)=\mathbf{d}(y,x)+\mathbf{f}(x,1)=2+8=10\\
&\mathbf{f}(x,3)=\mathbf{d}(x,y)+\mathbf{f}(y,2)=1+10=11\\
&\mathbf{f}(y,3)=\mathbf{d}(y,x)+\mathbf{f}(x,2)=2+6=8\\
&\quad\cdots
\end{aligned}$$

hence

$$\mathbf{f}(s,2)=\mathbf{min}\left\{\begin{array}{c}\mathbf{d}(s,x)+\mathbf{f}(x,1)\\ \mathbf{d}(s,y)+\mathbf{f}(y,1)\end{array}\right\}=\mathbf{min}\left\{\begin{array}{c}3+8\\ 5+5\end{array}\right\}=10$$

$$\mathbf{f}(s,3)=\mathbf{min}\left\{\begin{array}{c}\mathbf{d}(s,x)+\mathbf{f}(x,2)\\ \mathbf{d}(s,y)+\mathbf{f}(y,2)\end{array}\right\}=\mathbf{min}\left\{\begin{array}{c}3+6\\ 5+10\end{array}\right\}\begin{array}{l}\leftarrow\\ =9.\end{array}$$

(Since all distances are nonnegative, $\mathbf{f}(s,i)$, for $i\geq 4$, need not be evaluated; however, the reader should verify that $\mathbf{f}(s,4)=13$.) Finally,

$$\mathbf{f}(s)=\mathbf{min}_i\ \{\mathbf{f}(s,i)\}=\mathbf{min}\left\{\begin{array}{c}\mathbf{f}(s,1)\\ \mathbf{f}(s,2)\\ \mathbf{f}(s,3)\end{array}\right\}=\mathbf{min}\left\{\begin{array}{c}\infty\\ 10\\ 9\end{array}\right\}\begin{array}{l}\\ =9.\\ \leftarrow\end{array}$$

We conclude that the shortest path from s to t is of length 9. To 'reconstruct' this shortest path, we note that the number of branches in the path is 3, that the next state which minimizes $\mathbf{f}(s,3)$ is x, and in turn that the next state which minimizes $\mathbf{f}(x,2)$ is y, and again in turn the next state which minimizes $\mathbf{f}(y,1)$ is t. Hence, the optimal path traverses the nodes in the sequence $sxyt$. (The arrows ($\leftarrow$) in the above illustrate this reconstruction process.)

Suppose now that we have an acyclic graph G' which is G with branch (y,x) removed. A topological sort yields the sequence of nodes $\{t,y,x,s\}$. Hence we evaluate $\mathbf{f}(t)$, $\mathbf{f}(y)$, $\mathbf{f}(x)$, $\mathbf{f}(s)$ in that order:

$$\mathbf{f}(t)=0$$

$$\mathbf{f}(y)=\mathbf{d}(y,t)=5$$

$$\mathbf{f}(x)=\mathbf{min}\left\{\begin{matrix}\mathbf{d}(x,t)+\mathbf{f}(t)\\ \mathbf{d}(x,y)+\mathbf{f}(y)\end{matrix}\right\}=\mathbf{min}\left\{\begin{matrix}8+0\\ 1+5\end{matrix}\right\}=6$$

$$\mathbf{f}(s)=\mathbf{min}\left\{\begin{matrix}\mathbf{d}(s,y)+\mathbf{f}(y)\\ \mathbf{d}(s,x)+\mathbf{f}(x)\end{matrix}\right\}=\mathbf{min}\left\{\begin{matrix}5+5\\ 3+6\end{matrix}\right\}=9.$$

Again, the shortest path from s to t is of length 9. In this example of the use of the target-stage approach, $S_0=\{t\}$, $S_1=\{y\}$, $S_2=\{x\}$, $S_3=\{s\}$.

8.2.5 The Traveling-salesman Problem

A variation of the 'routing' problem of finding the shortest path in an N-node directed graph $G=\{S,R\}$, from any node to another, is the 'touring' problem of finding the shortest Hamiltonian path—i.e., the shortest simple cyclic path which traverses each node of the graph. We can solve this, the well-known *traveling-salesman* problem, by employing the dynamic programming approach (see Bellman [8.8]). In essence, we must incorporate into our prior formulation information about nodes which have already been traversed (so that we know which nodes remain to be traversed).

We proceed by defining the state of our process as the sequence of nodes which have already been traversed, denoted by $\langle s_1s_2...s_i\rangle$. Then the

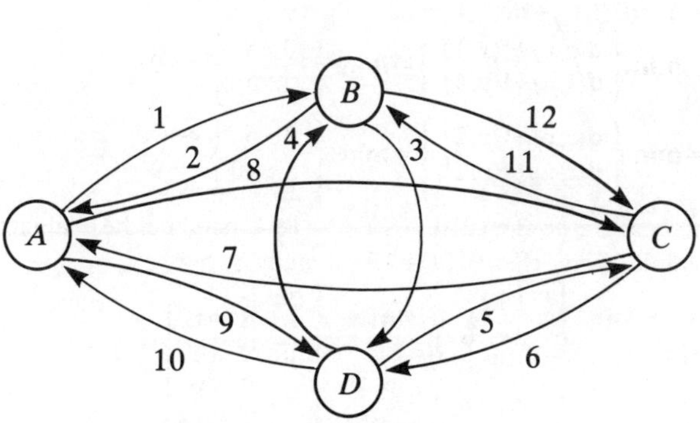

Figure 8.2.5

path that first traverses the nodes s_1, s_2, ..., s_i (in that order) and optimally continues to each of the other nodes before returning to s_1 has remaining length

$$\mathbf{f}(\langle s_1, s_2 \ldots s_i \rangle) = \mathbf{min}_{a \notin \{s_1, \ldots, s_i\}} \{\mathbf{d}(s_i, a) + \mathbf{f}(\langle s_1 s_2 \ldots s_i a \rangle)\}$$

with

$$\mathbf{f}(\langle s_1 s_2 \ldots s_N \rangle) = \mathbf{d}(s_N, s_1).$$

Since we seek cyclic paths, we may start at any node $a \in S$, and determine $\mathbf{f}(\langle a \rangle)$.

Example Consider the graph G shown in Figure 8.2.5. We wish to find the shortest Hamiltonian path in G. Arbitrarily starting at node A, we proceed as follows:

$$\mathbf{f}(\langle A \rangle) = \mathbf{min} \left\{ \begin{array}{l} \mathbf{d}(A,B) + \mathbf{f}(\langle AB \rangle) \\ \mathbf{d}(A,C) + \mathbf{f}(\langle AC \rangle) \\ \mathbf{d}(A,D) + \mathbf{f}(\langle AD \rangle) \end{array} \right\}$$

$$\mathbf{f}(\langle AB \rangle) = \mathbf{min} \left\{ \begin{array}{l} \mathbf{d}(B,C) + \mathbf{f}(\langle ABC \rangle) \\ \mathbf{d}(B,D) + \mathbf{f}(\langle ABD \rangle) \end{array} \right\}$$

$$\mathbf{f}(\langle AC \rangle) = \mathbf{min} \left\{ \begin{array}{l} \mathbf{d}(C,B) + \mathbf{f}(\langle ACB \rangle) \\ \mathbf{d}(C,D) + \mathbf{f}(\langle ACD \rangle) \end{array} \right\}$$

$$\mathbf{f}(\langle AD \rangle) = \mathbf{min} \left\{ \begin{array}{l} \mathbf{d}(D,B) + \mathbf{f}(\langle ADB \rangle) \\ \mathbf{d}(D,C) + \mathbf{f}(\langle ADC \rangle) \end{array} \right\}$$

$$\begin{aligned} \mathbf{f}(\langle ABC \rangle) &= \mathbf{d}(C,D) + \mathbf{f}(\langle ABCD \rangle) \\ &= \mathbf{d}(C,D) + \mathbf{d}(D,A) = 6 + 10 = 16 \\ \mathbf{f}(\langle ABD \rangle) &= \mathbf{d}(D,C) + \mathbf{d}(C,A) = 5 + 7 = 12 \leftarrow \\ \mathbf{f}(\langle ACB \rangle) &= \mathbf{d}(B,D) + \mathbf{d}(D,A) = 3 + 10 = 13 \\ \mathbf{f}(\langle ACD \rangle) &= \mathbf{d}(D,B) + \mathbf{d}(B,A) = 4 + 2 = 6 \\ \mathbf{f}(\langle ADB \rangle) &= \mathbf{d}(B,C) + \mathbf{d}(C,A) = 12 + 7 = 19 \\ \mathbf{f}(\langle ADC \rangle) &= \mathbf{d}(C,B) + \mathbf{d}(B,A) = 11 + 2 = 13. \end{aligned}$$

So

$$\mathbf{f}(\langle AB \rangle) = \mathbf{min} \left\{ \begin{array}{c} 12+16 \\ 3+12 \end{array} \right\} = \mathbf{min} \left\{ \begin{array}{c} 28 \\ 15 \end{array} \right\} \begin{array}{l} =15 \\ \leftarrow \end{array}$$

$$\mathbf{f}(\langle AC \rangle) = \mathbf{min} \left\{ \begin{array}{c} 11+13 \\ 6+6 \end{array} \right\} = \mathbf{min} \left\{ \begin{array}{c} 24 \\ 12 \end{array} \right\} = 12$$

$$\mathbf{f}(\langle AD \rangle) = \mathbf{min} \left\{ \begin{array}{c} 4+19 \\ 5+13 \end{array} \right\} = \mathbf{min} \left\{ \begin{array}{c} 23 \\ 18 \end{array} \right\} = 18$$

and

$$\mathbf{f}(\langle A\rangle) \quad = \min \left\{ \begin{array}{c} 1+15 \\ 8+12 \\ 9+18 \end{array} \right\} = \min \left\{ \begin{array}{c} 16 \\ 20 \\ 27 \end{array} \right\} \begin{array}{c} \leftarrow \\ =16. \\ \end{array}$$

We conclude that the desired path has length 16. To reconstruct this path, we note that the next state that minimizes $\mathbf{f}(\langle A\rangle)$ is B, and the next state that minimizes $\mathbf{f}(\langle AB\rangle)$ is D, from which the next state must be C before returning to A. Hence the optimal path traverses the nodes in the order $ABDCA$. (The arrows (←) in the above illustrate this reconstruction process.)

8.2.6 A Survey of Applications

In this chapter, dynamic programming is given preferential treatment over other optimization techniques (such as linear programming) because of its versatility. To illustrate this, we list below some of the many applications of dynamic programming to systems programming problems which we discuss in this book.

(a) finding the optimal code for a set of characters (Section 9.1.1);
(b) finding the best way to arrange data for linear search (Section 8.2.7) or for a binary tree search (Section 9.1.2);
(c) selecting a member from a queue for which service is to be provided next (Section 9.2.2);
(d) selecting the optimal page to be replaced in a paging system (Section 9.2.3);
(e) allocating computer system resources to computing processes, in general (Section 9.3.1);
(f) deciding how to partition a program that is too large for memory (Section 9.3.1);
(g) deciding how much memory, if any, competing processes should have (Section 9.3.2);
(h) determining the best way to arrange and to access files on external storage media (Section 9.3.3);
(i) deciding the order in which tests should be performed in a decision tree (Section 9.3.3).

Some other interesting applications include:

(j) finding a string in a language 'closest' to another string (in the sense of requiring a minimum number of transformations to go from one to the other; see Bellman [8.9] and Wagner [8.52]); this has applications to error-correcting parsing and spelling correction;

(k) parsing context-free languages by the 'Cocke-Younger-Kasami' algorithm (see Younger [8.58]);
(l) evaluating arithmetic expression, such as matrix products (see Godbole [8.25], or generating object code for operator trees (see Aho [8.2]);
(m) finding maximal common subsequences, e.g., for object code generation (see Wagner [8.53] and Hirschberg [8.27]);
(n) setting tabs in word processing output files (see Peterson [8.43]);
(o) compressing data which have redundancies (see Wagner [8.51]);
(p) synthesizing finite state transition systems (see Bellman [8.13]).

Numerous other applications may be found in the literature, many of which are discussed in Brown [8.17] and Lew [8.39].

8.2.7 Application: Linear Search

As an example of how dynamic programming can be applied, we return to the rather trivial problem mentioned in Section 7.2.3. Given a set of data $A=\{a_1,a_2,\ldots,a_N\}$, and given the probability p_i that a_i will be accessed, we wish to find the optimal ordering of A, $\hat{A}=(a_{i_1},a_{i_2},\ldots,a_{i_N})$, such that the expected number of accesses using linear search,

$$\mathbf{Exp}(I)=1\times p_{i_1}+2\times p_{i_2}+\cdots+N\times p_{i_N},$$

is minimal. This problem can be formulated as a multi-stage decision process as follows. At stage k, we must choose a member a_{i_k} of the set of data, which has not been previously chosen, as the kth member of the ordered set $\hat{A}$. The cost of making this kth choice, denoted $\mathbf{d}(a_{i_k},k)$, equals $k\times p_{i_k}$. Note that the additive cost assumption holds: the cost of making successive choices i_k and i_{k+1} is $\mathbf{d}(a_{i_k},k)+\mathbf{d}(a_{i_{k+1}},k+1)$, and the overall cost is $\Sigma_{k=1}^{N} k\times p_{i_k}$. The state S of the process, at stage k, is defined as the set of members of A which have not yet been chosen to be in $\hat{A}$. Then the problem reduces to that of solving the dynamic programming functional equation

$$\mathbf{f}(S,k)=\min_{a_{i_k}\in S}\{\mathbf{d}(a_{i_k},k)+\mathbf{f}(S-\{a_{i_k}\},k+1)\}$$

for $\mathbf{f}(A,1)$, given that $\mathbf{f}(\{a\},N)=\mathbf{d}(a,N)$ (or that $\mathbf{f}(\emptyset,N+1)=0$, which is equivalent).

It should be emphasized that, for any given problem, there may be many ways to formulate a multi-stage decision process, where the manner in which states and cost functions are defined differs. For example, we may also solve the above problem using the dynamic programming functional equation

$$\mathbf{f}(S)=\min_{a_{i_k}\in S}\{\mathbf{d}(S)+\mathbf{f}(S-\{a_{i_k}\})$$

where $\mathbf{d}(S)=\sum_{i\in\{j|a_j\in S\}} p_i$ and $\mathbf{f}(\emptyset)=0$. Note that $\mathbf{d}(S)$ does not depend upon the choice of $a_{i_k}\in S$, but, of course, $\mathbf{f}(S-\{a_{i_k}\})$ does. It can be shown that the $a_{i_k}\in S$ which has maximum probability p_{i_k} minimizes $\{\mathbf{d}(S)+\mathbf{f}(S-\{a_{i_k}\})\}$, hence the optimal ordering $\hat{A}$ has the elements in order of decreasing probability. (In other words, for this simple problem, we may just sort A instead of solving a dynamic programming functional equation; but, in general, we cannot *expect* such a simplification.)

8.3 OTHER OPTIMIZATION TECHNIQUES

Dynamic programming is just one of numerous optimization techniques. It is emphasized here because of its generality. As indicated above, dynamic programming can be used to solve a variety of software systems problems, ranging from optimal ways of representing, structuring, and processing data, to optimal ways of scheduling or allocating resources. Dynamic programming can be applied to problems related to data structures, program or algorithm design, compilers and interpreters, loaders, operating systems, and computer networks. (In addition to these software or systems programming applications, dynamic programming can also be applied in other computer science areas, such as hardware design and artificial intelligence, as well as to the business, economics, and engineering areas in which it evolved. See Bellman [8.10] for a general overview.)

Although we emphasize dynamic programming because it can be applied to so many different problems, it may be that, for a given problem, there may be other applicable techniques which are more efficient. There are numerous books which provide good surveys of some of these other optimization techniques: these books are usually classified under 'operations research', 'control theory', or 'applied mathematics' (rather than 'computer science'), and, accordingly, the applications they treat are not of direct relevance to software systems. In this section, we briefly discuss just some of the more prominent of these other optimization techiques. For example, see Adby [8.1]. In the next chapter, we discuss techniques which are restricted to only one major class of applications, because the class is such an important one.

8.3.1 Calculus-based Techniques

In Section 8.1.5, we introduced the 'classical' method of finding the minimum (or maximum) of a real function $\mathbf{f}$ of a single real variable x, which is explained in detail in any calculus text (e.g., Rudin [8.46]). In summary, to find $\mathbf{min}_x\mathbf{f}(x)$, we must find a point with zero slope by setting the first derivative of $\mathbf{f}(x)$ with respect to x equal to zero, and solving the resulting equation

$$\frac{d}{dx}[\mathbf{f}(x)]=0$$

for x; we denote a solution to this equation by x^*. (For this equation to hold, we must assume that $\mathbf{f}(x)$ is differentiable for x in some 'neighborhood' of x^* [where a neighborhood of x^* is an open interval of reals (x^*-h, x^*+h), for some $h>0$].) However, the satisfaction of the foregoing equation is not a sufficient condition for x^* to be a minimizing value of $\mathbf{f}(x)$ for several reasons: one is that x^* may be a maximizing value instead; another is that x^* may be only a relative minimizing value [i.e., it may be minimizing only in a neighborhood of x^*]; and yet another is that $\mathbf{f}$ may not be differentiable at its minimum point, for example, when its minimum value occurs at an end-point a_1 or a_2 of a restricted domain $\mathcal{D}_\mathbf{f}=[a_1,a_2]$. Of course, in principle, we can find the absolute minimizing value of $\mathbf{f}$ by first finding all solutions of $d/dx[\mathbf{f}(x)]=0$, denoted x_1^*, x_2^*,..., and then adding the values of $\mathbf{f}$ at any points a_1, a_2,..., where it is not differentiable, and finally determining

$$\min\ \{\mathbf{f}(x_1^*), \mathbf{f}(x_2^*),..., \mathbf{f}(a_1), \mathbf{f}(a_2),...\}.$$

Because this is such a poor procedure, we raise two important questions. Firstly, under what circumstances is a solution to $d/dx[\mathbf{f}(x)]=0$ a (relative) minimizing value, and secondly, under what circumstances is it an absolute minimizing value? We provide (without proof) answers to these questions in the form of two theorems:

Theorem 1 A solution x^* to $d/dx[\mathbf{f}(x)]=0$ is a relative minimum if $d^2/dx^2[\mathbf{f}(x^*)]>0$.

Theorem 2 A solution x^* to $d/dx[\mathbf{f}(x)]=0$ is an absolute minimum in an interval $[a,b]$ if $\mathbf{f}$ is *convex*, i.e., if

$$\mathbf{f}(\lambda x_1+(1-\lambda)x_2)\le\lambda\mathbf{f}(x_1)+(1-\lambda)\mathbf{f}(x_2)$$

for all x_1, $x_2\in[a,b]$, $0\le\lambda\le1$.

Consider now a generalization of the foregoing to a real function $\mathbf{f}$ of N real variables, $x_1,...,x_N$. Analogous theorems involving partial derivatives are then applicable.

Theorem 3 A necessary condition for a vector $X^*=(x_1^*,x_2^*,...,x_N^*)$ to minimize $\mathbf{f}(x_1,..., x_N)$ is that it satisfies the set of equations

$$\frac{\partial}{\partial x_i}[\mathbf{f}(x_1, x_2,..., x_N)]=0, \qquad i=1, 2,..., N.$$

This assumes that the partial derivatives exist in the neighborhood of X^*. The vector $(\partial \mathbf{f}/\partial x_1, \partial \mathbf{f}/\partial x_2, \ldots, \partial \mathbf{f}/\partial x_N)$ is called the ***gradient***, denoted $\nabla \mathbf{f}(X)$, which corresponds to the slope of $\mathbf{f}(X)$ when $N=1$; to find a minimum requires that a 'stationary point', i.e., one with zero gradient, be located.† Constraints of the form $\mathbf{g}(x_1, \ldots, x_N)=c$ or $\mathbf{g}(x_1, \ldots, x_N) \leq b$ may complicate matters further.

For example, suppose we wish to minimize a real function $\mathbf{f}(x_1, x_2, \ldots, x_n)$, with respect to n independent real variables $x_1, \ldots, x_n$, where these variables satisfy an equality constraint $\mathbf{g}(x_1, x_2, \ldots, x_n)=c$. Suppose also that $\mathbf{f}$ and $\mathbf{g}$ are continuous and have continuous derivatives. Then if the constraint equation has real solutions, and if $\partial^2 \mathbf{f}/\partial x_i^2 > 0$ at the minimizing value X^*, then we may proceed as follows. Let $\Phi = \mathbf{f} + \lambda(\mathbf{g} - c)$, where the variable λ, called the *Lagrange multiplier*, is independent of the x_i. Then the stationary points of $\mathbf{f}$ must satisfy

$$\frac{\partial \Phi}{\partial x_i} = 0, \qquad \text{for } i = 1, 2, \ldots, n.$$

These n equations plus the constraint equation $\mathbf{g}(x_1, x_2, \ldots, x_n)=c$, if independent, allow the $n+1$ variables $x_1, \ldots, x_n$ and λ to be evaluated.

Calculus-based optimization techniques have limited applicability to software system problems because the differentiability requirement is not satisfied for a discrete system. However, often a discrete system can be approximated by a continuous one, and calculus can then be applied to obtain approximate answers. Sometimes these answers are useful, and sometimes not.

Several interesting applications of the foregoing theory are discussed in the next section. Readers should also recall our use of calculus in Section 3.1.4.

8.3.2 Application: Buffering

Buffer Size Recall the method illustrated in Figure 3.2.6(d) for storing variable-length data (records) in fixed-size cells (blocks), and the discussion of buffer size in Section 7.3.4. We assume that cells have fixed portions reserved for data and for overhead. Let C denote the cell size, D denote the portion of a cell reserved for data, and E denote the portion reserved for overhead: then $C=D+E$. If a data record of length R is too large to fit within the data portion of one cell, then it 'overflows' to one or more additional cells. If these other cells are adjacent, we say the record 'spans' a sequence of

† A sufficient condition requires 'positiveness' of the matrix $(\partial^2 \mathbf{f}/\partial x_i \partial x_j)$, or 'convexity', concepts which we will not discuss herein. We simply note that they are generalizations of Theorems 1 and 2. See Saaty [8.47].

cells; the overhead portion of each cell contains a flag which indicates whether the data continues to the next cell or ends. Alternatively, overflowing portions of a data record can be placed at any location, provided a pointer to it is set; in this case, the overhead portion of the cells consists of linked list pointers. The question we address here is, given E and R, what is the best choice of C?

An obvious choice for C is $R+E$. However, let us assume that C cannot be too large, or in other words that we wish to store a record whose length R exceeds $D(=C-E)$. This record would occupy $\mathbf{ceiling}(R/(C-E))$ cells, with an amount of wasted space equal to

$$\mathbf{f}(C)=C\cdot\mathbf{ceiling}(R/(C-E))-R.$$

We would like to determine the value of C which minimizes this function of C (where R and E are constants). We cannot use calculus directly since **ceiling** is a discontinuous function. Thus, we proceed by observing that the amount of wasted space is the amount of overhead

$$E\cdot\mathbf{ceiling}(R/(C-E))$$

plus the unused space in the data portion of the last cell

$$D\cdot\mathbf{ceiling}(R/(C-E))-R.$$

The latter term, for unspecified R, is approximately $(C-E)/2$.. If we approximate the first term $E\cdot\mathbf{ceiling}(R/(C-E))$ by the lower quantity $ER/(C-E)$, then we should approximate the latter term by a comparably higher quantity, say, $C/2$. Thus, we have

$$\tilde{\mathbf{f}}(C)=\frac{ER}{C-E}+\frac{C}{2}$$

as an approximation to $\mathbf{f}(C)$. Using calculus, we find that the value of C which minimizes $\tilde{\mathbf{f}}(C)$ is $\hat{C}=E+\sqrt{(2RE)}$. Finally, if R is of variable rather than constant length, it is reasonable, but no longer optimal, to use its average in the formula.

In summary, for a given E,

$$E+\sqrt{(2E)}\cdot\sqrt{(R_{\text{avg}})}$$

is the 'best' choice for cell or buffer size, in the sense that space is then (approximately) minimal. This formula also applies to the situation where messages of variable length R are received and stored in buffers of fixed size C. (See Wolman [8.57].)

Buffer Pool Consider next the related problem where information is saved in a pool of buffers (of a fixed size) as it arrives, and is processed only after a fixed amount R has arrived (where R is measured in buffer units). Assume that the per unit time cost of saving information is a linear function ($=C_0+br$) of the amount r of buffer units occupied (because this space cannot be otherwise used), and that information arrives at a constant rate of λ buffer units per unit time. Then

$$\int_0^T (C_0+b\lambda t)\, dt=C_0T+b\lambda T^2/2$$

is the cost of saving information for T units of time. Assume also that there is a fixed cost C_1 associated with emptying the buffers (by processing the information). The cost for one cycle of saving and emptying the buffers is then

$$\left(C_0T+\frac{b\lambda T^2}{2}\right)+C_1.$$

The value, $\hat{T}$, of T which minimizes this cost, may be found by calculus. Dividing by T and differentiating with respect to T, we have

$$\frac{b\lambda}{2}-\frac{C_1}{T^2}=0.$$

Hence,

$$\hat{T}=\sqrt{\left(\frac{2C_1}{b\lambda}\right)},$$

and the 'optimal' size of the buffer pool is

$$\hat{R}=\lambda\hat{T}=\sqrt{\left(\frac{2C_1\lambda}{b}\right)}.$$

This is related to the 'inventory' problem; see Hillier [8.26].

Blocking Each physical block recorded on conventional magnetic storage media requires a nontrivial amount of 'overhead' space (equal to about 1/2 inch (13 mm) for standard magnetic tape). Consequently, when a file is written on such media, it is desirable to minimize the number of physical blocks into which the file is divided, or equivalently to maximize the number of bytes B in each physical block. In other words, for a file size of S bytes, we would like to choose B so as to minimize S/B. Of course, S/B is a minimum when $B=S$, but B is generally constrained by the relatively small amount of main storage space which can be used as file I/O buffers. Suppose there are C bytes of main storage which must be shared by N files. Let S_i denote the size of the ith file, and B_i denote the size of its blocks. Then we wish to minimize the total number of blocks

$$\sum_{i=1}^{N} \frac{S_i}{B_i}$$

subject to the constraint that

$$\sum_{i=1}^{N} n_i \times B_i = C,$$

where n_i is the number of buffers of size B_i allocated to the ith file. (It is common for n_i to have value two, for each i, when 'double buffering' is used.) Given $\{s_1,...,s_N\}$, $\{n_1,...,n_N\}$, and C, this optimization problem can be solved using the Lagrange multiplier method. (See Walker [8.54].)

8.3.3 Search Techniques

We formulate a general optimization problem as follows. Let $x_1,..., x_N$ be a set of N variables, where x_i can assume values from the set V_i, and let

$$\mathbf{f}\colon V_1 \times V_2 \times \cdots \times V_N \rightarrow R,$$

where R is the set of real numbers. Denote the domain of **f** by

$$\mathscr{D}_\mathbf{f} \ (\subseteq V_1 \times V_2 \times \cdots \times V_N),$$

and the range of **f** by

$$\mathscr{R}_\mathbf{f} \ (= \{\mathbf{f}(x_1,..., x_N) \mid (x_1,...,x_N) \in \mathscr{D}_\mathbf{f}\}).$$

Let R^* denote the minimum of the set $\mathscr{R}_\mathbf{f}$, which we assume exists;† R^* always exists when $\mathscr{D}_\mathbf{f}$ is finite, but need not exist otherwise. Our objective is to find a member $x^* \in \mathscr{D}_\mathbf{f}$ which yields the minimum value R^*, i.e., such that

$$\mathbf{f}(x^*) = R^* = \mathbf{min}_{(x_1,..., x_N) \in \mathscr{D}_\mathbf{f}} \{\mathbf{f}(x_1,..., x_N)\}.$$

There may be more than one such member of $\mathscr{D}_\mathbf{f}$; if so, any one will do. We say that the minimization of **f** with respect to the variables $x_1,..., x_N$ is 'constrained' by the choice of $\mathscr{D}_\mathbf{f}$. Two important ways in which constraints can be specified are in terms of either an equality $\mathbf{g}(x_1, x_2,..., x_N) = c$ or an inequality $\mathbf{g}(x_1, x_2,..., x_N) \leq b$; then

$$\mathscr{D}_\mathbf{f} = \{(x_1,..., x_N) \in V_1 \times \cdots \times V_N \mid \mathbf{g}(x_1,..., x_N) = c\}$$

†This assumption is necessary since R^* need not exist. For example, if $\mathbf{f}(x) = x$, and x may be arbitrarily negative, then $\mathscr{R}_\mathbf{f}$ has no minimum value. Or if $\mathbf{f}(x) = x$ and $\mathscr{D}_\mathbf{f} = (0,1)$ = the intervals of reals between 0 and 1, noninclusive, then the minimum value of $\mathscr{R}_\mathbf{f}$ (or actually its greatest lower bound) is 0, which is not in $\mathscr{R}_\mathbf{f} = (0,1)$.

or

$$\mathcal{D}_{\mathbf{f}}=\{(x_1,\ldots, x_N)\in V_1\times\cdots\times V_N \mid \mathbf{g}(x_1,\ldots, x_N)\le b\}.$$

Note that the imposition of a constraint reduces the domain of values for which **f** is defined. (For real functions of real variables, this may affect the differentiability of **f**.)

In situations where calculus-based methods are inapplicable, especially when we consider the discrete (as opposed to continuous) systems that arise in computer science, (optimizing) *search* techniques are commonly used to solve the general optimization problem posed here. When the domain is a finite set, an 'exhaustive search' is possible in principle (but not in practice, if the set is very large): this involves evaluating $\mathbf{f}(x_1,\ldots, x_N)$ for each of the members of $\mathcal{D}_{\mathbf{f}}$, and then finding the minimum of the finite set of resulting values. Although frequently impractical, exhaustive search may be the only possible way to find an absolute minimum for some problems. Where a relative minimum is sought, several 'guided search' techniques may be used instead, based on the assumption that the gradient ($\nabla\mathbf{f}$), as we approach a relative minimum, becomes closer and closer to zero. For a discussion of some guided search techniques, such as 'steepest descent', 'relaxation', and 'random search', see Bekey and Karplus [8.5]. Briefly, these techniques start by evaluating **f** at one (or two) arbitrarily chosen initial point(s), and then proceed to evaluate **f** at some other point chosen on the basis of the prior values of **f** or of its gradient $\nabla\mathbf{f}$. The basic idea is to choose next points so as to reduce **f** until no further reduction is possible, hopefully in a short enough time to be practicable.

8.3.4 Mathematical Programming

Dynamic programming is one of many optimization techniques known collectively by the name 'mathematical programming'. It has the advantage of being very general, but is often too inefficient. We briefly mention some other mathematical programming techniques below. For details, see Hillier and Lieberman [8.26] or other 'operations research' texts.

Suppose we wish to find

$$\mathbf{min}_{\{x_1, x_2,\ldots, x_N\}}\mathbf{f}(x_1, x_2,\ldots, x_N)$$

subject to inequality constraints of the form

$$\mathbf{g}_j(x_1, x_2,\ldots, x_N)\ge b_j$$

where **f** and $\mathbf{g}_j$ are *linear* functions, i.e., where

$$\mathbf{f}(x_1, x_2,\ldots, x_N)=a_1x_1+a_2x_2+\cdots+a_Nx_N$$

and

$$g_j(x_1,x_2,\ldots,x_N)=\beta_{1j}x_1+\beta_{2j}x_2+\cdots+\beta_{Nj}x_N.$$

Furthermore, suppose that the variables x_i must be nonnegative real numbers; $x_i \geq 0$. Under these assumptions, the optimization problem is called a *linear programming* problem, and can be solved using what is known as the 'simplex' method. This method is described in numerous elementary books, such as Barsov [8.3]; see Dantzig [8.19] for details. It is also the basis of standard mathematical programming packages such as Linpack [8.20].

In *integer programming,* the variables $\{x_1,\ldots,x_N\}$ are restricted to be integers. Sometimes these integers are further restricted to have Boolean values 0 or 1. For an application to paging, see Kral [8.34]. For an application to the assignment of files, see Dowdy [8.21].

A common approach to solving integer programming problems is the *branch-and-bound* methodology which is characterized by the task of bounding $f(x_1,\ldots,x_N)$ for various subsets of $\{x_1,\ldots,x_N\}$. If the lower bound for one subset is greater than the lower bound for another, then we may restrict our attention to the latter subset (provided, of course, that the lower bound is achievable). See Lawler [8.37] for a survey. Branch and bound techniques are commonly used in artificial intelligence applications; see Nilsson [8.42]. For an application to hierarchical memory allocation, see Ramamoorthy [8.44]. For an application to decision tables, see Reinwald [8.45].

EXERCISES

1. For the properties in Section 8.1.2, give their maximization analogues.
2. Show that property (P1) is not necessarily true for negative c.

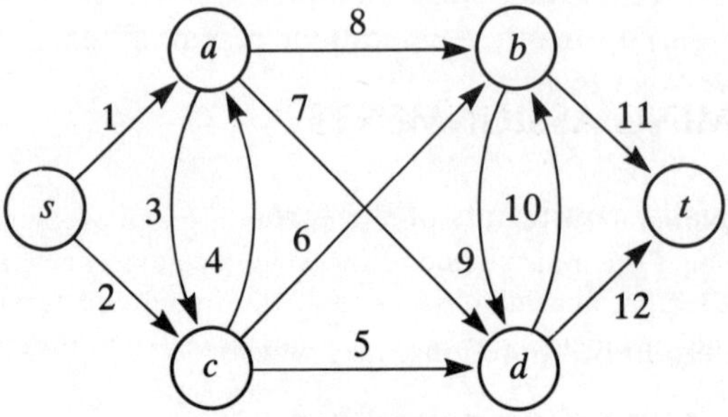

Figure 8A

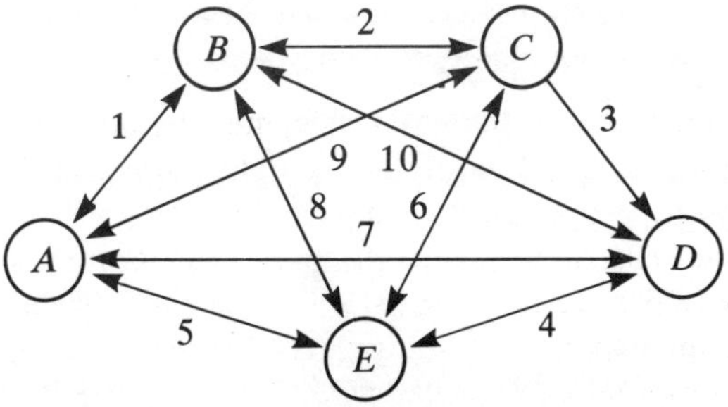

Figure 8B

3. Show that

$$|a| \geq a \geq \min\{b,c\} \qquad \text{for} \qquad c \leq a \leq b.$$

4. Given the graph shown in Figure 8A, find the shortest path from node s to node t using dynamic programming.

5. Given the graph shown in Figure 8B, solve the traveling-salesman problem.

***6.** Show how the index register allocation problem of Horowitz [8.28] can be solved using dynamic programming.

7. Recalling Theorem 1, what does

(a) $d^2/dx^2\ [\mathbf{f}(x^*)] < 0$, and
(b) $d^2/dx^2\ [\mathbf{f}(x^*)] = 0$

imply?

***8.** Solve the following linear programming problem:

$$\min_{\{a,b\}} 3a + 5b$$

subject to $4a + 2b \geq 6,\ a \geq 0,\ b \geq 0.$

***9.** Read one of the references cited in Section 8.2.6, and reformulate the problem in the notation of this Chapter.

PROGRAMMING ASSIGNMENTS

1. Write a program to solve the shortest-path-in-a-graph problem using dynamic programming. (The input should be the weighted adjacency matrix representation of the graph, and identifications of the initial and terminal nodes; the output should include the sequence of nodes traversed by the shortest path, and its length.)

2. Write a program which solves the traveling-salesman problem using dynamic programming.

3. Write a program which searches for a maximum of the function:

(a) $\mathbf{f}(x)=ax^2+bx+c$ for $0 \le x \le 1$.
(b) $\mathbf{f}(x,y)=x^2+y^2-2xy$ for $0 \le x \le 1$, $0 \le y \le 1$.

4. Use a computer to solve the linear programming exercise of preceding section. (Use a 'package' if available.)

***5.** Write a program which determines the 'best' parabolic (polynomial of degree 2) approximation to $\mathbf{F}(x)=e_x$ for $x \in [0,1]$, using one of the approaches mentioned in Section 8.1.6.

6. Verify computationally the dynamic programming solution to the linear search problem of Section 8.2.7; i.e., write a program which solves a specific numerical example.

***7.** Write a computer program which solves one of the applications cited in Section 8.2.6.

9 APPLICATIONS—OPERATING SYSTEMS

9.1 APPLICATIONS TO GRAPHS AND TREES

In this section, we discuss several applications of dynamic programming, as well as some relatively special-purpose optimization techniques, which have great importance to problems both within and without computer science. Within computer science, applications to operating systems are of major interest. However, we first consider problems which can be formulated in graph-theoretic terms.

9.1.1 Minimal Weighted-depth Binary Tree

Given a set of N objects with numerical weights $W=\{w_1, w_2,\ldots, w_N\}$, we wish to construct a full binary tree having the object weights as its leaves. We define the *weighted depth* $\hat{D}$ of the tree as the weighted average of the depths of its leaves;

$$\hat{D}=\sum_{i=1}^{N} w_i \times \textbf{depth}\ (a_i).$$

(If all of the weights are equal to $1/N$, then the weighted depth is just the average depth of the tree, as defined in Section 4.2; when $N=1$, $\hat{D}=0$.) Figure 9.1.1(a) shows the possible ways that weights $W=\{1, 2, 3\}$ can be placed as leaves in a full nonpositional binary tree; of these trees, the first has minimal weighted depth. Our objective is to determine the minimal weighted depth tree for any given set W. We shall proceed by constructing the tree from 'bottom up' (i.e., from leaf level to the root). The decision to be

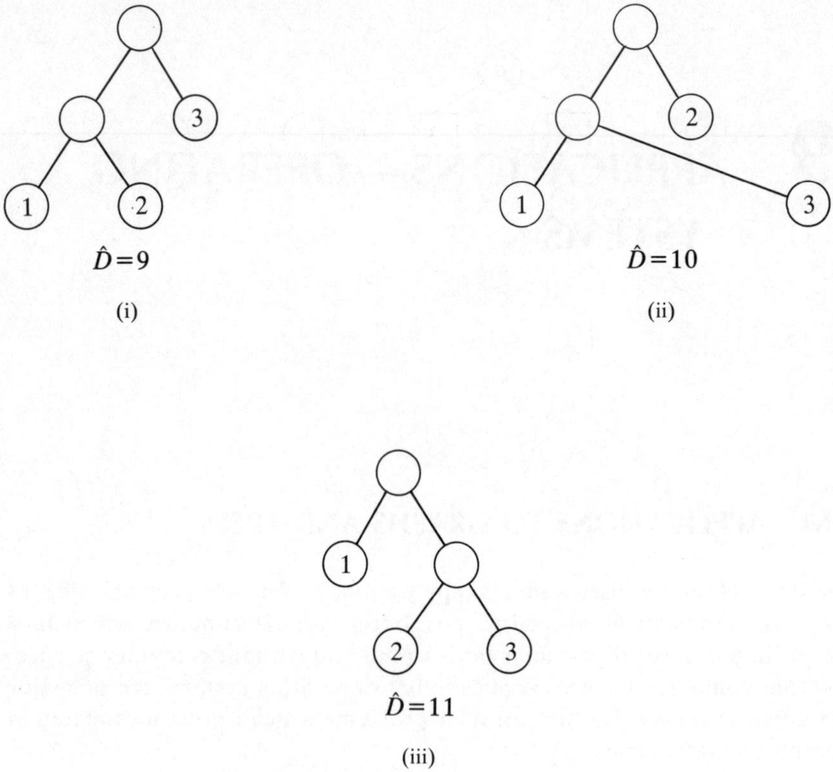

Figure 9.1.1(a)

made at each stage of the tree's construction is which two nodes (from a state set S) are to be combined next, thereby eliminating these nodes (from S), but introducing a new internal node (to S) in the process. Initially, S consists of the N leaf nodes corresponding to the given N weights. The procedure terminates when S consists of only one node (the root!).

Example Let $W=\{1, 2, 3, 4, 5\}$, and so, initially, $S=\{1, 2, 3, 4, 5\}$, as shown in Figure 9.1.1(b)(i). Suppose we first choose to combine leaf nodes 2 and 3; S then becomes

$$S-\{2,3\}\cup\{a\}=\{1, a, 4, 5\},$$

where a is introduced as the parent of nodes 2 and 3. The new set of nodes is shown in Figure 9.1.1(b)(ii), where square nodes indicate those that have been eliminated

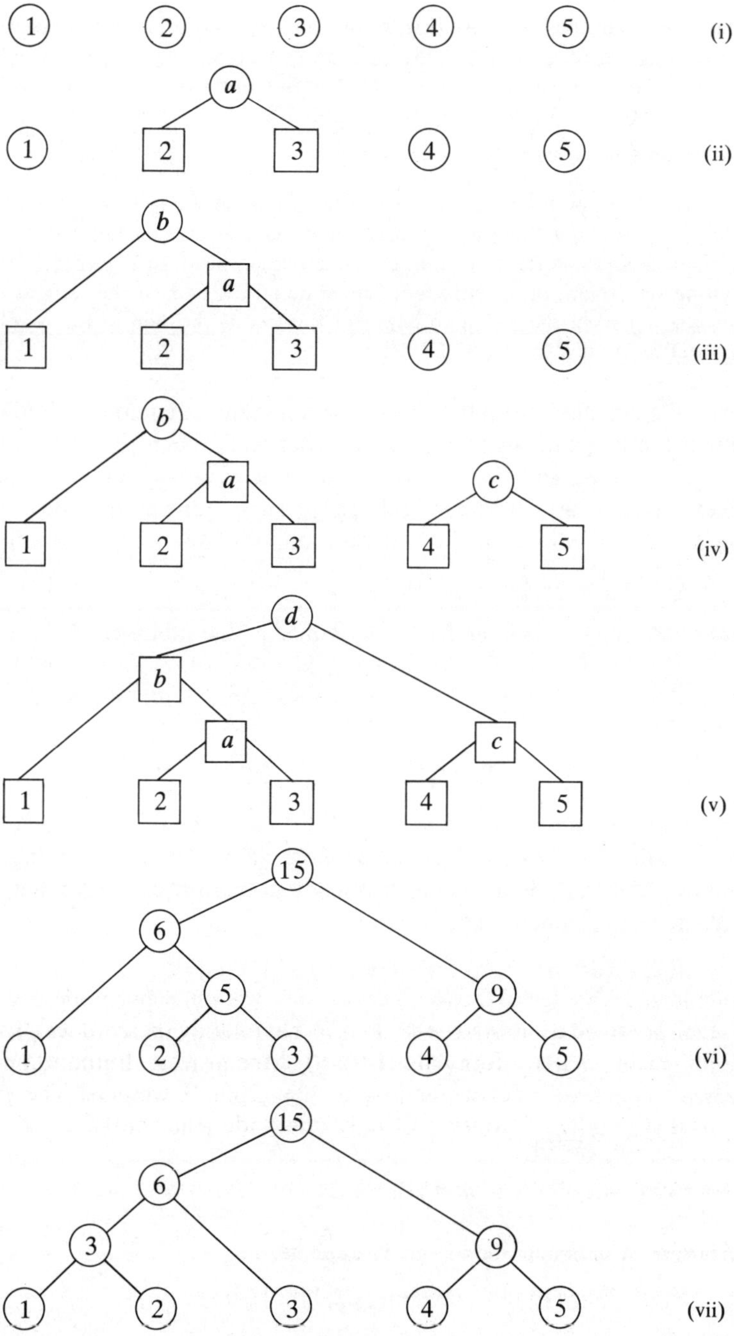
1
2
3
4
5
(i)
a
1
2
3
4
5
(ii)
b
a
1
2
3
4
5
(iii)
b
a
c
1
2
3
4
5
(iv)
d
b
a
c
1
2
3
4
5
(v)
15
6
5
9
1
2
3
4
5
(vi)
15
6
3
9
1
2
3
4
5
(vii)

Figure 9.1.1(b)

from consideration. Suppose that we next choose to combine leaf node 1 with new node a; S then equals $\{b, 4, 5\}$ as shown in Figure 9.1.1(b)(iii). Continuing, S becomes $\{b,c\}$ as shown in Figure 9.1.1(b)(iv), and finally $\{d\}$ as shown in Figure 9.1.1(b)(v). The resulting tree has weighted depth

$$1\times2+2\times3+3\times3+4\times2+5\times2=35.$$

In this example, the decisions as to which nodes are to be combined next were made arbitrarily, so the resulting tree cannot be expected to be optimal. Note that the weighted depth of the tree can be evaluated, as shown in Figure 9.1.1(b)(vi), by letting the weight of an introduced node equal the sum of the weights of its two successor nodes: the weighted depth of the tree is the sum of the weights of the introduced nodes $(5+6+9+15=35)$.

We conclude from the above example that, in combining and eliminating two nodes with weights w_i and w_j from set S, the introduced node should be given the weight w_i+w_j. Furthermore, the weighted depth of a tree T with leaves $\{w_1,\ldots, w_n\}$, where w_i and w_j have a common parent node α, is greater by the sum w_i+w_j than the weighted depth of the tree T' having leaves

$$\{w_1,\ldots, w_n\}-\{w_i,w_j\}\cup\{w_i+w_j\},$$

where T' can be obtained from T by replacing the subtree of T rooted at α by a single leaf having weight (w_i+w_j). In other words, w_i+w_j is the cost of deciding to combine w_i and w_j, and the weighted depth of the tree T is given by

$$\hat{D}(T)=(w_i+w_j)+\hat{D}(T').$$

If we let S denote the set of weights of the leaf nodes of T, and let $\mathbf{f}(S)$ denote the minimal weighted depth achievable among all trees having S for its leaves, then applying the principle of optimality we have the dynamic programming functional equation

$$\mathbf{f}(S)=\min_{\{w_i,w_j\}\subseteq S}\{(w_i+w_j)+\mathbf{f}(S-\{w_i,w_j\}\cup\{w_i+w_j\})\},$$

which is solved by starting with S equal to the original set of weights W, and ending when $\#(S)=1$ for which $\mathbf{f}(S)=0$. We may instead stop when $\#(S)=2$ since

$$\mathbf{f}(\{w,w'\})=(w+w')$$

for any two weights w and w'.

Example Continuing the foregoing example,

$$\mathbf{f}(\{1,2,3,4,5\})=\min_{(i,j)}\{(i+j)+\mathbf{f}(\{1,2,3,4,5\}-\{i,j\}\cup\{i+j\})\}$$

$$=\min\left\{\begin{array}{l}(1+2)+\mathbf{f}(3,4,5,3)\\(1+3)+\mathbf{f}(2,4,5,4)\\\dots\\(2+3)+\mathbf{f}(1,4,5,5)\\\dots\\(4+5)+\mathbf{f}(1,2,3,9)\end{array}\right\}$$

$$\mathbf{f}(3,4,5,3)=\min\left\{\begin{array}{l}(3+4)+\mathbf{f}(5,3,7)\\\dots\\(5+3)+\mathbf{f}(3,4,8)\end{array}\right\}$$

$$\mathbf{f}(5,3,7)=\min\left\{\begin{array}{l}(5+3)+\mathbf{f}(7,8)\\(5+7)+\mathbf{f}(3,12)\\(3+7)+\mathbf{f}(5,10)\end{array}\right\}$$

$$\mathbf{f}(7,8)=(7+8)$$

etc. We leave as an exercise the demonstration that

$$\begin{aligned}\mathbf{f}(1,2,3,4,5)&=(1+2)+\mathbf{f}(3,4,5,3)\\&=(1+2)+(3+3)+\mathbf{f}(4,5,6)\\&=(1+2)+(3+3)+(4+5)+\mathbf{f}(6,9)\\&=(1+2)+(3+3)+(4+5)+(6+9)\\&=3+6+9+15=33.\end{aligned}$$

The corresponding optimal tree is shown in Figure 9.1.1(b)(vii).

Remark Huffman [9.30] studied the minimal weighted-depth binary tree problem from another point of view and showed in effect that the two smallest values $\{w_i,w_j\}$ in S are always minimizing values. Utilizing this fact, $\mathbf{f}(S)$ can be found without performing any minimization operations at all. For example, in the foregoing example,

$\{1,2\}$ are the smallest values in $\{1,2,3,4,5\}$
$\{3,3\}$ are the smallest values in $\{3,4,5,3\}$
$\{4,5\}$ are the smallest values in $\{4,5,6\}$
$\{6,9\}$ are the smallest values in $\{6,9\}$.

Application: Coding A problem we mentioned in Section 7.1.1 is that of finding the 'best' (minimal mean word length) instantaneous binary code for 10 source characters having probabilities $\{p_1,p_2,\dots,p_{10}\}$; this is called a *Huffman* code. The problem can be solved by finding the minimal weighted depth tree with the probabilities as the weights. Such a tree, for equal probabilities ($p_i=0.1$), is shown in Figure 9.1.1(c)(i). If branches to the left are labeled 0 and those to the right are labeled 1, and the leaves are given corresponding path labels, as shown in Figure 9.1.1(c)(ii), then these leaf labels constitute the desired code (cf. Code 5 of Table 7.1.1). Note that the tree can be used for decoding by examining one bit at a time. A tree of this nature is also called a *radix* tree.

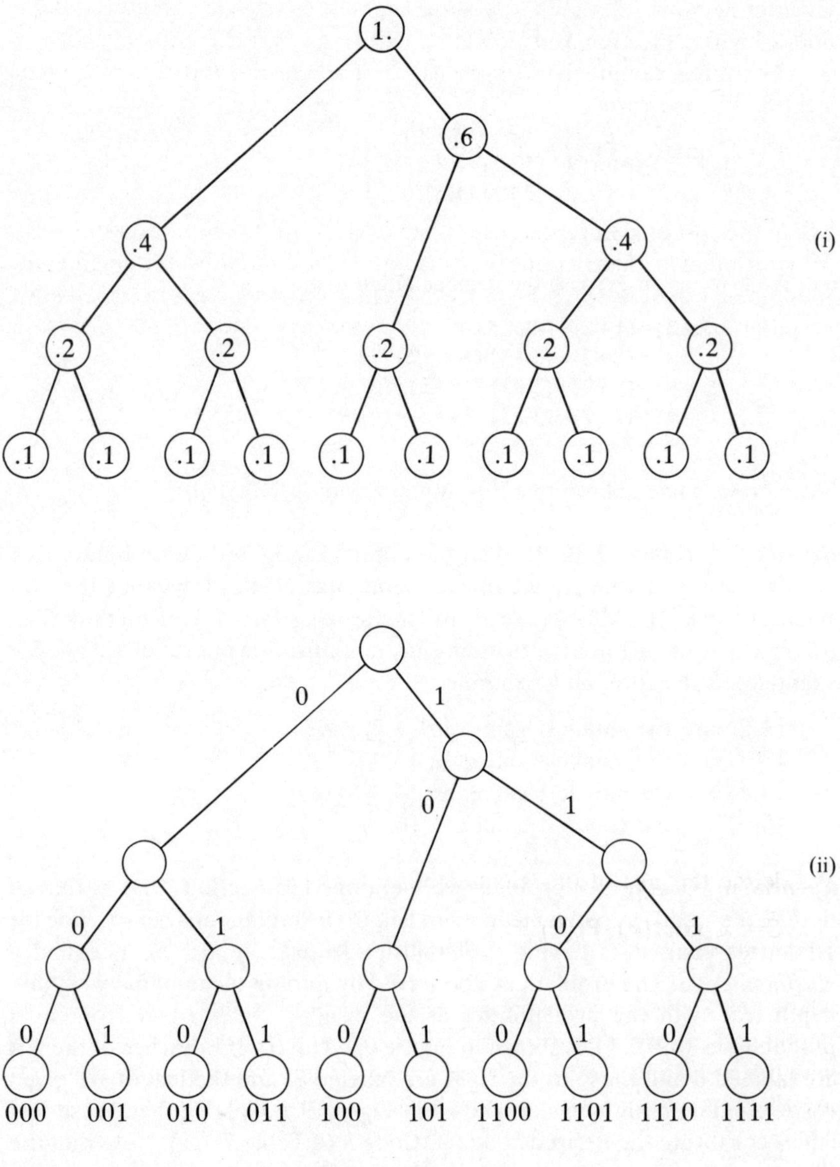

Figure 9.1.1(c)

9.1.2 Optimal Sorted Binary Tree

In this section, we consider the problem of constructing the optimal sorted binary tree for a set Δ of keyed data items. Recall the discussion of binary tree searching given in Section 5.3.3, and the analysis of the cost of a tree given in Section 7.2.4. While a good heuristic is to place high probability nodes high in the tree, our objective here is to find the optimal policy.

Let us re-examine the formula used to find the cost $\hat{C}$ of a sorted binary tree: it is of the form

$$\hat{C}=\sum_{\substack{\text{internal}\\ \text{node } i}} \mathbf{cost}(i)\cdot\mathbf{Pr}(i)+\sum_{\substack{\text{terminal}\\ \text{node } j}} \mathbf{cost}(j)\cdot\mathbf{Pr}(j)$$

where the cost of an internal (round) node is its depth plus one, and the cost of a terminal (square) node is its depth. Consider now the augmented (nonbinary) tree shown in Figure 9.1.2(a), where we have introduced additional terminal nodes (shown hatched) to represent data present in the tree.

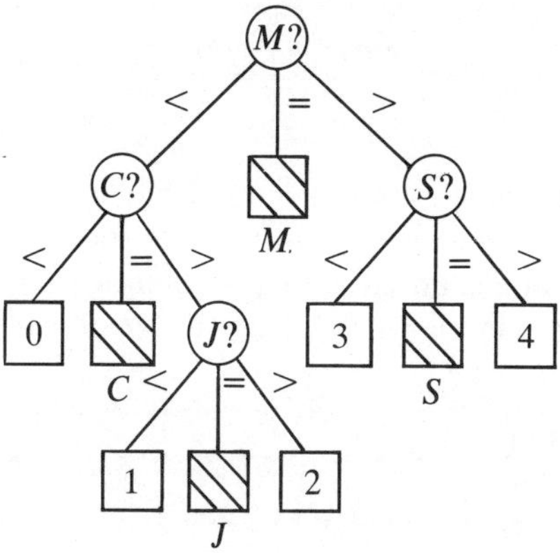

Figure 9.1.2(a)

We define the cost of this augmented tree as

$$\tilde{C}=\sum_{\substack{\text{terminal}\\ \text{node } k}} \mathbf{cost}(k)\cdot\mathbf{Pr}(k)$$

where the cost of any terminal (hatched or nonhatched) node equals its depth, and the probability of a hatched node equals that of its parent node; with this definition, $\tilde{C}=\hat{C}$. The costs of internal nodes and their probabilities are thereby subsumed by the new nodes, hence

$$\tilde{C}=2q(0)+2p(C)+3q(1)+3p(J)$$
$$+3q(2)+1p(M)+2q(3)+2p(S)+2q(4),$$

where $p(k)$ denotes the probability of a hatched terminal node (representing data in the tree), and $q(k)$ denotes the probability of a nonhatched terminal node (representing missing data).

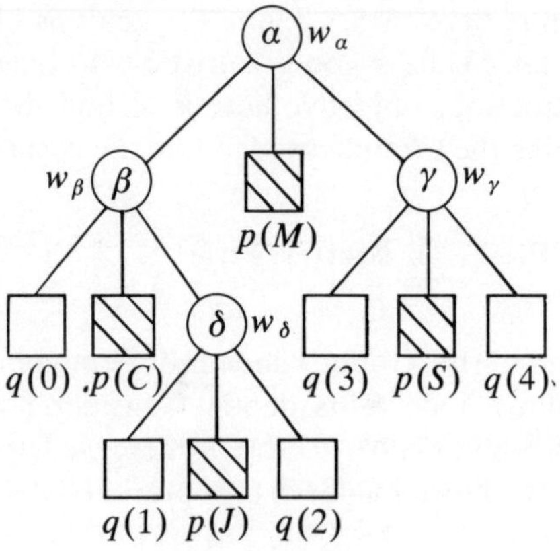

Figure 9.1.2(b)

Suppose now that each internal node i has an associated 'weight' w_i equal to the sum of the weights of its immediate successor nodes, where the weights of terminal nodes equal their probabilities. For the same tree as before, shown redrawn in Figure 9.1.2(b), the weights of the internal nodes are as follows:

$$
\begin{aligned}
w_\delta &= q(1)+p(J)+q(2) \\
w_\beta &= q(0)+p(C)+w_\delta \\
&= q(0)+p(C)+q(1)+p(J)+q(2) \\
w_\gamma &= q(3)+p(S)+q(4) \\
w_\alpha &= w_\beta+p(M)+w_\gamma \\
&= q(0)+p(C)+q(1)+p(J)+q(2)+p(M)+q(3)+p(S)+q(4).
\end{aligned}
$$

Note that the weight of an internal node i equals the sum of the weights of the terminal nodes in the subtree rooted at i. Note also that the sum of the weights of all of the internal nodes $(w_\alpha+w_\beta+w_\gamma+w_\delta)$ is the cost of the tree $\tilde{C}$. Unfortunately, unlike the case discussed in Section 9.1.1, it is not possible to construct the optimal tree 'bottom up' by combining leaves having minimal weights. The main reason for this is that the tree must be sorted, so minimal weight leaves which are not 'adjacent' cannot be combined. Instead, we employ a 'top down' approach: specifically, we initially seek the best choice for the root of the tree, then seek best choices for the roots of the possible subtrees, and so on.

Recall from above that the cost of a tree (e.g., the one in Figure 9.1.2(b)) equals the sum of the weights of its internal nodes. We may generalize this to define the cost of any of its subtrees. In other words,

$$\begin{aligned}
C_\alpha &= \text{cost of subtree rooted at } \alpha = w_\alpha + w_\beta + w_\delta + w_\gamma \\
C_\beta &= \text{cost of subtree rooted at } \beta = w_\beta + w_\delta \\
C_\gamma &= \text{cost of subtree rooted at } \gamma = w_\gamma \\
C_\delta &= \text{cost of subtree rooted at } \delta = w_\delta .
\end{aligned}$$

Note that

$$\begin{aligned}
C_\beta &= w_\beta + w_\delta = (q(0) + p(C) + w_\delta) + w_\delta \\
&= q(0) + p(C) + 2q(1) + 2p(J) + 2q(2);
\end{aligned}$$

as we would expect, terminal node probabilities are multiplied by their depths in the subtree.

For the tree shown in Figure 9.1.2(b), what is needed is a formula relating the cost of choosing (say) M for the root with the costs of its left and right subtrees, which we denote by $\mathbf{f}(\{C,J\})$ and $\mathbf{f}(\{S\})$, respectively. Employing dynamic programming, the cost of the optimal tree can be found by solving a functional equation of the form

$$\mathbf{f}(\{C,J,M,S\}) = \mathbf{min} \left\{ \begin{array}{l} \mathbf{cost}(C) + \mathbf{f}(\emptyset) + \mathbf{f}(\{J,M,S\}) \\ \mathbf{cost}(J) + \mathbf{f}(\{C\}) + \mathbf{f}(\{M,S\}) \\ \mathbf{cost}(M) + \mathbf{f}(\{C,J\}) + \mathbf{f}(\{S\}) \\ \mathbf{cost}(S) + \mathbf{f}(\{C,J,M\}) + \mathbf{f}(\emptyset) \end{array} \right\}$$

where $\mathbf{f}(\Delta)$ denotes the cost of the optimal sorted tree containing the data items in the set Δ. The basic problem is that of determining the 'decision cost' function $\mathbf{d}(k,\Delta)$ associated with choosing, as the root, node k from the set Δ, with all nodes in Δ less than k being in the left subtree and all nodes in Δ greater than k being in the right subtree. More formally, we write the dynamic programming functional equation

$$\mathbf{f}(\Delta) = \mathop{\mathbf{min}}_{k \in \Delta} \{\mathbf{d}(k,\Delta) + \mathbf{f}(\{k' \in \Delta \mid k' < k\}) + \mathbf{f}(\{k' \in \Delta \mid k' > k\})\}$$

where $\mathbf{f}(\emptyset) = 0$. To proceed, we must define $\mathbf{d}(k,\Delta)$ in a manner consistent with our means of evaluating the cost of a given subtree.

We conclude that, if Figure 9.1.2(b) is the optimal tree, then $\mathbf{d}(M,\Delta)$ must satisfy

$$\mathbf{f}(\{C,J,M,S\}) = \mathbf{d}(M,\Delta) + \mathbf{f}(\{C,J\}) + \mathbf{f}(\{S\})$$

or

$$C_\alpha = \mathbf{d}(M,\Delta) + C_\beta + C_\gamma .$$

Hence, $\mathbf{d}(M,\Delta)$ must equal w_α; note that $\mathbf{d}(M,\Delta)$ depends only upon Δ, and not upon the choice M. In general, $\mathbf{d}(k, \Delta)$ equals the sum of the weights of the terminal nodes (both hatched and nonhatched) in the subtree which contains the data items in the set Δ; i.e., we define

$$\mathbf{d}(k, \Delta) = \Sigma_{\kappa \in \Delta} p(\kappa) + \Sigma_j q(j)$$

where the second term is summed over those missing data nodes adjacent to the keys in Δ.

9.1.3 Shortest and Longest Paths (Optimal Routing)

Optimal routing refers to the problem of finding the shortest or longest path in a graph. Such problems can be solved by a variety of more efficient methods than dynamic programming, but none more general. While general methods are important, so are efficient specialized methods, such as that of Dijkstra [9.17]. However, here we shall only illustrate a dynamic programming solution. The decision table of Table 9.1.3 computes the length of the shortest path from node 1 to node N in a digraph whose branch lengths are given in the weighted adjacency matrix D.

$\lambda=?$	0	1	1	1	1
$I \geq 1$	–	**T**	**T**	**T**	**F**
$J \leq N$	–	**T**	**T**	**F**	–
$NU > D(I,J)+C(J)$	–	**T**	**F**	–	–
$\lambda:=1$; $XX:=\infty$; **get** N, $D(*,*)$	×	–	–	–	–
$C(N):=0$; $V(N):=0$; $I:=N$	×	–	–	–	–
$NU:=D(I,J)+C(J)$	–	×	–	–	–
$VE:=J$	–	×	–	–	–
$J:=J+1$	–	×	×	–	–
$C(I):=NU$; $V(I):=VE$	–	–	–	×	–
$I:=I-1$; $J:=I+1$; $NU:=XX$; $VE:=J$	×	–	–	×	–
put "$LEN=$" $C(1)$	–	–	–	–	×
exit	–	–	–	–	×

Table 9.1.3

One important generalization is the case where branch lengths may be negative and the graph is cyclic. Then shorter and shorter paths become possible by repeatedly traversing a loop of negative length. In this event, the shortest acyclic path may be found instead. Furthermore, the longest (acyclic) path may be found by simply negating all branch lengths. Other variations include finding the shortest path between each pair of nodes in a graph, finding the nth shortest path for $n \geq 1$, and finding the path with shortest expected length when branch lengths are stochastic, i.e., known only probabilistically. In each case, various constraints can also be taken into account. See Dreyfus and Law [9.19] and Loui [9.44] for further information.

Remark Numerous computer science problems have been solved by utilizing graph models and optimal routing and touring formulations. Most of the applications mentioned in Section 8.2.6 have such formulations.

9.1.4 Scheduling (PERT/CPM)

A scheduling problem arises when there are several processes (tasks, activities, jobs) which must be executed. A *schedule* is a specification of the times during which each process is executed. The *length* of a schedule is the time which elapses between the initiation of the first process(es) and the termination of the last process(es). An optimal schedule is one having minimal length subject to contraints of various kinds. For example, the order in which processes are executed may be subject to *precedence* and *resource* constraints. Precedence is a chronological constraint which requires that execution of some process i be terminated before execution of another process j can be initiated. A resource constraint prohibits simultaneous execution of two or more processes because their combined resource requirements exceed the supply of available resources. A processor is an example of a resource which may be in limited supply. In this section, we assume only precedence constraints.

If processes are represented as nodes in a digraph, then the constraint that process i must precede process j (or j must succeed i) can be represented as a branch from node i to node j. Such a graph is called a *precedence graph*. For example, Figure 9.1.4(a) represents the situation where process A must precede processes B and C, processes B and C must precede process D, process C must precede processes D and E, and processes D and E must precede process F. Precedence graphs must, of course, be acyclic.

An alternative is to represent processes as branches in a digraph. Then the constraint that process i must precede process j can be represented by having branch i incident *to* a node *from* which branch j is also incident. However, the case where j has another predecessor i' and i has another successor j', but where i' need not precede j', requires special treatment: a 'dummy' branch must be inserted between i and j. Such dummy branches are said to represent dummy processes which need not actually be executed, and

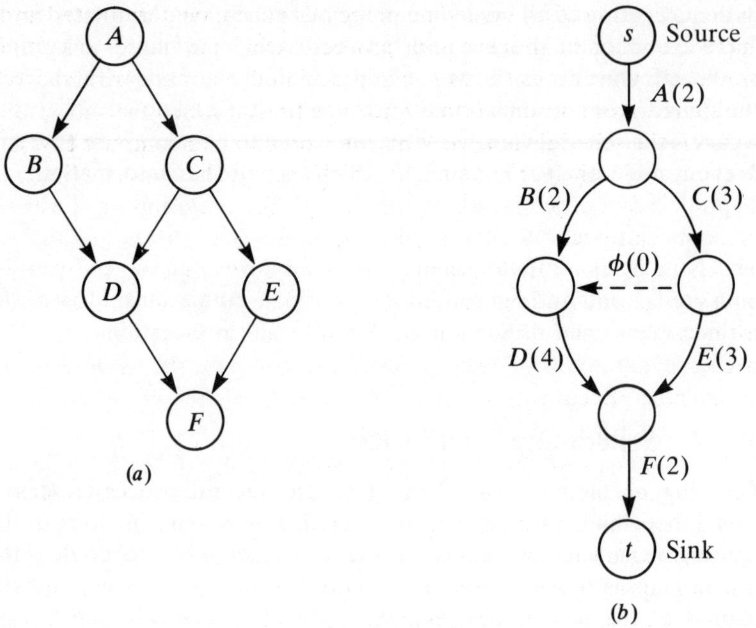

Figure 9.1.4

consequently need no resources. Figure 9.1.4(b) illustrates this alternative representation for the foregoing example. (A dummy branch ϕ, shown dashed, has been added between branches C and D.) Such a digraph is called a *PERT graph*.

PERT/CPM PERT and CPM are names given to certain approaches suitable for finding the longest or 'critical' path in a graph. We may, of course, adapt dynamic programming or other methods for this purpose, but the use of PERT/CPM yields additional information relevant to the scheduling applications for which they were designed. For an introduction, see Armstrong-Wright [9.5], and for further details, see Kaufmann [9.34].

By convention, PERT graphs have only one source and one sink node, and their branches are numerically labeled. A branch label is interpreted as the time required to execute the corresponding process (i.e., the process duration). Process durations are shown parenthesized in Figure 9.1.4(b); e.g., process C takes 3 units of time. A dummy process has zero duration. The single source and sink requirements are not restrictive since multiple sources and sinks can be merged into single ones.

The nodes of a PERT graph are said to represent *events,* i.e., the states at which the execution of all preceding processes must have terminated and at which the execution of all succeeding processes can be initiated. Since the source node has no predecessor, we use its associated event as our reference point: the source event occurs at time zero. The time at which the sink node event occurs is the schedule length. What we would like to compute first is, for each event node, the *earliest* time at which the event can occur (i.e., at which all predecessor processes have terminated). The earliest time at which the sink event can occur is of special importance; we call it the *critical* schedule length of the PERT graph. The critical schedule length is the minimum time required to execute all the processes, which is achievable if there are no delays other than those dictated by precedence constraints.

For each event node, we also would like to compute the *latest* time at which the event can occur without unnecessarily delaying the sink event, i.e., without increasing the schedule length. The difference between the latest and the earliest times is known as the event's *slack*. Of greater interest is the difference between the latest time associated with any event node and the earliest time associated with an immediate predecessor event node. If this time difference equals the duration of the process connecting the two event nodes, then the process is said to be *critical* (or to have a zero 'float'); if the time difference exceeds the duration, the balance is called the *float* of the process. A critical process must be initiated immediately upon completion of all of its predecessor processes, because at least one of its successor processes is also critical. A noncritical process can be delayed for a period of time not exceeding its float.

In summary, the critical schedule length of a PERT graph is achievable if critical processes are not delayed, and if noncritical processes are not delayed for longer than their floats. A sequence of branches associated with critical (zero float) processes traces a path from the source node to the sink node. Such a sequence is called a *critical path*. Critical paths can be found by computing the earliest and latest times associated with each event node, and then computing floats for each process branch.

Formalization A PERT graph with events E and processes P is defined as a weighted acyclic connected digraph (E,P), where $P(\subseteq E \times E)$ is a numerically labeled set, and where E contains a unique node (the source) with zero in-degree and a unique node (the sink) with zero out-degree. The branch labels or weights represent (nonnegative) process durations; we denote the duration of process (e,e') by $\mathbf{d}(e,e')$. Let $E=\{e_1, e_2, \ldots, e_N\}$ be ordered topologically, so that e_1 is the source, e_N is the sink, and $(e_i,e_j)\notin P$ for $j<i$. Finally, define $\mathbf{t}^-(e)$ as the earliest time at which event e can occur, and $\mathbf{t}^+(e)$ as the latest time at which event e can occur (without increasing the schedule length). Then, we may proceed by computing

$$\mathbf{t}^-(e_i) = \max_{(e_j,e_i)\in P}\{\mathbf{t}^-(e_j)+\mathbf{d}(e_j,e_i)\}, \qquad \mathbf{t}^-(e_1)=0$$

$$\mathbf{t}^+(e_i) = \min_{(e_i,e_j)\in P}\{\mathbf{t}^+(e_j)-\mathbf{d}(e_i,e_j)\}, \qquad \mathbf{t}^+(e_N)=\mathbf{t}^-(e_N).$$

Furthermore,

$$\textbf{slack}\ (e) \ = \mathbf{t}^+(e)-\mathbf{t}^-(e)$$
$$\textbf{float}\ (e_i,e_j) = \mathbf{t}^+(e_j)-\mathbf{t}^-(e_i)-\mathbf{d}(e_i,e_j).$$

We define the length of a path (from the source to the sink) as the sum of the durations of the processes which comprise the path. The longest path in the numerically branch-labeled PERT graph must be a critical path since delay of any of the processes comprising the path necessarily delays the sink event. Thus, the length of a longest path is the critical schedule length. In the above, a critical path is a sequence of branches (e_i,e_j), each of whose float is zero, with path length equal to $\mathbf{t}^+(e_N)$ or $\mathbf{t}^-(e_N)$.

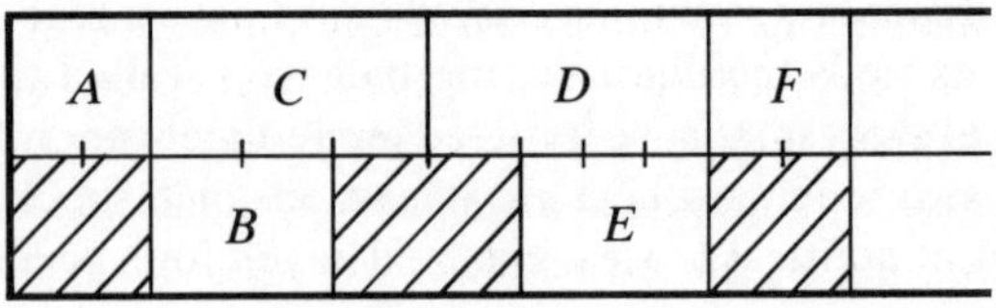

Figure 9.1.4(c)

Example Consider again the PERT graph and process durations given in Figure 9.1.4(b). The critical path consists of the sequence $\langle A,C,\phi,D,F\rangle$, which has length 11. Therefore, in an optimal schedule, processes *A, C, D,* and *F* would be performed in sequence (without intervening delays), process *B* would be performed concurrently with process *C,* and process *E* would be performed concurrently with process *D*. Such a schedule is depicted in Figure 9.1.4(c), where time is represented horizontally; a timing diagram of this sort is called a *Gantt chart*. Note that, in this example, process *B* is scheduled as early as possible, while process *E* is scheduled as late as possible.

The foregoing schedule assumes that concurrent processing is possible; if not, then a schedule with critical length is not realizable. (In fact, if only one process at a time can be performed, then every schedule has the same length.) We discuss scheduling with concurrency constraints in Section 9.2.2; in our discussion of PERT, we assumed that there are no such constraints. See Lloyd [9.43].

Application*: *Multiprocessing We describe one related problem here, that of determining the 'degree of concurrency' of a PERT graph—i.e., the maximum number of processes which may have to be executed concurrently in order for the critical schedule length to be achievable. It is simply the size of the largest directed cut (see Section 4.1.2) which disconnects the source from the sink, for it may be necessary for the processes in this largest cut to be performed concurrently. In the context of computers, this gives the minimum number of processing units required to handle a system of processes without unnecessary delays. While dummy processes must be considered in finding directed cuts, they are not counted in the sizes of the cuts (since they are not actually processed). In the foregoing example, $\{C, \phi, D\}$ is not a directed cut, and the size of the directed cut $\{B, \phi, E\}$ equals 2.

Remark In the foregoing, the assumption was made that process durations are known constants. In practice, these durations may be unknown variables. In order to proceed, of course, some information about the durations must be given. If expected values (means) or most likely values (modes) of process durations are known, then these values can be used directly in critical-path calculations to obtain an approximate answer. If upper and lower bounds on process durations are given, best and worst case schedules can be obtained (although those require a lot more work). The problem of finding a schedule with minimum *expected* length has also been studied for various assumptions on the probability distributions of the process durations. Normal, beta, and triangular distributions, which are easily characterized by minimum, average, and maximum values, are commonly assumed in PERT literature. See, for example, Archibald and Villoria [9.4] and MacCrimmon and Ryadec [9.45].

9.1.5 Network Flow

A directed graph may be used to model 'flows' between nodes along branches in a network. There may be limits to the size or rate of the flow through each branch. For example, in a model of a distributed processing network, nodes represent processors, branches represent channels of communication between processors, and information flows between processors through the channels at a limited data rate.

Formally, we define a *flow network* as a branch-labeled digraph, with a single source and single sink node, and whose numerical branch labels represent the flow *capacities* of the branches. The capacities must be nonnegative. The *flow* in a branch must be between 0 and its capacity; the *remaining* capacity of a branch is the difference between the flow capacity and the current flow in the branch. The flow in a path is limited by the minimum of the flow capacities of the branches in the path.

The branch flows must also satisfy a *conservation law*: at any node X (except for the source and sink), the sum of the flows in each branch incident *to* node X must equal the sum of the flows in each branch incident *from* node X; furthermore, the flow into the sink node must equal the flow out of the source node. If all branch flows were zero, the conservation law would be satisfied trivially. We are interested in finding the maximum flow out of the source node and into the sink node such that the conservation law holds and the branch flow capacities are not exceeded throughout the network.

One solution to this *maximal flow* problem is based on the observation that if the network is 'cut' into two pieces, with the source in one piece and the sink in the other, then the amount of flow from one piece to the other is limited by the forward flow capacities of the cut branches. (*Forward* capacities are those associated with branches directed from the piece containing the source to the piece containing the sink.) This is true for any cut, and in particular is also true for the *minimal cut,* i.e., one in which the sum of the forward capacities of the cut branches is a minimum. This is the basis for what is known as the 'max-flow min-cut' (Ford/Fulkerson [9.21]) theorem:

> The maximum flow out of the source node and into the sink node (such that the conservation law holds and the branch flow capacities are not exceeded) is equal to the minimal value of the forward capacities among all cuts which disconnect the source from the sink.

Unfortunately, this theorem does not provide a practical means for finding the maximal flow, since, in principle, all possible cuts must be examined. We may instead use the procedure described by Hillier and Lieberman [9.26].

We first define the capacity of a path as the capacity of its 'weakest link', i.e., that branch in the path having the least capacity. Each branch has a *reverse* capacity equal to the maximum amount of flow that can flow in the reverse direction along a branch; usually zero initially, the reverse capacity may equal the forward flow since in essence a reverse flow cancels the forward flow. We assume that, initially, there is no flow in the network. We then proceed as follows:

Do repeatedly:

Step 1 Find a path P from the source to the sink with a positive remaining capacity. (If none exists, stop.)

Step 2 Determine the branch in P having the minimum remaining capacity; denote this remaining capacity by C^*.

Step 3 In each branch of P,

- (a) increase the flow by C^*
- (b) decrease the remaining capacity by C^*
- (c) increase the reverse remaining capacity by C^*.

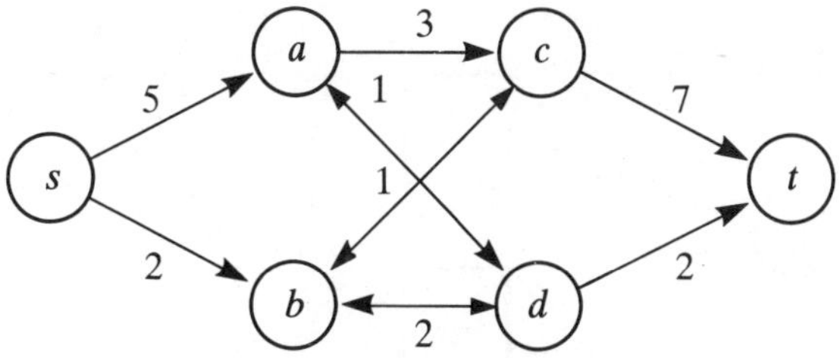

Figure 9.1.5

Using this algorithm for the flow network of Figure 9.1.5, we find that the maximal flow is 6. Other maximal flow algorithms are discussed in Cheung [9.13].

A related problem assumes that each branch has a cost in addition to a capacity. A given flow (generally less than the maximum permitted) from the source to the sink is then specified, and the objective is to find the minimal cost flow of the given size. For an application to scheduling, see Stone [9.51]. Additional applications to computer-communication networks appear in Abramson and Kuo [9.1] and Ahuja [9.3].

9.2 OPERATING SYSTEMS CONCEPTS

Of the numerous problems associated with the design of computer operating systems, we address only those related to the allocation of resources to concurrent processes, and to which the mathematics of the preceding chapters can be readily applied. The two main types of resources are processors and storage: CPU scheduling and main memory management are our primary concerns. When processes can execute concurrently while sharing resources, care must be taken to control their interactions. We discuss various 'control' problems first.

9.2.1 Concurrent Processes

Recall the process precedence graph of Section 9.1.4, which is shown again in the left-hand side of Figure 9.2.1(a)(i). We discussed previously how PERT can be used to find an optimal (minimum length) schedule, but this schedule may not be realizable if there is a limit on the number of processes which can be processed concurrently. We show how to find an optimal schedule given N processors in Section 9.2.2. In many cases, (chronological) precedence constraints are necessary only if processes share data. For exam-

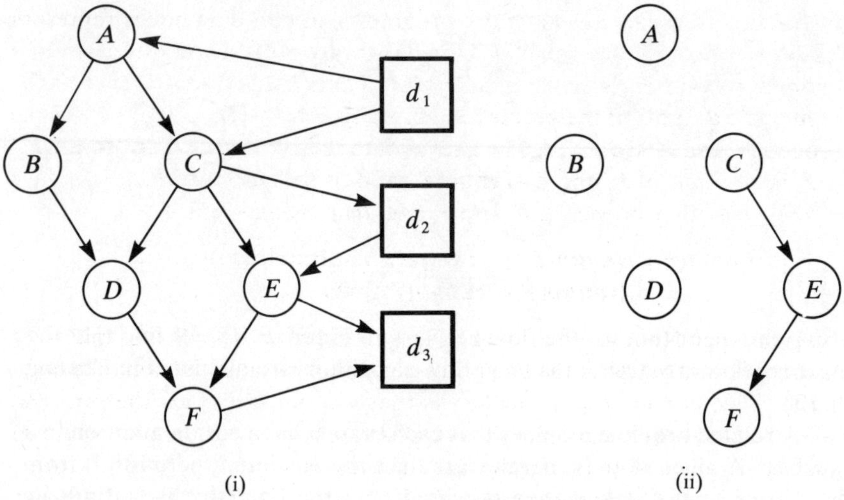

Figure 9.2.1(a)

ple, in the figure, if processes A and C do not share any data as inputs or outputs, then the branch from A to C is superfluous. Elimination of this branch increases the amount of parallelism since A and C can then execute concurrently, thereby enabling a reduction in the schedule length.

We emphasize that although A and C *can* execute concurrently, whether or not they *do* depends upon the operating system scheduler. Let $\overline{A}$ and $\underline{A}$ denote the initiation and termination, respectively, of process A. The constraint that A must precede C means that $\underline{A}$ must precede $\overline{C}$. Without such precedence constraints, any of the following sequences of events can take place:

(1) $\overline{A}...\underline{A}...\overline{C}...\underline{C}$ (2) $\overline{C}...\underline{C}...\overline{A}...\underline{A}$
(3) $\overline{A}...\overline{C}...\underline{C}...\underline{A}$ (4) $\overline{C}...\overline{A}...\underline{A}...\underline{C}$
(5) $\overline{A}...\overline{C}...\underline{A}...\underline{C}$ (6) $\overline{C}...\overline{A}...\underline{C}...\underline{A}$

We say A and B execute *in series* (or *sequentially*) in the case of (1) or (2), and *in parallel* in the cases of (3) to (6). Parallel execution of two or more processes on a single time-shared processor requires *interleaving* of the instructions of the process; each process is then said to alternate between the *running* state, when its instructions may execute, and the *idle* (but ready to be run) state, when its instructions may not execute.

Maximal Parallelism Recall the distinction between instruction and data nodes of an *I–D* flowgraph (see Section 7.3.1). Here, we assume that instruction nodes represent 'processes' which can be as small as a single instruction to as large as an entire program, and that data nodes represent 'cells' which can be as small as a single storage word or register within a computer to as large as an externally stored data file and which serve as the inputs and outputs of the processes. Formally, let $\mathcal{P}=\{P_1,\ldots,P_n\}$ be a set of processes, and $\mathcal{D}=\{d_1,\ldots,d_m\}$ be a set of data cells. We write $d_i \in \mathbf{domain}(P_j)$ if d_i is an input of P_j and $d_i \in \mathbf{range}(P_j)$ if d_i is an output of P_j.

We say that processes P_i and P_j are *data-independent* if

$$[\mathbf{domain}(P_i) \cap \mathbf{range}(P_j)] \cup [\mathbf{range}(P_i) \cap \mathbf{domain}(P_j)] \cup [\mathbf{range}(P_i) \cap \mathbf{range}(P_j)] = \emptyset.$$

Data independence can be tested by inspection of the *I–D* flowgraph. Note that it does not matter if the two processes share the same domain; however, if the range of one process can be the range or domain of another process, then the order of execution of the two processes can affect the computations and final values of the outputs of the processes.

Given a set of processes $\mathcal{P}=\{P_1,P_2,\ldots,P_n\}$, let $\lessdot \subseteq \mathcal{P} \times \mathcal{P}$ be a precedence relation on $\mathcal{P}$. Note that $\lessdot$ is transitive and asymmetric. In the process precedence graph, we only show the immediate predecessor/successor relationship. A system of concurrent processes $(\mathcal{P},\lessdot)$ is said to be *determinate* if, for each pair of data-dependent processes (P_i,P_j), a precedence constraint holds between the two processes (so that one must be executed before the other).

Given a determinate system, in order to maximize parallelism, it may be desirable to eliminate all precedence constraints which are superfluous (in the sense that their removal does not affect determinacy). Such an equivalent 'maximally parallel' system is given by $(\mathcal{P},\lessdot')$, where

$$\lessdot' = \{(P_i,P_j) \in \lessdot \mid P_i \text{ and } P_j \text{ are data-independent}\}^+$$

(where $^+$ denotes the positive transitive closure). In Figure 9.2.1(a)(i), the branch from A to C can be removed, but the branches from C to E and E to F cannot; all the other branches can also be removed, yielding the maximally parallel system shown in Figure 9.2.1(a)(ii). Note that its 'degree of concurrency' is 4 (compared with 2, originally). For further details, see Bernstein [9.10] and Baer [9.6]. For a survey of other aspects of parallel processing, see Jones [9.32].

Process States A common method of modeling systems of concurrent processes is in terms of the states of the individual processes. For example, if each process can be in one of three states, *running*(**R**), *blocked*(**B**) and *eligible*(**E**), then the 'configuration' of a system of N processes is an N-tuple specifying the states of the N processes; e.g., (**B,R,E,B**) represents a system of four processes where the second is running (or 'executing'), the third is eligible (or 'ready') to be run, and the other two are blocked (or 'waiting') until some event occurs (hence are not eligible to be run). A process makes a transition from state **R** to state **B** upon executing a **wait** operation (e.g., one associated with an I/O request), and from state **B** to state **E** upon a **signal** indicating the occurrence of the awaited event. A process makes a transition from state **R** to state **E** if its execution is *preempted* because it has been running for longer than its pre-assigned time limit (or 'slice'), or because some higher priority process (which has just become eligible) is to be run instead. We say a process is *scheduled* when it makes the transition from state **E** to state **R**. Figure 9.2.1(b)(i) shows the possible state transitions for a single process.

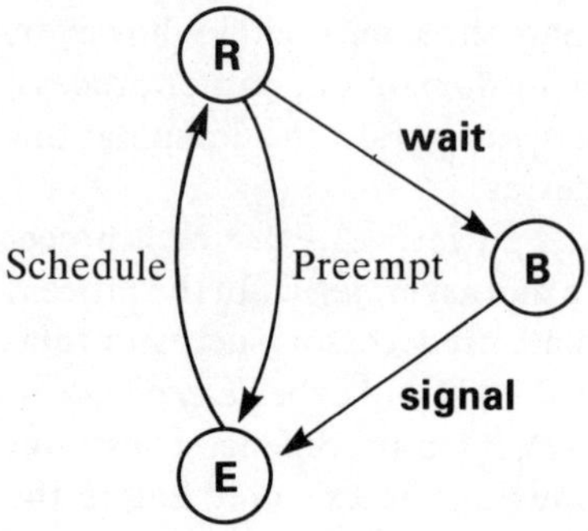

(i)

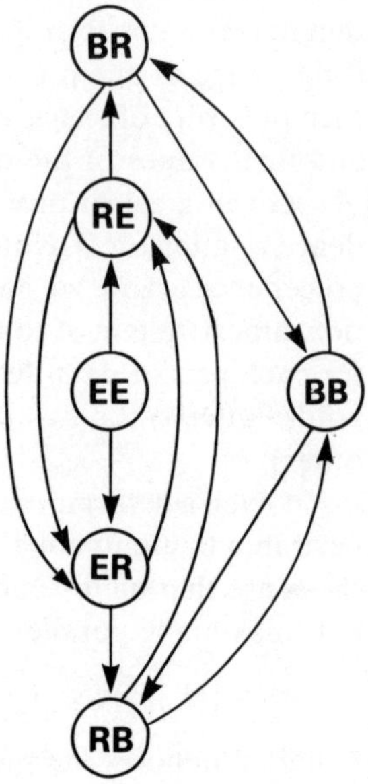

(ii)

Figure 9.2.1(b)

Figure 9.2.1(b)(ii) shows the possible configuration transitions for a two-process single-processor system. Note that configuration **(R,R)** [or **RR** for short] is not possible since we have assumed here that only one process can execute at a time. Furthermore, we have assumed that, whenever the processor becomes available, an eligible process is immediately scheduled, so that the transition of one process out of state **R** (to state **E** or **B**) is simultaneously accompanied by the transition of a process in state **E** (if any) to state **R**. This assumption accounts for the absence of configurations **(B,E)** and (**E,B**) from the figure, and for the unreachability of the 'initial' configuration (**E,E**) from any other configuration. (For simplicity, we have also assumed that some simultaneous transitions cannot occur; see Exercise 6.)

Synchronization One method of implementing the precedence constraints in a system of concurrent processes is to have each process, upon its termination, **signal** each of its successors. In turn, each process should **wait** for such a signal from each of its immediate predecessors before proceeding. For example, the processes of Figure 9.2.1(a)(i) would have the forms:

Process A:	Process B:	Process C:	Process D:
...	**wait**(s_1)	**wait**(s_2)	**wait**(s_3)
signal(s_1)	...	...	**wait**(s_4)
signal(s_2)	**signal**(s_3)	**signal**(s_4)	...
		signal(s_5)	**signal**(s_6)

Process E:	Process F:
wait(s_5)	**wait**(s_6)
...	**wait**(s_7)
signal(s_7)	...

where the branches have separate 'flag' variables used to distinguish the various events. We call a special flag variable used for synchronization purposes a *semaphore*. See Dijkstra [9.18].

For cyclic processes, where the results produced by one process are passed to another process to be consumed repetitively, using a circular list of buffers (recall Section 7.3.4), the consumer must wait for a 'filled' signal from the producer and the producer must wait for an 'emptied' signal from the consumer. Rather than attaching a two-valued buffer flag to each of K individual buffers, synchronization can be achieved by using two multiple-valued semaphores. One semaphore s_F is incremented each time the producer fills a buffer and is decremented (if positive) each time the consumer empties one, so its net value (initially zero) is the number of occupied buffers. The other semaphore s_E is incremented each time the consumer empties a buffer and is decremented (if positive) each time the producer fills one, so its net value (initially K) is the number of available buffers. The proviso 'if positive' in the above means that a process cannot decrement a semaphore whose value is zero; the process must wait instead for an incrementation signal from another process. Thus the producer and consumer processes have the form:

Producer:	*Consumer*:
do-forever	**do-forever**
wait(s_E)	**wait**(s_F)
fill buffer	empty buffer
signal(s_F)	**signal**(s_E)
end-do	**end-do**

where the **wait** and **signal** operations on a semaphore s may be defined as follows:

wait(s): **if** $s>0$ **then** $s:=s-1$ **else** **wait**(s)
signal(s): $s:=s+1$

Note that these definitions of **wait** and **signal** are consistent with the earlier example where each semaphore $\{s_1,\ldots,s_7\}$ is initially zero. Recalling Figure 9.2.1(b), a running process goes into the blocked state upon executing a **wait** operation, and becomes eligible upon a **signal** operation made by some other process.

Deadlock We say a process is 'blocked' if its execution cannot be initiated or continued until some event (external to the process) takes place. If this event can never occur, we say the process is *deadlocked*. If a process is deadlocked waiting for an event which can only be signaled by a second deadlocked process, which in turn is waiting for a signal from the first process, we say the processes are *mutually* deadlocked. For example, if process A holds resource I and is waiting for resource II, and process B holds resource II and is waiting for resource I, then A and B are mutually deadlocked. Deadlock of this sort can be prevented by requiring that the processes acquire the resources in a fixed order, e.g., I before II. Holt [9.27] discussed a digraph model and means of detecting deadlock of this sort.

A related problem arises when processes can be allocated various amounts of a given type of nonpreemptible resource. Nonpreemptibility means that each process need not release the resources it holds until it completes execution. A form of deadlock occurs if, at some point in time, each process requires more of the resource than is available. If the maximum resource requirements of each process is known, then, to be 'safe' from such deadlock, it is sufficient for the system to withhold enough resources to enable each process to complete execution. For a less conservative safety criterion, in an optimization context, see Section 9.3.3. Also see Llewellyn [9.42] for a state transition model.

Mutual Exclusion In an earlier section, we noted that it is generally desirable to permit two (or more) processes to execute concurrently whenever they have no precedence constraint. However, there are occasions when the *order* of execution of the processes may not matter, but their nonsequential execution can lead to trouble. For example, they may both contain an incrementation statement which adds one to a common counter c. Thus, when the processes execute in either order, c will be incremented by two, but if they execute in parallel, it is possible for c to be incremented by only one; the effect of one process can be lost. To avoid this problem, without adding an unnecessary precedence constraint, those sections in each process which cannot execute in parallel must be identified as *critical,* and some mechanism must exist to ensure the ***mutual exclusion*** of critical sections (i.e., to ensure that no more than one process can be executing its critical section at a time). We remark that while we may make critical sections arbitrarily large, e.g., so that an entire process is a critical section, doing so is at the expense of the efficiency gained by parallelism.

One proposal for ensuring mutual exclusion is to precede each critical section by a test of a flag which indicates whether or not some other process has entered a critical section; each process should therefore set this flag when it enters its critical section, and reset or clear the flag when it leaves the critical section. Observe that the test and set operations must be 'indivisible' in the sense that, between the beginning of the test and the ending of the set, other processes must be 'locked out'. Otherwise, exclusion may not be ensured, or else both processes may be locked out (i.e., blocked or deadlocked). For a state-transition model applicable to this problem, see Gilbert and Chandler [9.23]; they show how methods for finding paths in a digraph can be used to determine whether mutual exclusion (without blocking) holds.

9.2.2 Process Scheduling

Recall that, in our discussion of scheduling in Section 9.1.4, we assumed no limit on the degree of concurrency. In practice, and in particular for computer operating systems, there is an upper limit imposed by the number of available processors. Given a process precedence graph $(\mathcal{P},\lessdot)$, consider the problem of determining the minimum length schedule subject to the given precedence constraints and the additional constraint that no more than N processes can execute concurrently. The length of a schedule is defined as the time required to complete execution of all the processes. For simplicity (and for practicality), we assume each process has the same execution time, equal to *one* time unit; if a process has execution time equal to T, then it can always be viewed as a sequence of T unit-time processes. We solve the optimization problem by means of dynamic programming.

We define the 'state' S as the subset of processes in $\mathcal{P}$ which have not yet executed. We say $s \in S$ is 'eligible' if each of its predecessors has executed. The 'decision' at each stage is to choose (i.e., 'schedule') a subset S of eligible processes, of size less than or equal to N, to execute next. Since we have assumed unit-time processes, the 'cost' of each scheduling decision equals 1 time unit; the decisions differ by the number of future decisions which will be required in the 'next' state. Thus, we have the dynamic programming functional equation

$$\mathbf{f}(S)=\mathbf{min}_{\{S'\subseteq S \mid \#(S')\leq N \text{ and each } s\in S' \text{ is eligible}\}} \{1+\mathbf{f}(S-S')\},$$

where $\mathbf{f}(\emptyset)=0$. The optimal schedule then has minimal length $\mathbf{f}(\mathscr{P})$.

There are, of course, other important optimality criteria. For example, computer users are interested in minimum turnaround or response times. The *turnaround time* of a process is defined as the difference between the time at which a process completes execution and the time at which it arrives; this is the time during which the process waits for execution plus the time during which the process actually executes (cf. Eq. 6.4.2(c)). In a 'time-sharing' system, execution is not necessarily continuous: execution of a process can be *preempted,* to allow other processes to execute for a while, before its execution is continued. In this event, minimizing the time until the next execution of a process, sometimes called the *response time,* is often as important to a user as minimizing the time until completion of the process.

A significant problem is that the minimization of turnaround or response times for one process is generally at the expense of other processes. For the computer system as a whole, minimization of *average* (possibly weighted) turnaround times is a common system objective. Another common objective related to the efficiency of the system as a whole (rather than of individual processes) is the maximization of *throughput,* which we define as the rate (number per unit time) at which processes are completed.

For example, in *round-robin* scheduling, each process is given a fixed 'quantum' or 'slice' of time before preemption, and the next job to be scheduled is (say) the one which has not executed for the longest time. This policy favors short jobs since they will complete execution in the fewest rounds.

Priority Scheduling In the above, we assumed that all eligible processes have equal priority. In practice, some processes are given priority over others, based upon:

1. time of arrival ('first-come–first-serve')
2. service time requirement ('shortest job first')
3. resource requirement ('smallest job first')
4. urgency ('earliest deadline first')
5. importance of user ('VIP first').

A scheduling algorithm which selects for execution the first member of a linear list (or waiting queue) ordered in decreasing order of priority is called *priority scheduling* (or 'list scheduling'). The schedules which result, of course, would not necessarily be of minimal length.

One important special case is where the processes are independent, i.e., where $\lessdot = \emptyset$. In this event, shortest-job-first scheduling minimizes average turnaround time. On the other hand, longest-job-first scheduling is better for the criterion of minimal schedule length. In general, no scheduling policy is always best, even for a single criterion. The value of a scheduling policy depends upon numerous factors, such as precedence constraints (if any) and their structure, number of processors and their interchangeability, variation in process execution times, and process priorities. For further information, see Hellerman [9.25].

9.2.3 Memory Management (Paging)

Paging systems permit many users to share the limited main memory of virtual-memory computers in a time-sharing or multiprogramming environment without regard to memory size limitations. For a detailed discussion of paging and virtual memory concepts, see Coffman and Denning [9.16]. In essence, a program is partitioned into blocks, called *pages*, of fixed size and only a subset of these pages resides in main memory blocks, called *frames*, at any one time. The size of this subset is called the program's *allotment*. A reference to a page of a program which is not resident is called a *page fault*. The program then blocks until the 'demanded' page is loaded from auxiliary memory—a very inefficient operation.

We address ourselves first to the demand-paging *replacement* problem. Informally, this problem is that of deciding, upon the occurrence of a page fault, which page residing in main memory at that time should be replaced. The optimality criterion generally adopted is to minimize the number of faults or replacements. If future page references are known, the optimal policy is to replace that page whose time until next reference is maximum. An algorithm based upon this replacement policy, often called *Belady's algorithm*, is unrealizable, however, since it requires exact knowledge of future behavior.

The alternative is to estimate or predict the future. Common estimation algorithms utilize *a priori* probabilities of reference (cf. Markov chain program models) or measurements of past behavior (cf. working set models). The *working set* of a program at time t with 'window' parameter W is defined as the set of pages which are among the last W pages referenced. Assuming working sets change slowly with time, called the 'locality' property, future references will usually be made to those pages which have been recently referenced. Therefore, a reasonable replacement policy, called *Least Recently Used* (LRU), chooses that page which has not been referenced for the longest time. We remark that a desirable property for a replacement policy to possess is that increases in allotment should not lead to increases in page faults; unfortunately, not all policies possess this property.

State Transition Model We adopt, with slight variations, the notation and definitions given in Aho [9.2]. Let $N=\{1,...,n\}$ be the *pages* of a given n-page program $\mathscr{P}$, and $M=\{1,...,m\}$ be the page *frames* of memory, with $1 \le m \le n$. Let $M_m=\{X \mid X \subseteq N \text{ and } \#(X) \le m\}$. We shall call a subset $X \subseteq N$ a *resident set,* and the number m the page *allotment* given to $\mathscr{P}$. A *paging algorithm* (for M and N) is an algebraic system $\mathscr{A}=(Q, q_0, \mathbf{g})$ in which

(1) Q is a set of *control states* of the algorithm,
(2) $q_0 \in Q$ is the *initial control state,* and
(3) $\mathbf{g}$: $(M_m \times Q) \times N \to (M_m \times Q)$ is the *allocation map,* having the property that r is in X' whenever $\mathbf{g}((X,q),r)=(X',q')$.

If the allocation map is of the form

$$\mathbf{g}((X,q),r)= \begin{cases} (X,q') & \text{if } r \in X \\ (X+r,q') & \text{if } r \notin X,\ \#(X)<m \\ (X+r-y,q') & \text{if } r \notin X,\ \#(X)=m,\ y \in X, \end{cases}$$

we say $\mathscr{A}$ is a *demand paging* algorithm. The second case can be disregarded if resident sets have a fixed allotment, which we assume for simplicity hereafter. A demand-paging algorithm *processes* a page reference string $r_1...r_t...r_T$ of length T (each $r_i \in N$) from initial resident set x_0 by generating the sequence of 'configurations' $\{(x_t,q_t)\}_{t=0}^{T}$ such that $(x_t,q_t)=\mathbf{g}((x_{t-1},q_{t-1}),r_t)$ for $1 \le t \le T$. If $r_t \notin x_{t-1}$, a *page fault* is said to occur at time t.

We observe that paging algorithms may be regarded as deterministic finite-state automata. A nondeterministic version may be defined by eliminating the control states. Thus a nondeterministic (fixed allotment) demand-paging algorithm is a system $(N,X_0,\mathbf{g})$ in which N is a set of pages, X_0 is the initial resident set (of size m), and $\mathbf{g}$:$M_m \times N \to 2^{M_m}$, where $M_m=\{X \mid X \subseteq N \text{ and } \#(X)=m\}$, and $\mathbf{g}$ is a one-to-many map satisfying

$$\mathbf{g}(X,r)= \begin{cases} X & \text{if } r \in X, \\ \{X+r-y \mid y \in X\} & \text{if } r \notin X. \end{cases}$$

We can define an equivalent one-to-one map by extending $\mathbf{g}$, so that

$$\mathbf{g}(X,y,r)= \begin{cases} X & \text{if } r \in X, \\ X-y+r & \text{if } r \notin X \end{cases}$$

for every $y \in X$.

Control System Formulation We may regard a paging algorithm as a multistage decision or 'control' system governed by the map

$$x_{i+1}=\mathbf{g}(x_i,y_i,r_i), \qquad i\geq 0,$$

where x_i is the resident set (state) at stage i, y_i is the replacement decision (control), and r_i is a page reference, which may only be known probabilistically. Our objective is to select a sequence $\{y_0,y_1,\ldots\}$ so as to minimize a given 'cost' function

$$\mathbf{H}(\{x_0,x_1,\ldots\},\{y_0,y_1,\ldots\},\{r_0,r_1,\ldots\})$$

for a page reference string $\{r_0,r_1,\ldots\}$. We suppose $\mathbf{H}$ is separable, so that

$$\mathbf{H}=\mathbf{h}(x_0,y_0,r_0)+\mathbf{h}(x_1,y_1,r_1)+\cdots,$$

which simply asserts that the cost associated with going from state x_i to state x_{i+1} depends only on x_i, y_i, and r_i. This supposition is easily satisfied for paging systems, for example, by letting $\mathbf{h}$ represent page faults,

$$\mathbf{h}(x_i,y_i,r_i)=\begin{cases} 1 & \text{if } r_i\notin x_i, \\ 0 & \text{if } r_i\in x_i, \end{cases}$$

so that $\mathbf{H}$ is simply the total number of page faults encountered, a random variable.

We define the 'page replacement problem' as that of minimizing the expected value of the cost $\mathbf{H}$,

$$\min_{\{y_i\}}\{\underset{\{r_i\}}{\mathbf{exp}}\sum_i \mathbf{h}(x_i,y_i,r_i)\},$$

subject to $x_{i+1}=\mathbf{g}(x_i,y_i,r_i)$, $i\geq 0$, with initial condition x_0. The solution to this problem is a sequence of decisions $\{y_0^*,y_1^*,\ldots\}$, where y_i^* is the 'optimal' page to replace at stage i. We wish in fact to determine y_i^* as a function of x_i, for various assumptions on the page reference string. In general, the problem is quite difficult to solve without simplifying (and often unrealistic) assumptions. For example, if the page references $\{r_i\}$ are independent, dynamic programming can be employed to show that the optimal page to replace is the resident page whose probability of reference is least. See Lew [9.39]. This replacement policy can be approximated by choosing the least frequently used (LFU) page.

9.3 RESOURCE ALLOCATION

In this section, we discuss a variety of optimal resource allocation problems, each of which can be formulated in dynamic programming terms. Again, there may be more efficient techniques to solve some of these problems, but none more general.

9.3.1 Static Allocation

We formulate here the basic problem of optimally allocating a finite set of resources among independent parallel processes. Optimality must, of course, be relative to a specified quantitative criterion. Let $\mathbf{c}_k(q_k)$ denote a nonnegative measure of the expected performance of process P_k given an allocation of integral size q_k. We assume that the *allocation cost* $\mathbf{c}_k(q_k)$ for each process P_k is independent of $\mathbf{c}_j(q_j)$ for any other process P_j; this is a 'separability' assumption. Suppose there are N processes $\{P_i\}$ and M total units of a single resource type. Then the optimal (minimal-cost) *static* resource allocation problem may be formally stated as

$$\min_{\{q_k\}} \sum_{k=1}^{N} \mathbf{c}_k(q_k), \qquad \text{subject to} \qquad \sum_{k=1}^{N} q_k \le M,$$

where we require $0 \le q_k \le M$ for each k. We remark that the separability assumption permits us to allocate resources to processes sequentially in an arbitrary order.

For convenience, we introduce a dummy process P_0 representing unused resources, so that the inequality constraint above may be replaced by an equality $\sum_{k=0}^{N} q_k = M$; $\mathbf{c}_0(q_0)$ is thus a measure of inefficient utilization of resources. If, for any reason, all resources must be allocated, we may let $\mathbf{c}_0(q_0)=\infty$ for $q_0>0$, $\mathbf{c}_0(0)=0$. The introduction of P_0 permits withholding of resources from processes, as may be necessary to avoid deadlocks or other scheduling problems.

In order to solve the optimization problem we have posed, we first reformulate it as one of finding the shortest path in an acyclic graph with nonnegative labels. This routing problem can then be solved by dynamic programming.

Routing Problem Formulation Let $\mathcal{P}=\{P_0, P_1, \ldots, P_N\}$ be an augmented set, as described above, of N competing processes, and M be the number of units of available resources. Let $\mathbf{c}_k(q)$ be given, for each $q=0, 1, \ldots, M$, and $k=0, 1, \ldots, N$.

Definition An *allocation graph* $G=(S,B,C,\lambda)$ is a labeled digraph, where S is a set of nodes, B is a set of branches, C is a set of branch labels, and λ is a labeling function, defined as follows:

(a) $S=\{[k,i] \mid k\in\{0, 1, \ldots, N+1\}, i\in\{0, 1, \ldots, M\}\}$;
(b) $B=\{(a,b) \mid a=[k_a,i_a]\in S, b=[k_b,i_b]\in S, k_b=k_a+1, i_b\le i_a\}$;
(c) C=set of nonnegative real numbers (branch lengths);
(d) λ: $B\rightarrow C$ such that $\lambda(a,b)=\mathbf{c}_{k_a}(i_a-i_b)$.

Each node $[k,i]$ is associated with the *state* of having i units of available resources, some portion of which is to be allocated to process P_k, where allocations are made sequentially in order of the indices (justified by the separability assumption). A branch $([k,i],[k+1,i-q])$ represents the state transition resulting from allocating q units of resources to process P_k, at a cost $\mathbf{c}_k(q)$ given by the branch label. A path is a sequence of branches (or allocations) whose length (cost) is the sum of the individual branch labels.

Proposition The optimal static allocation problem is equivalent to the problem of finding the shortest path in the allocation graph from node $[0,M]$ to node $[N+1,0]$.

Example*: *Storage partitioning Let us suppose that the resource to be allocated is main memory space, so that each process (program) can be executed only a portion at a time. Our underlying assumption here is that each program can execute in a range of partition sizes, where the cost (efficiency) of its execution is a function of the partition size—as is the case in an *overlaying* environment. An 'overlay' is the replacement of a portion of a program by another portion. Let $\mathbf{c}_k(q_k)$ be defined as the expected number of overlay calls made by program P_k (proportionate to its expected execution time) given an allocation of size q_k. This function can be determined from a knowledge of the branch execution frequencies in the program and its overlay structure, e.g., using Markov chain analysis. See Kral [9.36]. This same information can be used to partition a program into pages to minimize interpage references or page-faulting; see Kernighan [9.35] and Lew [9.38].

9.3.2 Dynamic Allocation

In general, $\mathbf{c}_k(q_k)$ is a time-varying function, $\mathbf{c}_k(q_k,t)$. Consequently, reallocations should be made 'every so often', at a set of discrete reallocation times $\{t_1, t_2, t_3, \ldots\}$. In discretizing time, we assume that the functions $\mathbf{c}_k(q_k,t_i)$ accurately represent the allocation costs during the entire time interval (t_i,t_{i+1}). The time-varying allocation problem can be reduced to a sequence of (time-independent) static allocation problems, by re-solving the basic routing problem at each time t_i. That is, we may let $\mathbf{c}_k(q_k)=\mathbf{c}_k(q_k,t_i)$, obtaining the optimal allocation $q_k^*(t_i)$ for process P_k in the interval (t_i,t_{i+1}).

Example*: *Page allotment Let us suppose that the resource to be allocated is page frames in a paging system. Let $\mathbf{c}_k(q_k,t)$ be defined as the expected page-fault cost of program P_k given a page-frame allotment of size q_k at time t. Methods for estimating the function are discussed in Franklin [9.22] and Lew [9.38].

Reallocation In the foregoing, we have neglected many factors of importance, because they lead to much more complex models than we wish to consider in this book. One factor is the choice of times at which resources should be reallocated. In nonpreemptive systems, reallocations would only take place when a process runs to completion. In preemptive systems, however, reallocation times would be determined by scheduling considerations, depending upon, for example, such factors as time slices or process blockings.

An 'optimal' set of allocation times, which would minimize unnecessary reallocations, is generally impractical to find, but heuristic approximations may prove feasible. For example, we may monitor the performance of the system and make reallocations only when the performance drops below some prespecified level. See Chu [9.14].

We have also neglected the fact that each reallocation of resources can only be done at a price—in part because of the overhead required to change the amount of resources allocated to any process, and in part because of the calculations required to re-solve the routing problem.

Cost of Inactivity We say a process P_k is 'inactive' at the times when $q_k(t)=0$, at a cost $\mathbf{c}_k(0,t)$. We also say that execution of the process P_k is 'delayed' until (at least) time t_{i+1} if $q_k(t_i)=0$. This may not, of course, be tolerable for certain processes. If any or all of the processes are required to be active in the time interval (t_i,t_{i+1}), $\mathbf{c}_k(0,t_i)$ may be set arbitrarily high. When there are insufficient resources, then a cost of inactivity may be defined based upon scheduling criteria. Scheduling priorities may be assigned based upon processing times (completed or remaining), storage demands, and so forth. This formulation requires that the costs of inactivity $\mathbf{c}_k(0,t_i)$ be made commensurable with nonvacuous allocation costs $\mathbf{c}_k(q,t_i)$, $q>0$.

Example: CPU Scheduling Our discussion of resource allocation problems has mostly been general, without specification of the resource. CPU processing time is, of course, one very important type of resource, which formally can be allocated in the same manner as before. Suppose that there are M processors to be allocated among N independent processes. If processors cannot be shared, or cannot be allocated to more than one process at a time, then we may let $\mathbf{c}_k(0,t_i)$ be the cost of inactivity for process P_k for the interval (t_i,t_{i+1}), and $\mathbf{c}_k(1,t_i)=0$. The costs $\mathbf{c}_k(0,t_i)$ can be defined according to common scheduling priorities, so that the highest priority one will always be scheduled next whenever a processor is available. It should be emphasized that this approach minimizes the criterion $\Sigma_i\Sigma_k\ \mathbf{c}_k(q_k,t_i)$ (for separate costs), as opposed to the more common criteria of minimal schedule length and mean time or number in system.

We remark that by allowing $q_k(t)$ to be zero, this formulation automatically yields the optimal 'degree of multiprogramming' (number of active processes). After solving for the optimal set of allocations, $\{q_k^*(t_i)$

$|1 \leq k \leq N\}$, the optimal degree of multiprogramming for the interval (t_i, t_{i+1}) is simply the number of nonzero $q_k^*(t_i)$.

Stochastic Costs We assumed above that $\{\mathbf{c}_k(q,t_i)\}$ are precalculable deterministic independent functions. This may not be realistic for several reasons. Firstly, $\mathbf{c}_k(q,t_i)$ may not accurately 'predict' $\mathbf{c}_k(q,t)$ over the entire interval (t_i, t_{i+1}), especially since it must be calculated at time t_i or earlier. Secondly, the cost may depend upon nondeterministic factors—e.g., data dependencies or new process arrivals. Thirdly, the costs may be correlated in time, so that $\mathbf{c}_k(q,t)$ and $\mathbf{c}_k(q,t')$ are dependent for $t \neq t'$. To handle this more complex situation, we refer readers to the theory of Markovian decision processes (of 'D.H.' type, Kaufman [9.33]). See also Howard [9.28] and Ross [9.49].

Example: Page Replacement We remark that the page *replacement* problem may also be formulated as a Markovian decision process. (See Ingargiola [9.31].)

9.3.3 Scheduling Problems

Scheduling problems arise naturally when allocations of resources (other than CPUs) may vary with time. For example, if resources are withheld from one process and given to another, overall delays may result. Allocation and scheduling decisions are thus clearly interrelated; an optimal 'schedule' must minimize the sum of allocation costs over time, subject now to constraints of various types. In the remainder of this chapter, we discuss how certain 'control' aspects of scheduling may be treated in the context of non-CPU resource allocation. The problems we consider below no longer assume independent processes.

Process Coordination When parallel processes are not independent, so that they interfere if executed concurrently, allocation decisions should take into account the dependencies. For example, if processes are partially ordered according to a chronological precedence relation, for (say) determinacy reasons, then resources should be allocated only to 'eligible' processes whose predecessors have all been terminated. Again, if (cyclic) processes must satisfy synchronization or exclusion constraints, only certain processes are 'eligible' for resources. Formally, we may set the allocation costs for ineligible processes arbitrarily high, and use the foregoing procedures to distribute resources among the eligible processes. Allocation costs may be functions of time, or else functions of the 'state' of the system (e.g., values of semaphores), or both. For example, the costs may be made functions of how 'critical' processes are; a critical process is one on the longest path of a scheduling (PERT) network, which we discussed in Section 9.1.4.

When allocation of resources is constrained by 'soft' precedence relations among processes, for which there may be preferences rather than precedences (superseding other measures of allocation cost), a problem of a different nature arises. We may in this case associate a higher cost with one ordering of two processes rather than with another, where we assume that these costs are independent of the sizes of the allocations. Let $\{P_1,\ldots,P_N\}$ denote the set of processes. Given $\{C_{ij}\}$, the costs of P_i preceding P_j, the optimal ordering of the processes can be found by solving a 'traveling-salesman' touring problem. Let $\mathbf{f}_k(A,P_i)$ be the minimum cost of finally executing process P_k, preceded by all the processes in the subset A, starting with the execution of process P_i in A. Initial and terminal conditions can be added by letting a_i and b_i be the costs of starting and ending, respectively, with process P_i; these will be zero if there are no constraints or preferences.

Proposition The optimal ordering of processes can be found by solving the dynamic programming functional equation

$$\begin{aligned}&\mathbf{f}_k(\{P_{s_1},\ldots,P_{s_\ell}\},P_i)\\&\quad=\min_j\ \{C_{ij}+\mathbf{f}_k(\{P_{s_1},\ldots,P_{s_\ell}\}-\{P_i\},P_j)\},\end{aligned}$$

for $\mathbf{f}_k(\{P_1,\ldots,P_N\},P_k)$ given that $\mathbf{f}_k(\{P_i\},P_i)=b_i$. The $j\in\{s_1,\ldots,s_\ell\}$ which yields the minimum designates the process to be executed immediately after P_i.

The initial process is given by $\min_k\ [a_k+\mathbf{f}_k(\{P_1,\ldots,P_N\},P_k)]$, with succeeding processes determined by reconstruction.

Example*: *File Scheduling Given a queue of processes wishing to access a file, the dominant cost factor depends upon the device-dependent access times (such as seek times) required by the processes. The optimal ordering of processes can be determined as above, where $\{C_{ij}\}$ are the access times associated with servicing process P_j immediately after P_i, a_i is the cost of servicing P_i initially (based on the initial state of the device), and $b_i=0$, for each i. We note that if all C_{ij} are taken to be zero, the *first* process to be serviced is that having the shortest access time.

Suppose that P_{i_1}, P_{i_2}, ..., P_{i_N} is an optimally ordered sequence of processes. Then resources should be allocated to processes in that order. If the processes may execute concurrently, then resources may be allocated *in toto* first to P_{i_1}, then P_{i_2}, and so forth, until the supply is exhausted; the latter processes are then generally delayed. Alternatively, or if the processes must be concurrent, allocation costs may prescribe the apportionment as before. If two processes may not be concurrent, then the one with lower priority is ineligible until the higher priority process terminates.

Redundant Processes The situation may arise in parallel processing applications where a set of processes has 'redundancies'. A process, for example, may be rendered unnecessary by some outcomes of certain other processes, so that it would be wasteful to execute the former process first. If decisions as to whether or not to execute processes depend upon other processes, we have a new scheduling problem, which may be formalized as follows.

Given a set of processes $\{P_1,...,P_N\}$, let $V_i=\{v_{i1},...,v_{ir_i}\}$ be a set of 'outcomes' associated with P_i. Execution of P_i selects one element of V_i as its outcome, say v_{ij}; we denote this by writing $\mathbf{e}(P_i)=v_{ij}$. Let $\mathbf{C}(P_i|R)$ be the cost of executing P_i given that the predicate R is true. Define

$$\mathbf{f}_k(\{P_{s1}, P_{s2}, ..., P_{sk}\}|r_k)$$

as the minimum achievable cost of execution, where there are k remaining processes $\{P_{s1},...,P_{sk}\}$, given that R_k is true.

Proposition The optimal ordering of processes can be found by solving

$$\mathbf{f}_k(\{P_{s1},...,P_{sk}\}|R_k)$$
$$=\min_i\{\mathbf{C}(P_i|R_k)+\sum_{j=1}^{r_i}\mathbf{f}_{k-1}(\{P_{s1},...,P_{sk}\}-\{P_i\}\ |R_k\wedge\mathbf{e}(P_i)=v_{ij})\}$$

for $\mathbf{f}_N(\{P_1,...,P_N\}|\mathbf{true})$ given that $\mathbf{f}_1(\{P_i\}|R_1)=\mathbf{C}(P_i|R_1)$. The $i\in\{s_1,...,s_k\}$ which yields the minimum in the above designates the process to be executed next (at stage k; $k=N,...,1$) given that R_k is true.

We observe that R_k is the conjunction of predicates of the form $\mathbf{e}(P_i)=v_{ij}$ for P_i not in the set $\{P_{s1},...,P_{sk}\}$ of processes executed at the earlier stages, $k+1,...,N$. If $\{P_{s1},...,P_{sk}\}$ consists of redundant processes given the outcome R_k, then $\mathbf{C}(P_i|R_k)=0$ for each such process. Otherwise, we may let $\mathbf{C}(P_i|R_k)=c_i\times\Pi(R_k)$, where c_i is the execution cost (resource requirement) of process P_i, and $\Pi(R_k)$ is the probability that R_k holds. (The latter depends upon likelihoods of various outcomes; equilikelihood may be assumed if better probabilistic information is not available.)

With the above definition, the dynamic programming procedure yields the hierarchical order of process executions which minimizes the expected total cost associated with a set of redundant processes. It should be emphasized that the ordering is not linear, but is of tree form. We remark, in conclusion, that sequencing constraints (for precedence reasons, as in the prior section) can be incorporated in this new algorithm by restricting the minimum in the functional equation to be taken only over those P_i in $\{P_{s1},...,P_{sk}\}$ which can be executed prior to each of the remaining processes.

Example: Decision tables The algorithm presented above may be used to convert decision tables to optimal computer programs (cf. Bayes [9.7] and Lew [9.40]). To minimize time, let c_i be the time required to evaluate 'condition' P_i and $\mathbf{Pr}(R_k)$ be the probability that the specified combination of condition values R_k obtains.

$\mathbf{C}(P_i|R_k)=0$ if the 'subtable' associated with R_k has a common action set. To minimize space, let c_i be the space required to evaluate P_i, and $\mathbf{Pr}(R_k)=1$ if the probability of R_k is nonzero, else $\mathbf{Pr}(R_k)=0$. (The latter formulation also applies to the problem of minimizing weighted depths; see Section 9.1.1.)

Deadlock Avoidance If we assume nonpreemptible resources, or, more weakly, that there is a nonzero lower bound on the amount of resources a process can be allocated once activated (possibly a time-varying bound), then requests for additional resources by processes cannot be granted on demand without risk of deadlock. A deadlock avoidance criterion may rule out the shortest path in the allocation graph as an acceptable solution to the optimal resource allocation problem. In this event, a 'safety' constraint may be formally added to the optimization problem. For example, suppose that we require knowledge of the maximum resource requirement, M_k, of each process P_k. Then the set of allocations $\{q_k(t_i)\}$ is *safe* if

$$(\forall j)\ M_j - q_j(t_i) \leq q_0(t_i) + \sum_{k \in K_i} q_k(t_i) \qquad \text{where} \qquad K_i = \{k\,|\,q_k(t_i) = M_k\}.$$

We should therefore minimize $\Sigma_k \mathbf{c}_k(q_k, t_i)$ subject both to the safety criterion and to the constraint $\Sigma_k q_k = M$. The safety constraint is quite conservative; the universal quantifier can be replaced by a sequence of existential ones if the inequality is evaluated iteratively with $q_0(t_i)$ increased by $q_j(t_i)$ each time the inequality holds (cf. Dijkstra [9.18]). We remark, in addition, that while some overhead is required to evaluate constraint relations, the computational requirements of dynamic programming algorithms are generally reduced significantly by their addition.

An alternative approach is the direct utilization of dynamic programming to determine the kth shortest path, for $k=1,2,\ldots$; the smallest k for which the kth shortest path does not violate the safety criterion then yields the desired solution. If no such path exists, the system is initially deadlocked. In this event, one or more of the deadlocked processes must be 'aborted' to effect a recovery, at a cost, of course, which should be minimized.

One method of preventing deadlocks where there are different types of resources is to require that processes use the resources in a prescribed order. A dynamic programming solution of this 'flow-shop' problem for two resource types appears in Held and Karp [9.24].

EXERCISES

1. For 10 source characters having probabilities (weights) $p_1=p_2=0.2$, $p_3=p_4=0.05$, $p_5=p_6=0.1$, $p_7=p_8=0.05$, $p_9=p_{10}=0.10$, find the Huffman code.
2. For the 'equally likely letters' example of Section 7.2.4, find the optimal sorted binary tree.
3. Use the decision table program (Table 9.1.3) to find the shortest path from s to t in the graph of Figure 9.1.5.
4. (a) Convert the inverse of the precedence graph of Figure 9.1.4 to a PERT graph.
   (b) Use PERT to find the critical path in the resulting graph.
5. Find the maximum flow from s to t in the capacitated network of Figure 8A.
6 (a) Redraw Figure 9.2.1(b)(ii) assuming no unnecessary restrictions on simultaneous transitions.
   (b) Redraw Figure 9.2.1(b)(ii) under the assumption that two processors are available (so state **(R,R)** is possible).
7. Show that shortest-job-first scheduling does not minimize average turnaround time when there are precedence constraints.
8. Show that the FIFO replacement policy possesses the property that increases in allotment can lead to increases in page faulting.
9. Solve the optimal static resource allocation problem for $N=3$, $M=4$, and $\mathbf{c}_1(q)=20-5q$, $\mathbf{c}_2(q)=2^{5-q}$, $\mathbf{c}_3(q)=\mathbf{max}\{(2-q)^3,0\}$, for $q=0,1,2,3,4$.
10. Find the optimal decision tree for Figure 4.2.5(b) assuming equally likely rules and unit time condition tests.
11. Show that the shortest-access-time-first disk scheduling policy can lead to blocking of some file requests.
12. An allocation graph has $(N+2)$ $(M+1)$ nodes. How many branches does it have?

PROGRAMMING ASSIGNMENTS

1. Write a program which finds the optimal sorted binary tree for an arbitrary set of data, where the probabilities $\{p(k)\}$ and $\{q(k)\}$ are given.
2. Write a program which solves the scheduling example of Section 9.1.4.
3. Write a program which solves the flow example of Section 9.1.5.
4. Write a program which simulates a queueing system (cf. Chapter 6), and computes average turnaround times and throughputs for the FCFS and shortest-job-first scheduling policies.
5. Write a program which generates page reference strings, and compare the LRU and LFU replacement policies.
6. Write a program which determines whether or not a set of allocations is safe.

POSTSCRIPT

We have covered a wealth of material in this book. To regain some perspective, we summarize below the major relationships between software systems and mathematics which have provided our motivation. We have only hinted at many of these relationships, and hope that readers will pursue them further. It has been our intent to provide adequate preparation for this venture.

1. The term *software* is generally associated with computer programming or its product, programs. A *computer program* is simply a list of 'statements': program statements may be 'declarative', to specify *data* and their structure, or 'executable' (instructions) to specify operations to be performed on such data. A given finite sequence of operations is an *algorithm*. Instructions and data must be represented in some *language* to be useful. *Translators* must also be available to convert programs in one language to another. An *operating system* is required to actually 'run' a set of programs on a computer. 'Software systems' include all of the above.

2. A *computer program* can be thought of as a set of instructions operating on a set of data. Graphs can be used to model the structure associated with a set of instructions (cf. *flowcharts*), as well as *data structures* (e.g., linked lists, trees, arrays, strings). Relationships between instructions and the data they operate upon can also be modeled by a graph. The operation performed by a program on data 'computes' a *function* from input data to output data: in fact, since data have string representations (e.g., strings of digits), the function need not be numerical, but may be regarded as a string translation from an input string to an output string; furthermore, since all character strings have numerical (e.g., ASCII or EBCDIC) representations, and a set of numbers may be uniquely represented by a single (Gödel) number, the function may be regarded as a map from an integer to another integer.

3. A computer program can also be thought of as a string of characters from some alphabet, where the string (sentence) must satisfy given structural rules (to be 'syntactically' correct). A sentence is syntactically correct if it can be *parsed* according to a set of syntax (string substitution) rules, called a *grammar*. Grammars can be represented as directed graphs, called *syntactic charts*. A parse of a sentence (with respect to a grammar) can be represented as a tree. A *language* is a set of syntactically correct sentences. A *compiler* is a computer program which performs a translation from input source strings to output object strings; syntactically incorrect source strings are translated into diagnostic message strings.

4. The sequential operation of a computer program can be modeled by decomposing the (composite) function that it computes into an equivalent sequence of functions ('tasks'). The sequence of intermediate outputs is the 'semantics' (meaning or 'behavior') of the program. The same function may have different decompositions, each defining an algorithm for computing the function. (Algorithms are 'equivalent' if they compute the same function.) Algorithms have associated *complexities* (e.g., time and space), which may be used for comparing equivalent algorithms. Optimizing compilers attempt to find less complex equivalent algorithms, for example, by utilizing the structural information provided by a graph model.

5. A computer program is a concrete representation of an (abstract) algorithm. For a program to be correct, not only must it be a syntactically correct character string, but the behavior of its abstraction must be semantically correct (i.e., must coincide with *asserted* propositions about intermediate outputs). Some examples of assertions include sequences of bounds on intermediate data values and relations which hold between different values. 'Static' semantic errors can be found by compiler analysis of structural properties of programs, but 'dynamic' error diagnosis requires comparisons of 'assertions' with monitored program behavior. A correct (error-free) program is said to be completely 'debugged', but correctness can only be proven by exhaustive testing or by logical deduction.

6. A computer program is also a set of data, portions of which are *interpreted* (executed) by a computer as instructions operating on other portions. A *computer* is a machine (*automaton*) which moves from configuration to configuration according to a set of (fixed) transition rules. A configuration is the set of values of all data in the machine, one of which specifies the *state* of the machine; the transition rules map states and data values into new states and data values. A computer maps input data to output data via a sequence of transitions, this sequence being determined by the algorithm which is 'programmed' as part of the input data. (A 'microprogrammable' computer permits the transition rules to be modified.) This sequence of configurations is called a *trace*.

7. While a given computer program defines a deterministic sequence of configurations, nondeterministic models of programs are also useful. A computer program may of course be thought of as a fixed set of instructions operating on a fixed set of input data yielding a fixed set of output data. If the same set of instructions is the 'program' and different sets of data were 'input' to the program, then the 'output' and the sequence of configurations leading to that output would be *data-dependent*. If now the input were not known specifically, but instead a probability was associated with each set of input, then outputs and configurations would be 'random' variables. Markov chain models are useful for modeling programs probabilistically. The *complexity* of a program can be analyzed (estimated) in this fashion.

8. Complexity of programs often depends upon how data is represented or structured. To operate upon data requires first that the data be *accessed* (located and, if stored externally, retrieved). If data is stored sequentially but in no particular order, linear searching (sequential access) for an arbitrary item is simplest, but mapped (indexed, directoried, or 'hashed') searching (direct access) may be more efficient. If the data is linearly ordered and stored sequentially in that order (i.e., *sorted*), binary searching and other comparison techniques are possible. Linearly ordered data may also be stored in a tree structure (binary, balanced, radix, etc.). Mathematical programming techniques, for example, are applicable to problems of finding optimal data structures, i.e., structures which minimize expected access time when probabilities of requiring individual data items are known.

9. Data structures may be classified according to how access to data items can be made. In *sequential* structures (strings, stacks, queues) each item has a unique immediate predecessor and successor (possibly null) which can be found using a 'displacement' calculation. In *linked* structures (linked lists, trees, and graphs) each item's successors or predecessors can be found from explicit location 'pointers' (addresses). In *mapped* structures (arrays and tables) each item can be found 'directly' by evaluation of a function (possibly tabular or hashed).

10. Data base *processes* include accessing (search/retrieval) sets of data items, sorting or restructuring the data sets (or constructing them, in the first place, from sets of items), inserting new items and deleting old ones (possibly all), and traversing the data base. A 'traversal' is a sequence of accesses to each data item at least, but ideally exactly, once. Traversal algorithms for linked structures are related to path finding in a graph.

11. A computer program may be partitioned into a set of subprograms, called *tasks*. A 'job' is also a set of tasks. A 'task system' is a set of tasks together with a set of chronological precedence relations between tasks (which may be modeled by a directed graph). Execution of a task requires

allocation of limited resources (processor and memory, at least) to the task. An *operating system* allocates resources to tasks in some manner subject to precedence and limitation constraints.

12. Once a task has been allocated its resource requirements (possibly in part only), execution of the task may be initiated. Execution may then terminate autonomously, or else by force in the case of preemptibility of a resource. A task is *active* during the intervals of time between its initiations (activations) and terminations. When a task has no further resource requirements, it is completed; a task once activated is incomplete until then. A *parallel processing* (time-sharing) system is one where incomplete tasks may coexist (be concurrent in time). If active tasks may coexist, the system is termed *multiprocessing*; otherwise it is termed *multiprogramming*. A *sequential* system is one where a task, once activated, must be completed before another task can be activated.

13. A *scheduler* allocates resources (esp. CPUs) to tasks in some linear order, generally for a fixed 'quantum' of time. Allocation decisions are subject not only to precedence (e.g., priority) and limitation ('degree' of multi-tasking) constraints, but also to determinacy, deadlocking, mutual exclusion, and synchronization considerations. *Critical path* scheduling (cf. PERT) is a special case applicable to task systems with given precedence relations and durations of tasks. It may be modeled by a labeled directed graph. A *queueing* system is also a useful model for representing a set of tasks awaiting 'service' by a resource scheduler; a *feedback* queue, in particular, permits tasks to return to a servicer, as required in a time-sharing system. Markov chain techniques are also applicable to the analysis of queues.

14. A *memory manager* allocates 'regions' of memory, of limited size and specifically located, to tasks; a *loader* moves tasks from one location to another, a *linker* makes location-dependent connections between tasks. Related problems include the design of overlay structures, memory reallocation, and garbage collection. When many hierarchies of memory exist, allocation problems are greater.

15. A *paging* system is a memory management system where main memory is allocated in blocks of equal size (called page frames). Subsystems include a *paginator* which partitions tasks into segments (which fit within a page frame), an *allotter* which decides on the number of page frames to allocate to each active task (generally less than its total requirement), and a *fetch/replacer* which decides which segments of an active task are to be loaded into main memory page frames (generally varied dynamically). A task executing in a paging system can be modeled by a Markov chain and its performance analyzed thereby.

16. *Optimal* resource allocation requires quantification of 'cost' criteria associated with allocation decisions; an optimal allocation minimizes

total cost. Mathematical programming techniques may be employed to minimize such cost criteria subject to various constraints; one such technique, *dynamic programming,* is especially useful when costs associated with separate decisions are independent.

17. A computer *network* is a set of computers (or resource servers) linked together in some fashion. It may be modeled by a (labeled) directed graph. Information *flow* between computers is generally restricted both in type and in amount, depending on the links ('channels'). When channels have bounded 'capacity' (e.g., rate of information flow), the flow rate between points in the network is also bounded. Mathematical programming techniques may be employed to determine maximum flows. Delays in networks can be analyzed by the use of queueing models.

REFERENCES

(*Note:* **M** denotes mathematical orientation, **C** denotes programming orientation, and **A** denotes mathematical application; see Preface.)

CHAPTER 1

[1.7] is a general reference to the mathematics introduced in this book, and [1.20] is a general reference to the computer science topics. In addition to the references cited in the body of this chapter, we recommend for further reading [1.2], [1.6], and [1.1], for Sections 1.1, 1.2, and 1.3, respectively.

[1.1] Beckman, F.S., *Mathematical Foundations of Programming,* Addison-Wesley, Reading, Mass. (1980). **A**

[1.2] Booth, T.L., *Digital Networks and Computer Systems, 2nd Ed.,* Wiley, New York (1978). **C**

[1.3] Blum, E.K., *Numerical Analysis and Computation, Theory and Practice,* Addison-Wesley, Reading, Mass. (1972). **A**

[1.4] Copi, I.M., *Introduction to Logic, 2nd Ed.,* Macmillan, New York (1961). (6th Ed.: 1982.) **M**

[1.5] Forsythe, G.E., M.A. Malcolm and C. Moler, *Computer Methods for Mathematical Computations,* Prentice-Hall, Englewood Cliffs, NJ (1977). **A**

[1.6] Freeman, P., *Software Systems Principles: A Survey,* Science Research Associates, Chicago (1975). **C**

[1.7] Gellert, W., H. Kustner, M. Hellwich and H. Kastner (eds.), *The VNR Concise Encyclopedia of Mathematics,* Van Nostrand Reinhold, New York (1977). **M**

[1.8] Gries, D., *The Science of Programming,* Springer-Verlag, Berlin (1981). **A**

[1.9] Hoare, C.A.R., "An axiomatic basis for computer programming", *Comm. ACM,* **12**(10), 576–80 and 583 (1969). **A**

[**1.10**] IBM, *System/360 Scientific Subroutine Package, Version III, Programmer's Manual,* Form No. GH20-0205-4, IBM Corp., White Plains, NY (1970). **C**

[**1.11**] Isaacson, E. and H.B. Keller, *Analysis of Numerical Methods,* Wiley, New York (1966). **A**

[**1.12**] Knuth, D.E., *The Art of Computer Programming, Vol. I: Fundamental Algorithms,* Addison-Wesley, Reading, Mass. (1968). **A**

[**1.13**] Knuth, D.E., *The Art of Computer Programming, Vol. II: Seminumerical Algorithms,* Addison-Wesley, Reading, Mass. (1969). **A**

[**1.14**] Krutz, R.L., *Microprocessors and Logic Design,* Wiley, New York (1980). **C**

[**1.15**] Lew, A., "Decision tables for general-purpose scientific programming," *Software—Practice and Experience,* **13**(2), 181–188 (1983). **C**

[**1.16**] Mackenzie, C.E., *Coded Character Sets: History and Development,* Addison-Wesley, Reading, Mass. (1980). **C**

[**1.17**] Mendelson, E., *Number Systems and the Foundations of Analysis,* Academic Press, New York (1973). **M**

[**1.18**] Niven, I. and H.S. Zuckerman, *An Introduction to the Theory of Numbers,* Wiley, New York (1960). (4th Ed.: 1980.) **M**

[**1.19**] Quine, W.V.O., *Methods of Logic, Revised Ed.,* Holt, Rinehart and Winston, New York (1959). (4th Ed.: Harvard Univ. Press, 1982.) **M**

[**1.20**] Ralston, A. and E.D. Reilly, Jr. (eds.), *Encyclopedia of Computer Science and Engineering, 2nd Ed.,* Van Nostrand Reinhold, New York (1983). **A**

[**1.21**] Reynolds, J.C., *The Craft of Programming,* Prentice-Hall International, London (1981). **A**

[**1.22**] Rudin, W., *Principles of Mathematical Analysis, 2nd Ed.,* McGraw-Hill, New York (1964). (3rd Ed.: 1976.) **M**

[**1.23**] Sammet, J.E., *Programming Languages: History and Fundamentals,* Prentice-Hall, Englewood Cliffs, NJ (1969). **C**

[**1.24**] Standish, T. A., *Data Structure Techniques,* Addison-Wesley, Reading, Mass. (1980). **C**

CHAPTER 2

We recommend [2.5] as a general reference for the algebraic concepts introduced in this chapter. More advanced concepts appear in [2.16]. Among the numerous textbooks which emphasize computer science applications of algebra are [2.4], [2.6], and [2.9]. See also the textbooks on discrete mathematics cited in Chapter 4.

[**2.1**] Backus, J., "Can computer programming be liberated from the von Neumann style? A functional style and its algebra of programs," *Comm. ACM,* **21**(8), 613–641 (1978). **A**

[**2.2**] Barron, D.W., *Recursive Techniques in Programming, 2nd Ed.,* Macdonald and Jane's, London (1975). **C**

[**2.3**] Beckman, F.S., *Mathematical Foundations of Programming,* Addison-Wesley, Reading, Mass. (1980). **A**

[**2.4**] Birkhoff, G. and T.C. Bartee, *Modern Applied Algebra,* McGraw-Hill, New York (1970). **A**

[**2.5**] Birkhoff, G. and S. MacLane, *A Survey of Modern Algebra, 3rd Ed.,* Macmillan, New York (1969). (4th Ed.: 1977.) **M**

[**2.6**] Bobrow, L.S. and M.A. Arbib, *Discrete Mathematics: Applied Algebra for Computer and Information Science,* W.B. Saunders, Philadelphia, Pa. (1974). **A**

[**2.7**] Brady, J.M., *The Theory of Computer Science: A Programming Approach,* Chapman and Hall, London (1977). **A**

[**2.8**] Eilenberg, S., *Automata, Languages, and Machines,* Academic Press, New York and London (Vol. A, 1974; Vol. B, 1976). **M**

[**2.9**] Gill, A., *Applied Algebra for the Computer Sciences,* Prentice-Hall, Englewood Cliffs, NJ (1976). **A**

[**2.10**] Halmos, P.R., *Finite-Dimensional Vector Spaces, 2nd Ed.,* Van Nostrand, Princeton, NJ (1958). **M**

[**2.11**] Halmos, P.R., *Naive Set Theory,* Van Nostrand, Princeton, NJ (1960). **M**

[**2.12**] Henderson, P., *Functional Programming: Application and Implementation,* Prentice-Hall International, London (1980). **C**

[**2.13**] Iverson, K.E., "Notation as a tool of thought", *Comm. ACM,* **23**(8), 444–465 (1980). **C**

[**2.14**] Knuth, D.E., *The Art of Computer Programming, Vol. I: Fundamental Algorithms,* Addison-Wesley, Reading, Mass. (1968). **A**

[**2.15**] Liu, C.L., *Introduction to Combinatorial Mathematics,* McGraw-Hill, New York (1968). **M**

[**2.16**] MacLane, S. and G. Birkhoff, *Algebra,* Macmillan, New York (1967). **M**

[**2.17**] McCarthy, J., "Recursive functions of symbolic expressions and their computation by machine, Part I", *Comm. ACM.,* **3**, 184–195 (1960). **C**

[**2.18**] Scott, D., "Data types as lattices", *SIAM J. Computing,* **5**(3), 522–586 (1976). **A**

[**2.19**] Wand, M., *Induction, Recursion, and Programming,* Elsevier North-Holland, New York (1980). **A**

[**2.20**] Zadeh, L., "Fuzzy sets", *Information and Control,* **8**, 338–353 (1965). **M**

CHAPTER 3

In addition to the references already cited, we recommend for further reading [3.20] for Section 3.1, [3.3] for Section 3.2, and [3.14] for Section 3.3.

[3.1] Bellman, R., *Introduction to Matrix Analysis,* McGraw-Hill, New York (1960). (2nd Ed.: 1970.) **M**

[3.2] Bellman, R., "Dynamic programming and Lewis Carroll's game of doublets", *Bull. Inst. Math. and its Applics.,* **17**, 86–87 (1968). **M**

[3.3] Berztiss, A.T., *Data Structures: Theory and Practice, 2nd Ed.,* Academic Press, New York (1975). **A**

[3.4] Blum, E.K., *Numerical Analysis and Computation, Theory and Practice,* Addison-Wesley, Reading, Mass. (1972). **A**

[3.5] Brainerd, W.S. and L.H. Landweber, *Theory of Computation,* Wiley, New York (1974). **A**

[3.6] Cody, W.J. and W. Waite, *Software Manual for the Elementary Functions,* Prentice-Hall, Englewood Cliffs, NJ (1980). **A**

[3.7] Floyd, R.W., "A descriptive language for symbol manipulation", *J. ACM,* **8**, 579–584 (1961). **C**

[3.8] Galler, B.A. and A.J. Perlis, *A View of Programming Languages,* Addison-Wesley, Reading, Mass. (1970). **A**

[3.9] Ginsburg, S., *The Mathematical Theory of Context-Free Languages,* McGraw-Hill, New York (1966). **A**

[3.10] Griswold, R.E., J.F. Poage and I.P. Polonsky, *The SNOBOL4 Programming Language, 2nd Ed.,* Prentice-Hall, Englewood Cliffs, NJ (1971). **C**

[3.11] Hovanessian, S.A., *Computational Mathematics in Engineering,* D.C. Heath, Lexington, Mass. (1976). **A**

[3.12] Isaacson, E. and H.B. Keller, *Analysis of Numerical Methods,* Wiley, New York (1966). **A**

[3.13] Knuth, D.E., *The Art of Computer Programming, Vol. I: Fundamental Algorithms,* Addison-Wesley, Reading, Mass. (1968). **A**

[3.14] Kurki-Suonio, R., *Computability and Formal Languages,* Studentlitteratur, Lund (1971). **A**

[3.15] Lewis, H.R. and C.H. Papadimitriou, *Elements of the Theory of Computation,* Prentice-Hall, Englewood Cliffs, NJ (1981). **A**

[3.16] Maurer, W.D. and T.G. Lewis, "Hash table methods", *Computing Surveys,* **7**(1), 5–19 (1975). **C**

[3.17] Paige, L.J. and J.D. Swift, *Elements of Linear Algebra,* Blaisdell, New York (1965). (2nd Ed.: Wiley, 1974.) **M**

[3.18] Pearl, M., *Matrix Theory and Finite Mathematics,* McGraw-Hill, New York (1973). **M**

[**3.19**] Peterson, W.W., "Addressing for random-access storage", *IBM J. Res. Dev.*, **1**(2), 130–146 (1957). **C**

[**3.20**] Rice, J.R., *Numerical Methods, Software, and Analysis,* McGraw-Hill, New York (1983). **A**

[**3.21**] Rudin, W., *Principles of Mathematical Analysis, 2nd Ed.*, McGraw-Hill, New York (1964). (3rd Ed.: 1976.) **M**

[**3.22**] Simmons, G.F., *Introduction to Topology and Modern Analysis,* McGraw-Hill, New York (1963). **M**

[**3.23**] Trakhtenbrot, B.A., *Algorithms and Automatic Computing Machines,* D.C. Heath, Lexington, Mass. (1963). **A**

CHAPTER 4

Digraphs and trees are covered in numerous textbooks on discrete mathematical structures, such as [4.3] and [4.11]. See also the books on applied algebra cited in Chapter 2. The theory of graphs is treated in greater detail in [4.2], [4.5], and [4.14]. A variety of applications are presented in [4.1], [4.4], [4.8], and [4.10]. Computer science applications are emphasized in books concerned with data structures, such as [4.7], [4.9], and [4.16]; other relevant applications are given in Chapter 5.

[**4.1**] Bellman, R., K.L. Cooke and J.A. Lockett, *Algorithms, Graphs, and Computers,* Academic Press, New York (1970). **M**

[**4.2**] Berge, C., *The Theory of Graphs and its Applications,* Methuen, London (1962). **M**

[**4.3**] Berztiss, A.T., *Data Structures: Theory and Practice, 2nd Ed.,* Academic Press, New York (1975). **A**

[**4.4**] Deo, N., *Graph Theory with Applications to Engineering and Computer Science,* Prentice-Hall, Englewood Cliffs, NJ (1974). **A**

[**4.5**] Harary, F., *Graph Theory,* Addison-Wesley, Reading, Mass. (1969). **M**

[**4.6**] Harel, D., "And/or programs: a new approach to structured programming," *ACM Trans. Prog. Lang. Sys.*, **2**(1), 1–17 (1980). **A**

[**4.7**] Horowitz, E. and S. Sahni, *Fundamentals of Data Structures,* Computer Science Press, Potomac, Md. (1977). **C**

[**4.8**] Kaufmann, A., *Graphs, Dynamic Programming, and Finite Games,* Academic Press, New York (1967). **M**

[**4.9**] Knuth, D.E., *The Art of Computer Programming, Vol. I: Fundamental Algorithms,* Addison-Wesley, Reading, Mass. (1968). **A**

[**4.10**] Liu, C.L., *Introduction to Combinatorial Mathematics,* McGraw-Hill, New York (1968). **M**

[4.11] Liu, C.L., *Elements of Discrete Mathematics,* McGraw-Hill, New York (1977). **A**

[4.12] McCarthy, J., M.I. Levin, *et al., LISP 1.5 Programmer's Manual, 2nd Ed.,* MIT Press, Cambridge, Mass. (1965). **C**

[4.13] Moret, B.M.E., "Decision trees and diagrams", *Computing Surveys,* **14**(4), 593–623 (1982). **A**

[4.14] Ore, O., *Theory of Graphs,* American Mathematical Society, Providence, R.I. (1962). **M**

[4.15] Pearl, M., *Matrix Theory and Finite Mathematics,* McGraw-Hill, New York (1973). **M**

[4.16] Standish, T. A., *Data Structure Techniques,* Addison-Wesley, Reading, Mass. (1980). **C**

CHAPTER 5

The references of the preceding chapter are also relevant to the material covered in this chapter. A general reference for Sections 5.2 and 5.3 is [5.18]. Descriptions of digraph and tree algorithms may also be found in most books on data structures (such as [5.8]) or on the analysis of algorithms (such as [5.1]). A general reference for Section 5.4 is [5.7]. Additional structured programming concepts are discussed in [5.14]. Numerous textbooks cover complexity, computability, languages, and automata more thoroughly than here, as well as provide further references; for example, see [5.16] and [5.24]. Numerous applications of graphs and trees to compilers are described in [5.2]. A variety of other applications may be found in [5.11] and [5.47].

[5.1] Aho, A.V., J.E. Hopcroft and J.D. Ullman, *The Design and Analysis of Computer Algorithms,* Addison-Wesley, Reading, Mass. (1974). **A**

[5.2] Aho, A.V. and J.D. Ullman, *The Theory of Parsing, Translation, and Compiling,* Prentice-Hall, Englewood Cliffs, NJ (Vol. I: *Parsing,* 1972; Vol. II: *Compiling,* 1973). **A**

[5.3] Aho, A.V. and J.D. Ullman, *Principles of Compiler Design,* Addison-Wesley, Reading, Mass. (1977). **C**

[5.4] Allen, F.E., "Program optimization", *Annual Review in Automatic Programming* **5**, Pergamon, Oxford, 239–307 (1969). **C**

[5.5] Anderson, C., *An Introduction to ALGOL 60,* Addison-Wesley, Reading, Mass. (1964). **C**

[5.6] Backhouse, R.C., *Syntax of Programming Languages: Theory and Practice,* Prentice-Hall International, London (1979). **A**

[5.7] Beckman, F.S., *Mathematical Foundations of Programming,* Addison-Wesley, Reading, Mass. (1980). **A**

[5.8] Berztiss, A.T., *Data Structures: Theory and Practice, 2nd Ed.,* Academic Press, New York (1975). **A**

[5.9] Blum, E.K., "Towards a theory of semantics and compilers for programming languages", *J. Comp. Sys. Sci.,* **3**(3), 248–275 (1969). **A**

[5.10] Böhm, C. and G. Jacopini, "Flow diagrams, Turing machines, and languages with only two formation rules", *Comm. ACM,* **9**(5), 366–371 (1966). **A**

[5.11] Bowie, W.S., "Applications of graph theory in computer systems", *Intl. J. of Comp. and Info. Sci.,* **5**(1), 9–31 (1976). **A**

[5.12] Brown, P.J., *Writing Interactive Compilers and Interpreters,* Wiley, Chichester (1979). **C**

[5.13] Cohen, D.J. and C.C. Gottlieb, "A list structure form of grammars for syntactic analysis", *Computing Surveys,* **2**(1), 65–82 (1970). **C**

[5.14] Dahl, O.J., E.W. Dijkstra and C.A.R. Hoare, *Structured Programming,* Academic Press, New York (1970). **C**

[5.15] Davies, A.C., "The analogy between electrical networks and flowcharts", *IEEE Trans. Softw. Engrg.,* **SE-6**(4), 391–394 (1980). **A**

[5.16] Denning, P.J., J.B. Dennis and J.E. Qualitz, *Machines, Languages and Computation,* Prentice-Hall, Englewood Cliffs, NJ (1978). **A**

[5.17] Dennis, J.B., "Data flow supercomputers", *Computer,* **13**(11), 48–56 (1980). **C**

[5.18] Even, S., *Algorithmic Combinatorics,* Macmillan, New York (1973). **A**

[5.19] Ginsburg, S., *The Mathematical Theory of Context-Free Languages,* McGraw-Hill, New York (1966). **A**

[5.20] Ginsburg, S., *An Introduction to Mathematical Machine Theory,* Addison-Wesley, Reading, Mass. (1962). **A**

[5.21] Gries, D., *Compiler Construction for Digital Computers,* Wiley, New York (1971). **C**

[5.22] Harrison, M.A., *Introduction to Formal Language Theory,* Addison-Wesley, Reading, Mass. (1978). **A**

[5.23] Hoare, C.A.R. and D.C.S. Allison, "Incomputability", *Computing Surveys,* **4**(3), 169–178 (1972). **A**

[5.24] Hopcroft, J.E. and J.D. Ullman, *Introduction to Automata Theory, Languages, and Computation,* Addison-Wesley, Reading, Mass. (1979). **A**

[5.25] Howden, W.E., "Methodology for the generation of program test data", *IEEE Trans. Computers,* **C-24**(5), 554–559 (1975). **A**

[5.26] Jensen, K. and N. Wirth, *Pascal: User Manual and Report, 2nd Corr. Ed.,* Springer-Verlag, Berlin (1976). **C**

[5.27] Karp, R.M., "A note on the application of graph theory to digital computer programming," *Information and Control,* **3**, 179–190 (1960). **A**

[5.28] Knuth, D.E., *The Art of Computer Programming, Vol. I: Fundamental Algorithms,* Addison-Wesley, Reading, Mass. (1969). **A**

[5.29] Knuth, D.E., "Structured programming with go to statements", *Computer Surveys,* **6**(4), 261–301 (1974). **C**

[**5.30**] Lew, A., "Diagnostic compilers and debugging", *Proc. Internatl. Computer Symp., Vol. II,* 205–212 (1975). **C**

[**5.31**] Lew, A., "On the emulation of flowcharts by decision tables", *Comm. ACM,* **25**(12), 895–905 (1982). **C**

[**5.32**] Lucas, P. and K. Walk, "On the formal description of PL/I", *Annual Review in Automatic Programming* **6**, Pergamon, Oxford, 105–181 (1969). **A**

[**5.33**] McCabe, T.J., "A complexity measure", *IEEE Trans. on Software Engrg.,* **SE-2**(4), 308–320 (1976). **C**

[**5.34**] Minsky, M.L., *Computation: Finite and Infinite Machines,* Prentice-Hall, Englewood Cliffs, NJ (1967). **A**

[**5.35**] Nakata, I., "On compiling algorithms for arithmetic expressions", *Comm. ACM,* **10**(8), 492–494 (1967). **C**

[**5.36**] Pager, D., "A practical general method for constructing LR(k) parsers", *Acta Informatica,* **7**(3), 249–268 (1977). **A**

[**5.37**] Paull, M.C. and S.H. Unger, "Minimizing the number of states in incompletely specified sequential functions", *IRE Trans. Elect. Comp,* **EC-8**(3), 356–367 (1959). **A**

[**5.38**] Peterson, J.L., *Petri Net Theory and the Modeling of Systems,* Prentice-Hall, Englewood Cliffs, NJ (1981). **A**

[**5.39**] Peterson, W.W., T. Kasami and N. Tokura, "On the capability of while, repeat, and exit statements", *Comm. ACM,* **16**(8), 503–512 (1973). **A**

[**5.40**] Ramamoorthy, C.V. and S.B.F. Ho, "Testing large software with automated software evaluation systems", *IEEE Trans. Software Engrg.,* **SE-1**(1), 46–58 (1975). **A**

[**5.41**] Stoy, J.E., *Denotational Semantics: The Scott-Strachey Approach to Programming Language Theory,* MIT Press, Cambridge, Mass. (1977). **A**

[**5.42**] Tarjan, R., "Depth-first search and linear graph algorithms", *SIAM J. Computing,* **1**(2), 146–160 (1972). **M**

[**5.43**] Tarjan, R., "Enumeration of the elementary circuits of a directed graph", *SIAM J. Computing,* **2**(3), 211–216 (1973). **M**

[**5.44**] Tennent, R.D., *Principles of Programming Languages,* Prentice-Hall International, London (1981). **C**

[**5.45**] Trakhtenbrot, B.A., *Algorithms and Automatic Computing Machines,* D.C. Heath, Lexington, Mass. (1963). **A**

[**5.46**] Warshall, S., "A theorem on Boolean matrices", *J. ACM,* **9**(1), 11–12 (1962). **M**

[**5.47**] Yeh, R.T. (ed.), *Applied Computation Theory: Analysis, Design, Modeling,* Prentice-Hall, Englewood Cliffs, NJ (1976). **A**

CHAPTER 6

Textbooks on operations research are good general references for most of the material in this chapter. We recommend, for example, [6.14] and [6.32]. More specialized books include [6.10] for statistics (Section 6.1), [6.7] for probability (Section 6.2), [6.18] for Markov chains (Section 6.3), and [6.19] for queueing (Section 6.4). Books which cover a variety of computer science applications include [6.2] and [6.13]; books of related interest include [6.11], [6.16], and Vol. 3 of [6.35]. Probability and statistics are also applicable to most of the subjects covered in previous chapters. For example, see [6.20] for a discussion of statistical tests for random-number generators, [6.17] and [6.22] for statistical studies of computer arithmetic and roundoff errors, [6.24] for an introduction to random graphs, and [6.33] for a discussion of probabilistic languages.

[6.1] Afifi, A.A. and S.P. Azen, *Statistical Analysis : A Computer Oriented Approach, 2nd Ed.,* Academic Press, New York (1979). **A**

[6.2] Allen, A.O., *Probability, Statistics, and Queueing Theory with Computer Science Applications,* Academic Press, New York (1978). **A**

[6.3] Beizer, B., *Microanalysis of Computer System Performance,* Van Nostrand Reinhold, New York (1978). **C**

[6.4] Coffman, E.G., Jr. and P.J. Denning, *Operating Systems Theory,* Prentice-Hall, Englewood Cliffs, NJ (1973). **A**

[6.5] Cox, D.R. and W.L. Smith, *Queues,* Methuen, London (1961). **M**

[6.6] Everling, W., *Exercises in Computer Systems Analysis,* Springer-Verlag, Berlin (1972). **A**

[6.7] Feller, W., *An Introduction to Probability Theory and its Applications,* Wiley, New York (Vol. I, 2nd Ed., 1957; Vol. II, 1966). **M**

[6.8] Ferrari, D., *Computer Systems Performance Evaluation,* Prentice-Hall, Englewood Cliffs, NJ (1978). **A**

[6.9] Freiberger, W. (ed.), *Statistical Computer Performance Evaluation,* Academic Press, New York (1972). **A**

[6.10] Freund, J.E., *Modern Elementary Statistics, 4th Ed.,* Prentice-Hall, Englewood Cliffs, NJ (1973). **M**

[6.11] Graybeal, W. and U.W. Pooch, *Simulation : Principles and Methods,* Winthrop, Cambridge, Mass. (1980). **A**

[6.12] Hellerman, H., *Digital Computer System Principles,* McGraw-Hill, New York (1967). **C**

[6.13] Hellerman, H. and T.F. Conroy, *Computer System Performance,* McGraw-Hill, New York (1975). **A**

[6.14] Hillier, F.S. and G.J. Lieberman, *Introduction to Operations Research,* Holden-Day, San Francisco (1967). (3rd Ed.: 1980.) **M**

[6.15] Hoel, P.G., *Elementary Statistics, 4th Ed.,* Wiley, New York (1976). **M**

[**6.16**] Hovanessian, S.A., *Computational Mathematics in Engineering,* D.C. Heath, Lexington, Mass. (1976). **A**

[**6.17**] Hull, T.E. and J.R. Swenson, "Test of probability models for the propagation of roundoff errors", *Comm. ACM,* **9**(2), 108–113 (1966). **A**

[**6.18**] Kemeny, J.G. and J.L. Snell, *Finite Markov Chains,* Van Nostrand, New York (1960). **M**

[**6.19**] Kleinrock, L., *Queueing Systems,* Wiley, New York (Vol. I: *Theory,* 1975; Vol. II: *Computer Applications,* 1976). **M, A**

[**6.20**] Knuth, D.E., *The Art of Computer Programming, Vol. II: Seminumerical Algorithms,* Addison-Wesley, Reading, Mass. (1969). **A**

[**6.21**] Kobayashi, H., *Modeling and Analysis: An Introduction to System Performance Evaluation Methodology,* Addison-Wesley, Reading, Mass. (1978). **A**

[**6.22**] Kuki, H. and W.J. Cody, "A statistical study of the accuracy of floating point number systems", *Comm. ACM,* **16**(4), 223–230 (1973). **A**

[**6.23**] Liu, C.L., *Introduction to Combinatorial Mathematics,* McGraw-Hill, New York (1968). **M**

[**6.24**] Marshall, C.W., *Applied Graph Theory,* Wiley, New York (1971). **M**

[**6.25**] Moore, R.W., *Introduction to the Use of Computer Packages for Statistical Analysis,* Prentice-Hall, Englewood Cliffs, NJ (1978). **C**

[**6.26**] Papoulis, A., *Probability, Random Variables, and Stochastic Processes,* McGraw-Hill, New York (1965). **M**

[**6.27**] Pearl, M., *Matrix Theory and Finite Mathematics,* McGraw-Hill, New York (1973). **M**

[**6.28**] Ramamoorthy, C.V., "Discrete Markov analysis of computer programs", *Proc. ACM 20th Natl. Conf.,* 386–392 (1965). **A**

[**6.29**] Schucany, W.R., B.S. Shannon, Jr. and P.D. Minton, "A survey of statistical packages", *Computing Surveys,* **4**(2), 65–79 (1972). **C**

[**6.30**] Spirn, J.R., *Program Behavior: Models and Measurements,* Elsevier North-Holland, New York (1977). **A**

[**6.31**] Takacs, L., *Introduction to the Theory of Queues,* Oxford Univ. Press, New York (1962). **M**

[**6.32**] Wagner, H.M., *Principles of Operations Research with Applications to Managerial Decisions,* Prentice-Hall, Englewood Cliffs, NJ (1969). **M**

[**6.33**] Wetherall, C.S., "Probabilistic languages: a review and some open questions", *Computing Surveys,* **12**(4), 361–379 (1980). **A**

[**6.34**] Winkler, R.L., *Introduction to Bayesian Inference and Decision,* Holt, Rinehart and Winston, New York (1972). **M**

[**6.35**] Yeh, R.T., *et al.* (eds.), *Current Trends in Programming Methodology,* Prentice-Hall, Englewood Cliffs, NJ (4 volumes, 1977–78). **A**

CHAPTER 7

General references for much of the material in this chapter, especially as related to the processing of files, are [7.4] and [7.28]. For a comprehensive treatment of sorting and searching, see [7.20]. Sorting and searching are also discussed in books on the analysis of algorithms (e.g., [7.2]) and on data structures (such as those cited in Chapter 4). Information or data base systems are covered in detail in numerous books such as [7.13] and [7.35]. We especially recommend [7.12] and [7.30] for their algebraic approaches, [7.21] for some interesting matrix applications, and [7.33] for its statistical models.

[7.1] Abramson, N., *Information Theory and Coding,* McGraw-Hill, New York (1963). **M**

[7.2] Aho, A.V., J.E. Hopcroft and J.D. Ullman, *The Design and Analysis of Computer Algorithms,* Addison-Wesley, Reading, Mass. (1974). **A**

[7.3] Bernstein, A.J., "Program analysis for parallel programming", *IEEE Trans. Elect. Comp.,* **EC-15**(5), 757–767 (1966). **A**

[7.4] Berztiss, A.T., *Data Structures: Theory and Practice, 2nd Ed.,* Academic Press, New York (1975). **A**

[7.5] Chen, P.P., "The entity-relationship model: toward a unified view of data", *ACM Trans. Database Systems,* **1**(1), 9–36 (1976). **C**

[7.6] Codd, E.F., "A relational model of data for large shared data banks", *Comm. ACM,* **13**(6), 377–387 (1970). **C**

[7.7] Coffman, E.G., Jr. and P.J. Denning, *Operating Systems Theory,* Prentice-Hall, Englewood Cliffs, NJ (1973). **A**

[7.8] Comer, D., "The ubiquitous B-tree", *Computing Surveys,* **11**(2), 121–137 (1979). **C**

[7.9] Date, C.J., *An Introduction to Data-Base Systems, 3rd Ed.,* Addison-Wesley, Reading, Mass. (1981). **C**

[7.10] Flores, I., *Computer Sorting,* Prentice-Hall, Englewood Cliffs, NJ (1969). **C**

[7.11] Ganapathy, S. and V. Rajaraman, "Information theory applied to the conversion of decision tables to computer programs", *Comm. ACM,* **16**(9), 532–539 (1973). **A**

[7.12] Ghosh, S.P., *Data Base Organization for Data Management,* Academic Press, New York (1977). **A**

[7.13] Heaps, H.S., *Information Retrieval: Computation and Theoretical Aspects,* Academic Press, New York (1978). **A**

[7.14] Held, G., *Data Compression: Techniques and Applications, Hardware and Software Considerations,* Wiley, New York (1983). **C**

[7.15] Hellerman, H., *Digital Computer Systems Principles,* McGraw-Hill, New York (1975). **C**

[7.16] Hoare, C.A.R., "Quicksort", *Computer Journal,* **5**(1), 10–15 (1962). **C**

[7.17] Hsiao, D.K., *Systems Programming: Concepts of Operating and Data Base Systems,* Addison-Wesley, Reading, Mass. (1975). **C**

[7.18] Knott, G.D., "Hashing functions", *Computer Journal,* **18**(3), 265–278 (1975). **A**

[7.19] Knuth, D.E., *The Art of Computer Programming, Vol. I: Fundamental Algorithms,* Addison-Wesley, Reading, Mass. (1968). **A**

[7.20] Knuth, D.E., *The Art of Computer Programming, Vol. III: Sorting and Searching,* Addison-Wesley, Reading, Mass. (1973). **A**

[7.21] Langefors, B., *Theoretical Analysis of Information Systems, 4th Ed.,* Studentlitteratur, Lund (1973). **A**

[7.22] Lorin, H., *Sorting and Sort Systems,* Addison-Wesley, Reading, Mass. (1975). **C**

[7.23] Mackenzie, C.E., *Coded Character Sets: History and Development,* Addison-Wesley, Reading, Mass. (1980). **C**

[7.24] Martin, W.A., "Sorting", *Computing Surveys,* **3**(4), 147–174 (1971). **C**

[7.25] Maurer, W.D. and T.G. Lewis, "Hash table methods", *Computing Surveys,* **7**(1), 5–19 (1975). **C**

[7.26] Meadow, C.T., *The Analysis of Information Systems, 2nd Ed.,* Melville (Wiley), Los Angeles (1973). **C**

[7.27] Nievergelt, J., "Binary search trees and file organization", *Computing Surveys,* **6**(3), 195–207 (1974). **C**

[7.28] Peterson, W.W. and A. Lew, *File Programming and Processing,* Wiley, New York (to be published). **C**

[7.29] Peterson, W.W. and E.J. Weldon, Jr., *Error-Correcting Codes, 2nd Ed.,* MIT Press, Cambridge, Mass. (1972). **M**

[7.30] Rus, T., *Data Structures and Operating Systems,* Wiley, Chichester (1979). **A**

[7.31] Salton, G., *Automatic Information Organization and Retrieval,* McGraw-Hill, New York (1968). **A**

[7.32] Taylor, D.J., D.E. Morgan and J.P. Black, "Redundancy in data structures: improving software fault tolerance" and "Redundancy in data structures: some theoretical results", *IEEE Trans. Software Engrg,* **SE-6**(6), 585–594 and 595–602 (1980). **A**

[7.33] Teorey, T.J. and J.P. Fry, *Design of Database Structures,* Prentice-Hall, Englewood Cliffs, NJ (1982). **A**

[7.34] Ullman, J.D., *Principles of Database Systems,* Computer Science Press, Potomac, Md. (1980). **C**

[7.35] Wiederhold, G., *Database Design,* McGraw-Hill, New York (1977). **C**

CHAPTER 8

General introductions to optimization appear in [8.18] and [8.33]. Dynamic programming is discussed in greater detail in [8.6] and [8.12]. Other optimization techniques are treated in books on operations research (such as [8.26] and [8.40]), control theory (such as [8.48] and [8.30]), and applied mathematics (such as [8.47]). Computational aspects are emphasized in [8.4], [8.35], and [8.49]; see also the survey in [8.55]. For further information on the routing and traveling-salesman problems, see [8.23] and [8.14], respectively.

[8.1] Adby, P.R. and M.A.H. Dempster, *Introduction to Optimization Methods,* Chapman and Hall, London (1974). **M**

[8.2] Aho, A.V. and S.C. Johnson, "Optimal code generation for expression trees", *J. ACM,* **23**(3), 488–501 (1976). **A**

[8.3] Barsov, A.S., *What is Linear Programming?* D.C. Heath, Boston (1964). **M**

[8.4] Beale, E.M.L., *Mathematical Programming in Practice,* Pitman, London (1968). **C**

[8.5] Bekey, G.A. and W.S. Karplus, *Hybrid Computation,* Wiley, New York (1968). **C**

[8.6] Bellman, R., *Dynamic Programming,* Princeton Univ. Press, Princeton, NJ (1957). **M**

[8.7] Bellman, R., "On a routing problem", *Quart. Appl. Math.,* **16**, 87–90 (1958). **M**

[8.8] Bellman, R. "Dynamic programming treatment of the traveling salesman problem", *J. ACM,* **9**(1), 61–63 (1962). **M**

[8.9] Bellman, R., "Dynamic programming and Lewis Carroll's game of doublets", *Bull. Inst. Math. and its Applics.,* **17**, 86–87 (1968). **A**

[8.10] Bellman, R., *An Introduction to Artificial Intelligence: Can Computers Think?* Boyd and Fraser, San Francisco (1978). **A**

[8.11] Bellman, R., K.L. Cooke and J.A. Lockett, *Algorithms, Graphs, and Computers,* Academic Press, New York (1970). **A**

[8.12] Bellman, R. and S.E. Dreyfuss, *Applied Dynamic Programming,* Princeton Univ. Press, Princeton, NJ (1962). **M**

[8.13] Bellman, R., J. Holland and R. Kalaba, "On the application of dynamic programming to the synthesis of logical systems", *J. ACM,* **6**(4), 486–493 (1959). **A**

[8.14] Bellmore, M. and G.L. Nemhauser, "The traveling salesman problem", *Operations Research,* **16**, 538–558 (1968). **M**

[8.15] Blum, E.K., *Numerical Analysis and Computation, Theory and Practice,* Addison-Wesley, Reading, Mass. (1972). **A**

[**8.16**] Bonzon, P., "Necessary and sufficient conditions for dynamic programming of combinatorial type", *J. ACM,* **17**(4), 675–682 (1970). **M**

[**8.17**] Brown, K.Q., "Dynamic programming in computer science", Report No. CMU-CS-79-106, Dept. of Computer Science, Carnegie-Mellon Univ., Pittsburgh, Pa. (1979). **A**

[**8.18**] Courant, R. and H. Robbins, *What is Mathematics,* Oxford University Press, London (1941). **M**

[**8.19**] Dantzig, G.B., *Linear Programming and Extensions,* Princeton Univ. Press, Princeton, NJ (1963). **M**

[**8.20**] Dongarra, J.J., C.B. Moler, J.R. Bunch and G.W. Stewart, *Linpack User's Guide,* Society for Industrial and Applied Math., Philadelphia, Pa. (1979). **C**

[**8.21**] Dowdy, L.W. and D.V. Foster, "Comparative models of the file assignment problem", *Computing Surveys,* **14**(2), 287–313 (1982). **A**

[**8.22**] Draper, N.R. and H. Smith, *Applied Regression Analysis,* Wiley, New York (1966). (2nd Ed.: 1981.) **M**

[**8.23**] Dreyfuss, S.E., "An appraisal of some shortest-path algorithms", *Operations Research,* **17**(3), 395–412 (1969). **M**

[**8.24**] Dreyfuss, S.E. and A.M. Law, *The Art and Theory of Dynamic Programming,* Academic Press, New York (1977). **M**

[**8.25**] Godbole, S.S., "On efficient computation of matrix chain products", *IEEE Trans. Computers,* **C-22**(9), 864–866 (1973). **A**

[**8.26**] Hillier, F.S. and G.J. Lieberman, *Introduction to Operations Research,* Holden-Day, San Francisco (1967). (3rd Ed.: 1980.) **M**

[**8.27**] Hirschberg, D.S., "A linear space algorithm for computing maximal common subsequences", *Comm. ACM,* **18**(6), 341–343 (1975). **A**

[**8.28**] Horwitz, L.P., R.M. Karp, R.E. Miller and S. Winograd, "Index register allocation", *J. ACM,* **13**(1), 43–61 (1966). **A**

[**8.29**] Jacobs, O.L.R., *An Introduction to Dynamic Programming: The Theory of Multistage Decision Processes,* Chapman and Hall, London (1967). **M**

[**8.30**] Kalman, R.E., P.L. Falb and M.A. Arbib, *Topics in Mathematical System Theory,* McGraw-Hill, New York (1969). **M**

[**8.31**] Karp, R.M. and M. Held, "Finite-state processes and dynamic programming", *SIAM J. Appl. Math.,* **15**(3), 693–718 (1967). **M**

[**8.32**] Kaufmann, A., *Graphs, Dynamic Programming, and Finite Games,* Academic Press, New York (1967). **M**

[**8.33**] Kaufmann, A., *The Science of Decision-Making,* World Univ. Library (McGraw-Hill), New York (1968). **M**

[**8.34**] Kral, J., "The formulation of the problem of program segmentation in the terms of pseudo boolean programming", *Kybernetica,* **4**, 6–11 (1968). **A**

[**8.35**] Land, A.H. and S. Powell, *Fortran Codes for Mathematical Programming: Linear, Quadratic and Discrete,* Wiley, London (1973). **C**

[**8.36**] Larson, R.E. and J.L. Casti, *Principles of Dynamic Programming,* Marcel Dekker, New York (1982). **M**

[**8.37**] Lawler, E.L. and D.E. Wood, "Branch-and-bound methods: a survey", *Operations Research,* **14**(4), 699–719 (1966). **M**

[**8.38**] Lew, A., "Some results in differential approximation", *Intl. J. Comp. Math.,* **2**(3), 231–245 (1970). **A**

[**8.39**] Lew, A., "Optimal resource allocation and scheduling among parallel processes", *Parallel Processing* (ed., T. Feng), Springer-Verlag, Berlin, 269–279 (1975). **A**

[**8.40**] Moder, J. and S.E. Elmaghraby (eds.), *Handbook of Operations Research,* Van Nostrand Reinhold, New York (2 volumes: 1978). **M**

[**8.41**] Nemhauser, G.L., *Introduction to Dynamic Programming,* Wiley, New York (1966). **M**

[**8.42**] Nilsson, N.J., *Problem-solving Methods in Artificial Intelligence,* McGraw-Hill, New York (1971). **A**

[**8.43**] Peterson, J.L., J.R. Bitner and J.H. Howard, "The selection of optimal tab settings", *Comm. ACM,* **21**(12), 1004–1007 (1978). **A**

[**8.44**] Ramamoorthy, C.V. and K.M. Chandy, "Optimization of memory hierarchies in multiprogrammed systems", *J. ACM,* **17**(3), 426–445 (1970). **A**

[**8.45**] Reinwald, L.T. and R.M. Soland, "Conversion of limited-entry decision tables to optimal computer programs, I: Minimum average processing time", *J. ACM,* **13**(3), 339–358 (1966). **A**

[**8.46**] Rudin, W., *Principles of Mathematical Analysis, 2nd Ed.,* McGraw-Hill, New York (1964). (3rd Ed.: 1976.) **M**

[**8.47**] Saaty, T.L. and J. Bram, *Nonlinear Mathematics,* McGraw-Hill, New York (1964). **M**

[**8.48**] Sage, A.P., *Optimum Systems Control,* Prentice-Hall, Englewood Cliffs, NJ (1968). **M**

[**8.49**] Syslo, M.M., N. Deo and J.S. Kowalik, *Discrete Optimization Algorithms with Pascal Programs,* Prentice-Hall, Englewood Cliffs, NJ (1983). **A**

[**8.50**] Thesen, A., "Scheduling of computer programs for optimal machine utilization", *BIT,* **13**, 337–343 (1963). **A**

[**8.51**] Wagner, R.A., "Common phrases and minimum-space text storage", *Comm. ACM,* **16**(3), 148–152 (1973). **A**

[**8.52**] Wagner, R.A., "Order-*n* correction for regular languages", *Comm. ACM,* **17**(5), 265–268 (1974). **A**

[**8.53**] Wagner, R.A. and M.J. Fischer, "The string-to-string correction problem", *J. ACM,* **21**(1), 168–173 (1974). **A**

[**8.54**] Walker, E.S., "Optimization of tape operations", Appendix III in D. Van Tassel, *Programming Style, Design, Efficiency, Debugging, and Testing,* Prentice-Hall, Englewood Cliffs, NJ (1974). **A**

[**8.55**] White, W.W., "A status report on computing algorithms for mathematical programming", *Computing Surveys,* **5**(3), 135–166 (1973). **A**

[**8.56**] Williams, K., *Dynamic Programming: Sequential Decision Making,* Longmans, London (1970). **M**

[8.57] Wolman, E.J., "A fixed optimum cell-size for records of various lengths", *J. ACM,* **12**(1), 53–70 (1965). **A**

[8.58] Younger, D.H., "Recognition and parsing of context free languages in time n^3", *Information and Control,* **10**(2), 189–208 (1960). **A**

CHAPTER 9

The books on digraphs and trees cited for Chapters 4 and 5 are good general references for Section 9.1. We also recommend [9.11], [9.20], [9.29], and [9.48]. Additional binary tree applications may be found in artificial intelligence literature, such as [9.47] and [9.46]. (Other interesting aspects of artificial intelligence are presented in [9.8] and [9.52].) For an introduction to operating systems, see [9.12] and [9.53]; more mathematical treatments are given in [9.16] and [9.25]. Resource allocation and scheduling are discussed in greater detail in [9.9], [9.15], and [9.26]. See also the books cited for Chapter 6, especially those concerning system performance, and books on software engineering, such as [9.41] and [9.50]. Section 9.3 is based on [9.37].

[9.1] Abramson, N. and F.F. Kuo (eds.), *Computer-Communication Networks,* Prentice-Hall, Englewood Cliffs, NJ (1973). **A**

[9.2] Aho, A.V., P.J. Denning and J.D. Ullman, "Principles of optimal page replacement", *J. ACM,* **18**(1), 80–93 (1971). **A**

[9.3] Ahuja, V., *Design and Analysis of Computer Communication Networks,* McGraw-Hill, New York (1982). **A**

[9.4] Archibald, R.D. and R.L. Villoria, *Network-based Management Systems (PERT/CPM),* Wiley, New York (1967). **M**

[9.5] Armstrong-Wright, A.T., *CPM: Introduction and Practice,* Longmans, London (1969). **M**

[9.6] Baer, J.L., "A survey of some theoretical aspects of multiprocessing", *Computer Surveys,* **5**(1), 31–80 (1973). **A**

[9.7] Bayes, A.J., "A dynamic programming algorithm to optimise decision table code", *Australian Computer J.,* **5**(2), 77–79 (1973). **A**

[9.8] Bellman, R., *An Introduction to Artificial Intelligence: Can Computers Think?* Boyd and Fraser, San Francisco (1978). **A**

[9.9] Bellman, R., A.O. Esogbue and I. Nabeshima, *Mathematical Aspects of Scheduling and Applications,* Pergamon, Oxford (1982). **M**

[9.10] Bernstein, A.J., "Program analysis for parallel programming", *IEEE Trans. Elect. Comp.,* **EC-15**(5), 757–767 (1966). **A**

[9.11] Boffey, T.B., *Graph Theory in Operations Research,* Macmillan, London (1982). **M**

[**9.12**] Brinch Hansen, P., *Operating System Principles,* Prentice-Hall, Englewood Cliffs, NJ (1973). **A**

[**9.13**] Cheung, T.Y., "Computational comparison of eight methods for the maximal flow problem", *ACM Trans. Math. Softw.,* **6**(1), 1–16 (1980). **M**

[**9.14**] Chu, W.W. and H. Opderbeck, "The page fault frequency replacement algorithm", *Proc. AFIPS FJCC,* 597–609 (1972). **A**

[**9.15**] Coffman, E.G., Jr. (ed.), *Computer and Job-Shop Scheduling Theory,* Wiley, New York (1976). **A**

[**9.16**] Coffman, E.G., Jr. and P.J. Denning, *Operating Systems Theory,* Prentice-Hall, Englewood Cliffs, NJ (1973). **A**

[**9.17**] Dijkstra, E.W., "A note on two problems in connexion with graphs", *Numerische Mathematik,* **1**, 269–271 (1959). **A**

[**9.18**] Dijkstra, E.W., "Cooperating sequential processes", in F. Genuys (ed.), *Programming Languages,* Academic Press, London, 43–112 (1968). **C**

[**9.19**] Dreyfuss, S.E. and A.M. Law, *The Art and Theory of Dynamic Programming,* Academic Press, New York (1977). **M**

[**9.20**] Even, S., *Algorithmic Combinatorics,* Macmillan, New York (1973). **A**

[**9.21**] Ford, L. and D. Fulkerson, *Flows in Networks,* Princeton Univ. Press, Princeton, NJ (1962). **M**

[**9.22**] Franklin, M.A. and R.K. Gupta, "Computation of page fault probability from program transition diagram", *Comm. ACM,* **17**(4), 186–191 (1974). **A**

[**9.23**] Gilbert, P. and W.J. Chandler, "Interference between communicating parallel processes", *Comm. ACM,* **15**(6), 427–437 (1972). **A**

[**9.24**] Held, M. and R.M. Karp, "The construction of discrete dynamic programming algorithms", *IBM Systems J.,* **4**(2), 136–147 (1965). **A**

[**9.25**] Hellerman, H. and T.F. Conroy, *Computer System Performance,* McGraw-Hill, New York (1975). **A**

[**9.26**] Hillier, F.S. and G.J. Lieberman, *Introduction to Operations Research,* Holden-Day, San Francisco (1967). (3rd Ed.: 1980.) **M**

[**9.27**] Holt, R.C., "Some deadlock properties of computer systems", *Computing Surveys,* **4**(3), 179–196 (1972). **C**

[**9.28**] Howard, R.A., *Dynamic Programming and Markov Processes,* MIT Press, Cambridge, Mass. (1960). **M**

[**9.29**] Hu, T.C., *Combinatorial Algorithms,* Addison-Wesley, Reading, Mass. (1982). **A**

[**9.30**] Huffman, D.A., "A method for the construction of minimum redundancy codes", *Proc. IRE,* **40**, 1098–1101 (1952). **M**

[**9.31**] Ingargiola, G. and J.F. Korsh, "Finding optimal demand-paging algorithms", *J. ACM,* **21**(1), 40–53 (1974). **A**

[**9.32**] Jones, A.K. and P. Schwartz, "Experience using multiprocessing systems—a status report", *Computing Surveys,* **12**(2), 121–165 (1980). **A**

[9.33] Kaufmann, A., *Graphs, Dynamic Programming, and Finite Games,* Academic Press, New York (1967). **M**

[9.34] Kaufmann, A. and G. Desbazeille, *CPM: Applications of the PERT Method and its Variants to Production and Study Programs,* Gordon and Breach, London (1969). **M**

[9.35] Kernighan, B.W., "Optimal sequential partitions of graphs", *J. ACM,* **18**(1), 34–40 (1971). **A**

[9.36] Kral, J., "One way of estimating frequencies of jumps in a program", *Comm. ACM,* **11**(7), 475–480 (1968). **A**

[9.37] Lew, A., "Optimal resource allocation and scheduling among parallel processes", *Parallel Processing* (ed., T. Feng), Springer-Verlag, Berlin, 269–279 (1975). **A**

[9.38] Lew, A., "On optimal demand-paging algorithms", *Proc. 2nd USA–Japan Computer Conf.,* 205–212 (1975). **A**

[9.39] Lew, A., "Optimal control of demand-paging systems", *Information Science,* **10**(4), 319–330 (1976). **A**

[9.40] Lew, A., "Optimal conversion of extended-entry decision tables with general cost criteria", *Comm. ACM,* **21**(4), 269–279 (1978). **A**

[9.41] Lewis, T.G., *Software Engineering: Analysis and Verification,* Reston, Reston, Va. (1982). **A**

[9.42] Llewellyn, J.A., "The deadly embrace—a finite state model approach", *Computer Journal,* **16**(3), 223–225 (1973). **A**

[9.43] Lloyd, E.L., "Critical path scheduling with resource and processor constraints", *J. ACM,* **29**(3), 781–811 (1982). **M**

[9.44] Loui, R.P., "Optimal paths in graphs with stochastic or multidimensional weights", *Comm. ACM,* **26**(9), 670–676 (1983). **M**

[9.45] MacCrimmon, K.R. and C.A. Ryadec, "An analytical study of the PERT assumptions", *Operations Research,* **12**, 16–37 (1964). **M**

[9.46] Meisel, W.S. and D.A. Michalopoulous, "A partitioning algorithm with application in pattern classification and the optimization of decision trees", *IEEE Trans. Computers,* **C-22**(1), 93–103 (1973). **A**

[9.47] Nilsson, N.J., *Problem-solving Methods in Artificial Intelligence,* McGraw-Hill, New York (1971). **A**

[9.48] Papadimitriou, C.H. and K. Steiglitz, *Combinatorial Optimization: Algorithms and Complexity,* Prentice-Hall, Englewood Cliffs, NJ (1982). **A**

[9.49] Ross, S.M., *Applied Probability Models with Optimization Applications,* Holden-Day, San Francisco (1970). **M**

[9.50] Shooman, M.L., *Software Engineering: Design, Reliability and Management,* McGraw-Hill, New York (1983). **A**

[9.51] Stone, H., "Multiprocessing scheduling with the aid of network flow algorithms", *IEEE Trans. Softw. Engrg.,* **SE-3**(1), 85–94 (1977). **A**

[9.52] Watanabe, M.S., *Knowing and Guessing,* Wiley, New York (1969). **A**

[9.53] Welsh, J. and M. McKeag, *Structured System Programming,* Prentice-Hall International, London (1980). **C**

INDEX